Cognitive Systems Monographs

Volume 8

Editors: Rüdiger Dillmann · Yoshihiko Nakamura · Stefan Schaal · David Vernon

Henrik Iskov Christensen,
Geert-Jan M. Kruijff,
and Jeremy L. Wyatt (Eds.)

Cognitive Systems

Springer

Rüdiger Dillmann, University of Karlsruhe, Faculty of Informatics, Institute of Anthropomatics, Humanoids and Intelligence Systems Laboratories, Kaiserstr. 12, 76131 Karlsruhe, Germany

Yoshihiko Nakamura, Tokyo University Fac. Engineering, Dept. Mechano-Informatics, 7-3-1 Hongo, Bukyo-ku Tokyo, 113-8656, Japan

Stefan Schaal, University of Southern California, Department Computer Science, Computational Learning & Motor Control Lab., Los Angeles, CA 90089-2905, USA

David Vernon, Khalifa University Department of Computer Engineering, PO Box 573, Sharjah, United Arab Emirates

Editors

Henrik Iskov Christensen
Georgia Tech
RIM@GT
801 Atlantic Dr.
Atlanta, GA 30332
USA
E-mail: hichristensen@gmail.com

Jeremy L. Wyatt
School of Computer Science
University of Birmingham
Birmingham B15 2TT
UK
E-mail: jlw@cs.bham.ac.uk

Geert-Jan M. Kruijff
DFKI GmbH
Stuhlsatzenhausweg 3
66123 Saarbrücken
Germany
gj@dfki.de

ISBN 978-3-642-11693-3 e-ISBN 978-3-642-11694-0

DOI 10.1007/978-3-642-11694-0

Cognitive Systems Monographs ISSN 1867-4925

Library of Congress Control Number: 2010925225

Typeset & Cover Design: Scientific Publishing Services Pvt. Ltd., Chennai, India.

Printed on acid-free paper

5 4 3 2 1 0

springer.com

Quotes from the CoSy Science Advisors

> While there is still a long way to come close to the objective of the European Commission Cognitive Systems initiative "to construct physically instantiated or embodied systems that can perceive, understand, ... and interact with their environments and evolve in order to achieve human-like performance" this book is about one of the funded projects in this initiative. It gives and excellent insight into the challenges and benefits of working in an large interdisciplinary team to better understand the human mind and in order to build intelligent machines.

Heinrich Bülthoff, MPIK

> One of the great challenges of the 21st century is to build a robot that can perceive and act within its environment and communicate with people, while also exhibiting the cognitive capabilities that lead to performance like that of people. This book reports on the European Union project on Cognitive Systems. It offers detailed explanations of the exciting progress made on this challenge and serves as a foundation for the science of Cognitive Systems in the next part of this century.

Candance Sidner, BAE Systems

Preface

The present volume is a report on the research results generated by the project "Cognitive Systems for Cognitive Assistants" (CoSy), which was sponsored by the European Commission during the period 2004–2008.

The CoSy team was assembled to study the problem of embodied cognitive systems for domestic service tasks such as guidance, fetch and carry, etc. The main aim of the project has been to study the core technologies needed to build such systems rather than mere application development. The key competencies needed are: systems architectures, scalable knowledge representation, adaptive embodiment,categorical perception, planning and error recovery, learning, and situated dialog systems. All of these aspects were studied in the context of CoSy and exemplified using two "demonstrator scenarios" that were conceived to allow studies / evaluation in an integrated context.

The volume is organized into 4 parts. The introduction outlines the overall problem domain and the CoSy approach to the problem. The second part contains a number of chapters that detail progress on topical problems across architectures, perception, learning, planning and dialog systems. These competencies were integrated into systems as described in the third part of the book. The final section provides a perspective on the results obtained and considers some possible issues for future research.

The project has published extensively throughout its life and links to publications can be found at the project web facility `www.cognitivesystems.org`, where copies of associated project deliverables also can be retrieved. The CoSy web facility contains also a page with material that supplements the book. The page has pointers to published material, associated datasets, videos and open software. The electronic version of the book also has embedded links to the web facility and published papers. I.e., referenced material published by the consortium can be accessed through embedded links.

The consortium would like to express our gratitude for the support the European Commission has provided for this research. In addition we are grateful for the guidance and feedback we have received from our scientific advisors: Prof. Heinrich Bülthoff - MPIK, Prof. Benjamin Kuipers - UT Austin,

Dr. Candy Sidner - BAE Systems. We are also grateful for the support from the project reviewers: Prof. Igor Alexander - Univ. College London, Prof. Mark Steedman - Univ. of Edinburgh, Prof. John Tsotsos - York Univ. and Prof. Oliver Brock, UMASS. Finally, we appreciate the guidance from the associated EU project officer Cecile Huet.

Atlanta, Saarbrücken & Birmingham
November 2009

Henrik Iskov Christensen
Geert-Jan M. Kruijff
Jeremy L. Wyatt

List of Contributors

Mykhaylo Andriluka
TU Darmstadt
Multimodal Interactive Systems
Hochschulstrasse 10
D-64289 Darmstadt, Germany
andriluka@cs.tu-darmstadt.de

Trevor Benjamin
DFKI GmbH
Saarbrücken, Germany

Michael Brenner
Albert-Ludwigs Universität Freiburg
Department of Computer Science
Freiburg, Germany
brenner@informatik.uni-freiburg.de

Wolfram Burgard
Albert-Ludwigs Universität Freiburg
Department of Computer Science
Freiburg, Germany
burgard@informatik.uni-freiburg.de

Ian Calvert
University of Birmingham
School of Computer Science
Edgbaston
Birmingham, B15 2TT UK

Henrik I. Christensen
Georgia Institute of Technology
Robotics and Intelligent Machines
Atlanta, GA 30332-0280
hichristensen@gmail.com

Richard Dearden
University of Birmingham
School of Computer Science
Edgbaston
Birmingham, B15 2TT UK
rwd@cs.bham.ac.uk

Sanja Fidler
University of Ljubljana
Faculty of Computer and
Information Science
Visual Cognitive Systems Laboratory
Trzasaka 25
SE-1001 Ljubljana, Slovenia
sanja.fidler@fri.uni-lj.si

Mario Fritz
TU Darmstadt
Multimodal Interactive Systems
Hochschulstrasse 10
D-64289 Darmstadt, Germany
fritz@cs.tu-darmstadt.de

Nick Hawes
University of Birmingham
School of Computer Science
Edgbaston
Birmingham, B15 2TT UK
nah@cs.bham.ac.uk

Somboon Hongeng
University of Birmingham
School of Computer Science
Edgbaston
Birmingham, B15 2TT UK
S.Hongeng@cs.bham.ac.uk

Henrik Jacobsson
DFKI GmbH
Saarbrücken, Germany
henrikj@dfki.de

Patric Jensfelt
Royal Institute of Technology (KTH)
Center for Autonomous Systems
SE-100 44 Stockholm, Sweden
patric@csc.kth.se

Marek Kopicki
University of Birmingham
School of Computer Science
Edgbaston
Birmingham, B15 2TT UK
mzs@cs.bham.ac.uk

Matej Kristan
University of Ljubljana
Faculty of Computer and
Information Science
Visual Cognitive Systems Laboratory
Trzasaka 25
SE-1001 Ljubljana, Slovenia
matej.kristan@fri.uni-lj.si

Geert-Jan Kruijff
DFKI GmbH
Saarbrücken, Germany
gj@dfki.de

Ivana Kruijff-Korbayova
DFKI GmbH
Saarbrücken, Germany
ivana.kruijff@dfki.de

Ales Leonardis
University of Ljubljana
Faculty of Computer and
Information Science
Visual Cognitive Systems
Laboratory
Trzasaka 25
SE-1001 Ljubljana, Slovenia
ales.leonardis@fri.uni-lj.si

Pierre Lison
DFKI GmbH
Saarbrücken, Germany
Pierre.Lison@dfki.de

Nikodem Majer
TU Darmstadt
Multimodal Interactive Systems
Hochschulstrasse 10
D-64289 Darmstadt, Germany
majer@cs.tu-darmstadt.de

Oscar Martinez Mozos
Albert-Ludwigs Universität Freiburg
Department of Computer Science
Freiburg, Germany
omartine@informatik.
uni-freiburg.de

Bernhard Nebel
Albert-Ludwigs Universität Freiburg
Department of Computer Science
Freiburg, Germany
nebel@informatik.uni-freiburg.de

J. Kevin O'Regan
Laboratoire Psychologie
de la Perception,
Université Paris
Descartes and CNRS
Paris, France
jkrevin.oregan@gmail.com

David Philipona
Laboratoire Psychologie
de la Perception,
Université Paris
Descartes and CNRS
Paris, France

Christian Plagemann
Albert-Ludwigs Universität Freiburg
Department of Computer Science
Freiburg, Germany
plagem@informatik.
uni-freiburg.de

Andrzej Pronobis
Royal Institute of Technology (KTH)
Center for Autonomous Systems
SE-100 44 Stockholm, Sweden
pronobis@csc.kth.se

Bernt Schiele
TU Darmstadt
Multimodal Interactive Systems
Hochschulstrasse 10
D-64289 Darmstadt, Germany
schiele@cs.tu-darmstadt.de

Kristoffer Sjöö
Royal Institute of Technology (KTH)
Center for Autonomous Systems
SE-100 44 Stockholm, Sweden
krsj@csc.kth.se

Danijel Skocaj
University of Ljubljana
Faculty of Computer and
Information Science
Visual Cognitive Systems Laboratory
Trzasaka 25
SE-1001 Ljubljana, Slovenia
danijel.skocaj@fri.uni-lj.si

Aaron Sloman
University of Birmingham
School of Computer Science
Edgbaston
Birmingham, B15 2TT UK
axs@cs.bham.ac.uk

Mohan Sridharan
Texas Tech at Abilene
302 Pine Street
Abilene, TX 79601, USA
mohan.sridharan@ttu.edu

Michael Stark
TU Darmstadt
Multimodal Interactive Systems
Hochschulstrasse 10
D-64289 Darmstadt, Germany
stark@cs.tu-darmstadt.de

Jeremy Wyatt
University of Birmingham
School of Computer Science
Edgbaston
Birmingham, B15 2TT UK
jlw@cs.bham.ac.uk

Henrik Zender
DFKI GmbH
Saarbrücken, Germany
zender@dfki.de

Part I

Introduction

1

Cognitive Systems Introduction

Henrik I. Christensen[1], Aaron Sloman[2], Geert-Jan Kruijff[3], and Jeremy L. Wyatt[2]

[1] Robotics and Intelligent Machines, Georgia Institute of Technology, Atlanta, Ga. USA
hic@cc.gatech.edu
[2] Intelligent Robotics Lab, School of Computer Science, University of Birmingham, Birmingham, UK
{axs,jlw}@cs.bham.ac.uk
[3] DFKI GmbH, Saarbrücken, Germany
gj@dfki.de

1.1 Introduction

The CoSy project was setup under the assumption that the visionary FP6 objective

> *"To construct physically instantiated ... systems that can perceive, understand ... and interact with their environment, and evolve in order to achieve human-like performance in activities requiring context-(situation and task) specific knowledge"*

is far beyond the state of the art and will remain so for many years. From this vision several intermediate targets were defined. Achieving these targets would provide a launch pad for further work on the long term vision.

In particular it has been an objective to advance the *science* of cognitive systems through a multi-disciplinary investigation of *requirements*, *design options* and *trade-offs* for human-like, autonomous, integrated, physical (e.g. robot) systems, including requirements for architectures, for forms of representation, for perceptual mechanisms, for learning, planning, reasoning, motivation, action, and communication.

To validate science progress a succession of increasingly ambitious working systems are constructed to test and demonstrate the ideas. Devising demanding but achievable test scenarios, including scenarios in which a machine not only *performs* some task but shows that it *understands* what it has done, and why, is an integral part of the empirical study of cognitive systems.

In this chapter the basic objectives, expected results and organization of the project will be presented, whereas the remainder of the book present results that have been obtained during the CoSy project. The final chapters of the

H.I. Christensen et al. (Eds.): Cognitive Systems, COSMOS 8, pp. 3–48.
springerlink.com

book will provide reflections on progress in terms of new insight and major lessons.

1.2 Objective of Project

1.2.1 The Problem

Despite impressive progress in many specific sub-topics in AI and Cognitive Science, the field as a whole moves slowly. Most systems able to perform complex tasks that humans and other animals can perform easily, for instance robot manipulators, or intelligent advisers, have to be carefully crafted. Whatever intelligence they have could be described as 'insect-like' insofar as they have capabilities that they do not understand, they do not know why they do things one way rather than another, they cannot explain what they are doing, they cannot improve their performance by taking advice from a human, and they cannot give advice or help to someone else doing similar tasks. Part of the reason for this is that over the last few decades research has become fragmented: with many individuals and research teams focusing their efforts on narrowly defined problems in vision, or learning, or language understanding, or problem solving, or mobile robotics, for instance.

1.2.2 The Way Forward

A key part of the CoSy effort has been to try to overcome some of these limitations by using ideas from relevant disciplines to investigate an ambitious vision of a highly competent robot, combining many different capabilities in a coherent manner, for instance a subset of the capabilities of a typical human 4-5 year old child. The scientific importance of this objective is that such a robot requires generic capabilities providing a platform for many different sorts of subsequent development, since a child of that age can develop in any human culture and benefit from many forms of education. However, we do not underestimate the profound difficulties of this challenge.

The research makes use of and feeds results into the various component disciplines of AI and cognitive science, for instance, new results on perception, learning, reasoning, language processing, memory, plan execution, and studies of motivation and emotion. Perhaps more importantly: the project not only benefits from other disciplines but has also tried to provide new substantive contributions to those disciplines in the form of new theories and working models. The detailed tasks of developing working systems generate new research questions for the contributing disciplines.

1.2.3 Steps to Success

The goal of producing a robot with many of the capabilities of a human child is unrealistic for a five year research project: it is an significant long term challenge. However, by analysing the many requirements for moving in that direction, one can derive sets of successively less challenging sub-goals that provide steps towards the distant goal. Some of these sub-goals are achievable in the time-frame of the project and form the main deliverables of the project.

An important part of the effort has been to consider two main kinds of deliverables: *theory* and *implementation.*

1. **Theory Deliverables**
   A body of theory, at different levels of abstraction, regarding requirements, architectures, forms of representation, kinds of ontologies, types of reasoning, kinds of knowledge, and varieties of mechanisms relevant to embodied, integrated, multi-functional intelligent systems. The results are expected to be useful both for enhancing scientific understanding of naturally occurring intelligent systems (e.g. humans and other animals) and for the design of artificial intelligent systems.

   The theory results are built around the core idea of a self-modifying architecture comprising different sorts of capabilities which develop over time. The results cover both analysis of *requirements* for such an architecture and also *design options* with their trade-offs.

   Key ideas for the architecture are informed by biological considerations, e.g. the notion of an architecture combining components which developed at different evolutionary epochs, and which in humans operate concurrently, performing different categories of tasks, for instance:

   - *reactive* components controlling the details of execution of skilled behaviours (these are evolutionarily old and use mechanisms shared with many species)
   - *deliberative* components supporting thought and reasoning about what might happen next or what should be done at some later stage (these are found in fewer species and require highly specialised mechanisms and forms of representation – including human language in some cases)
   - *self-reflective, meta-management* components that monitor, evaluate, and (partially) control and redirect the reactive and deliberative processes (these are probably rare in animals because they require the ability to represent, compare and evaluate information-processing tasks and strategies, as opposed to merely implementing them).

   The requirements for perceptual and motor systems that operate concurrently with, and in close coordination with, processes in all the different architectural layers are analysed and architectures are proposed for both perception (e.g. [122]) and action mechanisms.

   Learning processes are different within and between different parts of the architecture and current theories of learning have to be substantially extended to explain how, for instance (a) kinds of learning that extend the individual's ontology for perceiving and thinking about the environment, and (b) kinds of learning that develop fluency and speed in motor performance, e.g. because the reactive layer is *trained* by processes in the deliberative layer.

Different varieties of communication and social interaction are related to the different architectural layers: for instance, (a) dancing, fighting and moving heavy objects require coupled *reactive* systems; (b) linked collaborative actions spanning spatial and temporal gaps, e.g. in building houses and bridges, require *deliberative* capabilities; (c) the ability to empathise, exhort, persuade, may require extensions of self-understanding in the *meta-management* system to support other-understanding. (All of these influences can go both ways: e.g. meeting requirements for social developments may enhance individual capabilities.)

As different sorts of architectures with these general features are possible it is important to consider an analysis of architectural options and trade-offs.

2. **Implementation Deliverables**
Several implementations of working systems demonstrating successful application of the theory, e.g. in a robot capable of performing a collection of diverse tasks in a variety of scenarios described in the project, including visual and other forms of perception, learning, reasoning, and communication (with another robot or a human) both in order to collaborate on some task and in order to explain what it is doing and why. Our aim has been to create robotic cognitive systems that have a kind of autonomy that will justify describing it as having its own goals, which may change over time.

The distinctive features of such a robot include integration of subfunctions (e.g. vision and other senses can be combined in making sense of a scene, vision can be used to disambiguate an utterance by looking at what the utterance refers to, and learning processes can enhance different kinds of capabilities, including linguistic, visual, reasoning, planning, and motor skills).

Integration does not imply *homogeneity*, such as use of the same type of representation throughout the system. For instance, low level visual mechanisms finding edge-features, optical flow patterns, etc. use different forms of representation from higher level mechanisms e.g. those recognizing and describing whole objects. Likewise, planning mechanisms will need different forms of representation from fine-grained motor control mechanisms.

Nature vs. Nurture: A major research issue concerns how much should be programmed into such a robot and how much will have to be learnt by interacting with the environment, including teachers and other agents. There are several AI projects aiming to develop intelligent systems on the basis of powerful and general learning mechanisms starting from something close to a "Tabula rasa" (e.g. the COG project at MIT, the Cyberlife Research "Lucy" Project, and the Dav project at Michigan State

University). Their hope is that the structure of the environment will cause the learning mechanisms to induce all required information about the nature of the environment.

Such projects are likely to be defeated by explosive search spaces requiring evolutionary time-scales for success.

Biological evolution enables individuals to avoid this problem by providing large amounts of "innate" information in the genomes of all species. In the case of humans this seems to include meta-level information about what kinds of things are good to learn, helping to drive the learning processes as well as specific mechanisms, forms of representation, and architectures to enable them to work.

Although such debates are most commonly associated with requirements for language learning (e.g. [118] the issues are deep and general. For instance, Kant [65] argued two centuries ago that notions of space, time and causation are presupposed by and cannot be learnt from perceptual experiences. Similar points could be made about notions of meaning, purpose and action.

Instead of taking a dogmatic stance on what needs to be innate CoSy has explored various alternatives for amounts and types of innate knowledge and included a first analysis of the trade-offs.

1.3 A Motivating Example

At the core of the CoSy project have been the methods for construction of embodied artifacts, such as robots, with advanced cognitive functions. Such systems should be endowed with facilities for automatic interpretation of the environment (in terms of mapping of the environment, recognition of a large number of objects, etc.), adaptive acquisition of new skills and tasks in cooperation with a human user, methods for advanced manipulation to allow the system to perform extended missions on behalf of the users and reasoning methods to ensure advanced autonomy. The interpretation facilities can be used both to ensure autonomy and to verbalise knowledge and mission parameters to a user. Advanced service robots can be used for a large range of tasks in our everyday environment ranging from vacuuming to clearing the table after dinner. The tasks can also be in terms of mobility support for elderly and handicapped. To be of true utility to an average citizen it is essential that the system has facilities for automatic adaptation to the environment of the user, it must also be able to adapt to the habits of the owner, it must be able to understand the importance and consequences of instructions, and it must be sufficiently flexible to learn new skills that it was not endowed with from the factory. An example scenario is outlined below.

1. The Hartmut family decides to acquire a CoSy system to assist them in their everyday life. A system is acquired from the local Robs-R-Us chain of stores

2. Upon arrival at home the system is unpackaged and booted up. Upon wakeup the CoSy system greets the owners and asks them to give their names (Helga and Oskar). The owners are asked to give it a name (Fido). The system then asks for a tour of the home. As they traverse the home it builds up a map of the house, and objects of particular interest are named to the system (this is the refrigerator), ...
3. When asked to clean the table after dinner, Fido already has a built-in strategy for cleaning of tables, but it needs to know the location of the dishwasher (Oskar shows him the dishwasher – a Bosch 4590). Upon recognition of the dishwasher Fido downloads an operations manual from the Cosy corporate site. It then goes about its task.

   Oskar notices that Fido is removing cutlery and dishes but moves glasses to one side of the table. He asks: 'Why are you leaving the glasses?' Fido answers: 'My movements can be clumsy, so I'll put them in the dishwasher after all the objects that might break them.'
4. After dishes, cutlery, glasses and bottles have been cleared, a single object is left on the table. Fido asks Helga: What am I to do about this object left on the table? This is an Orrefors vase please leave it on the table! Fido inquires if it is Ok to turn on the dishwasher.
5. A few months later the Hartmut family acquires a new GE refrigerator. When Fido detects the new item, he inquires about the new objects. He is told that it is a GE-Economy Fridge. Upon learning this he downloads the latest skills from the CoSy skill database to allow him to operate the unit in terms of opening of the door and placing objects on the shelves, etc.
6. Later they replace the door to the bathroom with a sliding door, a new object that Fido has never seen before. He asks Oskar to demonstrate to him how to operate the door. Initially Oskar shows him how to operate the door (learning by demonstration). Fido tries to replicate the action and is corrected when he pushes to hard.
7. One day Julie, the 8 year old grandchild of Helga and Oskar, comes to visit. At first Julie is a little afraid of Fido and keeps her distance. Later when she is alone with Fido she says to Fido: Can you play Ludo? Fido says: What is Ludo? Julie: A game - I'll teach you. Will you play? Fido understands the general category of game, does not know Ludo, but reasons that since Julie wants him to play, and he is not required for anything else at that time, he should oblige. He considers downloading information about Ludo from the Cosy corporate site, but reasons that if Julie offers to teach him she would probably like to do that, and he wants to please her. So he does not download the information and says: Yes please. Show me what to do. They both enjoy the ensuing interaction.

The above scenario is deliberately futuristic, but it does at the same time illustrate "automatic mapping", "recognition of objects", "recognition of limitations in knowledge or skills", "natural interaction using speech and gestures", "never ending dialogues" to resolve ambiguities, and affective states and processes.

These are key issues that have been studied in the CoSy projects. Through integration of perception, action generation, autonomy, flexible user interfaces it has, however, been possible to approach the problem of building truly flexible cognitive systems that can be used to implement scenarios as outlined above.

1.4 Organization of the Research/Research Questions

In the study of cognitive systems for cognitive assistants a number of key research challenges were identified: architectures, representations, learning, perception-action modelling, communication, and planning & failure handling. In addition to addressing these issues at a fundamental level there is also a need to consider the integration of the techniques into complete systems. Two scenarios were identified for the study of integrated systems: i) exploration/mapping of space, and ii) models of objects and concepts. Initially we will discuss the research challenges (Sections 1.4.1 – 1.4.7) followed by a discussion of the integrated scenarios (Sections 1.4.9 – 1.4.10).

The overall work has been organised according to two major milestones

1. "Using intermodality and affordances for the acquisition of concepts, categories and language"

   The goal of the first milestone was to develop an agent that was capable of exploring unknown spaces and objects in that space, learning representations of spaces and objects based on a known ontology. The agent can determine what it does not know (or about which it is not clear), and can carry out a dialogue with another agent to learn more. Fundamental to the latter effort was the acquisition of strategies for representing, coordinating and understanding multi-modal action & dialog acts.
2. "Introspection of models & representations; planning for autonomy – goal seeking"

   Based on the idea of an embodied cognitive agent that is able to explore its environment, and communicate with other agents about descriptions for spaces and objects in that environment, the goal of the second milestone iwas to develop an agent that can perceive of its environment in terms of affordances, acquire knowledge about such affordances, and use such knowledge to understand the intentions of other embodied agents in that environment. Based on such understanding, the agent is able to (re-)plan its own actions and dialogues to achieve its own goals, possibly requiring communication with other agents to request or negotiate cooperation.

1.4.1 Architecture

For many years, research in AI and computational cognitive science focused on forms of representation, algorithms to operate on them, and knowledge to be encoded and deployed or derived. In the last decade or two it has become clear that there is also a need to investigate alternative ways of putting pieces together into a complex functioning system, possibly including parts that operate concurrently and asynchronously on different sub-tasks, for instance, perception, action, reasoning and communicating.

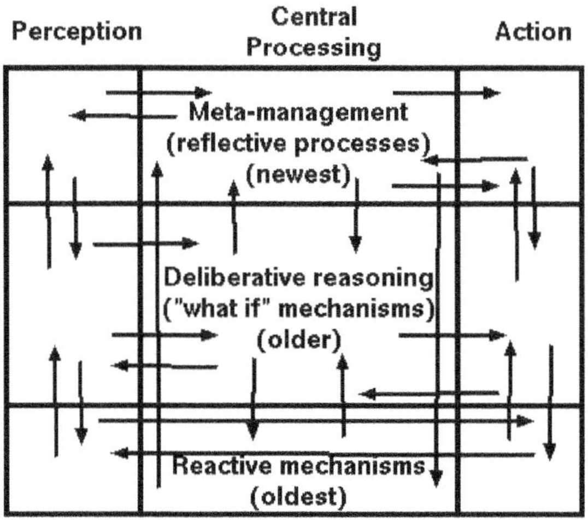

Fig. 1.1. The CogAff schema: superimposing 2 functional divisions, giving 9 categories of possible sub-mechanisms and many possible connections – not all shown.

Unfortunately this has led to a plethora of architectures being proposed including, for example SOAR, ACT (and its successors), PRODIGY, ICARUS, 3T, APEX, CLARION, CIRCA, EPIC, Subsumption architectures, H-COGAFF, and Minsky's emotion machine. Many of these were discussed at a DARPA/NSF-funded workshop in Stanford in March 2003.[1] One reason why this is a problem is that there is no agreement on what the space of possible architectures is like, nor on the terminology for describing architectures or on criteria for evaluating and comparing them.

One of the tasks for this project, therefore, has been to produce a framework for describing and comparing architectures. A first draft and relatively simple example of such a framework is the *CogAff schema* developed at the University of Birmingham, and described in [121], partly inspired by [89]. This

[1] Most of the presentations are online here: `http://www.isle.org/symposia/cogarch/`, including a useful survey paper by Langley and Laird. Another useful overview can be found in [40].

provides a way of classifying components of an architecture in terms of their functional role, starting with a crude three-way division between perceptual, central and action components, and another three-way division between components concerned with reactive, deliberative or meta-management functions. Superimposing these modes of division gives a grid of nine types of components which may or may not be present in an architecture, and which may be connected in various ways to other components as shown in Figure 1.1.

In recent years many researchers influenced by the work of Brooks [23] have attempted to design robots using only the bottom layer of the CogAff grid, the reactive layer, arguing that either features of the environment or emergent interactions between many individuals will produce effects that were thought to require deliberative and other mechanisms. Others have challenged this view arguing that it suffices only for simple organisms and insect-like robots.

Our approach has not been to engage in such battles but to try to understand under which conditions the various types of architectural components are useful. For example if building something at location A requires materials known to be at location B, and the agent does not have innately determined reactive behaviours that cause fetching of materials from B, then the deliberative ability to consider and evaluate a possible journey to B in advance of doing it will be useful. Of course this requires suitable forms of representation for learning and storing re-usable generalizations and perceptual mechanisms that can 'chunk' the environment into categories that support learning of useful generalisations. Without discretization, planning several steps ahead in a purely continuous space would be very difficult, although discretization of percepts may not be needed for certain kinds of reactive behaviours involving continuous control and no predictions.

One of the particularly interesting issues to explore is whether the kind of self-understanding that so many AI systems lack can be provided on the basis of a *meta-management* architectural layer permitting observation, classification, evaluation and possibly some control of internal states and processes, especially deliberative processes that are capable of getting stuck in loops, wasting resources be repeating sub-tasks or in other ways performing suboptimally. An important form of learning might include detecting such cases and finding out how to prevent them or reduce their effects. An early example of this sort of thing was demonstrated 30 years ago by Sussman [128], but never developed further. One of the issues to be explored is how our notion of meta-management relates to notions of "executive function" used in psychology and psychiatry. It is possible that empirical research on executive functions in humans can contribute design ideas for the construction of artificial cognitive systems. Likewise our design work may shed new light on puzzles about the nature of executive functions in humans.

Another interesting and important issue to consider is how such an architecture might develop. This relates to questions about the pros and cons of various degrees of complexity of the innate mechanisms given to a robot by its designers, or to animals by evolution.

Architecture Deliverables:

The study of architectures has continued throughout the project. It has three sorts of outputs:

- A succession of theoretical papers analysing the variety of possible architectures and their trade-offs in various contexts
- experimental designs for working systems to help investigate the theories and demonstrate the ideas

Software Tools:

A major requirement for the success of a project like this is the availability of tools that support construction of a complex architecture with many interacting, concurrently active components doing different things, e.g. concurrently active mechanisms involved in perception (concurrently processing at different levels of abstraction – eg. feature detection and perception of affordances), reasoning, motive generation, planning, plan execution, communication with other agents, evaluation of whatever's going on, learning of various kinds, etc.

Cosy has developed a platform for rapid prototyping and rapid deployment of intelligent systems combining multiple software and hardware resources. There were quite a lot of toolkits already available but mostly they tend either to be committed to a particular sort of architecture (e.g. SOAR, ACT-R, PRS, etc.) or else aimed at multi-agent systems composed of lots of relatively simple agents perhaps distributed over many machines. More general and open ended toolkits include the SimAgent toolkit developed at Birmingham [123], the Cogent toolkit developed at Birkbeck College and the Mozart toolkit developed in Saarbrücken and elsewhere. The project has evaluated these and other tools and developed a new improved tool - CAST that is presented in Chapter 2.

A cognitive systems project aimed at producing an integrated physically embodied agent with diverse but cooperative concurrently active and possibly strongly interacting capabilities needs tools that can be used to explore and test out ideas about different sorts of architectures – preferably rapidly and easily.

One of the requirements that arises in connection with architectures that include a meta-management layer is the need for mechanisms that allow self-observation during program execution. The SimAgent toolkit provides some support for that, used in work by Kennedy on self-monitoring intrusion detectors [66].

For all these explorations, merely using a powerful programming language, e.g. Prolog, or Lisp or Java is clearly not enough. For instance, different sorts of languages have been required for the different components (e.g. low-level vision, reasoning about where to go next, understanding or composing sentences, etc.) Facilities to support concurrency, self-observation and self-modification of running systems should if possible be general and re-usable. Rapid prototyping, i.e. rapidly producing new experimental implementations that can be run, is essential for research and exploration, as opposed to formal specification and verification, which are often the main concerns of software engineers: that only makes sense if the problem has already been understood, so that researchers know what they are trying to verify.

Often, discovering requirements is the hardest part of the research task in AI and exploratory design is a major tool for doing that (you soon find out that things you have built are no good for surprising reasons, narrowing the search space for what a working system needs to be able to do.)

Moreover, as the prototypes get more complex the human interface requirements for interactive debugging, testing and interrogating parts of the system, also get more complex and those requirements have to be met as well as support for whatever is being built.

Note on Representations:

There have been many debates in recent years about whether animals or robots need representations (E.g. see [23]). Our view on this is that anything that senses the environment and on the basis of what is sensed uses its own energy stores in order to select among possible actions is using information to determine what to do. Biological evolution discovered many variations on that theme depending on the kind of information acquired, how it is processed, how it is used, when it is used (e.g. long term storage may be required) how it is transformed, how it is combined with other information, and how it is communicated. In all cases there is some *medium* used for the information, but there are great differences between different media, including whether they are discrete or continuous, one-dimensional or multidimensional, what sorts of structures they can have, and so on. Some people may be inclined to argue about whether some of them are *really* representations or not, but we avoid such disputes by investigation *what kinds* of representations they are, and what their costs and benefits are to the organism.

1.4.2 Representations

A very important issue in the project has been the design of representations for objects, scenes, actions, dialogues and affordances. The representations must take into account visual, haptic, proprioceptive, and auditory — particularly linguistic — information, capturing temporal as well as static aspects. Many of these representations also need to be suitable for higher-level cognitive processes, such as deliberative and reflective processing. Others may be used only for reactive control. To accomplish the tasks, that formed for the basis for the project, the representations should

- enable integration of representations of objects, scenes, actions, events, causal relations and affordances; these representations should be linked together to form a consistent framework.
- allow incremental updating or sometimes correction.

- allow different types of learning (supervised, unsupervised, reinforcement).
- allow integration of various modalities, of very different input signals (visual, tactile, auditory, etc.), into a common framework of representations.
- deal with multiple cues, enable multiple representations of the same object, scene, event or action.
- be suitable for recognition and categorization in the presence of a cluttered background and variable illumination.
- take into account the context of an object, an action or a scene
- be scalable; they should enable efficient learning, storing of the representations, and recognition even when the number of objects, scenes and actions increases (in fact, with an increase in the number of entities, learning should become more efficient).
- accommodate semantics (in terms of language); they should be linked with symbolic descriptions.
- be suitable for acquiring, describing, and using spatial relationships, both quantitatively and qualitatively.
- be suitable for higher level cognitive tasks, such as reasoning or planning.
- enable the formation of concepts and hierarchies of concepts.
- allow introspection leading to refinement of the representations, and the construction of new representations at new levels of abstraction (e.g. capturing and exploiting temporal, or geometric regularities not explicitly programmed in by the designer).

Many of these issues have been addressed prior to CoSy. They were, however, addressed separately, considering only one, or a few aspects at a time. They have never been tackled together, in a common framework. This has been an important part of this project. The representations (and, consequently, the learning and usage of the representations) of objects, scenes, actions, events, causes and affordances should be linked together in a unifying framework. The representations may be obtained in different ways, but processes employing different representations need to able to exchange information, and to update some representations on the basis of others.

Task Specific vs. General Representations

For a cognitive system there is a need for representations which can, on the one hand, be quite general, to serve different tasks and goals, and, on the other hand, be tailored for particular tasks. The advantage with task specific representations is that they are efficient to use, and often compact. The problem with specific representations is that they can only be used for the task that they were trained for. One example is task specific recognition or classification e.g., a system trained to distinguish faces with glasses from faces without glasses would be inappropriate for other tasks e.g., to distinguish male from female faces. Another example would be a reactive controller for plugging a socket into a power point, which is not directly of use for another task such as grasping a cup.

A number of unsupervised learning methods produce generative representations which allow partial approximate reconstruction, or hallucination, of the input data. According to Grossberg [51], reconstruction is central for a two-way processing cognitive system. Memory traces in the short term memory could activate the long term memory forming a reconstruction, which gets compared to the momentary input. This mechanism allows categorization and establishment of a reference frame, forming a set of high-level expectations. These expectations influence further processing in a top-down manner, by limiting the contextual frame, checking for consistency, providing missing data etc. Cognition is therefore performed as a two-way, bottom-up and top-down process. Comparing the input to the "hallucination" can reinforce the good hypotheses in the interpretation and inhibit inconsistent parts such as noise and occlusions. A two-way interaction also enables the processing and organization of a huge amount of data, in contrast to the one-way processing with no feedback. While being significantly faster [104], the lack of interaction with the higher levels of a one-way processing schema could lead to a combinatorial explosion of complexity.

Nevertheless, supervised non-generative methods can produce simpler representations, which can be more efficient for specific tasks they are tuned for. Our goal should be to combine both, unsupervised generative methods and supervised non-generative methods, in order to obtain representations, which are general enough to enable a two-way processing and robust recognition, and to complement them with features, which are tailored for specific tasks.

1.4.3 Learning

Modes of Learning

Learning can be considered using a number of different modes:

Tutor Driven. A user (tutor) shows to the system an object or an action and explains to the cognitive system what he/she is showing or doing. The tutor can also provide figure-ground segmentation (e.g., showing the object on a black background), which facilitates the learning process and enables the creation of a reliable representation. In a similar manner, the tutor can guide the agent around the space and give it additional information (e.g. coordinates) to facilitate the creation of a map. In such a learning scenario, the user has a high degree of involvement.

Tutor Supervised. A cognitive system detects a new object, an action, event, affordance or a scene by itself and builds its representation in an unsupervised manner. Then it asks the user to provide additional information about the object or action (e.g., What is this object, action or room?). The involvement of the user in the learning process is now significantly reduced, yet it still assures that the produced representations are reliable/consistent.

Exploratory. A cognitive system detects a new instance of an object, action, event or scene, tries to recognize (categorize) it (using previously acquired knowledge, statistical properties, etc.), and then updates the representations accordingly. Such a learning scenario does not require any involvement on the part of the user.

To speed up the initial phase of the learning process and to enable development of consistent basic concepts, one could start with mainly tutor-driven learning with many user interactions. These concepts would consequently be used to detect new concepts with limited help from the user (an important role will be played by the dialogue between the cognitive system and a user). Later on in the process, when the ontology is sufficiently large, many new concepts could be acquired without user interaction.

Learning is embedded in the perception-action loop but must go beyond simple percept-react cycles involving deliberative processes such as reasoning and planning. This is particularly important in tutor supervised and exploratory modes of learning, where the agent must plan how to interact with the world so as to learn effectively. This will in turn require reflective processes to modify the learning strategy, as well as a model of motivation and attention.

Continuous Learning

As mentioned in Section 1.2.3 in order to avoid evolutionary time-scales in the development of the system, it has to initially encompass a certain level of pre-defined functionality and knowledge. It then has to be able to build new concepts upon these. It needs to do this throughout its lifetime.

It is therefore important that the representations employed allow the learning to be a continuous, open-ended, life-long process. The representations should not be learned once and for all in a training stage and then used in their fixed form for recognition, planning and acting. They should be continuously updated over time, adapting to the changes in the environment, new tasks, user reactions, user preferences, etc. So there will be no strict distinction between the activities of learning, planning, recognising, acting and introspection: these activities will be interleaved.

This is a non-trivial challenge. For example, most of the state-of-the-art algorithms for visual learning and recognition do not consider continuous learning. They follow the standard paradigm, dividing the off-line learning stage and the recognition stage [141, 139, 111, 79, 80, 63, 2, 42, 73]. Most of these approaches are not designed in a way which would enable efficient incremental learning, which is a basic prerequisite for continuous learning.

When having a hierarchical or multi-layered organization of representations, one has to ensure, when performing continuous learning, that the representations are updated on all levels and that also all mutual dependencies are updated accordingly, to assure the consistency of the representations. Some of these updates would need to be performed on-line (thus they have to be simple and fast) while some of them may be performed off-line from time to time (during the agent's "sleeping mode") by processing several updates at once.

There are two issues which are very important for reliable continuous learning. First, the representations have to be carefully chosen, such that they must allow efficient upgrading with the newly acquired data. Second, it is important how new data is extracted and prepared. When this is performed under the user's supervision, this operation is risk-free in the sense that the algorithm can update the model with high confidence. On the other hand, when the information, which is to be added into the representation, is autonomously extracted by the agent in an unsupervised manner (e.g., using a recognition procedure) there exists a possibility of propagating an erroneous extraction from the data through the learning process. Consequently, the representation could be corrupted with false data, resulting in poorer performance and a less reliable representation. Robust mechanisms, which prevent such propagation of errors, play an important role in the process of continuous learning.

Similarly, the already learned concepts should facilitate learning of new concepts. For instance, it should be easier to learn to recognise a new object after a hundred objects have previously been learned than only after a few objects have been considered. In the former case, information is captured in the representations of already learned objects, which should make learning of the new object easier. Such behavior (which is in accordance with some of the properties of human perception) is not a characteristic of many state-of-the-art algorithms for visual learning.

1.4.4 Perception-Action Modelling

In planning and acting approaches fall into two broad classes, although there are now points of contact. On the one hand there are abstract relational representations of the effects of actions as used in classical AI planning. On the other there are probabilistic models of action effects in continuous spaces as used in robot localisation and mapping. The former are general, and non-task specific, but assume observability of the world, and a hand-constructed abstraction from sensor data. Reasoning with them is also inefficient. The latter capture uncertainty in both action and observation, and are tractable for localisation and path planning in continuous spaces. They are, however, typically tied to geometric representations of space, and in some cases to quite strong assumptions about the sensors used. One of the tasks of this projects has been to try to connect these different types of representation in such a way that updates to one representation can be propagated to other representations.

In neither case are the representations of actions entirely suitable for continuous learning. Some attempts have been made to achieve this. Early techniques like chunking attempted to capture reusable patterns of behaviour, and probabilistic models of actions allow the natural integration of new experiences into existing action models. But our agents also need to learn to identify and acquire new action models on different time-scales, recognise new events, introspect about their causes and thus identify affordances for specific tasks.

Analogously in the literature on planning and learning to act there has been some work on automatic relevance detection when learning action models. This can be seen as a simple form of affordance learning, and has been studied in, for example, reinforcement learning. These algorithms learn action models that are more efficient because they are task specific. These are acquired by identifying features that are relevant to predicting the outcome on the task. The state-of-the-art techniques, however, work only with unstructured or propositional representations, and rely on statistical methods to detect the affordances. Another issue has been how knowledge be transferred from one task specific representation to another.

1.4.5 Continuous Planning and Acting in Dynamic Multiagent Environments

Continuous Planning

Planning and acting in realistic environments poses a number of difficulties to artificial agents many of which are due to the dynamic nature and partial observability of such environments: Other agents' actions as well as naturally occurring events (e.g. sunset) may change the agent's surroundings in ways it cannot foresee, control or even perceive. It is thus crucial that during plan execution agents react according to the perceived changes (e.g. stop moving when there is a human standing in your way). Unfortunately, on the one hand not all plans can easily be repaired during plan execution, on the other hand with increasing dynamics of the world an agent's knowledge will become less accurate and its plans more likely to be not executable or to achieve undesired effects. However, it is also not possible to give up planning at all or to use purely reactive forms of planning [3] since its lack of flexibility does not allow to account for diverse and varying goals and its lack of knowledge may prevent an agent from selecting a "good" action to perform without some lookahead planning.

Thus contingencies have to be taken into account already during the planning phase. Traditionally, research in AI Planning has addressed this problem in the subfields of *conditional planning* [105, 14, 56] and *probabilistic planning* [20, 77]. The problem solved by conditional and probabilistic planning algorithms is to find plans that will work not only for a given initial state but under all circumstances imaginable, a fact that makes the problem computationally hard [105, 77]. Considering the large number of unobservable features and of possible contingencies in dynamic multiagent environments it is clear that even small-sized problems will be hard to solve by conditional and probabilistic planners.

Luckily, in realistic domains (like those considered in the CoSy project) there is often no need to devise universally executable plans before acting as agents will continuously plan, act, monitor their actions, and learn in the same environment. Therefore, in this project techniques have been developed that allow agents to postpone the resolution of contingencies and handle them only when they actually occur. Ideally, an agent learns from past experiences which contingencies it can easily solve by re-planning at execution time. During the planning phase the agent can then simply "ignore its ignorance" about those contingencies. Instead of a conditional plan fragment, agents use so-called *assertions* in their plans, non-conditional pseudo actions achieving facts whose actual achievement will be planned for later. Assertions consist of preconditions and effects just like normal actions and special *planning trigger conditions* describing under which conditions in later phases of a planning-execution cycle detailed planning for the assertion's effect will be carried out. Trigger conditions mainly consist in a set of state variables whose values are unknown to the agent in the current state. As soon as the agent can observe the actual values of these variables a detailed sub-plan achieving the assertion is searched for.

The planning agent research goes beyond classical planning in integrating planning and execution into a continuous planning agent. It contrast to related approaches [142, 8, 144] the research goal of this project is not focused on plan repair or re-planning[2] but on planning under incomplete information and partial observability. Unlike the conditional and probabilistic planners presented earlier contingency resolution is based on postponing the solution of parts of the planning problem to later points in time. In their use of abstract pseudo actions that are decomposed into executable actions the resulting plans resemble hierarchical task networks [145, 38, 39]. In our approach, however, the abstraction hierarchy is not explicitly given, but the decompositions can be planned by the agent itself during a later phase of the planning-execution-monitoring cycle. Also the purpose of the abstraction is different from HTN planning: while HTN decompositions embody knowledge about how to solve subtasks, in our project assertions essentially represent a way to reason under imperfect knowledge.

The approach to automatically interleaved planning and execution taken in this project relies on explicit modeling of an agent's perception. Assertion achievement is planned for only when the observations specified by the triggering conditions can be made. It is therefore an important part of the proposed project to model different perceptors and sensing capabilities in a general way. There are two possibilities for modeling sensing: automatic sensing and active sensing by executing sensory actions [109] which both are allowed in our representation. Making perception explicit in this manner allows agents to plan their own sensing, i.e. to actively try to reach states in which observations can be made and thus to reduce their ignorance. Both kinds of sensing are modeled as *perceptive events*, active sensing events being triggered by explicit sensing actions of the agent while automatic sensing events "fire" similarly to

[2] Modern planners like FF [57] allow fast re-planning whenever it is necessary.

domain axioms [130] (also sometimes called derived predicates, causal rules, or indirect effects). In contrast to earlier approaches to planning with sensing [41, 67, 50, 9] the use of perceptive events allows to separate modeling of the physical effects of events from their perception by an agent. This separation is especially important for active failure diagnosis (cf. section 1.4.5), but also to describe when several agents with different perceptors are present in a common environment (cf. section 1.4.5).

Active Failure Diagnosis

In the recent years, tremendous advances have been achieved in the field of mobile robotics. The majority of the work, however, has focused on navigation aspects such as path planning, localization and mapping. In most approaches it is typically assumed that the sensors and actuators of the robot are reliable in the sense that their input always corresponds to the expected input and that there is no malfunction of the sensors or actuators. Accordingly, these systems are prone to failure in situations in which a sensor breaks and provides erroneous data that does not correspond to the assumptions.

Recently, several techniques have been developed to allow mobile robots to detect or reason about potential failures. For example, Murphy and Hershberger [84] present an architecture that allows a mobile robot to exploit causal models for generating tests about potential failures or changes in the environment. Leuschen et al. [76] use a mixture of methods to analyze the reliability of a robot. However, they do not provide means for fault detection. Scheding et al. [110] analyze the effects of faults to the innovation of a Kalman filter. They introduce a metric for determining the detectability of faults. Washington [138] uses a combination of Kalman filters and Markov decision processes to identify faults on a mobile platform. Whereas Kalman filters are efficient means for detecting faults, their drawback lies in the limited capabilities to represent ambiguous situations in which multiple failures could case the same effect. To also deal with ambiguous situations recently particle filters have been used with great success for fault detection in technical systems [31, 30, 83, 136, 137, 135]

One important problem in the context of applying particle filters for fault diagnosis in complex systems is caused by the high dimensionality of the state space. In practice, a huge number of particles is needed to track all potential fault states of the system. Therefore, recent approaches have especially addressed the problem of how to reduce the number of samples during the sampling process [83, 137, 135].

All these approaches, however, are passive in the sense that they do not exploit the actuators of the robot to identify potential faults. By generating appropriate actions, such as moving for example, the robot can actively generate measurements that make the robot more certain about potential failures, i.e. a malfunction of its laser range scanner. This way, the complexity of the state tracking problem can significantly be reduced. In particular, we have investigated techniques allowing mobile robots to identify and to deal with possible sensor failures. A crucial aspect in this context is active exploration, i.e. the search for actions that allow the robot to identify whether the current measurements are due to failures of the sensors or whether they correspond to the current state of the environment. Promising approaches for exploration of state spaces for active state estimation have recently been applied by various authors [24, 64, 60, 68, 108]. Most of the systems rely on approximations of partially observable Markov decision processes.

Throughout this project we have been especially interested in the integration of symbolic information into the process of active exploration for action generation. Especially we have been interested in triggering the generation of samples in the state estimation component based on assumptions represented at the higher-level planning system (cf. section 1.4.5). In many cases a fault will be detected by the planning system due to a time-out. The explicit representation of assertions (such as camera is functional, light is switched on) allows the system to generate samples in appropriate regions of the state space. This way we exploit high-level knowledge about the robot to efficiently generate actions that reduce the uncertainty at lower levels. To generate samples we have used techniques similar to those presented by Lenser and Veloso as well as Thrun et al. [74, 132]. Similar integrations of high-level and symbolic and probabilistic representations has been applied successfully in the past [55, 13]. Independently of actions triggered by the planning system, the robot permanently monitors its state using a particle filter. To efficiently detect failures, however, we exploit knowledge about the effects of typical faults and generate the samples according to likely states [74, 132].

Based on the sample-based representation of the posterior about the potential states of the system the robot can then generate actions that reduce its uncertainty about the potential failure conditions. In this context, techniques such as those mentioned above [24, 64, 60, 68, 108] are used. Once a fault(s) has been identified, the high-level system is notified so that appropriate actions can be generated at the planning level.

Collaborative Planning and Acting

The modeling and planning techniques suggested in section 1.4.5 help individual agents to act continuously in dynamic environments. However, those techniques do not account for the fact that the dynamic nature of most environments is a consequence of the presence of *multiple agents*. If this is the case, instead of being a handicap the dynamics can be exploited by means of *cooperation*. The notion of cooperation is at the heart of the CoSy project: artificial and human agents cooperate to solve problems, both by supplying each other with information and by executing parts of collaboratively developed plans. But cooperation is not limited to human-robot interaction. A robot must also interact with other "intelligent" household appliances. (For example, the Fido robot and the dishwasher may work out a plan together in which the dishwasher closes it door and starts cleaning the dishes when Fido has filled it). Sometimes it will even be necessary to collaborate with remote agents via the Internet (for example the CoSy repository might be such an agent, providing details of a sub-plan for Fido).

Collaborative action planning is not possible without *communication*. However, different groups of agents may have different ways to communicate. Groups of artificial agents can communicate using special-purpose languages with rigid protocols, while human-robot interactions should allow the human to use more convenient methods (cf. 1.4.6). For the sake of scientific clarity, we have adopted an abstract view of communication, assuming that a group of agents has decided on a common language and protocol that allows them to communicate beliefs and intentions by abstract "communicative acts" if they decide to do so. In this context the only focus is to investigate when and why agents should engage in communication, not how. For the case of human-robot interaction the latter question is dealt with when formulating how a concrete dialogue can be generated on the basis of the abstract communicative acts produced during the planning process.

Planning in multiagent systems or Multiagent Planning (MAP) is an ambiguous notion. It can describe both a planning process distributed among multiple *planning* agents as well as the concurrent execution of plans involving multiple *acting* agents. In most real-world domains both dimensions are directly coupled as agents are both searching for and executing plans. Traditionally however, subfields of AI have almost exclusively concentrated on either one of the two aspects. The field of AI Planning, for example, has developed formalizations of concurrent plans featuring durational, overlapping actions and nontrivial joint effects [96, 48, 16, 10, 45, 19], but to date only centralized planning algorithms have been developed for these expressive formalisms. In contrast, distributed planning algorithms developed in the fields of Distributed Artificial Intelligence (DAI) and Multiagent Systems (MAS) have mainly focused on coordination and merging of individually found plans [46, 47, 37, 114, 133], ignoring the specific issues of both planning for concurrent execution and planning with incomplete knowledge (cf. 1.4.5).

The example of planning with incomplete knowledge is typical for realistic multiagent applications like the ones studied in the CoSy project. It also shows how an artificial separation of single-agent planning, multiagent coordination, and individual execution has led previous research to neglect important properties of MAP domains: centralized planning algorithms, for example, have ignored the need for coordinative actions (synchronization, communication) in a multiagent plan; distributed planners have assumed that plans (or sets of possible plans) can be found individually before coordination starts; nontrivial effects or interferences of synchronously executed actions have been modeled and taken into account by neither approach. Just as the continuous planning agent presented in section 1.4.5 is intended to couple planning, monitoring, and execution its extension to the multiagent case in this project interleaves (and thus integrate) planning, coordination, and execution in a novel way.

A key element for this integration is a formalism that allows agents to reason about their distributed planning problem as well as about the concurrent execution of the partial plans found so far. Parts of such a formalism have already been developed [21, 22] and has now been extended in the CoSy project.

Several other formalizations of multiagent planning problems and plans rely on hierarchical task decompositions [35, 34, 52] which allow to circumvent some of the aforementioned problems. Most hierarchical MAP algorithms assume implicitly that these decompositions are predefined, i.e. explicitly formulated by the domain designers, and that they are commonly known among the agents [34, 32]. However, in general this cannot be assumed. In fact, dialogues often manifest an ongoing distributed search for a valid task decomposition [78]. *Assertions* as a new form of abstraction introduced in this project (cf. 1.4.5) allows agents to build and learn hierarchies automatically and interactively. In the multiagent setting individual planning and learning of assertions have been extended with the possibility to share assertions through communication or devise them collaborative through dialogue.

If planning and execution of partial plans are to be interleaved it is a necessary requirement that even the plan fragments produced are at least *executable*. Many classical and modern planning algorithms do not have this property. For example, regressive partial-order planning algorithms [81, 97, 88] search for solutions starting from the goals and may produce an executable plan only when the problem is completely solved. The same is true for many hierarchical planners [129, 143]. However, in recent years planners based on heuristic forward-search have been developed [11, 17, 57] that not only are more efficient than previous approaches but also produce executable plan fragments during the search process. Unfortunately, these algorithms produce totally-ordered sequences of actions which makes coordination (i.e. merging or synchronization) of several agents' plans difficult. In the CoSy project we have investigated techniques for *progressive partial-order planning* that allow one to combine executability and efficiency of forward-search techniques with the flexibility of partial-order planning. When used within the continuous framework of planning with assertions forward-search techniques can be also extended to hierarchical planning [86].

1.4.6 Models of Action and Communication for Embodied Cognitive Agents

Often, an agent finds itself in an environment that it cannot fully know, control, or introspect. It is here important to explore the possibilities that collaborative action and communication offer to an agent to explore its environment, either on its own or together with other agents. Particularly in the case of joint exploration and problem-solving, natural language is *the* most efficient and effective way to communicate between artificial agents and humans: Instead of learning an artifact-specific language, a human can use the communication means she is accustomed to with little or no learning curve. This perspective complements the other two settings, where we focus primarily on how an agent can explore its environment through *perception* of the environment's spatial organization, and the objects found in the environment.

An important aspect of natural communication is that it is inherently *multi-modal*. Humans communicate not only using natural language, but also through gesture and other non-verbal acts such as body positioning, movement, or pointing. There are long-standing efforts to construct multi-modal dialogue systems that are able to capture various of these aspects: Not only communication from system to human can use a variety of modalities (for example spoken and graphical output), but also that from human to system (for example spoken input and gesture, e.g. [27], or pen-based input, e.g. [61, 93, 94, 62]).

However, despite significant advances in the design and building of multi-modal dialogue systems over the past decades, these systems are still mostly confined to relatively simple tasks that require little or no flexibility in the kind of communication to be expected in such a setting. Most if not all of the system's behavior is structured to follow rigid predefined patterns. This does not allow such systems to carry out more flexible and adaptive, natural conversations between human and computer.

The literature on multi-modal dialogue systems [27, 7, 126] has raised several issues that need to be addressed, if we are to develop systems that *are* capable of flexible and adaptive, natural communication, thus going beyond the typical *master-slave* relation common to dialogue systems for information-seeking, moving to mixed-iniative systems as for example required in a collaborative problem-solving setting.

Integration of communication and action. Under a Gricean view of language, dialogue rests on coordinated reasoning about and determination of communicative intentions [5, 126]. To understand language, the issue is to build a representation of the communicated content (*language-as-product*) and relate that content to (communicative) intentions (*language-as-action*). This presents challenges both in utterance interpretation (for example when combining multiple modalities like speech and gesture, cf. e.g. [93, 62]) and generation (e.g. [127, 125]). We need to understand intentions as commitments in linguistic action, cf. [100], preferably separating generic strategies for (collaborative) problem-solving from domain-specific knowledge [5, 6, 15].

Recognition of intention, attention, and grounding/understanding. For an important part, communication is about communicating *intentions* [28, 100, 101], possibly in the context of planning and expressing complex (group) action [52, 78]. This is linguistically realized through different types of dialogue moves (e.g. Searle's speech acts, DAMSL, or the VerbMobil ontology) or generic problem-solving acts [6, 15], with the possibility to shift attention among intentions [53, 15].

Besides recognizing intentions, and shifts in attention, an important point is the grounding of *understanding*. An agent must be able to signal whether it understood what was communicated, and it must be able to recognize whether another agent has or has not understood. The agent needs to be able to ask for and provide feedback or further clarification, as necessary. This also relates to the problem of *belief*-modeling, possibly involving notions such as trust.

Mixed-initiative. An inherent aspect of natural dialogues is *mixed-initiative*. Agents must be able to lead or yield control in the conversation in ways that best accomplish their goals, i.e. they should be able to take or release the initative to speed up the solution when opportunities arise [4]. To be able to model mixed-initiative, *incremental* strategies are needed for understanding, planning, scheduling and execution/generation, cf. [126, 7]. This issue becomes even more apparent for embodied cognitive agents: Being set in a partially unknown and unpredictable environment, agents can hardly expect to work out the full details of a plan at the outset without ever having to change it in the light of the perceived results – new opportunities *will* arise (the ramification and qualification problems, cf. [102]).

The *embodiment* and the *cognitive* character of the systems we have considered in this project add several new issues and twists.

Cognitively plausible architecture. Multi-modal dialogue systems are engineered so as to optimally perform the task that they have been designed for. Resource-bounds, introduced to increase the naturalness of human-system communication, may appear ad-hoc unless they are motivated on cognitive grounds. For example, there is no reason for a system's memory and its reasoning capabilities to be bounded; it is humans who forget or are unaware of some logical consequences of their beliefs (cf. the logical omniscience problem).

One step in a more cognitively oriented direction is the introduction of incrementality and resource-boundedness into dialogue systems capable of flexible and adaptive, natural dialogues [7, 126]. The challenge is to let these mechanisms be based on models of *human sentence processing*, offering the possibility to handle complexity of communication in a way that is natural to a human language user.

Furthermore, multi-modal dialogue systems are normally endowed with precoded world knowledge and linguistic skills required for carrying out their tasks. From a cognitive perspective, the challenge is to develop models that could explain how agents develop such knowledge and skills, i.e. acquire them over time through interaction with the environment and other agents.

Finally, multi-modal dialogue systems usually have dedicated input channels. For example, the pen-input channel is dedicated to pen-input, and not to any other form of interacting with the environment. In a cognitively plausible setting, channels are more likely to be *multi-functional*: Human visual perception not only perceives gestures in communication, but for example also objects and events in the environment. This gives rise to various challenges for both recognition and realization: How to separate different types of information arriving over a multi-functional input channel? How could different multi-functional modalities be combined to aid, and correct, recognition (*cross-modal recognition*)? How can a particular type of information be sent through a multi-functional modality unambiguously?

Embodiment in an unknown environment. Most multi-modal dialogue systems developed so far are not embodied – either not at all, like information-seeking dialogue systems or more complex systems like TRAINS or TRIPS [5, 7], or they are set in a virtual environment, e.g. [26, 27]. Embodiment in a physical, unknown environment adds new challenges.

Because the environment itself may be dynamic, the agent needs incremental strategies to handle not only flexible dialogue, but also environment interrupts.

The qualification and ramification problems of the agent's knowledge of the world [102], coupled with the vagaries of sensoric-motoric control and perception, require the agent to actively perceive the environment and the outcome of its actions; it is not "said and thereby done": Communicated intentions, and the actions they represent, may work out differently in reality than expected.

1.4.7 Multi-Modal Recognition and Categorisation

The capacity to recognize and categorize objects plays a crucial role for cognitive systems in order to compartmentalize the huge numbers of objects it has to handle into manageable categories. In order to achieve this, the ability to learn from experience and adapt its object models must be at the core of any cognitive system. There is consequently a need for methods for automatic acquisition of models, and structuring those models into hierarchies to allow the system to operate in an open-ended fashion, i.e. beyond initial specifications. This is related to points made in other sections about the need for perceptual discretization or chunking to support deliberative processes of exploring branching possibilities, e.g. searching for plans or explanations, in order to avoid intractable branching continua.

Quite interestingly, for humans it was shown that entry-level categorization (i.e. "Is this a dog/cat?") is much faster in human vision than recognition or identification ("Is this my dog/cat?"). These findings suggest that humans do a sort of coarse to fine categorization and recognition of objects. This is also reflected in language by the way humans talk about objects and scenes. Evolution probably also developed the ability to categorise, e.g. food, prey, shelter, before the ability to identify individuals, which for many organisms is not a requirement.

In this project we have taken the approach that dialogue models and object models should cross-fertilize through cross-modal integration. Learning, and linguistic semantic ontologies have been tightly coupled with perceptual and planning ontologies. On the one hand language can serve as a trigger and provide guidance to learn and acquire new models but also to structure models and representations. On the other hand the role of object models and representations used in perception and action may serve as partial determinants of the semantics of the language. This is one example of what we refer to as cross-modal integration. As mentioned before, an important issue for any cognitive system is the ability to communicate about its internal knowledge and state so that a particular challenge will be also to communicate about the acquired models and model hierarchies of objects.

Recognition of objects is obviously one of the most fundamental problems of computer vision and consequently has been studied widely. However, one problem in recognition is that most approaches have been domain specific or tailored to narrow domains. In addition most approaches have been based on single visual cues such as invariants (points, lines, structures). It is here suggested that recognition must be based on use of multiple visual cues as no single cue is robust in a general setting [115, 82, 120]. In addition there is a need to exploit contextual information to constrain the set of potential objects [134]. Goal-oriented information has been utilized in a few recent efforts such as [29, 44]. In addition context information has been used in a number of other efforts as for example reported by [69, 85, 33].

In the context of categorization some have attempted to use the idea of functional representations for modeling of object categories such as chairs [124]. These efforts have, however, assumed access to a full CAD model of the objects of interest, which clearly is unrealistic. Recently there have been efforts to use simple geometric features for categorization of objects such as cars and tools [140, 43], but it is not immediately obvious that this effort will scale and generalize. In general conceptual models have been used as the basis for categorization, yet it is not immediately obvious that such collections of objects form visual categories in a natural fashion and there is thus a need to revise the concept of categorization to match visual categories, which may be combined into super-sets that form conceptual categories.

However by interacting with the objects and observing others using and interacting with these objects another route for cross-modal learning is possible and have been pursued in this project: Here, cross-modal integration is between perception and action where the system can start to represent and build models of object affordances and then can recognize and categorize objects based on those models.

Therefore, through construction of multi-cue and cross-modal representations that integrate these representations it is likely that more efficient and robust mechanisms for priming, verification and recognition can be derived. Another key ingredient is the use of task-oriented context driven methods for association in the context of multiple cues that each are distorted due to uncertainty. Therefore as part of this project the visual recognition a multi-cue and cross-modal basis for recognition and categorization have been studied.

Another important and notorious problem in computer vision in general, is figure-ground segmentation in which objects are separated from the background. A very rich set of studies have addressed this problem. A fundamental premise is that figure ground segmentation only makes sense when studied in a task context, i.e. the segmentation is task dependent. In this context the role of discriminant vs. representational model characteristics have also been studied. The former are required to recognize a large variety of objects, whereas the latter allow segmentation of those objects in novel scenes.

Our work on the scenario "Models for Objects and Concepts" combines the above ideas regarding the need for cross-modal and multi-directional processing in segmenting the environment in a context- and goal-dependent manner, including detection of regions of the environment about which the agent is unclear or confused. Depending on the agent's current goals, such regions may either be ignored, or else investigated more fully by re-directing attention to possible sources of disambiguation or possible cues to check doubtful partial interpretations. This not only helps to facilitate figure ground segmentation (combining bottom-up and top-down processing) but also allows testing, verification, and subsequent improvement of the models learned by the agent.

1.4.8 Scenario Driven Research

Sections 1.4.1 – 1.4.7 have discussed basic research challenges involved in the design of cognitive systems. Each of these key challenges are important by themselves but more importantly they also have to be studied in the context of integrated systems, where the *system level* issues can be studied as part of operational systems embedded in natural environments. To complement the studied of individual challenges two scenarios were defined as a basis for systems integration and empirical studies. The scenarios were:

Exploration/Mapping of Space. A fundamental competence of a cognitive agent is the ability to recognize its own interaction with the environment and to be able to generate an internal model of its environment as a prerequisite for operation and interaction with the environment. In these scenario the issue of self-recognition and insertion into the environment. This involves modelling of perception interaction and recognition of the embodiment. Once such a competence is available the agent can move about in the environment and generate an internal model of the environment. Without a complete/comprehensive method for categorisation of objects/structures in the environment the agent must be assisted by a tutor to recognize / disambiguate structures in the environment. The emphasis of this scenario is thus on spatial models and self-awareness.

Models of objects and concepts. For interaction with the environment, reasoning about scenes, and communication with other agents it is essential that the system has facilities for acquisition of models of objects, events and structures. This involves both static objects and dynamic phenomena. The number of objects present in a natural environment calls not only for recognition, but for categorisation of objects to place them in a spatio-temporal hierarchy. For operation and interaction with objects pure RE-cognition will facilitate a limited complexity of interaction and there is thus a need to recognize/classify affordances as a more generic level of modelling of structures and objects. In this scenario the emphasis has been on categorisation of objects, events and structures. Again the agent will cooperate with other agents in its environment to accomplish its missions/goals. In addition the agent will actively explore the environment to "discover" new objects and structures.

Each of these three scenarios discussed below:

1.4.9 Exploration/Mapping of Space

A fundamental aspect of any autonomous / intelligent system is the ability to navigate through semi-structured environments. Navigation involves three questions: i) Where am I? ii) How do I get to my destination? and iii) how do I detect that I have arrived at the destination? All of these questions are closely related to a representation of space and equally important the ability to reason about space. For autonomous operation in everyday life the system needs

- A self-image that models the relationship between perception and action, to understand how control actions influences perception of the world.
- Methods for localisation in the world (detection of present position)
- Methods for construction/acquisition of a map of the environment
- Methods to plan a sequence of actions (subject to the constraints /competencies) for task achievement

In terms of basic cognitive competencies (representation, recognition, learning and reasoning) each of these four issues have differing implications on the system architecture, models of the environments, interaction with other agents and operation.

Self-recognition and Spatial Insertion

Affordances and Newer Approaches
In AI and robotics, the classic approach to building an agent that is able to spatially explore its environment has been to assume that it is necessary to create, within the agent, an internal representation or cognitive map of the environment in which spatial locations (coded either in absolute or relative coordinate systems) are associated with objects situated at those locations.

In biological systems however such an approach is unrealistic because both the notion of location and the notion of object are not well defined.

As concerns the notion of location, biological agents grow and change, and so are often faced with modifications in their physical make-up which make it difficult to obtain accurate, repeatable measurements. An agent might for example be carrying a load or wearing interfering apparel, or moving in a situation where terrain characteristics like mud or water modify the characteristics of its motion and change the relation between its motor commands and the spatial displacement they produce.

Similarly, a biological agent has sensors whose properties may change over time, or which may be modified by environmental factors: an agent may be moving at night, in fog, under water, in a dust storm, in conditions of unusual lighting, where the response characteristics of its sensors might be severely modified.

Both these examples show that the notion of "location", simply conceived as a coordinate system related to agent displacements, as measured by its sensors and modified by its effectors, cannot be easily maintained in robust biological systems.

As concerns the notion of object, here again similar problems arise. An agent that establishes a map of its environment in terms of objects must necessarily rely on some way of describing and coding the objects. And yet in real biological life, the sensors used to identify objects may be damaged or affected by environmental conditions. Similarly, the objects themselves may suffer modifications or changes that make them no longer identifiable by simple pattern matching techniques. For example, houses may be repainted, tablecloths changed, furniture replaced, vegetation may grow, etc.

The notion of affordance [49] may be biology's answer to these problems, and a development of this notion, the notion of sensorimotor contingency [92], can be of additional help.

Take as an example the notions of "wall", "door", "table". In a classical approach to wayfinding, such entities may be stored in the cognitive map of a region by way of their particular visual characteristics: surface color, shapes, orientation, etc., and their associated locations. But exact identity or location of physical characteristics are generally unimportant as far as a biological agent is concerned. What is important is what the agent can do with respect to these entities. A wall is something which you can move along but not through. A door is a thing through which you can see things outside the room, and by which you can go out of the room. A table is a thing that you can put objects on.

To a large extent these properties are invariant with surface properties and visual aspect. The tablecloth can be changed, the size of the door can be modified, the wall can have a new picture hanging on it.

The trick is to find a way of coding such aspects of the agent's environment, a way which does not depend on the particular visual aspects. One partial answer to the problem is to use multimodal sensory cues. Making use of tactile, auditory, or other sense modalities may provide precious help.

But an additional, perhaps vital factor may be that in order to succeed in this task, biological agents are helped by use of their ability to move and manipulate their environments. For example, an open door can be detected by the fact that because of parallax, when the agent moves there is a well-defined patch of visual flow (coming from outside the door) which moves slower than the surrounding surfaces (corresponding to nearby walls). A table is a thing such that when the agent puts its hand on it, the table resists its pressure, and such that when the agent moves with respect to it, the flow field produced corresponds to that of a horizontal surface.

The World as an Outside Memory
A second strategy that appears to be used by biological agents in order to maximize their efficiency appears to be the idea of relying on the outside world as its own representation.

Instead of making a detailed internal representation of the world, with the risk that this representation will become out-of-date when the environment changes, biological agents often use crude global cues in their environments to allow them to home in to elements that they are interested in. For example, instead of coding the GPS coordinates of the coffee container, humans know it is in the cupboard, and that the cupboard is in the kitchen. Once they have moved into the kitchen they look for the cupboard and find the coffee.

This way of coding the environment has several advantages. The first is that it requires little memory. The second is that it is robust with respect to variations in object properties and locations. Changes in the exact position of the cupboard in the kitchen, the coffee in the cupboard make no difference to performance. The exact shape of the cupboard, the marks on the coffee package are of little importance, since these have not been stored. This way of coding information is also insensitive to changes in sensor and effector properties: Since it is not visual templates of objects which are coded but sensory and motor, code-independent aspects of the way the objects change as the agent moves, degradation or modification of the agent's sensory and motor apparatus will have little effect on behavior.

Clearly however this coding strategy is not always possible. There are many cases where surface markings, shapes etc. of objects, which cannot be coded in an intrinsic, code-independent sensorimotor fashion, must be retained. Furthermore there are cases where the exact identity of an object is important and must be stored, for example when an agent is required to detect small changes in the environment. In that case a supplementary encoding strategy must be used.

The problem of how to combine and coordinate both modes of representation must be solved. It is possible that this might be done within a Bayesian framework. It is possible that such an approach might provide an instantiation of the biological notion of attention, in which agents select useful parts of their environment to attend to, be it at the time of encoding or during everyday exploration.

Mapping of the Environment

Mapping of the environment [106] has been studied both in terms of data acquisition with subsequent map construction (off-line) [36],[70], [113] and more recently as part of Simultaneous Localisation and Mapping (SLAM) [103], [131], [54], [75] and [146]. The by far most dominating approach to localisation and mapping is based on use of geometric features encoded in an occupancy grid or as a metric map of features such as lines, corners, feature compositions, etc. The advantage of such maps is that they can be constructed using standard parametric estimation techniques such as Kalman filtering. Such maps do, however, have a number of problems:

- Geometry may not in an accurate fashion encode the concept of "place".
- It is not immediately obvious that the approach scales to natural large scale environments.
- The maps are unrelated to the embodiment of the agent.
- The maps may fail to contain crucial information for planning, spatial reasoning, etc.

While the geometry based methods are well suited for basic navigation there is obviously a need for a scalable, embodied, and richer representation for cognitive systems. The requirements for a model of the environment (for the purpose of moving about) are manifold:

- Encoding of position of objects/places
- Encoding of environmental topology
- Invariant to changes to perception system
- Invariant to changes in action system
- Facilitate spatial reasoning
- Allow localisation

In principle two different approaches can be considered to the problem of spatial exploration/mapping. i) direct mapping of space based on a sensori-motor map, ii) encoding of space as a hybrid topological, geometric/semantic map that facilitates reasoning. Both types of representation carry different types of advantages, which is the reason why both approaches have been pursued as part of the research in Cosy.

To enable operation in the vicinity of objects and allow interaction with them there is a need to consider the "affordances" of the objects, i.e. the type of interaction and operations that an object facilitates. Acquisition of such maps can be acquired through direct interaction with the objects through which the perception-action coupling can be determined and invariances that allow RE-cognition can be identified. The exploration can here use (at least) three modes of acquisition.

1. Autonomous exploration: in which the agents explores the environment on its own and gradually builds up a representation. The exploration can both both random or pseudo structured in terms of coverage
2. Operator Assisted: as the agent moves through the environment the various objects/structures are identified / labelled by the operator and the agent can query the operator about the objects detected.
3. Supervised: the robot is given a tour of the environment and all objects are directly identified to the agent.

In practical terms it is of interest to perform research on the use of all three modes of operation to consider the differences in generated representations and in terms of efficiency of learning.

In parallel to the generation of sensori-motor maps and their integration of structures in the environment it is of interest to consider more traditional hybrid representations of space in which a hybrid topological/graph based model is used for encoding of the over-all structure of the environment. To allow recognition of places there is a need to "categorize" objects into classes that allow relocalisation. At the same there there is a need to construct spatial maps in which the relations between places and objects are made explicit. Where traditional localisation and mapping has considered geometric maps the idea is here to utilize a semantic approach to localisation in which place recognition is based on recognition of objects specific to a location rather than pure geometric location. The places are then integrated into a graph structure that encode spatial relations. Such a representation will not only allow localisation and mapping but also reasoning about space. A crucial issue considered here is the handling of uncertainty to allow the representation to have convergence in terms of geometric accuracy and handling of topological events such as loop closing. In this approach there is also a need consider the efficiency of mapping in relation to each of the three interaction approaches outlined above.

1.4.10 Models for Object and Concepts

Generally, an agent lives and operates in an external world which is neither fully known to or predictable by the agent, nor fully introspectable – no representation is ever equal to the external world it represents. In order to understand, reason and act in such an unknown and 'hostile' environment the agent has to continuously learn and update its concepts and models about this world.

This scenario in particular concentrates on the formation of concepts about objects and object manipulation, as well as action, activities and events. A particular emphasis is on the joint exploration of, communication about, and dealing with unknown reality (objects & actions external to the embodied agents), whereby "unknown" means not only "not-seen-before" but also unknown in the sense of (ontologically) unknown affordances ("ontological novelty"; surprise & true abduction). In this context the different modes of learning (tutor driven, tutor supervised, and exploratory) are particularly important and their interplay have been explored in this scenario.

More specifically we have started from the exploration of "static" objects and simple action, yielding concepts based on a given ontology, objects about which the agent can obtain further information (introspection) using communication with other agents (requiring interpretation of deictic references). Also we have worked on the perception, representation, and interpretation of objects & actions in terms of affordances; introspection of an agent's notion of space in terms of (affect-)act-cause-effect, using abductive reasoning ("WHY don't I understand?") and communication with other agents.

In this scenario we deal with the following cognitive abilities of an agent:

- The agent should be able to learn which features of an object, or an arrangement of objects provide an affordance for achieving a certain goal. The agent should be able to classify or categorise objects by their affordances.
- The agent should be able to verbalise the properties of objects, the relationships between objects, and the presence (or lack) of affordances, for semi-novel scenes.
- The agent should be able to infer and verbalise its beliefs about the causes of events.
- The agent should be able to learn to categorise and recognize objects in cluttered visual fields under standard lighting conditions using time series of visual, haptic, verbal, and proprioceptive information.
- Representations of objects should encompass more than just static visual information; they should take into account also temporal aspects, other modalities, and higher-level cognitive processes in order to enable different modes of life-long learning.
- Robustness is an important aspect of a cognitive agent since it is equipped with imperfect sensors and effectors, and it is supposed to operate in a partially unpredictable real-world environment.
- The algorithms for visual learning and recognition should not require a user-specified figure-ground segmentation beforehand. Instead, the system should arrive at a segmentation on its own based on previous knowledge and expectations using a combination of multiple cues.
- Robust learning (not just robust recognition) is another very important issue and should be strongly intertwined with the process of continuous learning.
- Contextual information plays an important role during recognition and learning and should be therefore integrated into the representation of the objects and actions.

Learning of Spatial-Temporal Concepts and Causes

In the scenario on exploration and mapping of space we have devised representations suitable for expressing, among other things, the qualitative spatial relationships that exist between an agent and an environment. In the objects and concepts scenario this has been extended to deal specifically with the spatial relationships that exist between objects, and their other properties. In addition to static spatial relationships an important element of this scenario will be representations used to express, and learn the way that these relationships and properties may be altered. In effect we aimed to construct action models automatically, selecting the correct levels of abstraction to use in a hierarchy of representations.

Currently we have methods that are capable of expressing action models at either the sub-object (geometric) level or the object level. The former are probabilistic and essentially pose the problem of recognition of action sequences as being akin to the problem of localisation. They naturally have a time series formulation which is capable of representing the effects of actions, and there are also algorithms for learning action models. These types of representations are widely used in modelling visual behaviour [25]. They are flexible, but are data-driven, and prone to over-fitting. They have also been applied in limited cases to categorising time series information from robots. Open questions in this area include the issue of how to learn more structured models directly from data; how to identify events occurring over different time scales; and how to learn to segment robot time series into natural and meaningful chunks.

The abstraction required to model actions at an object level is typically done by hand, in for example the formulation of AI planning operators in cognitive robotics. These do not interface easily with sensory information. This work uses action models based on logics suitable for dealing with events. Some workers have now linked these directly to sensory processing to perform recognition and inference [116, 117, 119]. These systems currently allow recognition of events from visual or distance data and in some cases inference of their causes by abduction. There are also rigourous inference procedures, although these are typically intractable. Neither do they typically incorporate multiple modes.

Reasoning about causes at an object level is also important because it gives us a possible way of identifying affordances and expressing affordances in a relational manner. For example we can say that a teacup and teapot provide an affordance for pouring tea if the pot's spout is above the teacup. The exact vertical distance between them and the exact position of the cup relative to the spout are part of a geometric description which can be ignored at more abstract levels of reasoning.

In this scenario we will develop methods that connect the two levels of description: object and geometric. As in the case of object recognition we use context, so that recognition of behaviour and object will influence one another. In summary representations for reasoning about changes in objects and their relations that we have designed will:

- Efficiently infer causes of events.
- Give insights into natural phenomena such as change blindness and erroneous assignment of cause.
- Allow the recognition and categorisation of objects by the types of interaction that they may be engaged in.
- Allow reasoning about change at different levels of abstraction.
- Allow learning about the invariant properties of objects under changes to them.

Before even beginning to be useful, an agent has to understand its own body, and the relation of its body to the environment. In the imitation scenario for example, the agent has to be able to move its arm to a target position, do eye-hand coordination, and do these things irrespective of whether the arm is loaded with a load, impeded by some obstacle, or moving in unusual conditions such as injury or mechanical dysfunction, and irrespective of similar variations in its sensor characteristics.

Classical schemes for accomplishing these goals generally involve making precise models of the agent's motor and sensory apparatus, and generally use analytical means to calculate the required control sequences.

We propose an alternative scheme based on the notion of intrinsic sensorimotor laws. These are intrinsic invariants of the functions that link motor commands to the resulting sensory changes. The invariants are intrinsic in the sense that they are independent of the exact physical structure and characteristics of the agent, because they correspond to environmental affordances, and not to aspects of the agent itself. For example, moving the fingertip in a straight line is something that can be described in a way that does not depend on the sensory or motor code that is used by the agent [91, 98, 99].

Multi-modal Learning of Object Categories

Representations. Most of the state-of-the-art algorithms for visual learning and recognition represent objects and object categories as collections (or constellations) of descriptors or parts, where each part has its distinctive appearance and spatial position relative to the other parts [141, 139, 111, 79, 80, 63, 2, 42, 73].

These representations are, however, far too narrow, inefficient, and insufficient for accomplishing tasks which are envisioned in the project. The representations of objects should encompass much more than just static visual information; they should take into account also affordances, temporal aspects, other modalities, and higher-level cognitive processes. They should also enable different modes of life-long continuous learning. The representations of objects and concepts should be merged together with representations of places in a unifying framework.

From the point of view suggested in [91], [98] and [99] the intervention of action and multimodal intervention can be conceptualized and modelled naturally by appealing to the notion of intrinsic sensorimotor invariants. Like affordances, these presuppose, in an essential manner, active exploration on the part of the agent. For example, within this framework, an object is an entity that produces sensory changes which can be reproduced in other environments (that is, said more anthropocentrically: an object can be moved around). Object categories can also be conveniently described in this sensorimotor framework: a table is a thing whose top surface has certain intrinsic properties: the agent can glide its hand smoothly over it; the tabletop suffers certain visual transformations characteristic of flatness when the camera moves with respect to it.

Continuous Learning. An essential characteristic of a cognitive system is its ability to learn. The cognitive system has to be able to modify, adapt, and expand the ontology. Only by learning can a cognitive system develop and adapt to new situations enabling autonomous higher level reasoning and behaviors in a general real-world settings. It is very important that learning is a continuous, open-ended, life-long process.

Most of the state-of-the-art algorithms for visual learning and recognition do not consider continuous learning and follow the standard paradigm of the division on the off-line learning stage (characterized by a controlled environment, and slow, computationally demanding processing) and on the recognition stage, where the processing should be fast and should enable recognition and categorization (also) in unconstrained environments [141, 139, 111, 79, 80, 63, 2, 42, 73].

Most of these approaches are not designed in a way, which would enable efficient updating of the learned representations, as a basic prerequisite for continuous learning. In contrast, we have designed representations, which will not be learned once and for all in the training stage and then used in their fixed form at the recognition stage. They will be continuously updated over time, adapting to the changes in the environment, user reactions, user preferences, etc. A strict distinction between the learning and the recognition stage should diminish and learning and recognition should be performed intertwiningly.

The idea of continuous learning is natural within a sensorimotor, action-based framework, where the system is continuously evaluating the effect of its own actions on its sensory input, thereby effectively continuously recalibrating its own body schema and sensor properties.

Robustness. Robustness should be another important property of a cognitive agent, since it is equipped with imperfect sensors and effectors, and it is supposed to operate in a partially unpredictable real-world environment. The algorithms for object recognition and categorization should be able to robustly operate in such non-ideal conditions as well. They should not require an explicit and precise figure-ground segmentation beforehand. An implicit figure-ground segmentation should be carried out using the combination of bottom-up and top-down approaches, intertwined with the object recognition procedure [73, 18].

Furthermore, within an active, sensorimotor approach, the problem of figure-ground separation is radically simplified, since the agent can, by controlling its own motions, generate sensory changes which allow it to isolate objects from their surroundings.

Robust learning is even a greater challenge than robust recognition. Since during the learning process a model of an object is being built, there is no previous knowledge, which could be used to determine the relevance of local descriptors for characterizing the particular object. Therefore, robust learning should be strongly intertwined with the process of continuous learning, which could provide enough redundant information to determine statistically consistent data (this information can also be provided by a user acting as tutor). Only the consistent data would then be used to build the representations of objects, enabling robust learning (and updating of the representations) under non-ideal real-world conditions.

Context. Another issue, which has been addressed in the project, is contextual information. Traditional object recognition algorithms can, in general, recognize only isolated objects. Some newer object recognition algorithms are robust and enable recognition of objects in the presence of occlusions and cluttered background without explicit figure-ground segmentation. However, all these algorithms tend to discard background regions and use only the information contained in the foreground. However, the background (contextual) information can be very useful: objects usually appear in a specific environment (context). Thus, context can be very useful for reducing the number of objects, which can appear in a particular environment, it can increase the confidence about the identity of the objects (resolve possible ambiguities), and it can narrow the image area, which has to be explored (set the focus of attention). These facts have been considered very rarely in the algorithms for object recognition [134, 90]. Therefore, we aimed at incorporating contextual information into the representation of the objects and scenes. The contextual information should be continuously updated, reflecting the changes in the real environment.

Categorisation. Objects and concepts can be recognized/classified on different levels of abstraction, considering different properties and following different goals. Traditionally, computer vision research has frequently focused on the recognition of specific object instances. While there has been considerable interest in the more general problem of object categorisation from the beginning, research has only recently made progress on recognizing object class members in challenging real-world settings [112, 95, 141, 139, 2, 42, 72, 73].

These approaches try to implement basic-level object classification [107] using visual input. Having other modalities as well, one could also consider them for a more reliable object classification. For example, basic level categories are also specified by common motor programs for interaction with category members [107, 71]. Therefore functional categories could arise by grouping objects with similar affordances, ad-hoc categories [12] could result from objects' contextual information. Using higher level cognitive processes, the different types of higher level categorization (beyond basic level) could be possible.

Architecture. In this scenario the implications for the overall architecture have been investigated. It may for example turn out useful to have different perceptual and learning processes in a reactive layer, in a deliberative layer and in a meta-management layer that includes information about intelligent systems. All these mechanisms will of course share the same physical sensors and may share a lot of the low level visual analysis mechanisms.

However it is possible that outputs from those low level systems will go in parallel to different perceptual systems performing different sorts of tasks, including for instance feedback control of actions, rapid recognition of obstacles within reactive motion systems, classification of objects and events at a level of abstraction required for learning re-usable generalisations about the environment, and interpretation of mental states and intentions of other agents.

Communication and Language. One of the main necessities for an embodied cognitive agent is to be able to communicate about the environment it finds itself in, to be able to learn more about the environment or to coordinate actions among multiple agents so as to collaborate in solving a task.

Whether using communication about the environment for learning or for coordinating actions, a fundamental issue is how language can be used to refer to objects and locations in the environment, either by describing them directly or by referring to them. The long-term goal we want to explore in this project how such descriptions and references can be interpreted and realized in a cross-modal setting, integrating different modalities such as speech, gesture, vision, and movement ("body positioning").

Not only does a cross-modal integration of modalities often lead to more efficient and more natural communication. It is also an essential aspect of learning through language, where an agent uses communication to learn more about the environment it finds itself in. For example, the agent could communicate with an expert who guides the agent and provides all the necessary information. Learning through language involves acquiring new words as well as new ways of talking about and dealing with reality, and thus inherently requires the integration of information from various modalities.

In turn this leads to the interesting issue of how to ground the semantics learned for the new word. On the one hand, the semantics should reflect the sensory data, and the categorization of that data in relation to an ontology – the word meaning needs to be fit into the agent's existing knowledge of the environment. On the other hand, the semantics of the word puts it in semantic relations with other words, e.g. via synonymy (different words, same meaning), hypernymy (super/sub-concept) and hyponymy (sub/super-concept). This reflects a lexical organization which determines for example in what contexts it would be appropriate to use the word as a description. The interaction between linguistic knowledge and situational knowledge again stresses the broader understanding of cross-modality we develop in this project.

1.5 Consortium

(Embodied) Cognitive systems naturally involve research on robotics, perception, action modelling, reasoning, learning, language and representations. Consequently the CoSy consortium brings together core competence from each of these disciplines in addition to systems integration. The involves institutions are briefly summarized below.

The Centre for Autonomous Systems (CAS), KTH has been the coordinator of the consortium. CAS has a proven record of constructing (robot) systems for natural environments. The research involves computational vision, control methods, behaviour based systems, and systems integration. In CoSy the research has been on perception-action modelling, categorisation, integration and spatial mapping.

DFKI is the largest European R&D Center in Artificial Intelligence. Its main areas of research are language technology, intelligent user interfaces, automated reasoning, multi-agent technology, knowledge management and visualization. In CoSy DFKI has mainly contributed expertise and technology in language processing and dialogue modelling.

University of Freiburg has a long term record on representations, formal methods, planning and mobile agents. The group is well known for it basic research on artificial intelligence. In this project the group has in particular focused on planning, failure detection and multi-agent interaction.

The University of Birmingham has a strong record on general aspects of artificial intelligence, cognitive systems, architectures and robotics. The partners brought a strong AI / Cognition experience to the consortium.

The partner from TU Darmstadt has a solid background in computer vision, multi-cue integration and learning. In relation to CoSy TUD has in particular performed research on categorisation, object categorisation, cross modal representations, and learning.

University of Ljubjlana has a strong research effort on computational vision and robotics. The work has considered both integration, recognition, learning and visual representations. In CoSy the research has emphasized categorisation, learning and representations.

The Laboratory of Experimental Psychology, CNRS has long studied the role of perception action modelling and in particular the sensory characteristics of vision and it implications for perception in general. LPE-CNRS has in particular been involved in perception-action modelling, categorisation and multi-cue integration.

1.6 Organization of the Book

The CoSy project ended in August 2008 and the main results of the four year research program are reported in the following chapters.

Chapter 2 present the main results in terms of design of a common architecture for cognitive systems and corresponding considerations related to representations. A key aspect of design of system is the perception-action embedded or spatial insertion into the world. Aspects related to perception-action integration are presented in Chapter 3. The generation of spatial maps in terms of basic geometry, topology and semantics is presented in Chapter 4. Visual perception for mobility and manipulation is presented in Chapter 5. Cognition heavily relies on planning and recovery from failures as presented in Chapter 6. Another key aspect of a cognitive system is adaptation and learning across perceptual modalities, skills and tasks as presented in Chapter 7. Cognitive systems does not exist in isolation but interact closely with humans as part of task specification, skill learning and clarification. Human-robot interaction based on situated dialogue is thus a key aspect of a system as presented in Chapter 8. Chapters 9 and 10 present demonstration systems and the final two chapters present reflections on lessons learnt during CoSy and challenges for the future.

References

1. Proceedings of the 12th National Conference of the American Association for Artificial Intelligence (AAAI 1994), Seattle, WA, July 1994. The MIT Press, Cambridge (1994)
2. Agarwal, S., Roth, D.: Learning a sparse representation for object detection. In: Heyden, A., Sparr, G., Nielsen, M., Johansen, P. (eds.) ECCV 2002. LNCS, vol. 2353, pp. 113–130. Springer, Heidelberg (2002)
3. Agre, P., Chapman, D.: Pengi: An implementation of a theory of situated action (1987)
4. Allen, J., Byron, D., Dzikovska, M., Ferguson, G., Galescu, L., Stent, A.: An architecture for a generic dialogue shell. Journal of Natural Language Engineering 6(3), 1–16 (2000)
5. Allen, J., Byron, D., Dzikovska, M., Ferguson, G., Galescu, L., Stent, A.: Towards conversational human-computer interaction. AI Magazine 22(4), 27–37 (2001)
6. Allen, J., Ferguson, G.: Human-machine collaborative planning. In: Proceedings of the Third International NASA Workshop on Planning and Scheduling for Space, Houston TX (2002)
7. Allen, J., Ferguson, G., Stent, A.: An architecture for more realistic conversational systems. In: Proceedings of Intelligent User Interfaces 2001 (IUI 2001), pp. 1–8 (2001)
8. Ambros-Ingerson, J.A., Steel, S.: Integrating planning, execution and monitoring. In: Proceedings of the 7th National Conference of the American Association for Artificial Intelligence (AAAI 1988), Saint Paul, MI, August 1988, pp. 83–88 (1988)
9. Ashish, N., Knoblock, C.A., Levy, A.Y.: Information gathering plans with sensing actions. In: Steel, S. (ed.) ECP 1997. LNCS, vol. 1348, pp. 13–25. Springer, Heidelberg (1997)
10. Bacchus, F., Ady, M.: Planning with resources and concurrency: A forward chaining approach. In: IJCAI 2001 [58]
11. Bacchus, F., Kabanza, F.: Using temporal logic to control search in a forward chaining planner. In: Ghallab, M., Milani, A. (eds.) New Directions in AI Planning (EWSP 1995), Amsterdam, The Netherlands. IOS Press, Amsterdam (1996)
12. Barsalou, L.W.: Ad-hoc categories. Memory and Cognition 11, 211–227 (1983)
13. Beetz, M., Burgard, W., Fox, D., Cremers, A.B.: Integrating active localization into high-level robot control systems. Robotics and Autonomous Systems 23, 205–220 (1998)

14. Bertoli, P., Cimatti, A., Roveri, M., Traverso, P.: Planning in nondeterministic domains under partial observability via symbolic model checking. In: IJCAI 2001 [58], pp. 473–478
15. Blaylock, N., Allen, J., Ferguson, G.: Managing communicative intentions with collaborative problem solving. In: Smith, R., van Kuppevelt, J. (eds.) Current and New Directions in Discourse and Dialogue. Kluwer Academic Publishers, Dordrecht (2003)
16. Blum, A., Furst, M.L.: Fast planning through planning graph analysis. Artificial Intelligence 90(1-2) (1997)
17. Bonet, B., Geffner, H.: Planning as heuristic search. Artificial Intelligence 129(1-2), 5–33 (2001)
18. Borenstein, E., Ullman, S.: Class-specific, top-down segmentation. In: Heyden, A., Sparr, G., Nielsen, M., Johansen, P. (eds.) ECCV 2002. LNCS, vol. 2351, pp. 109–124. Springer, Heidelberg (2002)
19. Boutilier, C., Brafman, R.: Partial order planning with concurrent interacting actions. Journal of Artificial Intelligence Research (2001)
20. Boutilier, C., Poole, D.: Computing optimal policies for partially observable decision processes using compact representations. In: Proceedings of the 13th National Conference of the American Association for Artificial Intelligence (AAAI 1996), pp. 1168–1175. The MIT Press, Cambridge (1996)
21. Brenner, M.: Multiagent planning with partially ordered temporal plans. In: IJCAI 2003 [59]
22. Brenner, M.: Multiagent planning with partially ordered temporal plans. Technical report, Institut für Informatik, Univerisität Freiburg, Germany (2003)
23. Brooks, R.A.: Intelligence without representation. Artificial Intelligence 47, 139–159 (1991)
24. Burgard, W., Fox, D., Thrun, S.: Active mobile robot localization. In: Proc. of the International Joint Conference on Artificial Intelligence, IJCAI (1997)
25. Buxton, H.: Learning and understanding dynamic scene activity. In: European Conference on Computer Vision: ECCV Generative Model Based Vision Workshop, pp. 71–81 (2002)
26. Cassell, J., Bickmore, J., Campbell, I., Chang, K., Vilhjalmsson, H., Yan, H.: Requirements for an architecture for embodied conversational characters. In: Proceedings Eurographics 1999, Berlin, Germany. Springer, Heidelberg (1999)
27. Cassell, J.: Embodied conversational agents: Representation and intelligence in user interfaces. AI Magazine 22(3), 67–83 (2001)
28. Cohen, P.R., Levesque, H.J.: Intention is choice with commitment. Artificial Intelligence 42(2-3), 213–361 (1990)
29. Crowley, J.L., Christensen, H.: Vision as Process. ESPRIT BR Series. Springer, Heidelberg (1995)
30. de Freitas, N.: Rao-blackwellised particle filtering for fault diagnosis. In: Proc. of the IEEE Aerospace Conference (2001)
31. Dearden, R., Clancy, D.: Particle filters for real-time fault detection in planetary rovers. In: Proc. of the Thirteenth International Workshop on Principles of Diagnosis (2002)
32. Decker, K., Lesser, V.: Designing a family of coordination algorithms. In: Proceedings of the 1st International Conference on Multiagent Systems (ICMAS 1995), San Francisco, CA, pp. 73–80. MIT Press, Cambridge (1995)
33. Dickenson, S., Christensen, H.I., Tsotsos, J., Olofsson, G.: Active recognition integrating attention and view point control. CVIU 67(3), 239–260 (1997)

34. Durfee, E., Montgomery, T.: Coordination as distributed search in hierarchical behavior space. IEEE Transactions on Systems, Man, and Cybernetics (1991)
35. Durfee, E.: Coordination of Distributed Problem Solvers. Kluwer, Dordrecht (1988)
36. Elfes, A.: Sonar-based real-world mapping and navigation. IEEE Journal of Robotics and Automation RA-3(3), 249–265 (1987)
37. Ephrati, E., Rosenschein, J.S.: Divide and conquer in multi-agent planning. In: AAAI 1994 [1]
38. Erol, K., Hendler, J.A., Nau, D.S.: HTN planning: Complexity and expressivity. In: AAAI 1994 [1], pp. 1123–1129
39. Erol, K., Hendler, J.A., Nau, D.S.: Complexity results for hierarchical task-network planning. Annals of Mathematics and Artificial Intelligence 18, 69–93 (1996)
40. Ritter, F.E., et al. (eds.): Techniques for Modeling Human Performance in Synthetic Environments: A Supplementary Review. Defense Technical Information Center, Ft Belvoir, VA (2002)
41. Etzioni, O., Hanks, S., Weld, D., Draper, D., Lesh, N., Williamson, M.: An approach to planning with incomplete information. In: Nebel, B., et al. (eds.) [87], pp. 115–125
42. Fergus, R., Perona, P., Zisserman, A.: Object class recognition by unsupervised scale-invariant learning. In: CVPR 2003, pp. II:264–II:271 (2003)
43. Fergus, R., Zisserman, Z., Perona, P.: Object class recognition by unsupervised scale-invariant learning. In: CVPR. IEEE Press, Los Alamitos (June 2003)
44. Fink, G., Jungclaus, N., Kummert, F., Ritter, H., Sagerer, G.: A distributed system for integrated speech and image understanding. In: International Symposium on Aritifical Intelligence, Cancun, Mexico, pp. 117–126 (1996)
45. Fox, M., Long, D.: PDDL 2.1: an Extension to PDDL for Expressing Temporal Planning Domains (2002)
46. Georgeff, M.P.: Communication and interaction in multi-agent planning. In: Proceedings of the 3rd National Conference of the American Association for Artificial Intelligence (AAAI 1983), Washington, DC, August 1983, pp. 125–129 (1983)
47. Georgeff, M.P.: A theory of action for multiagent planning. In: Proceedings of the 4th National Conference of the American Association for Artificial Intelligence (AAAI 1984), Austin, TX, pp. 121–125 (1984)
48. Ghallab, M., Laruelle, H.: Representation and control in IxTeT, a temporal planner. In: Proceedings of the 2nd International Conference on Artificial Intelligence Planning Systems (AIPS 1994), Chicago, IL. AAAI Press, Menlo Park (1994)
49. Gibson, J.: The Perception of the Visual World. Houghton Mifflin, Boston (1950)
50. Golden, K., Weld, D.: Representing sensing actions: The middle ground revisited. In: Aiello, L.C., Doyle, J., Shapiro, S. (eds.) KR 1996: Principles of Knowledge Representation and Reasoning, pp. 174–185. Morgan Kaufmann, San Francisco (1996)
51. Grossberg, S.: The link between brains, learning, attention, and consciousness. Consciousness and Cognition 8, 1–44 (1998)
52. Grosz, B.J., Kraus, S.: Collaborative plans for complex group action. Artificial Intelligence 86(2), 269–357 (1996)

53. Grosz, B.J., Sidner, C.L.: Attention, intentions, and the structure of discourse. Computational Linguistics 12(3), 175–204 (1986)
54. Guivant, J., Masson, F., Nebot, E.: Simultaneous localization and map building using natural features and absolute information. Robotics and Autonomous Systems (2002)
55. Hähnel, D., Burgard, W., Lakemeyer, G.: GOLEX — bridging the gap between logic (GOLOG) and a real robot. In: Günter, A., Herzog, O. (eds.) KI 1998. LNCS, vol. 1504. Springer, Heidelberg (1998)
56. Hansen, E.A., Zilberstein, S.: LAO *: A heuristic search algorithm that finds solutions with loops. Artificial Intelligence 129(1-2), 35–62 (2001)
57. Hoffmann, J., Nebel, B.: The FF planning system: Fast plan generation through heuristic search. Journal of Artificial Intelligence Research 14, 253–302 (2001)
58. Proceedings of the 17th International Joint Conference on Artificial Intelligence (IJCAI 2001), Seattle, WA, August 2001. Morgan Kaufmann, San Francisco (2001)
59. Proceedings of the 18th International Joint Conference on Artificial Intelligence (IJCAI-03), Acapulco, Mexico, August 2003. Morgan Kaufmann, San Francisco (2003)
60. Jensfelt, P., Kristensen, S.: Active global localisation for a mobile robot using multiple hypothesis tracking. IEEE Transactions on Robotics and Automation 17(5) (2001)
61. Johnston, M.: Deixis and conjunction in multimodal systems. In: Proceedings of COLING 2000 (2000)
62. Johnston, M., Bangalore, S., Vasireddy, G., Stent, A., Ehlen, P., Walker, M., Whittaker, S., Maloor, P.: MATCH: An architecture for multimodal dialogue systems. In: Proceedings of the 40th Annual Meeting of the Association for Computational Linguistics (ACL 2002), pp. 376–383 (2002)
63. Jugessur, D., Dudek, G.: Local appearance for robust object recognition. In: CVPR 2000, pp. I:834–I:839 (2000)
64. Kaelbling, L.P., Cassandra, A.R., Kurien, J.A.: Acting under uncertainty: Discrete Bayesian models for mobile-robot navigation. In: Proc. of the International Conference on Robots and Intelligent Systems, IROS (1996)
65. Kant, I.: Critique of Pure Reason. Macmillan, London, 1781 (1929); Translated by Norman Kemp Smith
66. Kennedy, C., Sloman, A.: Autonomous Recovery from Hostile Code Insertion using Distributed Reflection. Journal of Cognitive Systems Research 4(2), 89–117 (2003)
67. Knoblock, C.A.: Planning, executing, sensing, and replanning for information gathering. In: Mellish, C. (ed.) Proceedings of the Fourteenth International Joint Conference on Artificial Intelligence, pp. 1686–1693. Morgan Kaufmann, San Francisco (1995)
68. Koenig, S., Simmons, R.: A robot navigation architecture based on partially observable Markov decision process models. In: Kortenkamp, D., Bonasso, R.P., Murphy, R. (eds.) Artificial Intelligence and Mobile Robots. MIT/AAAI Press, Cambridge (1998)
69. Koller, D., Daniilidis, K., Nagel, H.-H.: Model-based object tracking in monocular image sequences of road traffic scenes. IJCV 10(3), 257–281 (1993)
70. Kuipers, B., Byun, Y.-T.: A robot exploration and mapping strategy based on a semantic hierachy of spatial representations. Journal of Robotics and Autonomous Systems 8, 47–63 (1991)

71. Lakoff, G.: Women, Fire, and Dangerous Things – What Categories Reveal about the Mind. Univ. of Chicago Press, Chicago (1987)
72. Leibe, B., Schiele, B.: Analyzing appearance and contour based methods for object categorization. In: CVPR 2003, Madison, WI (June 2003)
73. Leibe, B., Schiele, B.: Interleaved object categorization and segmentation. In: Proceedings of British Machine Vision Conference (BMVC 2003) (September 2003)
74. Lenser, S., Veloso, M.: Sensor resetting localization for poorly modelled mobile robots. In: Proc. of the IEEE Int. Conf. on Robotics and Automation, ICRA (2000)
75. Leonard, J.J., Feder, H.J.S.: A computationally effecient mathod for large-scale concurrent mapping and localization. In: Proc. 9th International Symposium on Robotics Research (December 1999)
76. Leuschen, M., Walker, I., Cavallaro, J.: Robot reliability through fuzzy markov models. In: Proc. IEEE Annual Reliability and Maintainability Symposium (1998)
77. Littman, M.L., Goldsmith, J., Mundhenk, M.: The computational complexity of probabilistic planning. Journal of Artificial Intelligence Research 9, 1–36 (1998)
78. Lochbaum, K.E.: A collaborative planning model of intentional structure. Computational Linguistics 24(4), 525–572 (1998)
79. Lowe, D.G.: Towards a computational model for object recognition in IT cortex. In: First IEEE International Workshop on Biologically Motivated Computer Vision, May 2000, pp. 20–31 (2000)
80. Lowe, D.G.: Local feature view clustering for 3D object recognition. In: CVPR 2001, pp. I:682–I:688 (2001)
81. McAllester, D.A., Rosenblitt, D.: Systematic nonlinear planning. In: Proceedings of the 9th National Conference of the American Association for Artificial Intelligence (AAAI 1991), Anaheim, CA, pp. 634–639. The MIT Press, Cambridge (1991)
82. Minsky, M.L.: Steps towards artificial intelligence. In: Feigenbaum, E.A., Feldman, J. (eds.) Computers and Thought, pp. 406–450. McGraw-Hill, New York (1963)
83. Morales-Menéndez, R., de Freitas, N., Poole, D.: Real-time monitoring of complex industrial processes with particle filters. In: Proc. of the Neural Information Processing Systems Conference, NIPS (2002)
84. Murphy, R., Hershberger, D.: Handling sensing failures in autonomous mobile robots. International Journal of Robotics Research (2000)
85. Nagel, H.-H.: Image sequence evaluation: 30 years and still going strong. In: ICPR, Washington DC, pp. 148–158. IEEE Press, Los Alamitos (2000)
86. Nau, D., Cao, Y., Lotem, A., Munoz-Avila, H.: SHOP: Simple hierarchical ordered planner. In: Dean, T. (ed.) Proceedings of the 15th International Joint Conference on Artificial Intelligence (IJCAI 1999), Stockholm, SE. Morgan Kaufmann, San Francisco (1999)
87. Nebel, B., Swartout, W., Rich, C. (eds.): Principles of Knowledge Representation and Reasoning: Proceedings of the 3rd International Conference (KR 1992), Cambridge, MA. Morgan Kaufmann, San Francisco (1992)
88. Nguyen, X., Kambhampati, S.: Reviving partial order planning. In: IJCAI 2001 [58]

89. Nilsson, N.J.: Artificial Intelligence: A New Synthesis. Morgan Kaufmann, San Francisco (1998)
90. Oliva, A., Torralba, A., Castelhano, M.S., Henderson, J.M.: Top-down control of visual attention in object detection. In: Proceedings of the IEEE International Conference on Image Processing (2003)
91. O'Regan, J.K., Noë, A.: A sensorimotor account of vision and visual consciousness. Behavirol and Brian Sciences 24(4), 883–917 (2001)
92. O'Regan, J.K., Rensink, R.A., Clark, J.J.: Change-blindness as a result of "mudsplashes". Nature 398(34) (1999)
93. Oviatt, S.L.: Advances in the robust processing of multimodal speech and pen systems. In: Yuen, P.C., Tang, Y.Y., Wang, P.S. (eds.) Multimodal Interfaces B for Human Machine Communication. Series on Machine Perception and Artificial Intelligence, pp. 203–218. World Scientific Publisher, London (2001)
94. Oviatt, S.L.: Breaking the robustness barrier: Recent progress on the design of robust multimodal systems. In: Zelkowitz, M. (ed.) Advances in Computers, pp. 305–341. Academic Press, London (2002)
95. Papageorgiou, C., Poggio, T.: A trainable system for object detection. IJCV 38(1), 15–33 (2000)
96. Pednault, E.P.D.: Formulating multiagent, dynamic-world problems in the classical planning framework. In: Georgeff, M.P., Lansky, A. (eds.) Reasoning about Actions and Plans: Proceedings of the 1986 Workshop, Timberline, OR, pp. 47–82. Morgan Kaufmann, San Francisco (1986)
97. Scott Penberthy, J., Weld, D.S.: UCPOP: A sound, complete, partial order planner for ADL. In: Nebel, B., et al. (eds.) [87], pp. 103–114
98. Philipona, D., O'Regan, J.K., Nadal, J.-P.: Is there anything out there? inferring space from sensorimotor dependencies. Neural Computation 15(9), 2029–2050 (2003)
99. Philipona, D., O'Regan, J.K., Nadal, J.-P., Coenen, O.J.-M.D.: Perception of the structure of the physical world using unknown multimodal sensors and effectors. In: NIPS (2003) (in press)
100. Pollack, M.E.: The uses of plans. Artificial Intelligence 57(1), 43–68 (1992)
101. Rao, A.S., Georgeff, M.P.: Modeling rational agents within a BDI-architecture. In: Allen, J., Fikes, R., Sandewall, E. (eds.) Proceedings of the 2nd International Conference on Principles of Knowledge Representation and Reasoning (KR 1991), San Mateo CA, pp. 473–484. Morgan Kaufmann, San Francisco (1991)
102. Reiter, R.: Knowledge in Action: Logical Foundations for Specifying and Implementing Dynamical Systems. The MIT Press, Cambridge (2001)
103. Rencken, W.D.: Concurrent localisation and map building for mobile robots using ultrasonic sensors. In: IROS 1993, July 1993, vol. 3, pp. 2192–2197. IEEE, Los Alamitos (1993)
104. Riesenhuber, M., Poggio, T.: Models of object recognition. Nature Neuroscience 3, 1199–1204 (2000)
105. Rintanen, J.: Constructing conditional plans by a theorem-prover. Journal of Artificial Intelligence Research 10, 323–352 (1999)
106. Robinson, A., Petchenik, B.: The Nature of Maps: Essays toward Understanding Maps and Mapping. The University of Chicago Press, Chicago (1976)
107. Rosch, E., Mervis, C., Gray, W., Johnson, D., Boyes-Braem, P.: Basic objects in natural categories. Cognitive Psychology 8, 382–439 (1976)

108. Roy, N., Burgard, W., Fox, D., Thrun, S.: Coastal navigation: Mobile robot navigation with uncertainty in dynamic environments. In: Proc. of the IEEE Int. Conf. on Robotics and Automation, ICRA (1999)
109. Russell, S., Norvig, P.: Artificial Intelligence: A Modern Approach, 2nd edn. Prentice-Hall, Englewood Cliffs (2003)
110. Scheding, S., Nebot, E., Durrant-Whyte, H.: The detection of faults in navigation systems: A frequency domain approach. In: Proc. of the IEEE Int. Conf. on Robotics and Automation, ICRA (1998)
111. Schiele, B., Crowley, J.L.: Recognition without correspondence using multidimensional receptive field histograms. IJCV 36(1), 31–50 (2000)
112. Schneiderman, H., Kanade, T.: A statistical method of 3d object detection applied to faces and cars. In: CVPR 2000, pp. 746–751 (2000)
113. Schölkopf, B., Mallot, H.A.: View-based cognitive mapping and path planning. Technical Report 7, Max-Planck -Institut fur biologische Kybernetik (November 1994)
114. El Fallah Seghrouchni, A., Haddad, S.: A recursive model for distributed planning. In: Proceedings of the 2nd International Conference on Multi-Agent Systems (ICMAS 1996), Kyoto, Japan, pp. 307–314. MIT Press, Cambridge (1996)
115. Selfridge, O.: A paradigm for learning. In: Symposium on Mechanization of Thought Processes: National Physical Laboratory Symposium (1959)
116. Shanahan, M.: A logical account of the common sense informatic situation for a mobile robot. Electronic Transactions on Artificial Intelligence 2, 69–104 (1998)
117. Shanahan, M.: A logical account of perception incorporating feedback and expectation. In: Proceedings of Knowledge Representation 2002, pp. 3–13 (2002)
118. Shanker, S.G., King, B.J.: The emergence of a new paradigm in ape language research. Behavioral and Brain Sciences 25(5), 605–606 (2002)
119. Siskind, J.: Grounding the lexical semantics of verbs in visual perception using force dynamics and event logic. Journal of Artificial Intelligence Research 15, 31–90 (2001)
120. Sloman, A.: The Computer Revolution in Philosophy. Harvester Press/Humanities Press, Hassocks, Sussex (1978), `http://www.cs.bham.ac.uk/research/cogaff/crp`
121. Sloman, A.: Beyond shallow models of emotion. Cognitive Processing: International Quarterly of Cognitive Science 2(1), 177–198 (2001)
122. Sloman, A.: Evolvable biologically plausible visual architectures. In: Cootes, T., Taylor, C. (eds.) Proceedings of British Machine Vision Conference, pp. 313–322. BMVA, Manchester (2001)
123. Sloman, A., Logan, B.S.: Building cognitively rich agents using the Sim_agent toolkit. Communications of the Association for Computing Machinery 42(3), 71–77 (1999)
124. Stark, L., Boyer, K.W.: Generic object recognition using form and functions. Machine Perception and Artifical Intelligence 10 (1996)
125. Stone, M.: Modality in Dialogue: Planning, Pragmatics and Computation. PhD thesis, Department of Computer and Information Science, University of Pennsylvania (1998)
126. Stone, M.: Communicative intentions and conversational processes in human-human and human-computer dialogue. In: Trueswell, J., Tanenhaus, M. (eds.) World-Situated Language Use. The MIT Press, Cambridge (2003)

127. Stone, M., Doran, C.: Sentence planning as description using tree-adjoining grammar. In: Proceedings of ACL 1997, pp. 198–205 (1997)
128. Sussman, G.J.: A computational model of skill acquisition. American Elsevier, New York (1975)
129. Tate, A.: Generating project networks. In: Proceedings of the 5th International Joint Conference on Artificial Intelligence (IJCAI 1977), Cambridge, MA, August 1977, pp. 888–893. William Kaufmann, San Francisco (1977)
130. Thiebeaux, S., Hoffmann, J., Nebel, B.: In defense of axioms in PDDL. In: IJCAI 2003 [59]
131. Thrun, S., Fox, D., Burgard, W., Dellaert, F.: Robust monte carlo localisation for mobile robots. Artificial Intelligence 128(1–2), 99–142 (2001)
132. Thrun, S., Fox, D., Burgard, W., Dellaert, F.: Robust Monte-Carlo localization for mobile robots. Artificial Intelligence 128(1-2) (2001)
133. Tonino, H., Bos, A., de Weerdt, M., Witteveen, C.: Plan coordination by revision in collective agent based systems. Artificial Intelligence 142(2) (2002)
134. Torralba, A.: Contextual priming for object detection. IJCV 53(2), 153–167 (2003)
135. Verma, V., Gordon, G., Simmons, R.: Efficient monitoring for planetary rovers. In: International Symposium on Artificial Intelligence and Robotics in Space, iSAIRAS (2003)
136. Verma, V., Langford, J., Simmons, R.: Non-parametric fault identification for space rovers. In: International Symposium on Artificial Intelligence and Robotics in Space, iSAIRAS (2001)
137. Verma, V., Thrun, S., Simmons, R.: Variable resolution particle filter. In: Proc. of the International Joint Conference on Artificial Intelligence, IJCAI (2003)
138. Washington, R.: On-board real-time state and fault identification for rovers. In: Proc. of the IEEE Int. Conf. on Robotics and Automation, ICRA (2000)
139. Weber, M., Welling, M., Perona, P.: Towards automatic discovery of object categories. In: CVPR 2000, pp. II:101–II:108 (2000)
140. Weber, M., Welling, M., Perona, P.: Unsupervised learning of models for recognition. In: Vernon, D. (ed.) ECCV 2000. LNCS, vol. 1842, pp. 18–32. Springer, Heidelberg (2000)
141. Weber, M., Welling, M., Perona, P.: Unsupervised learning of models for visual object class recognition. In: Vernon, D. (ed.) ECCV 2000. LNCS, vol. 1843, pp. 18–32. Springer, Heidelberg (2000)
142. Wilkins, D.E.: Practical Planning. Morgan Kaufmann, San Francisco (1988)
143. Wilkins, D.E.: Can AI planners solve practical problems? Computational Intelligence 6(4), 232–246 (1990)
144. Wilkins, D.E., Myers, K.L., Lowrance, J.D., Wesley, L.P.: Planning and reacting in uncertain and dynamic environments. Journal of Experimental and Theoretical AI 7(1), 197–227 (1995)
145. Yang, Q.: Intelligent Planning: A decomposition and abstraction based approach. Springer, Heidelberg (1997)
146. Zunino, G., Christensen, H.I.: Simultaneous mapping and localisation in domestic environments. In: Dillmann, R. (ed.) Multi-Sensory Fusion and Integration for Intelligent Systems, Baden-Baden, DE, August 2001, pp. 67–72 (2001)

Part II

Component Science

2

Architecture and Representations

Nick Hawes[1], Jeremy L. Wyatt[1], Mohan Sridharan[1], Henrik Jacobsson[2], Richard Dearden[1], Aaron Sloman[1], and Geert-Jan Kruijff[2]

[1] Intelligent Robotics Lab, School of Computer Science, University of Birmingham, Birmingham, UK
{nah,jlw,mzs,rwd,axs}@cs.bham.ac.uk

[2] DFKI GmbH, Saarbrücken, Germany
{henrikj,gj}@dfki.de

2.1 Introduction

The study of architectures to support intelligent behaviour is certainly the broadest, and arguably one of the most ill-defined enterprises in AI and Cognitive Science. The basic scientific question we seek to answer is: "What are the trade-offs between the different ways that intelligent systems might be structured?" These trade-offs depend in large part on what kinds of tasks and environment a system operates under (*niche space*), and also what aspects of the *design space* we deem to be architectural. In CoSy we have tried to answer that question in several ways. First, by thinking about the requirements on architectures that arise from our particular scenarios (parts of niche space). Second, by building systems that follow well-defined architectural rules, and using these systems to carry out experiments on variations of those rules. Third, by using the insights from system building to improve our understanding of the trade-offs between different architectural choices, i.e. between different partial designs. Our objective in CoSy has not been to come up with just another robot architecture, but instead to try to make some small steps forward in a new science of architectures.

When the project started we had several specific scientific goals for our work. First, we wanted *to identify a space of possible architectures that includes most of the good ones for our scenarios.* Our methodology is to define this space of possible architectures using an *architectural schema*, which is a simply a set of constraints and rules on what architectures we allow. We refer to specific points within it as *architectural instantiations.* Second *we aimed to develop a toolkit and an experimental methodology to explore this space, identifying the trade-offs as we move through it.* Finally, our third goal was to use our toolkit *to quickly and easily implement robotic systems that adhere to certain architectural principles.* Those systems are described in Chapters 9 and 10. In this chapter we will describe our schema; some specialisations of

H.I. Christensen et al. (Eds.): Cognitive Systems, COSMOS 8, pp. 51–93.
springerlink.com

it; experiments showing the trade-offs between some of those specialisations; and the toolkit we devised. In addition we will describe how our work sits in between the two major camps of work on architectures for intelligence. The specific contributions are:

- a set of requirements for embodied cognition as a way of generating an architectural schema.
- a new architectural schema that draws on important work from both cognitive architecture and robotic architecture traditions.
- a toolkit for implementing architectures within this schema.
- a series of robotic implementations that utilise the schema and provide a proof of concept (covered in Chapters 9 and 10).
- a set of well defined problems in architectures, posed if the schema is accepted: binding; filtering; management; action fusion.
- experimental profiling of the effects of steps through schema space, specifically with respect to the filtering problem.

The contributions are sequenced in the chapter as follows. First we make some introductory remarks about the science of architectures, our methodology, and define our terminology. Second we describe run-time and design-time requirements that arise from the needs of the CoSy scenarios, and of which our architectural theory must take account. After this we describe our architectural schema (CAS) and show how it meets these requirements. Fourth, we describe how the schema poses at least four further problems that we refer to as the *binding problem*, the *filtering problem*, the *processing management problem*, and the *action fusion problem*. Solutions to these problems provide a second level of constraints and thus a more specific architectural schema. Fifth, we describe the relationship between CAS and other schemas. We follow these contributions with a brief discussion of future research directions.

2.2 Architectures and the Science of Cognitive Systems

We start by clarifying the role of work on "architectures". AI is a science that attempts to understand intelligent systems partly through the process of synthesising them, and partly through analysis of the systems that result. In most areas of AI while the problems are technically difficult, the methodology is clear. When considering integrated systems, however, the methodology itself is a stumbling block. How can we objectively compare the architectures for two systems that might perform different tasks, in different environments, using different processing content? The short answer is that we can't. Yet most examples of work on architectures in robotics are precisely systems that combine architectural choices with task specific processing content, and that are evaluated in different environments from one another. From the point of view of a science of architectures for embodied intelligence this is no good. For a science of architectures we need to separate the processing content from the architectural bones on which it is hung. In other words we want to assess how the changes in the architecture alter the run time properties of that system, independent of changes in the information processing content.

This is tricky, since in both natural and artificial systems the shape of the architecture is intimately related to the shape of the pieces it pulls together. In nature, moreover, the information processing architecture is intimately related to the hardware implementation and the information processing functions implemented in that hardware. On the engineering side, complete cognitive systems are still so difficult to construct that the architecture is essentially almost always bespoke. Separation of architecture and content in a theory is made harder by the fact that the processing in an intelligent system can be described at a variety of levels of abstraction. We may speak of the neural or machine architecture, of information processing architecture, or of the cognitive architecture. In a complete theory of architectures for intelligence we need a clear account of how these different levels of description are related. At the current time we do not even have a clearly defined terminology to distinguish the different uses of the word.

There is an important distinction in the literature between architectures that are entirely specified in advance and those that are partially specified. In CoSy, following [1, 2], we use the notion of an *architectural schema* to refer to architectures of the latter type. In simple terms we can think of an *architectural schema* as a set of constraints that defines a space of possible architectures and ways of moving through that space. We can add constraints to a schema to create sub-schemas, i.e. sub-spaces. If we add enough constraints we can create a fully specified architecture, and these we refer to as *architectural instantiations*. Hereon, we frequently use the terms *schema* and *instantiation* as shorthand for these terms.

If we have a software toolkit that implements a particular schema, it can be used to implement specific working systems that follow the schema. These implemented systems require us to make both a complete set of architectural choices, and to have processing content. As described above, the two are very hard to separate. Our approach must therefore be based on a judicious division of our design choices into architectural ones (i.e. ones we can easily vary in the schema), and those concerned with content (i.e. which vary outside the schema).

In CoSy the specific schema we have designed is called CAS (CoSy Architecture Schema). Our toolkit implementing this schema is called CAST. It allows researchers to implement systems that fall within the space defined by CAS. CAST is particularly powerful because it allows us to implement a system, and then change some of the architectural choices independent of the processing content of that specific system. Thus CAST allows us to take steps through the architectural space, and to isolate the effects on system behaviour due to architectural changes. This provides an important element necessary for a science of architectures as described above.

Overall our architectural work in CoSy is neither concerned with trying to model humans or any other specific type of animal, nor with trying to compete on performance on a specific set of tasks. Rather it is concerned with trying to understand the possibilities and trade-offs involved in different designs in relation to different sets of requirements. In our approach we have several stages through which we must go:

- We identify sets of scenarios (tasks and environments) that we want our systems to solve. We then identify the requirements on architectures arising from these scenarios.
- These requirements lead us to hypothesise an initial architectural schema, together with a toolkit implementing the schema.
- Using the scenarios and the schema toolkit we produce one or more systems based on one or more instantiations of the schema. In these designs there is, as far as possible, a clear separation between the architectural choices and the processing content.
- We analyse our architectural choices, both empirically, and by introspecting on the flaws in our designs. We use the toolkit to make architectural changes to these designs, while holding the processing content fixed.
- Using the insights gained we can refine our schema.

In the rest of this document we will step through this process, stopping along the way to reflect on the choices we made, and to compare our schema with previous work. We now turn to the two scenarios driving CoSy, and to the requirements arising from them.

2.3 Requirements for Architectures for Cognitive Robots

As stated previously our starting point is to analyse scenarios in order to define requirements for architectures. This is an example of backward chaining research in which we consider goals beyond our immediate capabilities, and work backwards to define sub-problems that are more easily tackled — while also understanding a little of the way their solutions might be composed. In the CoSy project we picked scenarios based on robot systems that are able to interact with humans, either while mobile in an office or home setting (the Explorer); or while manipulating objects in a space shared with a human (the PlayMate). In the Explorer scenario typical tasks might be human augmented mapping of an office environment, and using natural language utterances to name and describe places. For example in this scenario the robot can ask the human what type of room it is in, if it is unsure, and must understand questions, instructions to learn, and instructions to act. In the PlayMate scenario the tasks include the robot answering questions about the identity and properties of objects on the table and also about the spatial relationships between them. The robot may be given instructions to manipulate the objects to alter their spatial relationships, and to learn not just word labels for objects, but also the meanings of the property words used to describe an object (e.g. what visual features correspond to the word red).

What sorts of requirements do these scenarios and other typical robot scenarios place upon the robot designs, and in particular upon the manner in which the pieces of the intelligent system are put together? There are a number of well-known properties of the robot-environment interaction that any robot system operating in human environments typically has to deal with. Our run-time requirements therefore include the ability to deal with the following properties of this interaction:

- *Dynamism:* The world changes frequently, rapidly and independently of the robot.
- *Uncertainty:* Sensors are inaccurate, and the actions of the robot often fail to have the planned-for consequences.
- *Multiple modalities:* Many robots — but especially the robots in our two scenarios — must use information from multiple sensory modalities in order to make decisions. In our case these include simple haptics, vision, proprioception and speech. All but the most trivial robot also has multiple modes of action (manipulation, looking, facial expression, utterances, locomotion) enabled by multiple motor systems and the actions of these must be coordinated.
- *Re-taskability:* In our scenarios our robots must be re-taskable, either by others, or autonomously. Behaviour should be goal directed, but not to the extent that it cannot be re-tailored to the context, switched to a new task, or support interleaving of new tasks with old ones.

These are quite general *run-time requirements* on the interaction between a robot and its environment, i.e. requirements on the system during performance. Out of these arise run-time and *design-time requirements* on the architectural schema itself. Architectures for artificial cognition do not just structure the way components work together during a system run, but structure the engineering efforts of the designers. A good software implementation of an architecture or architecture schema should therefore assist the design process. This means that the architecture should make it easier to design a cognitive system, and easier to evaluate a system and understand the causes of its particular behaviour. A proper engineering science of cognition will also require the architecture to have a number of other properties:

- *Understandability:* Cognitive systems of the order of complexity we are describing must be understandable. This means that at least some of the tokens within the system need to be semantically transparent to the designers at some level of abstraction. If we are to understand why our robots succeed or fail in a task, and how they can be re-engineered, then we must have the ability to look at the tokens within the system, and to assign them meaning as designers. An architecture schema for cognitive systems therefore needs to provide a variety of clear ways by which tokens can be related to one another.

- *Incrementality in design:* Large, bug-free, complex software systems need to be constructed and tested incrementally. This requires that a schema allows new sub-systems to be added without completely redesigning the existing sub-systems.
- *Multiple specialist representations:* the field of AI has fragmented, and the sub-disciplines have developed their own specialist representations for inference and decision- making. In designing robots with multiple forms of sensing and acting we need to bridge the gaps between them.
- *Parallel processing:* many of the algorithms employed in vision, language processing, planning, and learning are computationally demanding. A serial model of processing is thus unworkable. There is a need for perceptual components to run in parallel, so that the system may respond rapidly to change. Action components must also run in parallel so that the robot can do more than thing at a time, e.g. looking and reaching.
- *Asynchronous updating:* information arrives in different modalities at different rates. In addition,processing in some modalities is slower than in others. This requires us to accept that updating of information will occur asynchronously across the system.

Given these constraints, the management of information flow in the robot system becomes key. How should information from one sub-system be communicated to others? How should decisions to act be combined and sequenced? How should we determine whether separate pieces of information are related? These are questions to which we should provide architectural answers. There are a number of properties a robot control system needs in order to satisfy the above requirements together with the specification of the Explorer and PlayMate scenarios (though they may not be the only way of satisfying them). An architectural schema should take account of these properties. In order to satisfy our requirements in our scenarios we will assume the following principles:

1. Our robots will have representations of entities in the world. These representations will be built up from sensory information and feature a variety of levels of abstraction.
2. Some of these representations will have roots in multiple sensory modalities.
3. There will be many concurrently running sub-systems in our robots refining and using these representations.
4. Both the PlayMate and the Explorer must represent and reason about hypothetical future states, in order to be able to plan, answer questions, etc. They will therefore require representations that support this kind of reasoning.
5. A large part of what our architectural solution will be concerned with is the refinement, sharing and transmission of these representations by and to these different sub-systems.

6. The system will not be able to draw all possible inferences from the available sensory information, and thus will have to make judicious choices about which processing to perform.
7. There will multiple modes of action which have to be coordinated.

In the next section we describe the architectural schema CAS. CAS encompasses systems that satisfy our assumptions above, and also satisfies the requirements that precede them. In the following sections we also explain why.

2.4 A New Architectural Schema

The requirements and assumptions described in the previous section give rise to a large space of possible architecture schema designs. In order to produce a single schema to constrain our research work we must design a schema that satisfies all of these requirements, whilst still being general enough to capture a selection of the space of possible designs. It is also important that any design reflects both our previous experiences as system designers (i.e. we have knowledge about what works and what doesn't work), and the experience of the wider research community (i.e. what concrete designs have proven successful in the past). Given all of these (interacting) constraints, it is not possible to claim that the following design is the best possible schema design for our scenarios. Instead we put it forward as an initial attempt at producing a schema to satisfy our requirements given our experience.

2.4.1 Key Features of CAS

To quickly convey the features of CAS we summarise them below. More detail is given in the following sections.

- Distributed shared memory: The schema contains subarchitectures each of which has a working memory. These subarchitectures are loosely coupled to reduce complex inter-dependencies. Systems could contain subarchitectures for motor control, vision, action, planning, language etc.
- Parallel refinement of shared representations: Each subarchitecture contains a number of processing components which run in parallel and that asynchronously read and update shared information via the subarchitecture specific working memory.
- Control of information and processing: Information flow is controlled by goals generated within the architecture at run time, allowing it to deal with new problems and opportunities. This allows the schema to support different approaches to processing (e.g. incremental processing, forward chaining, backward chaining etc.). The schema distinguishes between two classes of goal: global goals (that require coordination across subarchitectures), and local goals (that are dealt with inside a single subarchitecture).

- Knowledge management by ontologies: The knowledge that can be used within a subarchitecture is defined by an ontology for that subarchitecture. Relationships between the ontologies in different subarchitectures are defined by a set of general ontologies. These ontologies can also be used to define knowledge at an architecture-general level.

2.4.2 Subarchitecture Design

Components

Our schema starts on the level of a collection of processing components. Every component is concurrently active, allowing them to process in parallel. This satisfies our requirement of supporting concurrent processing. We do not specify any constraints on the contents of components: they could have behave like a node in a connectionist network, an activity in a behaviour based system [3], or a unit of processing in a decomposition by information processing function. Components can take input either directly from sensors, or from working memory. They can also directly control actuators in the manner of closed loop controllers, or initiate fixed action patterns. Components can have processing triggered by the appearance of certain information on the shared working memory, and can modify structures on that memory. Components may also have their own private memory. Finally, components are of two types: managed and unmanaged. *Unmanaged components* are low-latency processes that run all the time. They are useful for several types of processing. They can be used for low-latency early processing of information coming directly from sensors at a high rate. In a visual system, for example, they could correspond to the earliest stages of pre-attentive processing, pulling high bandwidth data from cameras at frame rate and pumping the processed frames onto the working memory. Alternatively they could implement reflexes, providing rapid reaction to sensory information; or they could implement monitors on signals that raise alarms or actions elsewhere in the system, or modify global parameters. *Managed components* by contrast are typically computationally expensive processes, and the schema assumes that there are not the computational resources available to run them all. They therefore post requests to run to the *task manager* associated with the subarchitecture.

Shared Working Memories

Rather than exchange information directly, processing components are connected to a *shared working memory*. The content of the working memory is solely composed of the outputs of processing components. Working memories also connect to and exchange information with other working memories in other subarchitectures. In our implementation of CAS the communication method between the working memory and the components determines the efficiency of the model. But for now let us consider simply that the schema itself allows reading and writing to working memories, and transfer of information between them.

This use of shared working memories is particularly well suited to the *collaborative refinement of shared structures.* In this approach to information processing, a number of components use the data available to them to incrementally update an entry on shared working memory. In this manner the results of processing done by one component can restrict the processing options available to the others in an informed way. As all components are active in parallel, the collective total processing effort (i.e. the amount of work done by all the components in solving a problem) may be reduced by sharing information in this way. This feature turns out to be a very powerful aspect of the schema.

Processing Management

Our previously discussed requirements included the requirement that any design should support the explicit control of processing. Although control strategies could be implemented using communication via working memory entries, failing to support control requirements in the schema would mean that they would fall outside of the regions of design space we could explicitly manipulate with it. It would also mean that system designers would have to reinvent the control strategies they required with each new instantiation. The schema therefore supports control strategies by including a dedicated control component, the *task manager*, in each subarchitecture. In addition to the usual component connections to working memory, each task manager has a dedicated control connection to each component in its subarchitecture. Task managers are also connected across subarchitectures, allowing control strategies to be coordinated across entire instantiations. The task manager can operate in either a demand driven mode or in request mode. In the demand driven mode components request permission to perform a particular task and then have this request accepted or rejected by the task manager. In request mode, the task manager sends task requests to components which can then be accepted or rejected.

2.4.3 System Wide Design

While a system could be composed of a single subarchitecture we intend that there should typically be several subarchitectures in operation. Support for multiple subarchitectures is required in the schema for a number of reasons. It allows the designer of an instantiation to include separate modules in their system, where each subarchitecture plays a specialised role in processing. This contributes towards satisfying the requirement of supporting the multiple representations required to develop a robotic system from currently available technology. Multiple subarchitectures also, if implemented correctly, improve support for concurrency.

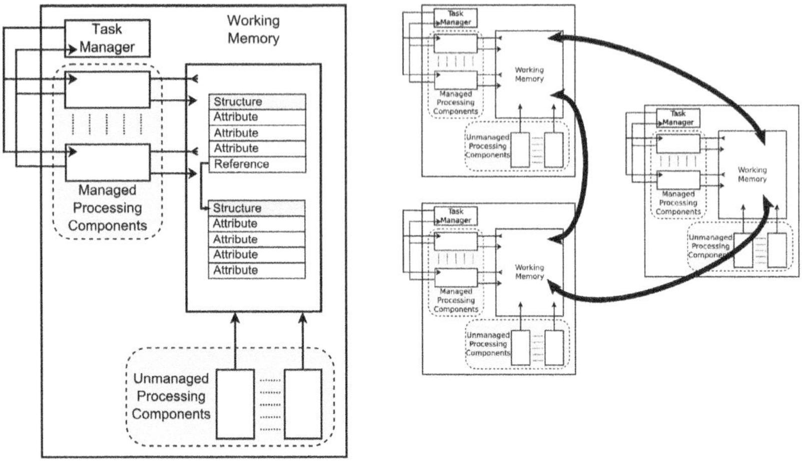

Fig. 2.1. Left panel: A single subarchitecture within CAS. There are components, which run in parallel, asynchronously updating shared structures on a common working memory. They can also take input from sensors or give output to actuators. The task manager determines whether managed components are able to run, while unmanaged components always run. **Right panel:** subarchitectures are coupled to make an overall system. Local changes are transmitted globally between working memories, not components directly.

In the integrated systems we describe in Chapters 9 and 10 we have between three and nine subarchitectures. The schema, note, makes no assumptions about whether system decomposition should be predominantly according to behaviour or information processing function. What is important is that the decomposition groups components that commonly exchange information via shared structures. When events occur on a working memory computation linear in the number of components must be performed to trigger processing by the right components. Early on in our system building experiences we discovered that doing this for all components in the whole system meant the system was paralysed by information change. One of the main benefits of having distributed working memories is precisely that the working memory can act as a gate-keeper for its local components, only letting them become aware of events elsewhere in the system when necessary. We show why this arrangement is beneficial in our exploration of different sub-schemas later in the chapter. Having loosely coupled subarchitectures also allows us to explore architecture sub-schemas where there are global controllers that utilize very abstract representations. We explore some specific architectural instantiations in Chapter 10.

In the next subsection we describe in more detail the way that CAS is implemented by CAST. This includes discussion of its memory and communication models. Following that we define four problems that are raised by CAS, and indeed by any architecture schema which satisfies our assumptions from Section 2.3. After describing these we describe sub-schemas for CAS that address each of the four problems.

2.4.4 CAST: A Toolkit Implementing CAS

In Section 2.4 we described how CAS works at an abstract, conceptual level. However, the details of the implementation of our schema in CAST determine a great deal about its efficiency, and how easy (or hard) it is for it to be specialised in one way or another — i.e. the kinds of moves the implementation supports through the space of architectures. In particular the communication, filtering and memory access models employed by CAST are key in how efficient particular architectural instantiations of the schema tend to be.

In our schema, a working memory is an associative container that maps between unique identifiers and *working memory entries*. Each entry is an instance of a *type*, which can be considered as analogous to a class description. A working memory entry can be any information that can be encapsulated into a single object class. A system that includes visual components may, for example, include entries that describe regions of interest and visually determined objects. A system that must navigate through a building may have entries representing maps of various floors, and objects that have been detected.

Components can add new entries to working memory, and overwrite or delete existing entries. Components can retrieve entries from working memory using three access modes: id access, type access and change access. For id access the component provides a unique id and then retrieves the entry associated with that id. For type access the component specifies a type and retrieves all the entries on working memory that are instances of this type. Whilst these two access modes provide the basic mechanisms for accessing the contents of working memory, they are relatively limited in their use for most processing tasks. Typically most component-level processing can be characterised by a model in which a component waits until a particular change has occurred to an entry on working memory, before processing the changed entry (or a related entry). To support this processing model, components can subscribe to *change events*. Change events are generated by the working memory to describe the operations being performed on the entries it contains. Different instantiations of the schema may support different content for change events, but a minimum set of useful information is the unique id and type of the changed entry, the name of component the made the change, and the operation performed to create the change (i.e. whether the entry was added, overwritten or deleted).

As stated previously an instantiation of the schema can be composed of one or more subarchitectures. The addition of multiple subarchitectures requires that the single subarchitecture schemes for working memory access are extended. We assume that a design would not include multiple subarchitectures unless necessary for imposing modularity on the processing, and so use subarchitecture boundaries to impose restrictions on cross subarchitecture communication. By default a change event is only broadcast within the subarchitecture in which it was generated. This restricts knowledge about the changes on working memory to within a subarchitecture. If a component should require information about changes on a working memory in a different subarchitecture, it can choose to subscribe to these changes. This action opens up a connection between the two working memories in question (the working memory where the change was generated, and the working memory local to the component requesting the information), down which all requested changes are sent as they are generated. The receiving working memory then includes any changes it receives from other subarchitectures in the list of changes it broadcasts within its subarchitecture. The schema allows restrictions to be placed on which working memories a component can access. The default is that a component can read entries from any working memory, but only write to the working memory in its own subarchitecture. Any variation of this scheme can be specified by an instantiation of the schema, allowing components to write to working memories as required.

As described in Section 2.4 support for multiple subarchitectures is required in the schema for a number of reasons. One of these is that it increases the support for concurrency in the system. In CAST, although components are concurrently active, their parallel interactions via working memory have the potential to become a bottleneck in processing. By distributing processing across multiple working memories, CAST allows designers to avoid delays caused by unrelated processing.

2.5 Four Problems

From our first year of system building experience in the PlayMate, and our attempts to build systems using other architectures (e.g. OAA [32]) we realised that there were a number of problems that must be addressed by all embodied intelligent systems that exchange representations between different sub-systems, and which must satisfy the requirements and assumptions we listed in Section 2.3. We refer to these as the *binding problem*, the *filtering problem*, the *processing management problem* and the *action fusion problem*. We define them as follows.

The *binding problem* arises as soon as we have a system with two pieces of information in different places that refer to the same entity. In some instances, to produce coherent decision-making and action execution we need to relate those pieces of information to one another. For example, if I have several blue objects in front of me, and someone refers to an entity as being "blue" how do I decide which entity that I can see is the object of the referent? In general, given many pieces of information residing in different sub-systems, how should the overall system efficiently decide which pieces are related to which other pieces and in what ways? In short *how do we match information from one component with information from another?* The binding problem occurs in many forms, and is a well studied phenomenon in the visual system [4], and in neural architectures [5].

The *filtering problem* arises as soon as a piece of information is created in one sub-system. How should the system decide where else that piece of information is needed? The key issue here, as stated above, is that we do not want to share all information with all sub-systems. The problem is that where information is needed depends both on the context and the problem the system is trying to solve [6]. We call this the filtering problem because we think of it precisely as the problem of deciding what information to send, and what information to filter away, i.e. to hold back from other sub-systems. Filtering mechanisms need to be cheap, context sensitive and to generate few false positives, and very few false negatives.

The *processing management problem* is the problem that arises because we do not have enough computational resources to draw all possible inferences from the sensory data. It is acknowledged that many animals utilise some form of attention to limit processing. On the one hand we can limit what information is processed, and on the other we can limit the processing that is done to the information selected. In other words we want to manage the processing according to the task. We want to investigate the different possible mechanisms and the trade-offs between them. In some solutions or models of human processing the solutions to the binding, filtering and processing management problems are intimately related.

The *action fusion* problem arises when different sub-systems recommend actions to motor systems that need to be fused or otherwise coordinated. Perhaps two behaviours are vying for control of a motor subsystem, or perhaps the activities of two separate motor systems need to be coordinated. In either case the architectural schema must have mechanisms for enabling this coordination. Behaviour based systems use arbitration or fusion mechanisms, which are limited to a very small number of tasks. Planners use action models to coordinate activity, and can produce controllers on the fly for many different tasks, but each of which has only limited feedback. The architectural question is again what the trade-offs are between different approaches. In CoSy were not able to investigate this question in detail, but review some related existing work at the end of the chapter.

We now go on to describe, for each of the first three problems, the solution spaces that we have investigated. We have captured the best of these solutions and incorporated them into CAS sub-schemas, i.e. as more specific parts of our architectural schema.

2.5.1 Binding

Requirements on Binding

Earlier, we stated that systems suitable for the PlayMate and Explorer scenarios often need to connect related information across sub-systems. We referred to this as the binding problem[1] [7, 8]. As a simple example take a system that can discuss and manipulate objects in a tabletop scene. Perhaps it receives the instruction "put the blue things to the left of the red thing". To carry this out it must be able to relate disparate pieces of information. The object identities and their properties from vision must be connected with corresponding information from the utterance. The robot also needs to connect not just information about physical entities, but also about relations between them — there is information in this example about current spatial relationships from both vision and language. Finally the goal state from language must be related to a possible spatial configuration, and the objects in this hypothetical state related to the objects in the current state. In fact the binding problem exists in a much larger range of designs than those necessary for the PlayMate or Explorer or indeed those covered by CAS. It exists in *any* system where information from multiple sub-systems must be *fused* in order to make decisions. The binding problem is related to theory tethering[2] and symbol grounding [9], in that some kind of binding must underlie either approach. After four years of investigating different approaches we have included a method for solving the binding problem in CAS. In this section we describe our solution as it currently stands.

There are two constraints from our scenarios that have influenced our approach to binding. The first is that one of the main features of both our scenarios is the need for an ability to perform deliberative reasoning, by which we mean processes that explicitly represent and reason about hypothetical world states. The second is that because the systems are also embodied, deliberation relies on representations derived from perceptual subsystems that are unreliable and update at unpredictable rates. Embodiment also requires that representations are interpretable by effector sub-systems.

These constraints lead us to impose the following three requirements on our solution to the binding problem. First, *binding must produce representations that are stable* for the duration of the deliberative processes they are involved in. For example, a representation of an object from vision should remain stable across multiple image frames if that object is to be involved in a planning process. Second, these *representations produced by binding must be at a level of abstraction appropriate for the processing they will be involved in.* For example, the positions of objects on a tabletop could be represented metrically or as predicate relations. The first is necessary for visual servoing, the second for understanding utterances about the scene. These two requirements are linked: the level of detail influences temporal stability, in that more abstract representations are typically more durable. Our third requirement arises because the symbols to be bound come from concurrently active, asynchronously updating sub-systems, binding cannot happen in a synchronous

[1] We realise that there are other binding problems in other fields, e.g. neuroscience, but they are somewhat different to the problem here.

[2] See `http://www.eucognition.org/wiki/index.php?title=Symbol_Tethering`

manner. To keep a representation of the state as current as possible, it is important that perceptual information is processed as soon as it is generated. Therefore it is important that any representation generated by binding can be incrementally and asynchronously extended as soon as new information is available. To summarise, the requirements on our approach to binding are:

- The representations it produces should be stable across the lifetime of the processing for which they are necessary.
- The representations should have the appropriate level of abstraction for the processing for which they are necessary.
- The process of binding must be proceed in an asynchronous and incremental manner.

The Binding Solution: Overview

Implicit Binding

Our solution to the binding problem within CAS relies on the separation of representations into two levels of abstraction. At the low level we have subarchitecture-specific representations (Figure 2.2). Within a subarchitecture the representations that reside on the working memory can be structured, they could for example be slot and filler structures. These structures allow the results of components to be bound together: two components can fill in different slots in the same structure. This could be because the different pieces of information in the two slots were derived from the same original data (e.g. the same region of interest (ROI) in an image), or from two different data items that the system designers deem to be related (e.g. a ROI tracked across two frames). This kind of binding therefore relies entirely on the structure of the representations, and the relationship of the processing components as decided at design time. The binding is not explicitly decided by the system, and thus we call it *implicit binding*. Implicit binding has turned out to be a very powerful feature of CAS. In addition since structures on the working memories may contain links to other structures — possibly on other working memories — there is the ability to use implicit binding to bind information along a single *processing trail*. This kind of book-keeping turns out to be another powerful feature of CAS. The reason for this is that it allows us to separate information that changes slowly from information that changes rapidly, e.g. the identity of an object from the visual description of that object. We can then store the slowly changing information in a new structure that has a reference back to the fast changing structure. So, in our example, the pose of the object could be recovered without re-abstraction by using the references in the processing trail. This is another kind of implicit binding.

Implicit binding, while powerful, is not suitable in all situations. If the decision to bind two pieces of information from two different subarchitectures must be made at run-time then we need *explicit binding*. In explicit binding the two pieces of information are translated into a general representation, and in this form are written to the working memory of a specialised binding subarchitecture. A third comparison component decides whether they should be bound into a set. This set is written to the binding working memory and is globally readable by the whole system.

The motivation for this centralised approach rather than any other is one of minimising the effort of translation. Suppose we have no general representation. This means that if every one of N subarchitectures needs information from all the others $N \cdot (N-1)$ pair-wise translation processes are required. Whereas if we have a general representation only at most $N \cdot 2$ translators are needed[3]. From a systems engineering viewpoint, translation to a single general representation makes the translation process more modular and less redundant (although this depends on implementation details). It is more modular in the sense that all translations from a collection of subarchitecture specific representations happen in one place in the system, rather than (potentially) in $N-1$ places (this also supports extensibility). It reduces redundancy because the $N-1$ translations may feature many similar operations, whereas these can all be grouped into a single translation into the general representation.

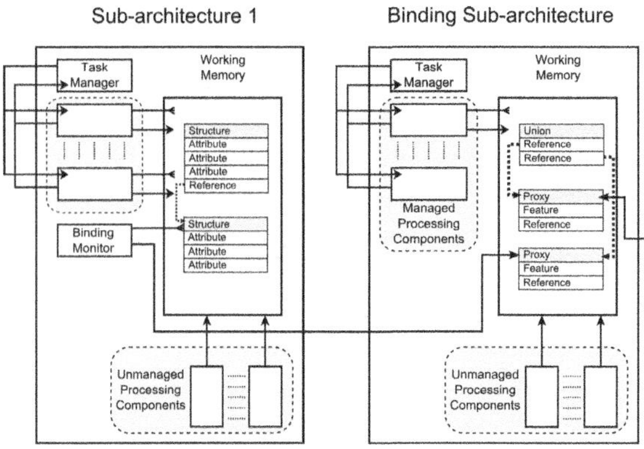

Fig. 2.2. The binding subarchitecture

To implement this scheme (see Figure 2.2) each subarchitecture has a special component called a *binding monitor* that translates any structures on its working memory into *binding proxies* that are written to the binder's working memory. A proxy is simply a set of *binding features*. Features are simply attribute-value pairs. For example, a visual subarchitecture may create proxies for visible objects that have feature attributes for colour, shape, ontological category etc. The role of a binding monitor is to keep its proxies linked to the source information in its subarchitecture, and to update them as that source

[3] This assumes separate steps for translating into and out of the general representation, which may be reduced into a single step in practice.

changes. This link allows other components to operate on the proxies in place of the source data.

All the binding monitors in the system write their proxies to the working memory in the *binding subarchitecture*. There a collection of components, which we will refer to as *the binder*, groups these proxies into *binding unions* based on whether their features match. Unions also have minimal internal structure, but are instead composed of the union of the sets of features from its bound proxies. The set of unions on binding working memory represents the current best architecture-wide hypothesis of the current state of the things the subarchitectures can represent about the world (including its own internal processing). This is based on the assumption that the underlying proxies and features are also the best hypotheses from the corresponding subarchitectures. The comparison of features is performed by feature *comparators*. Each comparator compares features to determine whether they could refer to the same underlying item of information (e.g. whether the colour or spatial location of two objects is the same). The collected results of these comparisons is used to determine whether a proxy could be bound into a union with other proxies. We now give a formal description of binding.

The Binding Solution: Details

The set of possible features is broad:

Definition 1. A feature space $\Phi^x \in \mathbf{\Phi}$ is any data format in the space of all possible data formats, $\mathbf{\Phi}$. $\phi_i^x \in \Phi^x$ denotes an instantiation of a particular representation where x should be interpreted as any feature space name.

For example, $\phi_{red}^{ColourLabel} \in \Phi^{Colour}$ denotes the colour "red" in the representation space of colours. In our CAST instantiation, $\mathbf{\Phi}$ corresponds to any representation that may inhabit a working memory.

Information from the subarchitectures (SAs) is shared as a collection of proxies:

Definition 2. A binding proxy is a structure $p = \langle F_p, u_p \rangle$ where F_p is a set of instantiated features of different types (i.e. $F_p = \{\phi_1^{x_1}, \phi_2^{x_2} \ldots \phi_n^{x_n}\}$) and u_p refers to a binding union with which the proxy is bound (see below).

The unions should express information from proxies that, by all accounts (cf. Algorithm 2.3), refer to the same entity. Unions simply inherit the features of the bound proxies and are defined as:

Definition 3. A binding union is a structure $u = \langle F_u, \mathbf{P}_u \rangle$ where $\mathbf{P}_u$ refers to the subset of proxies unified by the union u and F_u is defined as the union of the features in all proxies in $\mathbf{P}_u$.

The problem for the binder is to assess whether two proxies are matching or not. By matching we mean that they should refer to the same thing. To do this, all new or updated proxies are compared to all unions on the basis of their respective features. The basis of this comparison is that each pair of feature types has an associated comparator function:

Definition 4. A feature comparator is a function $\Delta : \Phi^x \times \Phi^y \rightarrow \{true, false, indeterminate\}$ returning a value corresponding to whether two feature instances are equivalent (or similar enough) or not. The comparator can also choose to not return a definite answer if the answer is undefined, or the uncertainty is too big (i.e. *indeterminate*).

Obviously, *indeterminate* is the only answer most such comparators can return, e.g. the comparison of a Φ^{Colour} and a $\Phi^{Position}$ is likely undefined[4]. However, for many pairs of features there exist informative comparators. For example, features such as linguistic concepts can be compared to other concepts (with ontological reasoning) or physical positions can be compared to areas.

Definition 5. Two feature spaces (Φ^x, Φ^y) are *comparable* iff $\exists(\phi_i^x, \phi_j^y) \in (\Phi^x, \Phi^y)$ such that $\Delta(\phi_i^x, \phi_j^y) \neq indeterminate$.

The more pairs of features from different SAs that are comparable, the more likely it is that proxies from these SAs will be accurately matched.

To compare a proxy and a union, the corresponding feature sets are the basis for scoring:

Definition 6. The binding scorer is a function $S^+ : \mathcal{P} \times \mathcal{U} \rightarrow \mathbb{N}$ where $\mathcal{P}$ and $\mathcal{U}$ denote the set of all proxies and unions respectively and

$$S^+(p,u) = \sum_{\phi_i^x \in F_p} \sum_{\phi_j^y \in F_u} \begin{cases} 1 \text{ if } \Delta(\phi_i^x, \phi_j^y) = true \wedge \phi_i^x \neq \phi_j^y \\ 0 \text{ otherwise} \end{cases}$$

where F_p and F_u are the feature sets of p and u respectively.

Note that identical features are not counted. This to prevent a union getting a higher score just because it is compared to one of its member proxies (this would sometimes prevent a proxy switching to a better union). The number of feature mismatches is also counted (i.e. with *true* replaced with *false* in S^+). That function is here denoted $S^- : \mathcal{P} \times \mathcal{U} \rightarrow \mathbb{N}$.

S^+ and S^- are the basis for selecting the best among all unions for each new or updated proxy. This is conducted by the function `bestUnionsforProxy` described in Algorithm 2.3. The result of $best$ = `bestUnionsforProxy` is a set of zero, one or more unions. If $|best| = 0$ then a new union will be created for the proxy p alone (i.e. with all the features of p). If $|best| = 1$, then the proxy is bound to that union.

When $|best| \geq 2$ we are faced with a *ambiguity*. To avoid deadlocks in such cases the binder can select a random union from *best* for binding. However,

[4] Of course, in the implementation, such undefined comparators are never invoked. Mathematically, however, this is exactly what happens.

```
bestUnionsforProxy(p,𝒰)
Input: A proxy, p, and the set of all unions, 𝒰.
Output: Best union(s) with which a proxy should bind.
begin
    best := ∅;
    max := 0;
    for ∀u ∈ 𝒰 do
        if S⁻(p, u) = 0 ∧ S⁺(p, u) > max then
            best := {u};
            max := S⁺(p, u);
        else if S⁻(p, u) = 0 ∧ S⁺(p, u) = max then
            best := best ∪ {u};
        end
    end
    return best;
end
```

Fig. 2.3. The algorithm which computes the set of best candidate unions for being bound with a new or updated proxy (see definitions 1-6 for an explanation of the notations)

bindings are *sticky*, meaning that if an already bound proxy subsequently matches a union in a larger "best"-list, then it will not switch to any of those unions. This to avoid excess processing in, and signaling from, the binder. This also helps to satisfy our requirement for symbols to be stable as far as possible. Ambiguities cannot be solved by the binder itself, but it can request help from other SAs. This may result, in principle, in the communication SA initiating a clarification dialogue with a human tutor.

Relations and Groups

The proxies and unions described so far have been assumed to roughly correspond directly to physical objects. They may also correspond to more abstract entities as well. To support this, two special proxy types are implemented in a slightly different manner: proxies denoting groups of proxies, and proxies denoting relationships between proxies.

Since proxies contain features that are of any representable type, proxies can also have features attributable to groups and relations, e.g. cardinality and relative metric information respectively, and explicit references to relating proxies. Currently we handle groups in a fairly simple yet direct way: a special kind of "group proxy" is created exactly like an ordinary binding proxy with all the features that the members of the group have in common (e.g. "the blue balls to the left of the mug" creates a group with features $\phi_{ball}^{Concept}$ and $\phi_{blue}^{ColourLabel}$ and with a spatial relation $\phi_{left\_of}^{SpatialRel}$-proxy to the $\phi_{mug}^{Concept}$-proxy. A separate process in the binding SA (the "group manager") then spawns off individual proxies which inherit the features of the group proxy.

Every time an individual is bound to something, a new proxy is spawned[5]. To all the other processes, the individuals appear as an endless supply of normal proxies.

Relation proxies are implemented in a similar way as standard proxies, but with additional features indicating the other proxies involved in the relation. Features of relation proxies are thus compared using the same mechanism that compares the features of standard proxies. For example, spatial metric features, e.g. $\phi^{\mathbb{R}^3}_{\mathbf{(x,y,z)}}$, could in principle be compared to a linguistic feature describing the same relation, e.g. $\phi^{SpatialRel}_{left\_of}$. It has turned out that features that link relations to normal proxies and vice versa make the scoring inefficient. Therefore, a separate scoring scheme similar to that in definition 6 is used to assess how well proxies match to unions w.r.t. their relational context.

Assume that union u_1 has a relation (union) $u_{1\to 2}$ to union u_2 If a now three additional proxies are added, arranged internally as two proxies and a relation proxy between them, p_a, p_b and $p_{a\to b}$ respectively, it is possible that they will be bound with the existing unions. First of all, if $S^+(p_a, u_1) = 0$ and $S^-(p_a, u_1) = 0$, then p_a may be unified despite no convincing score if $p_{a\to b}$ is unified with $u_{1\to 2}$. In other words, the relational context of an ordinary proxy, as defined by the relational proxies, can tip over the balance and favour that it binds with an existing union despite that there are no features that match. The relational proxy $p_{a\to b}$ may also be unified under similar conditions (i.e. where p_a is unified with u_1). If $S^-(p_a, u_1) > 0$ or $S^-(p_b, u_2) > 0$, however, the relation will not bind even if $S^+(p_{a\to b}, u_{1\to 2}) > 0$ and $S^-(p_{a\to b}, u_{1\to 2}) > 0$. The reason is that two relations that are between entities that cannot be the same, can also not be the same relation.

Visual & Spatial Reference Resolution

To illustrate how our binder supports a behaviour typically required of robots that interact with humans we now present present an example taken from the PlayMate system. For further examples see Chapters & .

Perhaps the most common use of information fusion systems is to interpret linguistic references in terms of visual information. Our binder handles this task as an instance of a more general problem of information fusion. We will here consider the simple situation where we have a red object and two blue objects on the table. The objects are arranged in a straight line of alternating colours. The human then asks the robot to "put the blue objects to the left of the red objects".

We will start our example in the visual subarchitecture, where change detection, tracking and segmentation components create representations of the objects in the scene. These objects have 3D poses and bounding boxes and a number of slots for visual properties such as colour, shape and size. These slots are filled by a recogniser that has been previously trained (see

[5] With some obvious limitations to allow finite groups and to prevent excess proxies being generated when members of different groups merge.

Chapter 7) using input from a human trainer [10]. For this example we assume the recogniser correctly extracts the colours of the objects as red and blue. When the scene becomes stable (determined by the change detector) the visual subarchitecture binding monitor creates a proxy for each of the currently visible objects. As the visual property recogniser processes the objects, the monitor updates the proxies with features reflecting these properties. This is an incremental process, so the visual proxies are updated asynchronously as the objects are processed. At this point only the visual proxies are present in the binding working memory, each one is bound to its own union.

The presence of objects in the visual working memory is also noticed by the components in the spatial subarchitecture. These abstract the objects as points on the tabletop, and the spatial binding monitor creates a proxy for each. These proxies are tagged with the ID of the visual proxy for the corresponding object so they are bound correctly[6]. Concurrently with the proxy creation, qualitative spatial relations between the spatial objects are added to working memory. These are generated by components using potential-field-based models of spatial relations [11]. In our example the two blue objects are to the left and to the right of the red object respectively. They are both also near the red object (but not near each other). As these relations are added, the spatial binding monitor reflects them on the binding working memory as relation proxies between the spatial proxies. The binder uses these as the basis of relations between the unions featuring the spatial proxies. This provides our basic model of the current state.

When the human speaks to the robot, a speech recognition module in the communication subarchitecture is triggered. The resulting speech string is written to the communication working memory. This triggers a cycle of deep sentence analysis and dialogue interpretation, yielding a structured logical description of the utterance's content (see Chapter 8 for more information about this processing). From this structure the communication binding monitor generates communication proxies for the discourse referents from the utterance and the relations between them. These proxies include features that can match against both those attached to visual proxies (colour, shape and size), and those attached to spatial proxies (relations based on spatial preposition). In the example two proxies are generated: one normal proxy for the red object with feature $\phi_{red}^{ColourLabel}$, and one group proxy for the blue objects with feature $\phi_{blue}^{ColourLabel}$. The binder uses the features of these communication proxies to bind them into unions with the visual and spatial proxies. In the example the $\phi_{red}^{ColourLabel}$-proxy is bound together with the visual and spatial proxies relating the red object, and the $\phi_{blue}^{ColourLabel}$-proxies spawned from the corresponding group proxy (see Section 2.5.1) are bound with the remaining proxies for the blue objects. This provides the system with an interpretation of the utterance in terms of the visual scene.

---

[6] A similar, but more general, functionality could be generated by matching location-derived features.

In this example, the process of reference resolution involves simply ensuring that the communication proxies referring to visual entities (i.e. those referring to objects in the tabletop scenario) are bound to unions that have a visual component. If the utterance contains spatial language, then relation proxies are generated by the communication binding monitor. This causes the binding process to bind proxies via the relations between proxies as well as the features of single proxies. Failure to bind proxies can trigger a number of different processes.

Binding Summary

In this section we have described two main mechanisms to tackle the binding problem: implicit binding and explicit binding. Implicit binding is essentially a design time choice, while explicit binding is a run time decision by the system itself. We have described how implicit binding also allows us to implement stable abstract features that are linked to rapidly changing features via the creation of *processing trails*. Finally we have described how explicit binding occurs. In addition to these basic mechanisms, elsewhere we have also explored the problem of early binding. This is when possible bindings can be used to cut down the number of possible interpretations in some subarchitecture specific process. In other words tentative bindings across subarchitectures can prune hypotheses within those subarchitectures. This kind of approach we term *incremental binding* and it is described in Chapter 8. The key issue with binding is whether or not a centralised approach to the problem is the right one. We have shown that it is possible, not for which kinds of niches it is the right choice.

2.5.2 Filtering

Previously we described the *filtering problem* as being how to decide efficiently where a piece of information that arises in one sub-system needs to be sent. In other words it is a problem of efficient information sharing. In CAS our atomic information generating sub-systems are components. There is a space of possible models for information sharing between components, ranging from point-to-point communication (i.e. that used by our OAA-based first PlayMate system) to a broadcast model. Between these two extremes exist a range of possible systems in which components share information with a subset of components. Which model is chosen can have a great impact on the final system behaviour. In this section we use the shared memory-based design of CAS to explore the effects of varying the information sharing patterns between components empirically. Specifically we achieve this by altering the ratio of components to subarchitectures in a subset of the PlayMate system.

We start with an $n{:}m$ design where n components are divided between m subarchitectures, where $n > m > 1$. The design is part of the PlayMate system described in Chapter 10, in which components are assigned to subarchitectures based on functionality (vision, binding or qualitative spatial reasoning), although for this experimental work arbitrary n:m assignments are also possible (and would explore a wider area of design space). We then reconfigure this system to generate architectures at two extremes of the design space for information sharing models. At one extreme we have an n:1 design in which all n components from the original system are in the same subarchitecture. At the other extreme of design space we have an n:n design in which every component is in a subarchitecture of its own. Each of these designs can be considered a schema specialisation of the CAS schema from which a full instantiation can be made.

These various designs are intended to approximate, within the constraints of CAS, various possible designs used by existing systems. The n:1 design represents systems with a single shared data store to which all components have the same access. The n:m design represents systems in which a designer has imposed some modularity which limits how data is shared between components. The n:n design represents a system in which a no data is shared, but is instead transmitted directly between components. In the first two designs a component has do to extra work to determine what information it requires from the available shared information. In the latter two designs a component must do extra work to obtain information that is not immediately available to it (i.e. information that is not in it's subarchitecture's working memory).

In order to isolate the effects of the architectural alterations from the other runtime behaviours of the resulting systems, it is important that these architectural differences are the *only* differences that exist between the final CAS instantiations. It is critical that the systems are compared on the same task using the same components. CAST was designed to support this kind of experimentation: it allows the structure of instantiations to be changed considerably, with few, if any, changes to component code. This has allowed us to take the original implementation described above and create the n:1, n:m, and n:n instantiations without changing component code. This means that we can satisfy our original aim of comparing near-identical systems on the same tasks, with the only variations between them being architectural ones.

To measure the effects of the architecture variations, we require metrics that can be used to highlight these effects. We previously presented a list of possible metrics that could be recorded in an implemented CAS system to demonstrate the trade-offs in design space [12, 13]. Ultimately we are interested in measuring how changes to the way information is shared impacts on the external behaviour of the systems, e.g. how often it successfully completes a task. However, given the limited functionality of our experimental system, these kind of behaviour metrics are relatively uninformative. Instead we have chosen to focus on lower-level properties of the system. We have compared the systems on:

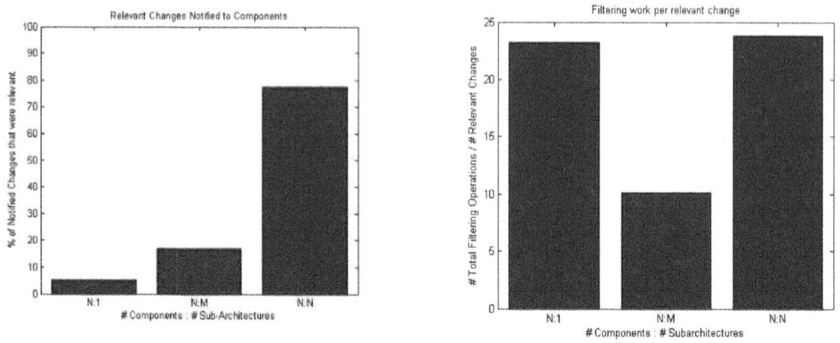

Fig. 2.4. Left panel: Average number of relevant change events received per component. Right Panel: Average filtering effort per relevant change event received.

1. variations in the number of **filtering operations** needed to obtain the change events necessary to get information to components as required by the task.
2. variations in the number of **communication events** required to move information around the system.

As discussed previously communication and change events underlie the behaviour of almost all of the processing performed by a system. Therefore changes in these metrics demonstrate how moving through the space of information sharing models supported by CAS influences the information processing profile of implemented systems.

We studied the three different designs in two configurations: one with vision and binding subarchitectures, and the second with these plus the addition of the spatial subarchitecture. This resulted in six final instantiations which we tested on three different simulated scenes: scenes containing one object, two objects and three objects. Each instantiation was run twenty times on each scene to account for variations unrelated to the system's design and implementation.

The results for the filtering metric are based around the notion of a *relevant event*. A relevant event is a change event that a component is filtering for (i.e. an event that it has subscribed to). Figure 2.4 demonstrates the percentage of relevant events received per component in each instantiation. 100% means that a component only receives change events it is listening for. A lower percentage means that the connectivity of the system allows more than the relevant change events to get the component, which then has to filter out the relevant ones. This is perfectly natural in a shared memory system. The results demonstrate that a component in an n:1 instantiation receives the lowest percentage of relevant events. This is because within a subarchitecture, all changes are broadcast to all components, requiring each component to do a lot of filtering work. A component in an n:n instantiation receives the greatest percentage of relevant changes. This is because each component is shielded by a subarchitecture working memory that only allows change events that are relevant to the attached components to pass. In the $n{:}n$ case because only a single component is in each subarchitecture this number is predictably high[7]. This figure demonstrates the benefits of a directly connected instantiation: components only receive the information they need.

However, this increase in the percentage of relevant changes received comes at a cost. If we factor in the filtering operations being performed at a subarchitecture level (which could be considered as "routing" operations), we can produce a figure demonstrating the total number of filtering operations (i.e. both those at a subarchitecture and a component level) per relevant change received. This is presented in Figure 2.4. This shows a striking similarity between the results for the $n{:}1$ and $n{:}n$ instantiations, both of which require a larger number of filtering operations per relevant change than the $n{:}m$ instantiations. In the $n{:}m$ systems, the arrangement of components into functionally themed subarchitectures results in both smaller numbers of change events being broadcast within subarchitectures (because there are fewer components in each one), and a smaller number of change events being broadcast outside of subarchitectures (because the functional grouping means that some changes are only required within particular subarchitectures). These facts mean that an individual component in an $n{:}m$ instantiation receives fewer irrelevant change events that must be rejected by its filter. Conversely a component in the other instantiations must filter relevant changes from a stream of changes containing *all of the change events in the system*. In the $n{:}1$ instantiations this is because all of these changes are broadcast within a subarchitecture. In the $n{:}n$ instantiations this is because all of these changes are broadcast between subarchitectures. Figure 2.5 (left panel) shows that these results are robust against changes in the number of objects in a scene. Also, the nature of the results did not change between the systems with vision and binding components, and those with the additional spatial components.

Figure 2.5 (centre panel) demonstrates the average number of communication events per system run across the various scenes and configurations for the three different connectivity instantiations. This shows that an $n{:}n$ instantiation requires approximately 4000 more communication events on average to perform the same task as the $n{:}1$ instantiation, which itself requires approximately 2000 more communication events than the $n{:}m$ instantiation. Figure 2.5 (right panel) demonstrates that this result is robust in the face of changes to the number of objects in a scene. The nature of the results also did not change between the systems with vision and binding components, and those with the additional spatial components.

This result is due to two properties of the systems. In the $n{:}n$ system, every interaction between a component a working memory (whether it's an

[7] The events required by the manager component in each subarchitecture mean the relevant percentage for the $n{:}n$ instantiations is not 100%.

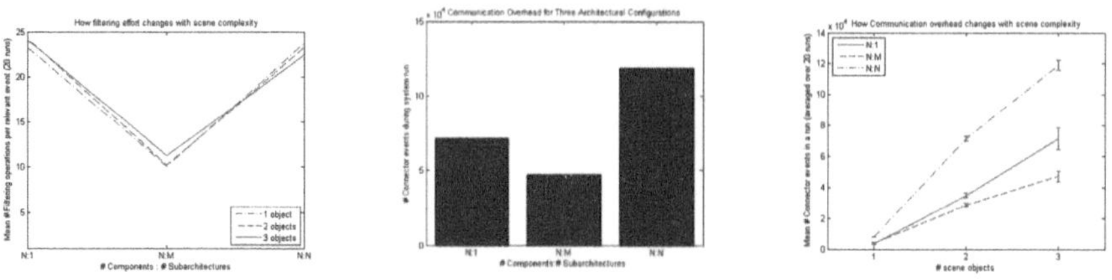

Fig. 2.5. Left panel: Average filtering effort per relevant event compared to scene complexity. Centre panel: Average total communication events per instantiation run. Right panel: Average total communication events per instantiation run compared to scene complexity.

operation on information or the propagation of a change event) requires an additional communication event. This is because all components are separated by subarchitectures as well as working memories. In addition to this, the number of change events propagated through the systems greatly effect the amount of communication events that occur. In the n:n and n:1 instantiations, the fact that they effectively broadcast all change events throughout the system contributes significantly to the communication overhead of the system.

2.5.3 Filtering Summary

From these results we can conclude that a functionally-decomposed n:m CAS instantiation occupies a "sweet spot" in architectural design space with reference to filtering and communication costs. This sweet spot occurs because having too much information shared between components in a system (the n:1 extreme) means that all components incur an overhead associated with filtering out relevant information from the irrelevant information. At the other extreme, when no information is shared by default (the n:n extreme) there are extra communication costs due to duplicated transmissions between pairs of components and (in CAS-derived systems at least) the "routing" overhead of transmitting information to the correct components (i.e. the filtering performed by working memories rather than components).

In addition we have demonstrated here an empirical approach to comparing different points within design space, where we have held the content of the system constant, while making architectural changes. In this way we have also shown how to use CAST to help carry out the empirical part of the science of architectures we discussed previously. We now proceed to discuss the third of our problems, that of managing processing across the system.

2.5.4 Processing Management

In this section we discuss the problem of how a complex artificial cognitive system such as the Explorer or PlayMate systems we describe in Chapters 9 and 10 should manage their internal activity. In particular, when the processing possibilities exceed the processing resources, how should the robot choose what kind of processing to do? We refer to this as the *processing management problem*. We sketch several possible solutions to it, and then discuss the solution we have been exploring, which relies on technologies for planning under uncertainty.

To begin with, consider a visual system that contains some of the many algorithms and representations described in Chapters 4 and 7. Each of these requires considerable computation to run, even in their classification (or non-learning) mode. In a robot with multiple competences we will need all of those vision algorithms and many more besides. For any natural visual scene running all such algorithms on all parts of the image is not feasible. This is not a problem in so far as we never need to perform all visual processing on a scene: the vision we need is determined by the task we are performing. To tackle this there has been much work on attention, and in particular the use of visual saliency models to identify which parts of the scene to process. There has been little or no work, however, on how to select which algorithms to run. It is this issue that we address here.

In our PlayMate domain, both a robot and human can converse about and manipulate objects on a table top (see [14]). Typical visual processing tasks in this domain require the ability to find the colour, shape identity or category of objects in the scene to support dialogues about their properties; to see where to grasp an object; to plan an obstacle free path to do so and then move it to a new location; to identify groups of objects and understand their spatial relations; and to recognize actions the human performs on the objects. Each of these vision problems is hard in itself, together they are extremely challenging. The challenge is to build a vision system able to tackle all these tasks. One early architectural approach to robot vision was to attempt a general purpose, complete scene reconstruction, and then query this model for each task. This is still not possible and in the opinion of many vision researchers will remain so. An idea with a growing body of evidence from both animals and robots is that some visual processing can be made more effective by tailoring it to the task/environment [15, 16, 17].

Consider the scene in Figure 2.6, and consider the types of visual operations that the robot would need to perform to answer a variety of questions that a human might ask it about the scene: "is there a blue triangle in the scene?", "what is the colour of the mug?", "how many objects are there in the scene?". In order to answer these questions, the robot has at its disposal a range of information processing functions and sensing actions. But, in any reasonably complex scenario (such as the one described above), it is not feasible (and definitely not efficient) for the robot to run all available information processing functions and sensing actions, especially since the system needs to respond to human queries/commands in real-time.

There are many approaches that could be taken. The key choices concern how bottom up and top down processing are mixed. Within the constraints

Fig. 2.6. A picture of the typical table top scenario—ROIs detected for processing are bounded by rectangular boxes

provided by CAS we have explored at least three approaches. In the first two approaches, processing opportunities can be identified as data arrives, and requests for processing resources be posted by the components to a local task manager. The task manager may simply have a policy which is unvarying, e.g. it permits all requests, or grants them up to some load threshold. This is what we have done for most our visual systems to date, because they have been very small. Alternatively the allocation policy could change according to the mode that the task manager is in. This is the approach we took in the communication sub-system for the Explorer and PlayMate systems. In this sub-architecture the task manager has modes, which it can switch between, and which are associated with different resource allocation policies. The third approach, and the one we detail here is to drive the processing in a largely top down way. In reality a mix of both top down and bottom up processing will be required. A top down approach essentially takes the current robot task, and uses this to determine which processing will be performed. There are several ways that we can conceive of the top-down determination of processing. In principle a (perhaps learned) task specific visual routine could be invoked from a library. The problem with this approach is that it is not at all obvious how to generalise from one task to another from a set of learned instances. A different approach would be to use planning to compose a completely new visual routine on the fly. The problem with this approach is that the planning process is itself expensive. However, we explore this approach, and show that it can work.

There already exists a body of impressive work on planning of image processing ([18, 19, 20, 21]). However, it is largely used for single images, requires specialist domain knowledge to perform re-planning or plan repair, has only been extended to robotic systems in the most limited ways, and poses the problem in an essentially deterministic planning framework or as a MDP ([21]). In our approach we push the field of planning visual processing in a new direction by posing the problem as an instance of probabilistic sequential decision making. We pose it as a Partially Observable Markov Decision Process (POMDP), thereby taking explicit, quantitative account of the unreliability of visual processing. Our main technical contribution is that we show how to contain one aspect of the intractability inherent in POMDPs for this domain by defining a new kind of hierarchical POMDP. We compare this approach with an earlier formulation based on the Continual Planning (CP) framework of [22]. Using a real robot domain, we show empirically that both planning methods are quicker than naive visual processing of the whole scene, even taking into account the planning time. The key benefit of the POMDP approach is that the plans, while taking slightly longer to execute than those produced by the CP approach, provide significantly more reliable visual processing than either naive processing or the CP approach. We give an overview of the POMDP approach here, describe the results, and relate it back to the CAS framework.

A POMDP Formulation of Visual Processing

In robot applications, typically the true state of the world cannot be observed directly. The robot can only revise its belief about the possible current states by executing actions, for instance one of the visual operators.

As stated previously, we pose the problem as a POMDP where we explicitly model the unreliability of the visual operators/actions. This probabilistic formulation enables the robot to maintain a probability distribution (the *belief state*) over the true underlying state. To do this we need an observation model that captures the likelihood of the outcomes from each action. In this paper, we only consider actions that have purely informational effects. In other words, we do not consider actions such as poking the object to determine its properties, with the consequence that the underlying state does not change when the actions are applied. However, the POMDP formulation allows us to do this, which is necessary if we wish to model the effects of operators that split ROIs, move the camera, or move the objects to gain visual information about them.

Each action considers the true underlying state to be composed of class labels (e.g. *red(R), green(G), blue(B)* for color; *circle(C), triangle(T), square(S)* for shape; *picture, mug, box* for sift), a label to denote the absence of any object/valid class—*empty* (E), and a label to denote the presence of *multiple* classes (M). The observation model for each action provides a probability distribution over the set composed of the normal class labels, the class label *empty* (E) that implies that the match probability corresponding to the normal class labels is very low, and *unknown* (U) that means that there is no single class label to be relied upon and that multiple classes may therefore be present. Note that U is an observation, whereas M is part of the underlying state: they are not the same, since they are not perfectly correlated.

Since visual operators only update belief states, we include "special actions" that answer the query by "saying" (not to be confused with language-based communication) which underlying state is most likely to be the true state. Such actions cause a transition to a terminal state where no further actions are applied. In the description below, for ease of explanation (and without loss of generality) we only consider two operators: *color* and *shape*, denoting them with the subscripts c, s respectively. States and observations are distinguished by the superscripts a, o respectively.

Consider a single ROI in the scene—it has a POMDP associated with it for the goal of answering a specific query. This POMDP is defined by the tuple $\langle \mathcal{S}, \mathcal{A}, \mathcal{T}, \mathcal{Z}, \mathcal{O}, \mathcal{R} \rangle$:

- $\mathcal{S} : \mathcal{S}_c \times \mathcal{S}_s \cup \text{term}$, the set of states, is a cartesian product of the state spaces of the individual actions. It also includes a *terminal* state (term). $\mathcal{S}_c : \{E_c^a, R_c^a, G_c^a, B_c^a, M_c\}$, $\mathcal{S}_s : \{E_s^a, C_s^a, T_s^a, S_s^a, M_s\}$
- $\mathcal{A} : \{color, shape, sRed, sGreen, sBlue\}$ is the set of actions. The first two entries are the visual processing actions. The rest are special actions that represent responses to the query such as "say blue", and lead to *term*. Here we only specify "say" actions for color labels, but others may be added trivially.
- $\mathcal{T} : \mathcal{S} \times \mathcal{A} \times \mathcal{S} \rightarrow [0, 1]$ represents the state transition function. For visual processing actions it is an identity matrix, since the underlying state of the world does not change when they are applied. For special actions it represents a transition to *term*.
- $\mathcal{Z} : \{E_c^o, R_c^o, G_c^o, B_c^o, U_c, E_s^o, C_s^o, T_s^o, S_s^o, U_s\}$ is the set of observations, a concatenation of the observations for each visual processing action.
- $\mathcal{O} : \mathcal{S} \times \mathcal{A} \times \mathcal{Z} \rightarrow [0, 1]$ is the observation function, a matrix of size $|\mathcal{S}| \times |\mathcal{Z}|$ for each action under consideration. It is learned by the robot for the visual actions (described in the next section), and it is a uniform distribution for the special actions.
- $\mathcal{R} : \mathcal{S} \times \mathcal{A} \rightarrow \Re$, specifies the reward, mapping from the state-action space to real numbers. In our case:

$$\begin{aligned} \forall s \in \mathcal{S},\ \mathcal{R}(s, shape) &= -1.25 \cdot f(\text{ROI-size}) \\ \mathcal{R}(s, color) &= -2.5 \cdot f(\text{ROI-size}) \\ \mathcal{R}(s, \text{special actions}) &= \pm 100 \cdot \alpha \end{aligned}$$

For visual actions, the cost depends on the size of the ROI (polynomial function of ROI size) and the relative computational complexity (the *color* operator is twice as costly as *shape*). For special actions, a large +ve (-ve) reward is assigned for making a right (wrong) decision for a given query. For example, while determining the ROI's color:

$\mathcal{R}(R_c^a T_s^a, \text{sRed}) = 100 \cdot \alpha, \mathcal{R}(B_c^a T_s^a, \text{sGreen}) = -100 \cdot \alpha$

but while computing the location of *red* objects:

$\mathcal{R}(B_c^a T_s^a, \text{sGreen}) = 100 \cdot \alpha$. The variable α enables the trade-off between the computational costs of visual processing and the reliability of the answer to the query.

Our visual planning task for a single ROI now involves solving this POMDP to find a policy that maximizes reward from the initial belief state. Plan execution corresponds to traversing a policy tree, repeatedly choosing the action with the highest value at the current belief state, and updating the belief state after executing that action and getting a particular observation. In order to ensure that the observations are independent (required for POMDP belief updating to hold), we take a new image of the scene if an action is repeated on the same ROI.

Actual scenes will have several objects and hence several ROIs. Attempting to solve a POMDP in the joint space of all ROIs soon becomes intractable due to an exponential state explosion, even for a small set of ROIs and actions. For a single ROI with m features (color, shape, etc.) each with n values, the POMDP has an underlying space of n^m; for k ROIs the overall space becomes: n^{mk}. Instead, we propose a *hierarchical decomposition*: we model each ROI with a lower-level (LL) POMDP as described above, and use a higher-level (HL) POMDP to choose, at each step, the ROI whose policy tree (generated by solving the corresponding LL-POMDP) is to be executed. This decomposes the overall problem into one POMDP with state space k, and k POMDPs with state space n^m. Space does not permit us to give the full details of the hierarchical POMDP formulation here, but these can be found in [23]. The key technical point is that in the HL-POMDP the observation function and the cost/reward specification for each action is based on the policy tree of a LL-POMDP that corresponds to that action. An example of the type of policy tree found for a LL-POMDP is given in Figure 2.7 where the LL-POMDP's policy tree has the root node representing the initial belief when the visual routine is called. At each node, the LL-POMDP's policy is used to determine the best action, and all possible observations are considered to determine the resultant beliefs and hence populate the next level of the tree.

Once the observation functions and costs are computed, the HL-POMDP model can be built and solved to yield the HL policy. During execution, the HL-POMDP's policy is queried for the best action choice, which causes the execution of one of the LL-POMDP policies, resulting in a sequence of visual operators being applied on one of the image ROIs. The answer provided by the LL-POMDP's policy execution causes a belief update in the HL-POMDP, and the process continues until a terminal action is chosen at the HL, thereby answering the query posed. For the query "Where are the blue objects" it provides the locations of all *blue* objects in the scene. For simpler occurrence queries (e.g. "Is there a blue object in the scene?") the execution can be terminated at the first occurrence of the object in a ROI. Both the LL and HL POMDPs are query dependent. Solving the POMDPs efficiently is hence crucial to overall performance.

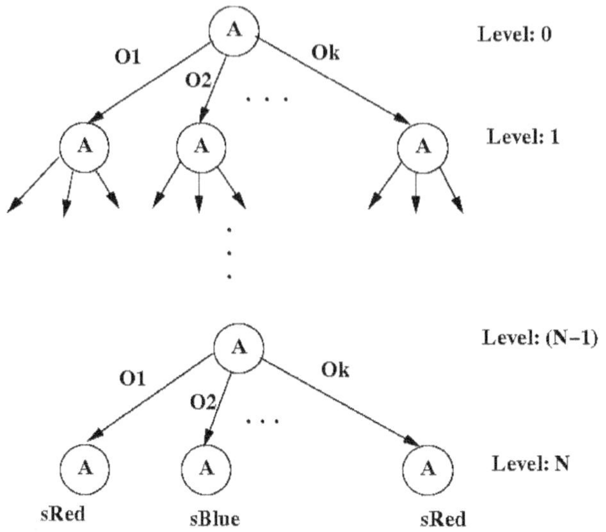

Fig. 2.7. Policy Tree of an ROI—each node represents a belief state and specifies the action to take

A Continual Planning Formulation

The Continual Planning (CP) approach of [22] interleaves planning, plan execution and plan monitoring. Unlike classical planning approaches that require prior knowledge of state, action outcomes, and all contingencies, an agent in CP postpones reasoning about unknowable or uncertain states until more information is available. It achieves this by allowing actions to assert that the preconditions for the action will be met when the agent reaches that point in the execution of the plan, and if those preconditions are not met during execution (or are met earlier), replanning is triggered. But there is *no* representation of the uncertainty/noise in the observation/actions. It uses the PDDL ([24]) syntax and is based on the FF planner of [25]. Consider the example of a *color* operator:

```
(:action colorDetector
:agent (?a - robot)
:parameters (?vr - visRegion ?colorP - colorProp )
:precondition (not (applied-colorDetector ?vr) )
:replan (containsColor ?vr ?colorP)
:effect (and
      (applied-colorDetector ?vr)
      (containsColor ?vr ?colorP) )   )
```

The parameters are a color-property (e.g. *blue*) being searched for in a particular ROI. It can be applied on any ROI that satisfies the precondition i.e. it has not already been analyzed. The expected result is that the desired color is found in the ROI. The "replan:" condition ensures that if the robot observes the ROI's color by another process, replanning is triggered to generate a plan that excludes this action. This new plan will use the *containsColor* fact from the new state instead. In addition, if the results of executing a plan step are not as expected, replanning (triggered by execution monitoring) ensures that other ROIs are considered. Other operators are defined similarly, and based on the goal state definition the planner chooses the sequence of operators whose effects provide parts of the goal state—the next section provides an example. The CP approach to the problem is more responsive to an unpredictable world than a non-continual classical planning approach would be, and it can therefore reduce planning time in the event of deviations from expectations. But, while actions still have non-deterministic effects, there is no means for accumulating belief over successive applications of operators. We show that the HiPPo formulation provides significantly better performance than CP in domains with uncertainty.

An Example Query

Figs 8(a)-8(d) show one execution example for an image with two ROIs.

The example query is to determine the presence and location of one or more *blue circles* in the scene (Fig 8(a)). Since both ROIs are equally likely target locations, the HL-POMDP first chooses to execute the policy tree of the first ROI (action u_1 in Fig 8(b)). The corresponding LL-POMDP runs the color operator on the ROI. The outcome of applying any individual operator is the observation with the maximum probability, which is used to update the subsequent belief state—in this case the answer is *red*. Even though it is more costly, the color operator is applied before shape because it has a higher likelihood of success, based on the learned observation functions. When the outcome of *red* increases the likelihood (belief) of the states that represent the "Red" property as compared to the other states, the likelihood of finding a blue circle is reduced significantly. The dynamic reward specification ($\alpha = 0.2$) ensures that without further investigation (for instance with a shape operator), the *best* action chosen at the next level is a terminal action associated with the "Red" property—in this case it is *sRedSquare*. The HL-POMDP receives the input that a red square is found in R_1, leading to a belief update and subsequent action selection (action u_2 in Fig 8(c)). Then the policy tree of the LL-POMDP of R_2 is invoked, causing the color and shape operators to be applied in turn on the ROI. The higher noise in the shape operator is the reason why it has to be applied twice before the uncertainty is sufficiently reduced to cause the choice of a terminal action (*sBlueCircle*)—the increased reliability therefore comes at the cost of execution overhead. This results in the belief update and terminal action selection in the HL-POMDP—the final answer is $(s\neg R_1 \wedge R_2)$, i.e. that a *blue circle* exists in R_2 and not R_1 (Fig 8(d)).

In our HiPPo representation, each HL-POMDP action chooses to execute the policy of one of the LL-POMDPs until termination, instead of performing

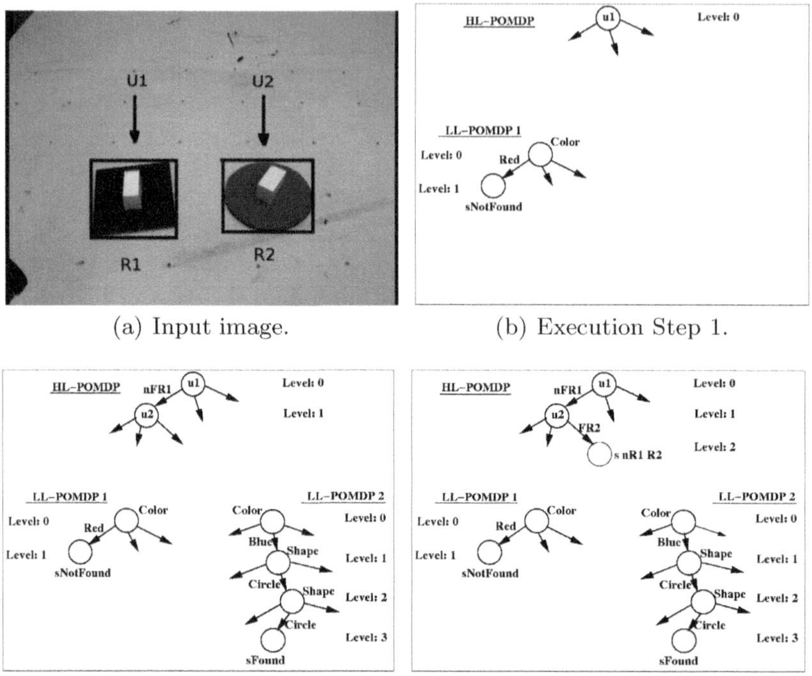

(a) Input image. (b) Execution Step 1.

(c) Execution Step 2. (d) Execution Step 3.

Fig. 2.8. Example query: "Where is the Blue Circle?" Dynamic reward specification in the LL-POMDP allows for early termination when negative evidence is found.

just one action. The challenge here is the difficulty of translating from the LL belief to the HL belief in a way that can be planned with. The execution example above shows that our approach still *does the right thing*, i.e. it stops early if it finds negative evidence for the target object. Finding positive evidence can only increase the posterior of the ROI currently being explored, so if the HL-POMDP were to choose the next action, it would choose to explore the same ROI again.

If the same problem were to be solved with the CP approach, the goal state would be defined as the PDDL string:

(and (exists ?vr - visRegion) (and (containsColor ?vr Blue) (containsShape ?vr Circle)))

i.e. the goal is to determine the existence of a ROI which has the color *blue* and shape *circle*. The planner must then find a sequence of operators to satisfy the goal state. In this case it leads to the creation of the plan:

(colorDetector robot vr0 blue)
(shapeDetector robot vr0 circle)

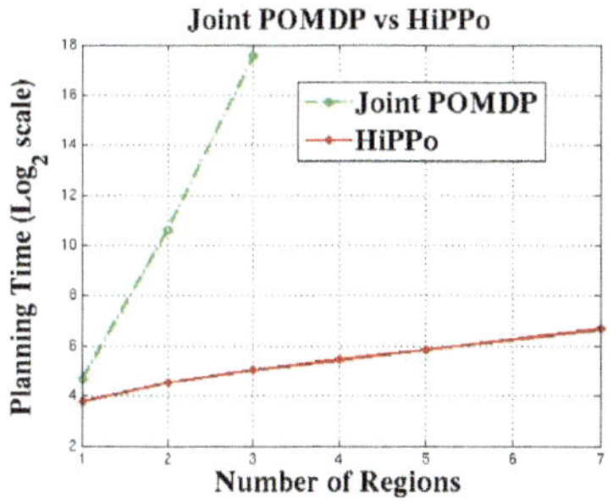

(a) HiPPo vs. joint POMDP. Joint POMDP soon becomes intractable.

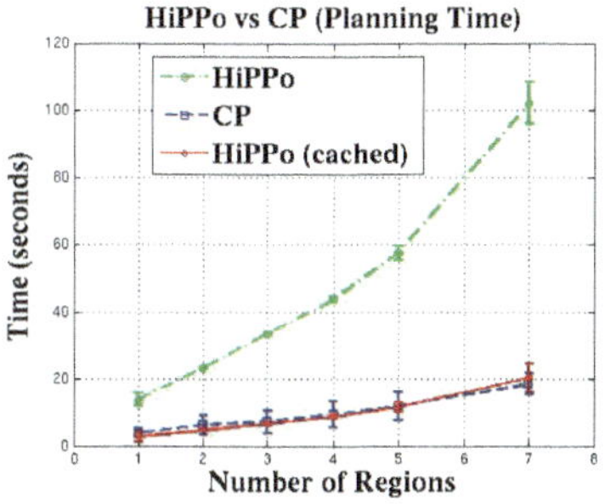

(b) Planning times of HiPPo vs. CP. Policy-caching makes results comparable.

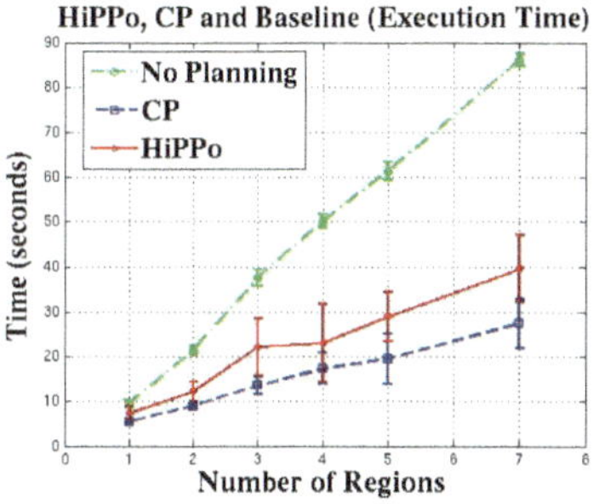

(c) Execution times of HiPPo, CP vs. No planning. Planning makes execution faster.

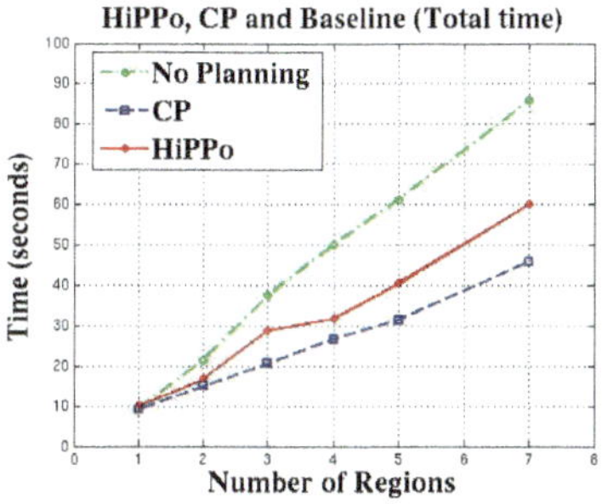

(d) Planning+execution times of HiPPo, CP vs. No planning. Planning approaches reduce processing time.

Fig. 2.9. Experimental Results—Comparing planning and execution times of the planners against no planning

i.e. the robot is to apply the color operator, followed by the shape operator on the first ROI. There is a single execution of each operator on the ROI. Even if (due to image noise) an operator determines a wrong class label as the closest match with a low probability, there is no direct mechanism to incorporate the knowledge. Any thresholds will have to be carefully tuned to prevent mis-classifications. Assuming that the color operator works correctly in this example, it would classify the ROI as being *red*, which would be determined in the plan monitoring phase. Since the desired outcome (finding *blue* in the first ROI) was not achieved, replanning is triggered to create a new plan with the same steps, but to be applied on the second ROI. This new plan leads to the desired conclusion of finding the *blue circle* in R_2 (assuming the operators work correctly).

An Experimental Study

In considering the trade-offs between the different types of solution to the processing management problem we consider several hypotheses that we can test. Specifically we hypothesise that:

- The hierarchical-POMDP planning (HiPPo) formulation is more efficient than the standard POMDP formulation.
- HiPPo and CP have comparable plan-time complexity.
- Planning is significantly more efficient than blindly applying all operators on the scene.
- HiPPo has higher execution time than CP but provides more reliable results.

In order to test these hypotheses we ran several experiments on the robot in the tabletop scenario. Objects were placed on the table and the robot had to analyze the scene to answer user-provided queries. Query categories include:

- Occurrence queries: Is there a red mug in the scene?
- Location queries: Where in the image is the blue circle?
- Property queries: What is the color of the box?
- Scene queries: How many green squares are there in the scene?

For each query category, we ran experiments over ~ 15 different queries with multiple trials for each such query, thereby representing a range of visual operator combinations in the planning approaches. We also repeated the queries for different numbers of ROIs in the image. In addition, we also implemented the naive approach of applying all available operators (color, shape and sift in our experiments) on each ROI, until a ROI with the desired properties is found and/or all ROIs have been analyzed.

Unlike the standard POMDP solution that considers the joint state space of several ROIs, the hierarchical representation does not provide the optimal solution (policy). Executing the hierarchical policy may be arbitrarily worse than the optimal policy. For instance, in the search for the *blue* region, the hierarchical representation is optimal *iff* every ROI is blue-colored. But as seen in Figure 9(a) that compares the planning complexity of HiPPo with the standard POMDP solution, the non-hierarchical approach soon becomes intractable. The hierarchical approach provides a significant reduction in the planning time and (as seen below) still increases reliability significantly.

Next, we compare the planning times of HiPPo and CP approaches as a function of the number of ROIs in the scene—Figure 9(b). The standard hierarchical approach takes more time than CP. But, the computationally intensive part of HiPPo is the computation of the policies for the ROIs. Since the policies computed for a specific query are essentially the same for all scene ROIs, they can be cached and not repeated for every ROI. This simple adjustment drastically reduces the planning time and makes it comparable to the CP approach.

Table 2.1. Reliability of visual processing

Approach	% Reliability
Naive	76.67
CP	76.67
HiPPo	91.67

Figure 9(c) compares the execution time of the planning approaches against applying all the operators on each ROI until the desired result is found. The HiPPo approach has a larger execution time than CP because it may apply the same operator multiple times to a single ROI (in different images of the same scene) in order to reduce the uncertainty in its belief state. In all our experiments the algorithms are being tested on a robot which has multiple modules to analyze input from different modalities (vision, tactile, speech) and has to bind the information from the different sources. Hence, though the individual actions are optimized and represent the state-of-the-art in visual processing, they take some time to execute on the robot.

A key goal of our approach is not only to reduce overall planning and execution time, but to improve the reliability of the visual processing. In these terms the benefits are very clear, as can be seen in Table 2.1. The direct application of the actions on all the ROIs in the scene results in an average classification accuracy of 76.67%, i.e. the sensing actions misclassify around one-fourth of the objects. Using CP also results in the same accuracy of 76.67%, i.e. it *only reduces the execution time* since there is no direct mechanism in the non-probabilistic planner to exploit the outputs of the individual operators (a distribution over the possible outcomes). The HiPPo approach is designed to utilize these outputs to reduce the uncertainty in belief, and though it causes an increase in the execution time, it results in much higher classification accuracy: 91.67%. It is able to recover from a few noisy images where the operators are not able to provide the correct class label, and it fails only in cases where there is consistent noise. A similar performance is observed if additional noise is added during execution. As the non-hierarchical POMDP approach takes days to compute the plan for just two ROIs we did not compute the optimal plan for scenes with more than two ROIs, but for the cases where a plan was computed, there was no significant difference between the optimal approach and HiPPo in terms of the execution time and reliability, even though the policy generated by HiPPo can be arbitrarily worse than that generated by the non-hierarchical approach.

A significant benefit of the POMDP approach is that it provides a ready mechanism to include initial belief in decision-making. For instance, in the example considered above, if R_2 has a higher initial belief that it contains a *blue circle*, then the cost of executing that ROI's policy would be lower and it would automatically get chosen to be analyzed first leading to a quicker response to the query.

Figure 9(d) shows a comparison of the combined planning and execution times for HiPPo, CP, and the naive approach of applying all actions in all ROIs (no planning). As the figure shows, planning is worthwhile even on scenes with only two ROIs. In simple cases where there are only a couple of operators and/or only one operator for each feature (color, shape, object recognition etc) one may argue that rules may be written to decide on the sequence of operations. But as soon as the number of operators increase and/or there is more than one operator for each feature (e.g. two actions that can find color in a ROI, each with a different reliability), planning becomes an appealing option.

Summary of Processing Management Work

In this sub-section we have explored the implications of the fact that within a complex cognitive system with many goals it will not be possible to perform all processing. We have studied this problem within the context of vision, specifically the kind of visual sub-architecture we use for the PlayMate scenario. Architecturally there are many possible solutions. Bottom up, data-driven processing is implemented naturally in CAS. In this section we have shown how to augment it with top down control, achieved using techniques for planning under uncertainty, and continual planning.

2.6 The Relationship of CAS to Previous Work on Architectures

2.6.1 Cognitive Architectures

There have been several attempts at unified theories of intelligence from within cognitive science. At least two of these emphasise the role of production systems. In SOAR Newell and Laird [26] proposed a production system model in which serial application of rules, written in a common form, modified representations held in a workspace shared by those rules. An important component of the theory was that there was a single unified representational language within which all data held in the shared workspace was expressed. Another key idea was that a set of meta-rules controlled the serial application of these productions. These three key ideas: a single shared workspace, serial application of processing elements, and a common representational language have been extremely influential. They are both simple to comprehend and allow the construction of effective systems. In ACT-R [27] John Andersen and colleagues have taken some of the elements of production systems and used them to produce models of aspects of human cognition that produce testable predictions. In ACT-R productions now represent the serial actions of processing in the thalamus and connect to information in buffers. Together these simulate the behaviour of multiple thalamic-cortical loops. ACT-R has been used to construct models that give impressively accurate predictions for human performance on a range of tasks, including reading and mental arithmetic. ACT-R models rely heavily on the provision of timing information about delays in each stage of processing. Both SOAR and ACT-R have in common the fact that they have widely available languages that allow researchers to implement computational models. Finally in Global Workspace Theory (GWT), Baars and Shanahan [6] have proposed the idea that conscious thought is explainable at an abstract level by the idea of a global workspace. The key idea in GWT is that local processes propose items to be posted onto a single global workspace, and that mechanisms exist that select one collection of items that are in turn re-transmitted to all the local processes. It can be seen that all three theories emphasise the idea of a single shared workspace for parts of cognition. There is evidence however, at least from robotics, that such a single workspace is an incomplete architectural account of intelligence. We now describe ideas from robotics on architectures, in order to compare and contrast them with the ideas from work on cognitive architectures.

2.6.2 Robotic Architectures

The first significant attempt to implement what might be loosely called a cognitive robot was the Shakey project [28]. Detailed examination of their approach bears fruit. The architecture in Shakey was dominated by a central workspace within which all data about the contingent state of the world was written in a single representational language. In the case of Shakey this was a form of first order predicate logic minus quantification. Sensory processing was essentially a business of abstracting from the raw sensor data to this predicate description. Typing of entities was captured using predicates, and the representation also captured some metric information for the highly simplified world. Qualitative action effects were captured using STRIPS operators, which addressed some of the difficulties previously encountered by the situation calculus. This declarative knowledge about action effects was used in a planning process that had available all knowledge in the world model. In addition to this the robot had fixed routines that would update the world model when sufficient uncertainty had accumulated about its state. This was the way that gross error recovery occurred: through re-sensing and then recalculating the world model. Simpler errors were handled using Intermediate Level Actions (ILAs). These were essentially discrete closed loop controllers that relied upon the world to settle between each step and thus couldn't deal easily with ongoing rapid change. Overall, Shakey shares, at an architectural level, some of the assumptions of SOAR and ACT-R. It relies on serial application of planning operators to simulate trajectories through the state space and selects courses of action based on those. It uses a single representational language, although it does reason with that representation using two different kinds of reasoning. It has completely serial control of execution: only one ILA is in control of the robot at a time. It collects all contingent knowledge about the world in a single shared workspace. It handles error recovery in two ways: re-sensing leading to model updating and re-planning, and closed loop recovery from errors without planning. Finally it does not provide an architectural answer to the problem of sensory interpretation: perceptual routines existed in Shakey, but there are no architectural constraints or aids to how they operate, communicate, or share information. They serve only to provide information to a central model in a unified language. The classic story about Shakey given by behaviour based roboticists, is that it could never have worked outside of its carefully controlled environment, and that even within it performance was unreliable. Shakey was able to perform different tasks, but it relied upon an accurate world model. It was able to construct a sufficiently accurate representation under benign sensory conditions, but robot vision researchers unsuccessfully spent the decade following the Shakey project trying to extend this approach of scene reconstruction to more natural visual environments. Thirty-five years later our ability to perform scene or surface reconstruction is still poor, although it has improved. Of course, to be fair to the designers of Shakey it is not clear that they took a principled stance on whether all aspects of a scene should be recovered, only that attempts to extend their approach often sought to do this, and have largely failed to date. A strong reaction to this paradigm that occurred in the mid 1980s was exemplified by the behaviour based approach to robotics [29]. The approach is characterised by a number of authors including proponents and sceptics. One key idea is that the system is almost entirely representation free. The meaning of this statement depends on what is meant by representation. Kirsch [30] describes three types of representation: data items the values of which co-vary with features in the world; declarative statements which release their information when queried; and predicate descriptions that allow generalisation by type, leading to the ability to reason about inheritance. Brooks' early robots were certainly representation free on the basis of either of the last two definitions. Furthermore, while modules may have representations of the first kind, they do not typically transmit these representations to other modules for further processing or consumption. In other words there is no sharing of information, only competition for control of the robot. Other important aspects of the approach are that many controllers run in parallel, that each is relatively simple, and that their action recommendations are fused through a single global mechanism. The obvious weakness of the behaviour based approach is the lack of evidence of its ability to scale to higher cognitive functions, despite nearly a quarter of a century of effort. Instead, roboticists typically use behaviours as the lowest level of control in a hierarchical system. Three tiered architectures such as 3T [31] employ behaviours at the lowest level, and link these to a symbolic planning level via a sequencing level in which transitions are made from behaviour to behaviour using a finite state machine like representation. Representations have made a re-appearance via advances in filtering and statistical approaches imported from machine learning. Behaviours, rather than truly behaviour based approaches have thus been merged into the tool-box of techniques employed by most roboticists within architectures, rather than being an architectural choice in their own right.

2.7 Summary of Contributions and Conclusions

In this chapter we have explored a small part of the space of designs for a particular part of niche space. Working from run- and design-time requirements imposed by our combination of task and environment (our two scenarios) we have suggested one architectural schema (CAS). We have argued that CAS includes a large number of architectural instantiations and sub-schemas that meet those requirements. From our experience of building real robot systems using the software implementation of the schema (CAST) we have identified four problems which are common to a very large range of systems, and which we argue warrant architectural solutions. These are the problems of *binding*, *filtering*, *processing management* and *action fusion*. We have gone on to detail our work within CAST to create solutions to each of these problems, and have tried to show that an empirical science of architectures is possible. The work in this aspect of CoSy has been exceptionally useful in allowing us to integrate the work of many of the other chapters. In particular we believe that any true systems approach to AI must include an architectural theory. We believe that CAS represents a significant step forward in architectures for embodied cognitive systems because of the way that it combines the parallelism and incrementality of behaviour based systems with the use of representations that should lie at the heart of any cognitive system.

References

1. Sloman, A., Scheutz, M.: A framework for comparing agent architectures. In: UK Workshop on Computational Intelligence, pp. 169–176 (2002)
2. Minsky, M., Singh, P., Sloman, A.: The st. thomas common sense symposium: Designing architectures for human-level intelligence. AI Magazine Summer (2004)
3. Arkin, R.C.: Behavior-Based Robotics. Intelligent robots and autonomous agents. MIT Press, Cambridge (1998)
4. Wolfe, J., Cave, K.: The psychophysical evidence for a binding problem in human vision. Neuron 24(1), 11–17 (1999)
5. van der Velde, F., de Kamps, M.: Neural blackboard architectures of combinatorial structures in cognition. Behavioral and Brain Sciences 29, 37–70 (2006)
6. Shanahan, M.P., Baars, B.J.: Applying global workspace theory to the frame problem. Cognition 98(2), 157–176 (2005)
7. Kruijff, G.-J.M., Kelleher, J.D., Hawes, N.: Information fusion for visual reference resolution in dynamic situated dialogue. In: André, E., Dybkjær, L., Minker, W., Neumann, H., Weber, M. (eds.) PIT 2006. LNCS (LNAI), vol. 4021, pp. 117–128. Springer, Heidelberg (2006), `http://www.cognitivesystems.org/cosybook/chap2.asp#Kruijff/etal:2006`

8. Jacobsson, H., Hawes, N., Kruijff, G.-J., Wyatt, J.: Crossmodal content binding in information-processing architectures. In: HRI 2008: Proceedings of the 3rd ACM/IEEE International Conference on Human Robot Interaction, pp. 81–88. ACM, New York (2008), http://www.cognitivesystems.org/cosybook/chap2.asp#Jacobsson/etal:2008a
9. Harnad, S.: The symbol grounding problem. Physica D: Nonlinear Phenomena 42, 335–346 (1990)
10. Skočaj, D., Kristan, M., Leonardis, A.: Continuous learning of simple visual concepts using incremental kernel density estimation. In: International Conference on Computer Vision Theory and Applications, Funchal, Madeira, Portugal, pp. 598–604 (2008), http://cognitivesystems.org/cosybook/chap2.asp#skocajVISAPP08
11. Brenner, M., Hawes, N., Kelleher, J., Wyatt, J.: Mediating between qualitative and quantitative representations for task-orientated human-robot interaction. In: Veloso, M.M. (ed.) IJCAI 2007, Proceedings of the 20th International Joint Conference on Artificial Intelligence, pp. 2072–2077 (2007), http://www.cognitivesystems.org/cosybook/chap2.asp#Brenner/etal:2007
12. Hawes, N., Sloman, A., Wyatt, J.: Towards an empirical exploration of design space. In: Kaminka, G.A., Burghart, C.R. (eds.) Evaluating Architectures for Intelligence: Papers from the 2007 AAAI Workshop, pp. 31–35. AAAI Press, Vancouver (2007), http://www.cognitivesystems.org/cosybook/chap2.asp#Hawes/etal:2007c
13. Hawes, N., Wyatt, J., Sloman, A.: Exploring design space for an integrated intelligent system. In: Bramer, M., Coenen, F., Petridis, M. (eds.) Research and Development in Intelligent Systems XXV: Proceedings of AI-2008, The Twenty-eighth SGAI International Conference on Innovative Techniques and Applications of Artificial Intelligence. Springer, Cambridge (2008), http://www.cognitivesystems.org/publications/cosyBib2008.asp#Hawes/etal:2008
14. Hawes, N., Sloman, A., Wyatt, J., Zillich, M., Jacobsson, H., Kruijff, G.-J., Brenner, M., Berginc, G., Skočaj, D.: Towards an integrated robot with multiple cognitive functions. In: Holte, R.C., Howe, A. (eds.) Proceedings of the Twenty-Second AAAI Conference on Artificial Intelligence (AAAI 2008), pp. 1548–1553. AAAI Press, Vancouver (2007), http://www.cognitivesystems.org/cosybook/chap2.asp#Hawes/etal:2007a
15. Lee, D.N.: The Optical Flow-field: The Foundation of Vision. Philosophical Transactions of the Royal Society London B 290, 169–179 (1980)
16. Horswill, I.: Polly: A Vision-Based Artificial Agent. In: AAAI, pp. 824–829 (1993)
17. Land, M.F., Hayhoe, M.: In What Ways do Eye Movements Contribute to Everyday Activities. Vision Research 41, 3559–3565 (2001)
18. Clouard, R., Elmoataz, A., Porquet, C., Revenu, M.: Borg: A knowledge-based system for automatic generation of image processing programs. PAMI 21
19. Thonnat, M., Moisan, S.: What can program supervision do for program reuse? In: IEE Proc. Software
20. Moisan, S.: Program supervision: Yakl and pegase+ reference and user manual, Rapport de Recherche 5066, INRIA, Sophia Antipolis, France (December 2003)
21. Li, L., Bulitko, V., Greiner, R., Levner, I.: Improving an Adaptive Image Interpretation System by Leveraging. In: Australian and New Zealand Conference on Intelligent Information Systems (2003)

22. Brenner, M., Nebel, B.: Continual planning and acting in dynamic multiagent environments. In: International Symposium on Practical Cognitive Agents and Robots, Perth, Australia (2006), http://www.cognitivesystems.org/cosybook/chap6.asp#Brenner/etal:2006a
23. Sridharan, M., Wyatt, J., Dearden, R.: HiPPo: Hierarchical POMDPs for Planning Information Processing and Sensing Actions on a Robot. In: International Conference on Automated Planning and Scheduling, ICAPS (2008), http://www.cognitivesystems.org/cosybook/chap2.asp#Sridharan/etal:2008a
24. McDermott, D.: PDDL: The Planning Domain Definition Language, Technical Report TR-98-003/DCS TR-1165. Tech. Rep., Yale Center for Computational Vision and Control (1998)
25. Hoffmann, J., Nebel, B.: The FF Planning System: Fast Plan Generation Through Heuristic Search. Journal of Artificial Intelligence Research 14, 253–302 (2001)
26. Laird, J.E., Newell, A., Rosenbloom, P.S.: Soar: An architecture for general intelligence. Artificial Intelligence 33(3), 1–64 (1987)
27. Anderson, J., Bothell, D., Byrne, M., Douglass, S., Lebiere, C., Qin, Y.: An integrated theory of the mind. Psychological Review 111(4), 1036–1060 (2004)
28. Nilsson, N.: Shakey the robot, Tech. Rep. Tech. Note 323, AI Center, SRI International (April 1984)
29. Brooks, R.A.: Intelligence without representation. Artificial Intelligence 47, 139–159 (1991)
30. Kirsch, D.: Today the earwig, tomorrow man? Artificial Intelligence 47, 161–184 (1991)
31. Gat, E.: On three-layer architectures. In: Artificial Intelligence and Mobile Robots (1997)
32. Cheyer, A., Martin, D.: The Open Agent Architecture. Journal of Autonomous Agents and Multi-Agent Systems 4(1), 143–148 (2001)

3

The Sensorimotor Approach in CoSy: The Example of Dimensionality Reduction

David Philipona and J. Kevin O'Regan

Laboratoire Psychologie de la Perception, Université Paris Descartes and CNRS
jkevin.oregan@gmail.com

3.1 Introduction

This chapter presents an algorithm implementing a basic requirement for any interface between sensory data and cognitive functions: dimensionality reduction. The algorithm extends the classical framework of dimensionality reduction to the case where sensory data are acquired through an embodied agent, by grounding the metric that is at the basis of the dimensionality reduction in the sensorimotor abilities of the agent. The final objective (which was not realized because of time constraints) within CoSy was to build on this to provide a demonstration of some basic unsupervised learning of interactions with space and objects, as would be required in an explorer-type scenario.

The algorithm provides several functionalities for the local analysis of sensory data: it can estimate a code independent distance between sensory data points, suggest actions to be taken so as to link neighboring sensory configurations, and decompose possible sensory variations into sets of variations with structurally different properties. Additionally, it can provide a local flattening of the sensory manifold and therefore it can be used in the way classical dimensionality reduction is used.

The work described is situated within an approach called the "sensorimotor approach" ([1]). The approach derives its motivation from the idea that biological and artificial systems that have a variety of different types of sensory inputs, and that have diverse and complicated effector mechanisms, are best described in a way that does not depend in an essential way on the particular sensory or motor systems involved, but is instead code-independent. Note that the sensorimotor approach, as understood here, is quite a radical departure from classic views of natural or artificial perception. It should be distinguished from other approaches where the term "sensorimotor" is used more loosely (cf. e.g. [2]), and from approaches like those of' 'Active Vision" (c.f. [3, 4, 5, 6, 7]) where action plays an important role only to the extent that it increases the amount of information available to the system. Here the sensorimotor approach is taken to imply the idea that the only type of knowledge

H.I. Christensen et al. (Eds.): Cognitive Systems, COSMOS 8, pp. 95–130.
springerlink.com

concerning the environment that a system can access is knowledge about the laws that link the system's actions to the resulting changes in sensory input. These *sensorimotor laws* are the only basic entities that a system can ever have access to.

The idea applies in an obvious way to geometrical notions: A system's understanding of the dimensionality and structure of the (probably Euclidean) space in which it is embedded is constrained by the particular movements the system can effect, and by the system's ability to observe the sensory changes these movements produce. But the idea must also apply to all other sensory notions that a system possesses: for example the notion of its own body characteristics, or its tactile perception or its perception of sound or color.

Indeed, a dditional work performed during the CoSy project using the sensorimotor approach concerned color. This work, though very interesting from a fundamental point of view, was of less direct concern to the CoSy platforms because using biologically inspired color perception was not considered a main priority in the project. The work can be referred to in [8].

3.2 Artificial Agents and Human Perception

Sensory data to be dealt with by embodied agents are complex, high dimensional and grounded in the continuous domain. On the other hand cognitive functions like navigation, planning and communication usually are considered algorithmic in nature. For this reason, most research in the field (although for an exception see e.g. [9]) considers that actions, sensory configurations and possible states of the system are taken from finite sets (this is for instance the standard framework in machine learning, topological map building, language, dynamic logic, qualitative reasoning: see for instance [10, 11, 12]). The complexity of these algorithms is often exponential in the number of such states, and consequently they crucially require tractable information as inputs. This requires modules to be implement that interface low level data acquisition and higher cognitive processing: for instance, modules that convert visual flow into a small collection of coordinates and attributes that identify the position and properties of objects in the field of view.

Evolution has endowed biological organisms with neuronal processing schemes adapted to the organisms' sensors. By analogy it makes sense for artificial systems to consider creating interfacing modules specifically engineered for the agent's sensory devices and the data format used by cognitive functions. On the other hand, while device dependent solutions will ensure a tractable level of efficiency and robustness, they also constrain a system's ability to adapt as task requirements and sensorimotor competences change, or as the agent's body suffers evolution, modifications, changes in environmental conditions, or damage. Such issues might presumably be handled at a cognitive level, but if information is erroneously discarded or even merely impoverished by early perceptual processing then the task may be very hard. In the case of humans, this seems essentially unrealistic in the long term as it would mobilize too many of the resources required by cognitive functions. Finally, if modifications occur at the sensorimotor level then it seems that methods considering a continuous framework will be more adaptable than methods conceptually based on the manipulation of a limited set of entities.

These arguments all suggest investigating the question of an unsupervised and multimodal approach to sensory data analysis. Insofar as global interactions with space and objects are based on the coupling of cognitive functions with the mastery of local interactions [13], such an approach would provide the required basic tools. It would ultimately lay the ground for unsupervised algorithms converging to the flexible exploitation of the sensorimotor abilities available to the agent.

The issues faced by research in the field of embodied articial agents are also related to research in the field of human perception. There is a long history of theories about the sensorimotor grounding of perceptual experience, especially as concerns space: Helmholtz at the end of the 19th century had already suggested a general formulation of the notion of displacement for a sensorimotor system [14], and Poincaré expressed his thoughts on the relation between compensability and space a few decades later [15]. More recently Piaget developed the idea that sensorimotor interactions were fundamental in structuring an individual's cognitive and perceptual abilities [16], and Gibson founded a theory of perception tightly linked with action [17]. As a natural outcome of this history several issues addressed today in developmental psychology revolve around the interplay between sensorimotor skills, cognitive/symbolic abilities and perception.

In the field of psychophysics, the importance of sensorimotor interactions has been recently pushed to quite an extreme limit in the idea that perception should be envisioned in terms of cognitive access to the mastery of sensorimotor abilities, rather than in terms of passive sensory processing decoupled from both cognitive and motor constraints [1]. Though this is more of philosophical importance than critical to a concrete project like CoSy, it is worth noting that the basic motivation for such a radical shift, as compared to the standard view of sensory processing, lies in a central difficulty faced by theories of perception: the difficulty in explaining the existence and character of the qualia of sensations [18]. Thus, an approach to perception emphasizing neuronal processing will be faced with the a priori impossible task of finding an explanation of qualia within properties of this processing or properties of the neuronal substrate underlying this processing [19, 20, 21]. On the other hand an approach emphasizing sensorimotor interactions has instead the easier task of matching qualia to properties of sensorimotor interactions.

3.3 Dimensionality Reduction

To address the mismatch between the complexity existing at the sensor level and the requirement that cognitive processing should be tractable, a common hypothesis is that perception is preceded by a process that extracts low dimensional representations of the sensory information. The so-called Swiss Roll gives a clear illustration of the situation. Given data points in the "high"-dimensional space $\mathbb{R}^3$ that really lie on a rolled plane, metrical reduction algorithms will search for a mapping from $\mathbb{R}^3$ to $\mathbb{R}^2$ that does not alter geodesic distances between points.

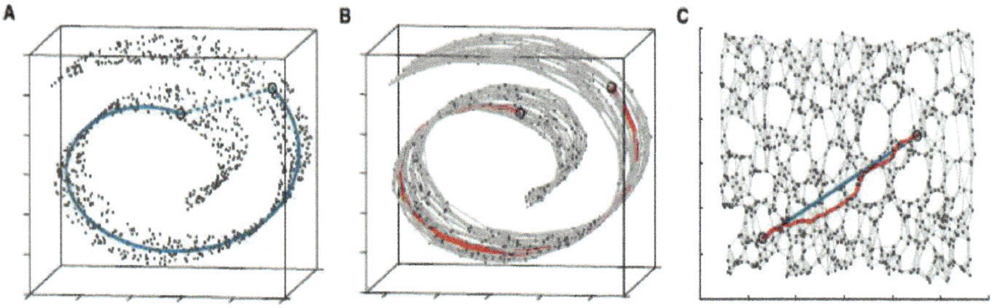

Fig. 3.1. Unfolding the Swiss Roll. *A*, the manifold on which the data lies; *B*, the data points; *C*, the resulting embedding. From [22].

However, as will be shown in the following overview, current research on methods of dimension reduction has mainly focused on technical aspects of dimensionality reduction itself without questioning the metric used on the data to be reduced. It is usually assumed that the format of data representation should itself suggest a metric. But according to the sensorimotor approach, we expect that actions, rather than the peculiarities of particular data representations, are what should essentially determine the metric to be used on sensory data.

3.3.1 Review of Classical, "Passive" Approaches

Variance Based Methods

Principal Component Analysis (PCA) is a ubiquitous method when data lie in a vector space, and it exists in several flavors: variance analysis and reduced rank analysis [23], Probabilistic PCA [24], incremental PCA [25], recursive PCA [26], etc. The basic idea is to provide the d-dimensional linear approximation of a dataset yielding minimal error with respect to the n-dimensional original data ($n \geq d$). Approximation is obtained by retaining from the data only their orthogonal projection on a d-dimensional subspace accounting for as much as possible of data variance. Actually, PCA provides an ordered set of n approximations of dimensions $d = 1 \ldots n$ having this property. It can be performed algebraically or using neural networks. Note that both the notion of minimal error and orthogonal projection follow from the specification of a norm on the data.

Several generalizations of PCA have been developed. Projection Pursuit (PP) searches for axes in input space that optimize a user defined index describing the "interestingness" of the axis [27, 28]. If the index of an axis is the variance of data along this axis then PP equates with PCA, but other indexes such as data kurtosis or multimodality can also be considered. Generalization of the Gaussian assumption underlying PCA to the more general assumption of an exponential family of probabilities has also been investigated [29].

Practical attempts towards a nonlinear version of PCA are found in Local PCA [30] or mixture of principal component analysers [31], which independently analyses cells of a partition of the input space, and Locally Weighted PCA (LWPCA), which interpolates with radial functions between different local models [32]. Kernel PCA [33] provides a nonlinear version of PCA through an implicit mapping of data in a higher dimensional space on which PCA is performed, computations being optimized by the fact that only scalar products of mapped data are actually required. Principal Curves (PC) and Principal Surfaces provide a more geometrical viewpoint on a nonlinear version of PCA: instead of searching for subspaces fitting the data, PC searches for surfaces fitting the data [34]. By analogy with the property satisfied by principal subspaces in the case of data drawn from a Gaussian distribution, each point of a principal surface is expected to be the mean of the data whose projection on the surface is this point. This Riemannian approach is quite general in extent, and there are developments for the more constrained (hence richer) case of Lie groups [35].

Distance Matching Methods

The methods of MultiDimensional Scaling (MDS) originally developed for psychophysics [36] can also be used for dimension reduction, by retaining from the data only their distances, and then searching from scratch for low dimensional representations satisfying the same set of distances. The simplest approach uses a least square citerion and is purely algebraic, but there are numerous variants. Notably, nonmetric MDS aims at replicating the rank orders for distances instead of the actual distances [37, 38], and is based on minimization of a complex cost function for the mismatch of distances instead of using the least square criterion. Sammon's NonLinear Mapping (NLM) follows the same philosophy as nonlinear MDS but uses yet another criterion for the mismatch of distances [39].

Laplacian eigenmaps can be seen as a form of distance matching [40]. The idea is to find real functions of unit norm over the data manifold that preserve locality as much as possible (i.e. close points are mapped onto close real values). An embedding can then be found from the fact that an orthonormal basis of such functions is the solution of a general eigenvalue problem, with the size of the eigenvalues quantifiying the goodness of locality preservation for the associated eigenfunctions.

Vector Quantization Methods

Self-Organizing Maps (SOM) represent a somewhat different trend of research, based on the idea of Vector Quantization (VQ) with topological constraints [41]: SOM provides a feature map from a specified set of representatives (called 'neurons" in the SOM nomenclature or "codebooks" in the VQ nomenclature) in a topological space to a set of prototypes in input space, so as to yield minimal quantization error in input space. Curvilinear Component Analysis (CCA) circumvents the requirement of a predefined neighborhood relation between representatives by adapting both the set of prototypes, using standard VQ methods, and their representatives, by a gradient descent minimizing the mismatch between the set of local distances for input data and the set of local distances for representatives [42]. Neural Gas (NG) and Growing Neural Gas (NNG) provide efficient VQ methods able to adapt to data with different dimensionalities at different locations [43, 44]. Approaches to dimension reduction based on Shannon information theory [45] should also be mentioned as they are closely related with VQ.

Geometric Methods

In recent years, two methods of dimension reduction for data lying on a nonlinear manifold have been very popular: Isomap [22] and Locally Linear Embedding (LLE) [46]. Isomap first computes geodesic distances between points within the input manifold using shortest path algorithms, and then applies MDS to represent the data in a space such that Euclidean distances correspond to the computed geodesic distances. Curvilinear Dimension Analysis (CDA) is very similar in spirit, its main difference being to use CCA instead of MDS [47]. As for LLE, it is based on the preservation of local linear relationships instead of the preservation of local distance relationships: LLE first expresses each point as a linear combination of its neighbors in input space and then computes a low dimensional representation satisfying identical linear relationships. A fundamental difference between Isomap and LLE is that the first tries to preserve global geodesic distances over the whole set of data while the second attempts to preserve local linear relationships between neighboring points: the first is global, the second local.

3.3.2 Standard Issues

Algebraic versus Gradient Based/Neural Network Methods

Approaches providing algebraic solutions such as PCA, classical MDS, Isomap and LLE profit from a clear analytical framework for studying the robustness of the algorithms and from highly optimized routines developed on computers for linear algebra. On the other hand approaches with gradient based solutions are more biologically appealing and suggest the existence of very efficient implementations in physical dynamical systems, that are presumably better behaved than algorithmic processing. It is therefore difficult to judge these approaches on the basis of this algebraic versus gradient based distinction, even though this point is often discussed in the arguments supporting each approach.

On the other hand, what may have an impact is the possible existence of strong discontinuities in the behavior of an algorithm with respect to its data and parameters. The behavior of Isomap, for instance, has been argued to be strongly dependent on the size of the neighborhood that is considered to estimate geodesic distances [48]: a neighborhood which is too large might cause collapse of nearby parts of the manifold on which the data lie and a neighborhood which is too small might lead to conjecture a myriad of unconnected components. This is a very unfortunate behavior, but it seems unavoidable in a numeric context which spoils the notion of continuity on which the geometric intuition of such algorithms is grounded.

Linear versus Nonlinear Approaches

PCA definitely possesses numerous advantages: it is easy to implement, robust (in the sense that it will not often yield absurd results), its memory and computational requirements scale reasonably with respect to the dimensionality of the data, and for about the same cost as a unique approximation it actually provides a series of approximations with increasing degree of accuracy. It must however be recalled that principal components are defined only on the basis of the variance of the data along a family of orthonormal axis. This means that the extracted components will not necessarily have an intuitive meaning. Indeed PP demonstrates that other criteria can equally well be considered, and may be more relevant from a statistical point of view. The linear assumption underlying PCA will also lead to an overestimation of the number of coordinates needed to approximate the data if the input dataset lies in a nonlinear manifold. Further, distances between approximations may be substantially different from distances in input space if the latter are geodesic distances. These two points are the main reasons that drove research from linear approaches to nonlinear ones.

Topological and Geometrical Constraints in the Nonlinear Case

Nonlinear approaches suffer from more subtle difficulties than might be imagined from the local notion of a smooth manifold: in particular, there may be a global topological incompatibility between the d-dimensional submanifold considered for reduction and the Euclidean space $\mathbb{R}^d$ in which a low dimensional representation is sought. Consider a circle in $\mathbb{R}^2$: any *global* attempt to represent it in the Euclidean space $\mathbb{R}$ having its intrinsic dimension will fail to preserve such an elementary feature as neighbor relations between its points.

There is research aiming to specifically address this issue either by cutting the data manifold so as to achieve an embedding in a Euclidean space having the intrinsic dimension d of the manifold [49], or by using Whitney's theorem to preserve the global topological structure of the data manifold but achieving an embedding in a Euclidean space of dimension $2d + 1$ [50].

As for metrical aspects, it may sometimes be the case that there is an incompatibility between the intrinsic curvature of the data manifold and the flatness of the representational space: consider the case of data lying on a fishbowl in $\mathbb{R}^3$. Such data cannot be represented in $\mathbb{R}^2$ without introducing a systematic distortion of distances as compared to geodesic distances on the data. A possible way to overcome this problem in some cases is to relax the assumption of a Riemannian metric on the data manifold, and use a conformal metric (i.e. undetermined by a scale factor at each point of the manifold) [51]. This can however only solve very specific cases.

3.3.3 The Central Issue of the Metric

All the approaches reviewed above have a fundamental problem. They assume a particular metric specified in input space, but in general they give no reason why this metric should be prefered over another one. Yet clearly the metric used for dimension reduction is crucial in determining 1) aspects of the data that can be neglected in the reduction process, 2) the kind of distance relationships that have to be preserved in representational space, and 3) the kind of coordinates yielded by dimension reduction.

Metrics Based on Noise

Dimensionality reduction presumably takes from its oldest solution, PCA, the conceptual framework that lets the Euclidean norm be considered as a natural default choice for a metric on data. Consider the very simple example of two-dimensional data obtained from one-dimensional data by a noisy process (the standard situation underlying PCA), so that

$$Y = u \cdot X + \epsilon$$

where $u \in \mathbb{R}^2$ is some vector, $X \in \mathbb{R}$ are the "true" data and ϵ is from white normal noise. The choice of norm $||y||^2 = y^T y = y_1^2 + y_2^2$ will equate statistical concerns with a geometrical picture: the estimation for X that results from maximum likelihood can be read as the orthogonal projection of data on the line yielding minimum distance expectation between data and their projections. This expectation equates in turn the geometrical picture with PCA since it can be shown that minimization of distance expectation is achieved by using the line yielding maximum projected variance. Here it must be emphasized that "orthogonal projection" and "distance" both *depend on the norm just defined.* Hence if one assumes that data are corrupted by independent and identical noise on each coordinate, the Euclidean norm is a natural criterion to minimize as regards approximation concerns.

Note of course that this is already no longer the case if the noise sources are not identical and independant. In order for the minimization of distance expectation to yield the maximum likelihood estimate for X, a coordinate system whitening the noise has to be considered: put $\tilde{Y} = Cov(\epsilon)^{-1/2}Y$, then

$$\tilde{Y} = v \cdot X + \tilde{\epsilon}$$

with $\tilde{\epsilon}$ white noise takes us back to the previous situation. But if $||\tilde{y}||^2 = \tilde{y}^T\tilde{y}$, then we have $||y||^2 = y^T Cov(\epsilon) y$ which shows that the noise covariance indeed defines the norm that should be used as a minimization criterion if we want to perform a maximum likelihood estimation of “true” data.

This kind of approach may therefore provide arguments for use of a specific metric, but the point is: methods such as Isomap are *not* really concerned by the issue of the noise. In fact, their simplest examples involve data without any noise. Thus the question is: what is that metric whose preservation is the main virtue of a "good" dimension reduction algorithm? This is discussed for instance in [52] from the perspective of standard tasks to be fulfilled in image processing. We propose here a somewhat different viewpoint.

Metrics Based on Invariance

Since images are a classical example where dimension reduction is often required, we could search as a starting point for arguments leading to consider specific metrics on image manifolds. For that purpose, we will distinguish between the continuous field amplitude $I(x, y)$ that represents the information with which sensors interact, and I_{ij} the pixel array delivered by the sensors of a camera. The idea is that the pixel representation should be considered merely as a choice of coordinates for the physical entity $I(x, y)$ facing the sensors. This coordinate system is obviously not neutral on general grounds: first, it will determine the kind of information loss induced by noise on these representations, and second, if computations performed by modules receiving such representations are limited then this representation will imply specific possibilities and impossibilities as regards what can be performed by such modules. While these two points are very important they are not usually mentioned explicitly in the framework of dimension reduction. From the viewpoint of this framework, coordinate systems should be neutral, and the choice of a metric has no reason to be based on it. This means that we must search within external constraints for ideas about what metric to use.

We might first try to find whether there is a natural metric on amplitude fields $I(x, y)$. As for infinitely extended fields, a presumably important requirement is that this metric should yield a notion of distance between fields independent of a frame of reference in the 2D plane, otherwise we would need to take into account some "absolute" position and orientation of our own viewfield to estimate the distance between two fields. Note $p = (x, y)$ for convenience, then I observed in a frame of reference with change of position

$t \in \mathbb{R}^2$ and change of orientation $r \in SO(1)$ is the field $(t, r) * I : p \mapsto I(r \cdot p + t)$, and therefore we want

$$d(I, J) = d((t, r) * I, (t, r) * J) \quad \forall t \in \mathbb{R}^2, r \in SO(1) \tag{3.1}$$

If d is a norm on square integrable functions, then it derives from a scalar product of the form

$$\langle I, J \rangle = \int k(p, q) I(p) J(q) dp dq$$

and we can easily show the constraint 3.1 to be equivalent to

$$k(p, q) = \tilde{k}(||p - q||)$$

where $||p||$ is the Euclidean norm on the 2D plane. Hence the norm

$$||I||^2 = \int I(p)^2 dp$$

is a simple case of such invariant norms obtained by using a delta function as a kernel: $\tilde{k}(v) = \delta(v - 0)$. This suggests that this norm on pixel arrays yields an intuitive notion of approximation (for instance when used to perform a PCA) because cameras have been built so that this norm satisfies invariance properties that may underlie our own perceptual analysis.

The previous simple formalization is instructive on three points:

- First, it shows that the standard norm used on pixel arrays has the very desirable property of being approximately invariant to changes of viewpoint on fields. This is however so only because of the regularity of the sensor array, and because all sensors are assumed identical and linear with respect to fields[1,2].
- Second it suggests a whole class of metrics on fields that are invariant and therefore could be considered on pixel arrays. It is easy to construct a whole family of invariant norms once we have one. This can be done using linear transformations A such that $A((t, r) * I) = (t, r) * A(I)$. Convolutions with kernels of the form $h(||p - q||)$ satisfy this, hence isotropic blurring or banks of linear invariant filters for instance obviously do not destroy invariance.
- And third, it suggests a generic criterion, invariance, to deal with the case where sensors no longer satisfy favorable constraints of regular layout,

[1] This shows that the physical architecture of a device *is part of its computational abilities*.

[2] As a consequence of this invariance we have the following result. Take $T = (dt, dr)$ a tangent vector to the identity in $SE(2)$, and put $I(s) = \exp(Ts) * I$ a trajectory in the space of pictures obtained by the action of the exponential of T: then $||\dot{I}||^2 = const.$

similarity and linearity of sensors. In the case of humans, it must be recalled for instance that photoreceptors are organized on the retina in very inhomogenous way, and furthermore that blood vessels in the eye obstruct a large number of photoreceptors. In the case of an artificial agent, the use of an inhomogeneous optical filter is equivalent to a distortion of sensor layout and sensor responsiveness. If such a filter is added to the camera, then the pixel norm and invariant norm on physical fields will not match anymore. The invariance property of the pixel norm will break down and the distance between pairs of pictures will for instance vary when both pictures are translated by the same amount. We may wonder why optical filters should be added to the camera: the point is merely to emphasze that important assumptions about the sensory device are made when considering the Euclidean norm on arrays of pixels. The role of these assumptions has to be kept in mind when conceiving an unsupervised approach to sensory processing.

The goal of section 3.4.1 will be to provide a more general formalization for the notion of invariance, so as to provide the definition in the most relevent way for a given sensorimotor system.

3.4 Dimension Reduction in the Context of Sensorimotor Interactions

Having considered a general overview of known dimension reduction methods and acknowledged the need for them to provide arguments for the choice of a metric, we try in this section to put forward sensorimotor arguments. We will first focus on the construction of a mathematical conception for what can be meant by a sensorimotor system. We need a picture that is constrained enough so that we avoid getting lost in wide-ranging considerations about dynamical systems, and yet a picture that is general enough so that we leave substantial room for applicability and unsupervised learning: we want to carefully weigh constraints to be added and put them in a general form so that they can be instantiated by an agent involved in sensorimotor interactions that may not resemble our own.

For the reasons we have already discussed, we deliberately make the choice of considering a continuous and dynamical picture of a sensorimotor system. Further we make the choice of a geometrical picture, that is: we want to think in a coordinate-free way about the mathematical objects considered in the model. This means in particular that we do not want models that depend on a specific encoding of sensory data and motor commands. This is the condition to find device independent algorithms, but it is also motivated by considerations at the foundations of the sensorimotor approach to perception developed in [1]: namely the assumption of a functional, rather than dualist or reductive physicalist explanation for perceptual experience.

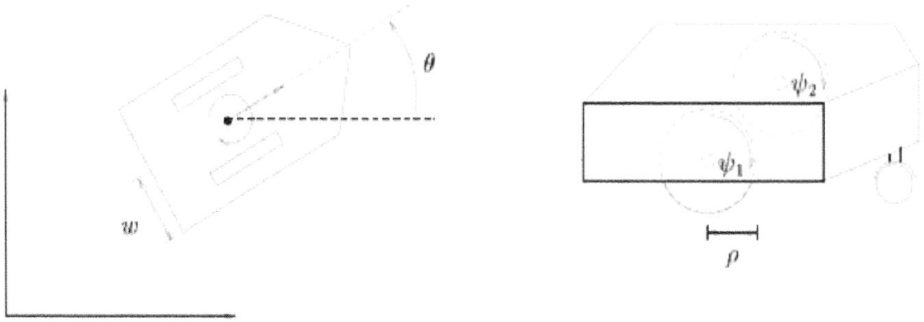

Fig. 3.2. Two wheeled vehicle on a 2D plane with various variables defining its physical state

3.4.1 Finding a Mathematical Framework

To ground the intuition of this framework we will often go back to a standard example of robotic system interacting with space: that of a two wheeled vehicle on a 2D plane (Figure 3.2).

Model for Sensorimotor Dynamics

A quite general model for sensorimotor interactions of an embodied agent could be

$$\begin{cases} \dot{\omega} = F(\omega, m) \\ s = G(\omega) \\ m = H(s, a) \end{cases} \tag{3.2}$$

where ω are physical states of the agent-and-environment, m are motor commands, s are sensory inputs, and a are action commands imposed by higher level cognitive functions. F describes the possible dynamics of the physical state under different commands, G essentially describes perceptual aliasing (the fact that a given sensory input most often can result from a whole set of physical states), and H describes feedback control laws associated with different actions a. This is for instance a model equivalent to that of [53], apart for one important point: actions a will be here considered *a priori* to constitute a continuous set rather than a finite set.

Our motivation to consider a continuous set of actions is that we want to investigate sensorimotor arguments for abstracting a finite number of actions so that the global sensorimotor interactions of the organism driven by cognitive functions still exploit the whole range of low level abilities[3]. This actually

[3] For instance, it seems intuitive that most of the time neighboring actions will lead to similar sensorimotor loops and therefore only one representative action for the whole set needs to be retained by cognitive functions. But there will also be some actions leading to very different sensorimotor loops, or bearing specific relationships with other loops, and these will need to be differentiated by cognitive functions if the overall structure of the sensorimotor dynamics is to be preserved.

means that we will overview some arguments for turning the previous system into a hybrid system [54] without losing too many of the possibilities offered by the underlying continuous dynamics. As we will see, the issue of isolating a finite number of actions will be related, in the algorithm that we suggest, to the question of finding a metric on sensory input. But more essentially the issue of the interface between a continuous dynamical system and cognitive functions (e.g. algorithms for Turing machines) is at the heart of our general approach in the scope of this work.

Notation: We will refer to the first equation of the model (3.2) as the *system dynamic* or *external dynamic*. We will refer to the dynamic involving feedback control as the *low level sensorimotor dynamic*, that is:

$$\dot{\omega} = F(\omega, H(G(\omega, a))) = \varphi(\omega, a) \tag{3.3}$$

The term low level is in contrast to the *global* dynamic which also involves slower cognitive feedback. This gives the picture of a fast dynamic (the low level sensorimotor dynamic) driven by a slow one (the cognitive dynamic), as also suggested in synergetics [55] and hybrid systems [54].

Overview of Structural Aspects

Model (3.3) is typically far too general to provide much insight about the possible structure of low level sensorimotor interactions. Constraints to consider in order to obtain a mathematically tractable framework are suggested by viewpoints that have arisen in the mathematical study of dynamical systems. We evoke these viewpoints as they can provide hints to ground selection of actions by specifying 'structures" that we may want to see preserved by cognitive processing.

Catastrophe Theory

An example of a possible and natural constraint to consider on the model, if it is to refer to a physical system, is the existence of a potential from which the dynamic derives. If the sensorimotor dynamic depends on a smooth potential then catastrophe theory details conditions under which slight changes of a given action will result in a very different organization of the associated low level sensorimotor dynamic: changes in the number of fixed points, the structure of stable, unstable and repelling sets, structure of cycles, etc. This could be used to partition the action space into areas associated with qualitatively similar dynamics, and identify low level sensorimotor dynamics that are singular in the sense that they require a very precise tuning of the parameters to occur. This would be the viewpoint of structural stability [56]: identify and represent dynamics that are robust to slight perturbations in the choice of action. While such general aspects of dynamical systems must be kept in mind, they do not seem to express sufficiently strong constraints to account for the kind of physical regularities that we need to ground a notion of space and objects.

Geometrical Nonlinear Systems

Another possible constraint is the one considered in standard approaches to nonlinear control theory [57], namely: a vector space structure for the set of all possible actions, and external dynamics affine with respect to actions i:

$$\dot{\omega} = \varphi_0(\omega) + \sum_i \varphi_i(\omega) \cdot a_i \tag{3.4}$$

This model can be applied to many robotic systems, and is further compatible with physiological work about the linear combination of motor primitives in humans [58]. The affine assumption is a very natural one from a geometrical perspective, as will be sketched later, and it will be retained in the following approach.

Example. To fix ideas, we can slightly detail the case of an ideal vehicle on the 2D plane. Put

$$\omega = (x, y, \theta, \psi_1, \dot{\psi}_1, \psi_2, \dot{\psi}_2)$$

where these variables describe, respectively, the position of the middle of the two wheels' contact points, the vehicle's heading direction, and the angles of a fixed point on each wheel with respect to the vertical. Then standard mechanical principles show that

$$\begin{aligned}(\dot{x}, \dot{y}, \dot{\theta}) &= f_0(\theta, \dot{\psi}_1, \dot{\psi}_2)\\ (\ddot{\psi}_1, \ddot{\psi}_2) &= J_1 \cdot \tau_1 + J_2 \cdot \tau_2\end{aligned}$$

where control variables τ_1 and τ_2 are the torques applied to drive the wheels. We then see that the dynamic is indeed linear in the motor commands.

In the context of nonlinear control theory, reachability is the issue of determining which configurations of the system can be reached from a given starting configuration. This is an issue well understood in the geometrical picture of equations (3.4), and it naturally suggests that an important and tractable requirement to be considered on the choice of a finite set of actions is that this choice should not reduce reachability, at least in the limit of infinite sequences of such actions. Here we will consider that issues about observability and stability are about available low level sensorimotor abilities, and so we will not discuss them.

Mechanical and Geometrical Symmetries

A last important constraint that may be considered is the case where the physical world possesses a very strong regularity, namely the existence of a symmetry group for system dynamics. This is the case we detail in the next section.

The existence of a symmetry group can be interpreted as the situation where an underlying space imposes that system dynamics are invariant when physical states are merely moved in this space. Consider lifting up the vehicle from the 2D plane then putting it at another place: There is no change in the dynamic ruling the system, there is no way to infer from the dynamic where the vehicle is.

If we take as our goal to investigate algorithms allowing a system to locate itself in space, we will need to define a constraint that will let us consider models of sensorimotor interactions in which there is indeed a space to find. The existence of a symmetry group will allow us to introduce on general grounds notions such as displacement, and relative configuration. This provides further ideas to select specific actions: we can expect some actions to bear very specific relationships with all other actions, and these specific relationships could provide a simplification for the set of policies to be considered[4]. In any case the presence of symmetries is known to simplify the complexity of optimal planning in the paradigm of machine learning [59].

In this chapter the symmetry group will suggest constraints on metrics to be used on sensory inputs: we wish to search for metrics respecting this symmetry.

Space: System Symmetries and Sensorimotor Symmetries

Symmetries

In order to formulate the assumption of the existence of a symmetry group, we make a slight shift of viewpoint towards a geometric framework: we rewrite the first equation of model (3.2) as

$$\dot{\omega} = F_m(\omega)$$

and we view F_m as a family of vector fields on the state manifold Ω parameterized by motor outputs m. We can see here that the assumption of a dynamic which is affine in the motor commands is natural, since the family F_m can be seen as a submanifold of the vector space of vector fields: the affine assumption can therefore be seen either as a local approximation or as a general frame inside which smooth constraints will be considered later.

The existence of a symmetry group Σ on Ω for the dynamics can be written

$$T^* F_m = F_m$$

Here T^* denotes the pull-back application for the diffeomorphism T of Ω (see for instance [60]).

We already have cited the common example of a vehicle moving on a 2D plane. Other examples could include a vehicle moving on 2D manifolds of

[4] Consider the case of commutative actions a and b. The set of possible action sequences to consider for planning reduce to $\{a^n b^m, n \in \mathbb{N}, m \in \mathbb{N}\} \sim \mathbb{N}^2$ while if actions do not commute then the set of possible action sequences to consider is $\mathbb{N}^{\{a,b\}} \sim \mathbb{R}$.

constant curvature[5]. To fit with the physics of our world, we will assume the existence of a smooth symmetry group, i.e. a Lie group.

Relative Configurations

Under suitable assumptions of regularity for the action of symmetries[6], the symmetry group induces a partitioning of the state manifold Ω into equivalence classes defining a manifold of *relative configurations*[7] Ω/Σ. This suggests a quite general aspect of an agent body in model (3.2), expressed in an abstract framework: the agent's own configuration resides in what remains among physical states when symmetries of the system dynamics have been taken into account.

Example. In the case of the two wheeled vehicle, physical states equivalent up to Euclidean symmetries of the 2D space are states with equal wheel configurations, hence relative configurations can be written $c = (\psi_1, \dot{\psi}_1, \psi_2, \dot{\psi}_2)$.

If further there is no physical state left invariant by symmetries[8] (any Euclidean displacement of the vehicle can be identified straightforwardly by reading out changes in ω), then the geometrical picture of model (3.2) is that of a principal bundle

$$\pi : \Omega \to \Omega/\Sigma$$

with group structure Σ.

Locomotion

This was for the constraints on the external dynamic. As for the sensorimotor model, we will take into account the control law of the agent in order to define its displacement abilities. Here also we operate a slight shift toward geometry by writing sensorimotor dynamics as

$$\dot{\omega} = \varphi_a(\omega)$$

[5] The case of manifolds of non-constant curvature intuitively calls for a distinction between a sensorimotor agent in a 3D symmetrical world constrained to move for a while on an arbitrary manifold (physical symmetries still exist, but are symmetries of the 3D world), and an agent in some sense living "inside" this manifold (there are no symmetries there as the agent could distinguish some places on the basis of a different relation between sequences of movement controlled by identical sequences of commands).

[6] The action must be properly continuous.

[7] We prefer to use this term rather than the one more classically used in this context of *shape*, because in our framework relative configuration could *a priori* include more than shape in the classical sense: for instance changes in the sensitivity of the sensors. Furthermore, if there are objects in the environment their configuration *relative* to the agent enters into Ω/Σ.

[8] In other words: if symmetries act freely.

and view them as a family of vector fields on the state manifold parameterized by actions.

Sensorimotor dynamics such that $T_\omega \pi \varphi_a(\omega) = 0$, $\forall \omega$ (i.e. *vertical* vector fields) define *sliding locomotion*: this means that such dynamics correspond to a displacement of the agent that occurs without any change of relative configuration. This is hardly expected to happen in our physical world[9], but is actually often considered as an approximation when abstracting details of a sensorimotor system.

Example. If we forget about the wheel configuration of the vehicle on the 2D plane, then the relative configuration manifold is reduced to a point (no change other than position and orientation is left in the model) and locomotion will therefore necessarily be sliding locomotion. Indeed, assuming high gain feedback controllers that let wheel rotation velocity quickly converge to specified velocities $a = (v_1, v_2)$, we can write

$$(\dot{x}, \dot{y}, \dot{\theta}) \approx \frac{\rho}{2l}(l(v_1 + v_2)\cos\theta, l(v_1 + v_2)\sin\theta, v_1 - v_2) \tag{3.5}$$

which provides a sensorimotor model with sliding locomotion. Note the requirement to involve control schemes to obtain such a model: on physical grounds the configuration of the wheels *cannot* be abstracted from the model.

On the other hand, some sensorimotor dynamics will correspond to locomotion associated with changes of relative configuration. This suggests focusing on the link between changes in relative configuration and locomotion. It turns out that both geometrical constraints of the form "no slipping contact" and mechanical constraints (that is, stemming from a Lagrangian approach), can be expressed as a connection on the principal bundle $\pi : \Omega \to \Omega/\Sigma$ [61, 62]. This has been shown to be valid for various artificial systems on a rigid ground [63] or in an ideal fluid [64], as well as for swimming microscopic organisms when viscous forces dominate inertial forces [65]. The fact that the physics can be written as a connection implies that only the trajectory of relative configurations affect the agent's displacement, not the velocity at which the trajectory is traversed. This is a strong constraint on the external dynamics of a sensorimotor system. It may actually be valid only piecewise, for instance when considering manipulation [66] or legged locomotion [67]. At the level of the external dynamics this connection allows a holonomy group to be defined, that is, the group of displacements for that dynamic[10]. It also allows a notion of curvature to be defined on the principal bundle π that can

[9] We mean here: *controlled* sliding locomotion is hardly expected to happen, since any system left in outer space with an initial nonzero momentum will suffer sliding locomotion.

[10] Note that this group need not be a Lie group: in particular it might be a discrete subgroup of the symmetry group (even though the whole framework is continuous). We will however assume a Lie group.

then be used to highlight a distinction between *stationary changes of relative configuration* (whose sensorimotor dynamics are within flat submanifolds of Ω and correspond to changes of relative configuration that do not interfere with locomotion) and other changes that can be called *locomotive changes of relative configuration.*

3.4.2 Back to Dimensionality Reduction

Constraints on a Spatial Metric for Sensory Data

An invariant metric g on the state space Ω is a metric that is invariant under the action of symmetries:

$$T^* g = g$$

It must therefore satisfy, using the symmetry of the metric :

$$\begin{aligned} g_{|\omega}(F_m(\omega), F_n(\omega)) &= T^* g_{|\omega}(F_m(\omega), F_n(\omega)) \\ &= g_{|T(\omega)}(D_p T F_m(\omega), D_p T F_n(\omega)) \end{aligned}$$

and using the symmetry of system dynamics:

$$g_{|\omega}(F_m(\omega), F_n(\omega)) = g_{|T(\omega)}(F_m(T(\omega)), F_n(T(\omega))) \tag{3.6}$$

In particular, we find the intuitive result:

Land Survey Lemma: The norm of tangent state change $\dot{\omega}$ associated with a given motor command m is the same *for all states associated with the same relative configuration.*

In the case of system dynamics that are linear in motor commands we have

$$g(F_m, F_m) = \sum_{i,j} m_i \cdot m_j \cdot g(F_{m_i}, F_{m_j})$$

Therefore if for each state ω the system dynamics span the tangent space $T_\omega \Omega$ then the choice of an invariant *metric* reduces to the choice for each relative configuration of a *scalar product* on the finite dimensional vector space of system dynamics. Note that there are *a priori* several invariant metrics, still leaving room for the consideration of other constraints such as biinvariance or energetic costs of the different dynamics.

No Local Aliasing

Now of course the point is that physical states are not accessible to the agent, and we would like to actually define a metric on sensory inputs. If there is no *local* perceptual aliasing then we can still apply the land survey lemma: the norm of tangent sensory input change $\dot{s}$ associated with a given motor command m must be the same for all sensory inputs associated with the same relative configuration.

This will be the case considered in section 3.5.

Local Aliasing: A Suggestion

In the more difficult case where there *is* local aliasing, then the previous viewpoint could be used to search for a metric on sensory inputs that satisfy the land survey lemma "as much as possible". We will not follow this line here, but we present the suggestion anyway: gradient descent could be used to minimize over a parameterized set of metrics g_θ the invariance criterion:

$$\epsilon^2(\theta) = \sum_{i,j,k} \left[g_\theta(\dot{S}_{k,i}, \dot{S}_{k,i}) - g_\theta(\dot{S}_{k,j}, \dot{S}_{k,j}) \right]^2$$

where for each k, $S_{k,i}(t), i = 1 \dots n$ is a set of sensory evolutions resulting from motor command m_k such that the relative configuration of the agent at time $t = 0$ is the same for all evolutions. The minimizer will be a metric such that, in expectation, the norm of tangent sensory inputs associated with identical spatial displacements is the same.

Note that the use of the metric obtained could be interpreted as Bayesian inference, since the metric learned could then be read as agent beliefs on physical states associated with the sensory inputs used to infer the distance between the those physical states.

Nonlinear Sensory Manifolds

Submanifold of Euclidean Spaces

Nonlinear approaches to dimensionality reduction usually define a metric on a data manifold by considering the metric induced on a submanifold of some space $\mathbb{R}^d$ by a norm on that space. Yet it is not clear on general grounds why we should consider the sensory manifold as a submanifold of some vector space. Of course, this makes sense from the computational viewpoint since enforcing such a hypothesis allows use of linear algebra and limits the number of parameters to estimate, therefore leading to simple algorithms with fast generalization. But it imposes a constraint that may reduce the possible number of distance-preserving transformations compatible with such a model. Invariance to spatial displacements suggests for instance considering a metric that is preserved when pictures are zoomed, so that the distance between two object pictures does not depend on the depth from which the pictures are taken[11]. Yet when taken together with invariance to 2D translations and rotations, this cannot be enforced using a norm. This therefore suggests investigating more general models for the metric than those deriving from a norm in a Euclidean embedding space. This will be the framework of the algorithm presented in section 3.5.

[11] That is, the same change of depth for two pictures should not change the distance between them. Independent changes of depths for the two pictures may (will) still affect the distance.

Unfolding and Beyond

As already discussed, the standard framework of nonlinear dimension reduction focuses on the unfolding of data manifolds and therefore essentially deals with flat manifolds such as the swiss roll. The issue that data manifolds can possess intrinsic curvature (such as a sphere that cannot be flattened out without introducing large distortions like those observed on earth maps) is actually often discarded as a second order problem because the metric with respect to which this curvature has been defined anyway did not rely on strong justifications. In our case however, since we are searching for metrics grounded on sensorimotor interactions, we expect curvature to reflect important differences between sensory data variations with respect to these interactions and we do not want to discard this information.

The goal of unfolding the data however usually stems from the use of dimension reduction algorithms to visualize high dimensional data in a 2D vector space. But in the case of an embodied agent this is not a natural requirement and we only want to exploit the geometrical structure as it is, therefore leaving room for a more open framework. For instance, we may want to compute distances between sensory inputs to know whether two sensory configurations are more or less similar so as to perform quantization on this basis. Or, we may want to know what actions can link two neighboring sensory inputs so that we can actively try to perform alignment in a navigation context. Or we may want to identify points of highest or lowest curvature in a given sensory trajectory, so as to consider a possible segmentation of sensorimotor interactions. All these tasks can be performed without having to flatten out the data.

3.5 An Embodied Algorithm for Dimension Reduction

Here we finally present a general attempt to build an invariant metric for sensory inputs delivered by an unknown sensorimotor device (satisfying a certain linearity constraint described in 3.5.1). To ground the intuition and explain simulations run in section 3.5.2, here is the example we have in mind. Consider an agent exploring a 2D world that is an infinitely extended 2D picture. Assume that the agent is able to move above the picture, drifting ahead, sideways, and turning to different extents so that its action space is continuous. Assume that the agent's sensory inputs are constituted by a small, say 10×10 pixel, snapshot taken of the picture below the agent. We would like the agent to build a metric on its sensory inputs that is invariant with respect to symmetries of its sensorimotor interactions in this world. We would also like the agent to discover that there are some actions (namely translations) that bear a very special relationship with other possible types of actions.

We will make the assumption that there is no relative configuration change (see section 3.4.1) and that there is no local perceptual ambiguity (i.e. perceptual aliasing). As for ambiguities, we will then show with examples that the algorithm does not suffer significantly from them.

3.5.1 Description of the Algorithm

Invariant Sensorimotor Metrics

Transformation Based Metrics

With the example of our 2D agent in mind, we note that each displacement of the agent corresponds to modifications of sensory inputs that are approximately[12] linear in sensory inputs. The reason is that sensory inputs result from sampling of a field I on space and that the viewpoint transformation $(t, r) * I : p \mapsto I(r \cdot p + t)$ (see section 3.3.3) is linear in I. Therefore if each sensor output is linear in the field (but without requiring that all sensors are identical nor that they have a uniform layout), then we can expect that each motor command resulting in a displacement (t, r) is associated with a global linear transformation $P(m)$ of sensory inputs:

$$\dot{s} = P(m) \cdot s \tag{3.7}$$

In this case, the Land survey lemma suggests the invariance criterion:

$$g_{|s}(P(m) \cdot s, P(m') \cdot s) = const$$

for all points s of the sensory submanifold and all motor commands m and m'.

We recall that a simpler way to describe metrics than providing an explicit formula is by specifying an orthonormal basis $\{X_i(s)\}$ at each point of the sensory manifold. Evaluation of the metric is then performed by computing tangent vector coordinates in this basis:

$$\dot{s} = \sum_i x_i X_i(s) \Rightarrow ||\dot{s}||^2 = \sum_i x_i^2 \tag{3.8}$$

As a consequence, if we consider the case where there is a *finite* number of displacements triggered by commands m_i and that the family of vectors $X_i(s) = P(m_i) \cdot s$ forms a basis of the tangent space of the sensory manifold at each s, then we can use transformations $P(m_i)$ to define a norm by positing that the family of $X_i(s)$ should be an orthonormal basis. Then obviously we will have $g_{|s}(P(m_i) \cdot s, P(m_j) \cdot s) = \delta_{ij}$ which is indeed independent of s.

The specific example of the 2D agent in a picture world therefore suggests that invariant metrics can be built from linear predictors of sensory changes

[12] Only approximately, as modifications at the edges of view field *cannot* be predicted from what is in the view field.

induced by actions. The whole problem then is: first choose appropriate actions, i.e. appropriate predictors, and second compute geodesic distance. The first point will be explained in section 3.5.1, here we present the approximation scheme that will be used for computing geodesics.

Approximation Scheme for Local Distances

If the motor command m is held fixed for a duration Δt, then equation 3.7 can be integrated using the matrix exponential:

$$s_2 = \exp\left(P(m)\Delta t\right) s_1 \Rightarrow s_2 - s_1 = \left[\exp\left(P(m)\Delta t\right) - Id\right] s_1$$

Therefore the approximation scheme in time for linear predictors $\bar{P}(m)$ of sensory changes resulting from holding motor command m fixed for a time Δt reads:

$$s_2 - s_1 = \bar{P}(m)s_1, \quad \bar{P}(m) \approx P(m)\Delta t$$

Herein the notation P will refer to the differential transformation of sensory inputs expressed in equation 3.7, while the notation $\bar{P}$ will refer to the numerical estimation of this transformation (i.e. to a predictor) based on sensory changes between a time t and a time $t + \Delta t$.

Given a finite family of predictors $\bar{P}(m_i)$, a linear approximation for the evaluation of distances between neighboring points (i.e. equation 3.8) is consequently given by

$$s_2 - s_1 = \sum_i x_i \cdot \bar{P}(m_i)s_1 + \epsilon \Rightarrow d(s_2, s_1)^2 = \sum_i x_i^2$$

This approximation is depicted in Figure 3.3.a.

To provide a more robust estimation for the case of strong nonlinearities, we however consider in the algorithm a slight improvement of this scheme. This improvement requires to have, in addition to the family of actions m_i, a family of actions n_i such that sequences of commands (m_i, n_i) yield no change of sensory inputs. The way to learn such a family will be explained in the section on action choice, here we give the expression for the improved approximation scheme depicted in Figure 3.3.b:

$$\begin{array}{l} s_2 - s_1 = \sum_i x_i \cdot \bar{P}(m_i)s_1 + \sum_i y_i \cdot \bar{P}(n_i)s_1 + \epsilon \\ \quad \text{with} \quad x_i \geq 0\,, y_i \geq 0 \end{array} \Rightarrow d(s_2, s_1)^2 = \sum_i x_i^2 + \sum_i y_i^2 \tag{3.9}$$

Justification for this scheme stems from the fact that if we have a family of commands n_i such that:

$$\exp\left(P(m_i)\Delta t\right) \exp\left(P(n_i)\Delta t\right) s = s \; \forall s$$

then to first order

$$Id + P(m_i)\Delta t + P(n_i)\Delta t = Id$$

and therefore $P(m_i) = -P(n_i)$, which can be used in equation 3.8 to split $\sum_i x_i P(m_i) \cdot s$ into terms with positive and negative coordinates, substitute $P(m_i)$ with negative coordinates into $-P(n_i)$ with positive coordinates, and find the scheme 3.9 involving only positive coordinates.

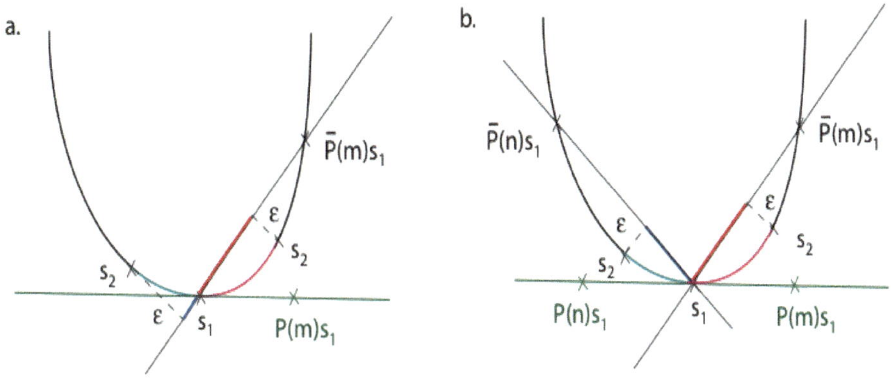

Fig. 3.3. Approximation scheme for geodesic distance between neighbors. Green axis shows the tangent space at s_1, and $P(m)s_1$ shows a normal basis for the transformation based metric. Assume the length of the segment joining s_1 to $\bar{P}(m)s_1$ has been normalized to 1, then red and blue segments show the coordinate of s_2 in the basis defined by predictor $\bar{P}(m)$, and in this one dimensional example the length of these segments actually is the approximation for the geodesic distance $d(s_2, s_1)$. a) Approximation of geodesic distances based on the sole predictor $\bar{P}(m)$ using both positive and negative coordinates of s_2 on $\bar{P}(m)s_1$. There is a big distortion in the estimation of distances depending on the sign of the coordinate. b) approximation of geodesic distances based on predictors $\bar{P}(m)$ *and* inverse predictors $\bar{P}(n)$ with positivity constraint: the previous distortion is removed (see text)

Computation of Geodesic Distances

Using a set of predictors $\bar{P}(m_i)$ and $\bar{P}(n_i)$, the algorithm analyses sets of data points $\{s_t\}$ with the following procedure. The algorithm considers each point, computes the distance with its k nearest neighbors using the scheme 3.9 (positive coordinates x_i and y_i are computed using Matlab's routine `lsqnonneg` for linear least squares with nonnegativity constraints), and finally computes shortest path distances between points that are far apart using Floyd's algorithm, in exactly the same way Isomap does. In this way, the algorithm is able to compute a distance between sensory data that is dependent of its action abilities and not on the sensory code. Basically, the distance $d(s_1, s_2)$ gives an indication of the time the agent has to hold the appropriate motor command that will take it from sensory configuration s_1 to sensory configuration s_2.

If the transformation based metric on the sensory manifold is flat, then the matrix $M_{ij} = d(s_i, s_j)$ of geodesic distances can be used with MDS to represent the data in a vector space having the intrinsic dimensionality of the data.

Choice of Actions

Ultimately the choice of motor commands to be isolated should be based on a general analysis of the holonomy group, such as provided by the analysis of the group root system (see for instance [68]). So far however we limit ourselves to a much simpler approach, namely: *identify* commuting displacements, *choose* a generating family for this set, then complete this family into a basis for the agent's displacement abilities. Within these constraints, we will choose actions associated with changes of sensory inputs as different as possible with respect to the noise-norm on sensory inputs.

Identification of Commuting Displacements

The identification of commuting displacements is done through a piecewise bilinear approximation for the composition of predictors resulting from two successive motor commands, as explained now.

In the first stage the system draws randomly a number of motor commands m_i, $i = 1 \ldots N$ ($N = 30$). It must be noted that if the model described by the equation 3.7 were exact, then we could write the expression for the predictors associated with successive motor commands from transformations $P(m_i)$ because integrating the equation with $m = m_1$ for Δt and then $m = m_j$ for another Δt would yield

$$s_2 = \exp\left(P(m_j)\Delta t\right) \exp\left(P(m_i)\Delta t\right) s_1$$

But to be robust to underdetermination of the sensory system, which imply that equation 3.7 only holds approximately, it is better to build an estimation of the predictor associated with successive motor commands using direct experimentation of sequences of motor commands. The algorithm therefore performs a linear regression over a sufficient number of pairs of {after/before} sensory inputs to learn predictors $\bar{Q}_{ij} = \bar{Q}(m_i, m_j)$ associated with sensory changes resulting from each sequence of two moves.

If the mapping $\bar{Q}$ is smooth, then we can consider a piecewise bilinear approximation $\hat{Q}$ of $\bar{Q}$ based on the sample of learnt predictors $\bar{Q}_{ij}$. $\hat{Q}$ is defined by the constraints to have value $\bar{Q}_{ij}$ at grid nodes (m_i, m_j) and to be linear along each edge of grid cells (see Figure 3.4).

In equations, this means that we put

$$\hat{Q}(m, m') = \sum_{i \in I, j \in J} \alpha_i \beta_j \bar{Q}_{ij}$$

where the α_i and β_j express the linear relationship between m, m' and the experimented commands closest to them $\{m_i, i \in I\}$ and $\{m_j, j \in J\}$, i.e. $m = \sum_{i \in I} \alpha_i m_i$ and $m' = \sum_{j \in J} \beta_j m_j$. $\hat{Q}$ can then be used to infer that some

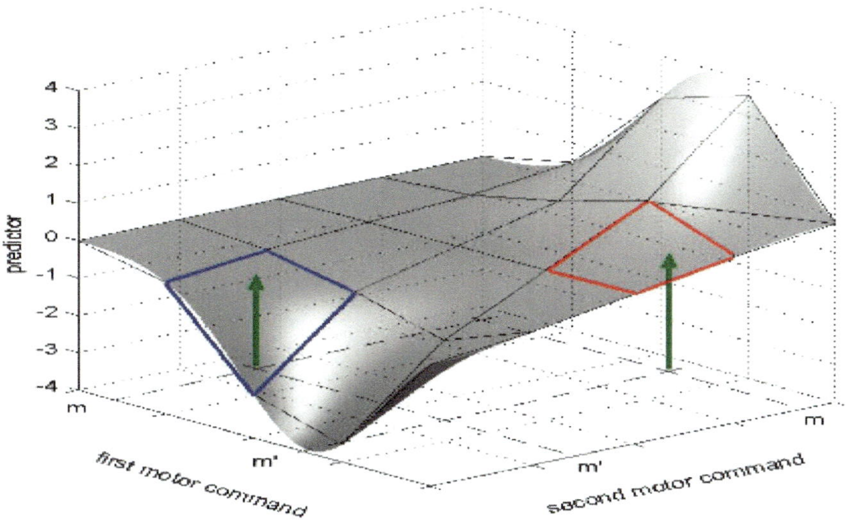

Fig. 3.4. Computing commuting predictors (illustrative sketch). Transparent gray surface is the predictor mapping $\bar{Q}$, black wireframe is the piecewise bilinear approximation $\hat{Q}$ from sampling $\bar{Q}$ on a grid with nodes (m_i, m_j). Red cell correspond to the choice of a neigborhood for motor sequences (m, m'), blue cell shows neighborhood for the motor sequences preformed in reverse order (m', m), green arrows shows a sequence that is expected from the approximation $\hat{Q}$ to yield the same predictor in both orders.

(not experimented) motor commands should lead to commuting predictors: this is done by minimizing the criterion

$$\Delta_{I,J}(\alpha, \beta) = ||\hat{Q}(m, m') - \hat{Q}(m', m)||^2 = || \sum_{i \in I, j \in J} \alpha_i \beta_j \left[\bar{Q}_{ij} - \bar{Q}_{ji}\right] ||^2$$

for neigborhoods (I, J) of each motor command sequences (m_i, m_j) (see Figure 3.4). In this equation, we consider the norm on matrices defined by $||A||^2 = Tr(A^T A)$ (recall that $\bar{Q}_{ij}$ are predictors, i.e. matrices).

Choose a Generating Family for Translations and Complete It

The algorithm retains a defined number of linearly independent (α, β) having the smallest Δ value (commuting commands, i.e. translations) and largest Δ value (non-commuting values, i.e. rotations). For now, the number of commands retained is given to the algorithm and not estimated. Predictors $\bar{P}(m_k)$ associated with the corresponding motor commands m_k are then learnt to be used in the scheme 3.9.

Inference of Inverse Actions

For an improved approximation scheme for local distances, we also rely on the inference of a family of actions n_k such that sequences of motor commands (m_k, n_k) do not yield any change of sensory inputs. This inference can be done from $\bar{Q}$ by looking for commands n_k such that $\bar{Q}(m_k, n_k) = 0$, and therefore an approximation can be computed from $\hat{Q}$ by minimizing for each k the criterion:

$$\Delta_k(\beta) = ||\hat{Q}(m_k, n)||^2 = ||\sum_{i \in I, j \in J} \alpha_i \beta_j \bar{Q}_{ij}||^2$$

where $m_k = \sum_{i \in I} \alpha_i m_i$ and $n = \sum_{j \in J} \beta_j m_j$. Predictors $\bar{P}(n_k)$ associated with the corresponding motor commands n_k are then learnt, to be used in the scheme 3.9.

3.5.2 Results

Dimension Reduction for a Flat Sensory Manifold

To test the whole procedure, we consider first the case where the agent in a 2D picture world is limited to translational moves (Figure 3.5). The reason to do so is that this should theoretically correspond to a case where the sensory manifold is flat with respect to the predictor metric, and therefore geodesic distances computed by the algorithm can be checked by embedding sensory data in a two dimensional vector space using MDS.

Note, as already emphasized several times, that in this system there actually *is* local aliasing since sensory change resulting for instance from forward translation cannot be predicted without errors from the content of the patch. The algorithm however shows no excessive sensitivity to this underdetermination.

Comparison with Isomap

The previous check done, Figure 3.6 shows the effect of a linear modification of sensory encoding. This naturally has a large impact on Isomap since this alteration induces a change in the metric (hence of geodesic distances) that Isomap aims to preserve in the embedding procedure. By comparison, our method is by design independent of such alterations.

Curved Sensory Manifold

To conclude, we consider the initial example where the agent can freely translate *and* rotate (Figure 3.7). This will theoretically result in a sensory manifold that is curved with respect to the transformation based metric, and the MDS procedure cannot therefore avoid a fundamental distortion of distances. Nonetheless, if we consider only a limited area of the sensory manifold then

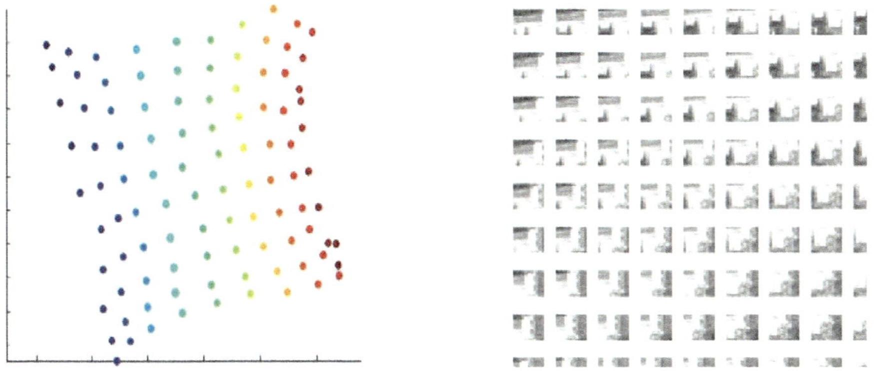

Fig. 3.5. The data is constituted with a grid of translated patches, whose coordinates in the 2D-world are of the form $\{x_0 + n, y_0 + m\}$ with n and m positive integers. x_0 and y_0 are randomly chosen. The camera does not rotate at all. *Right*, a sample of such patches is shown, each patch being translated slightly with respect to its neighbors. *Left*, the output: the embedding of the data points, which is two-dimensional since the only existing dimensions are vertical and horizontal translations. The curving is a border effect, due to the small quantity of data. Our algorithm successfully exhibits an approximatively square grid. Note that the algorithm does not necessarily find axes corresponding to orthonormal translations.

distortions should be kept reasonnable (as in the case for maps of limited areas of the earth).

This additionally allows us to check that the algorithm is indeed able to learn by itself the difference between commuting and noncommuting motor commands (i.e. flat and curved directions in the sensory manifold).

3.5.3 Related Work

Non-trivial Metrics on Data

In recent work oriented towards MRI, [52] and [69] hint that classical methods for dimensionality reduction produce parameters without intuitive meaning. Starting with pictures of an object undergoing unknown deformations, they aim at characterizing these deformations and tracing back the journey of the object. They propose to introduce an appropriate distance measure in Isomap using complex Gabor filters applied on image samples. Arguments given for the use of such a distance measure however only revolve arround the specific task of image processing, while we have tried to build a more general perspective. It could be interesting to make the link with this approach, as the metric based on Gabor filters actually turns out to be invariant.

In [70], it is shown how a discrete-valued auxiliary variable dependent on the original data induces a statistical metric on these data (namely: Fisher distance between the conditional probabilities of the auxiliary variable). There

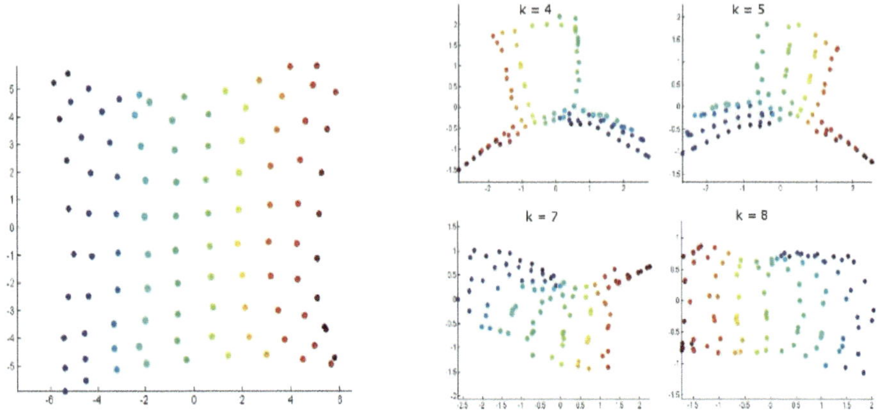

Fig. 3.6. Again, the camera producing the data does not rotate, and the grid of patches is similar to the one in Fig. 3.1. *Left*, our algorithm is not affected and outputs a regular embedding. *Right*, Isomap, tested with different radial parameters (number of neighbors=4,5,7,8), is affected by the distortion introduced in the image.

is an obvious similarity with the information bottleneck method [45]. This auxiliary variable could however be argued to constitute an indirect way of supervising these algorithms, as it is not clear to what they could correspond in the general framework of an embodied agent. Nonetheless, this is potentially of considerable interest if considered for instance in interaction with the communication system: linguistic interaction with a user may indeed provide such auxiliary variables. This approach is therefore complementary to ours, as we may envision the existence of several metrics on sensory data depending on the task the agent is involved in: a sensorimotor metric for self insertion in space and basic sensorimotor tasks, a side-information related metric for tasks requiring to match sensory data with auxiliary data.

More generally, information theory, Bayesian frameworks, statistical considerations on the input space also provide ways to build or improve a metric (see for instance [71]) and define a distance relevant to the problem.

Lie Groups on Sensory Data

[72] propose an algorithm able to estimate the generator matrix G of a one dimensional Lie group, and already suggested that this could be the ground for object recognition methods robust to change of viewpoint. Their work is extended by Bengio et al. in [73] where they develop a method for non-local manifold learning. Their unsupervised algorithm is designed to learn and infer, even at a large scale, the transformation on which it is trained. In our case such large scale predictions turned out to be unreliable, which is the reason why we based our work on direct learning (rather than inference)

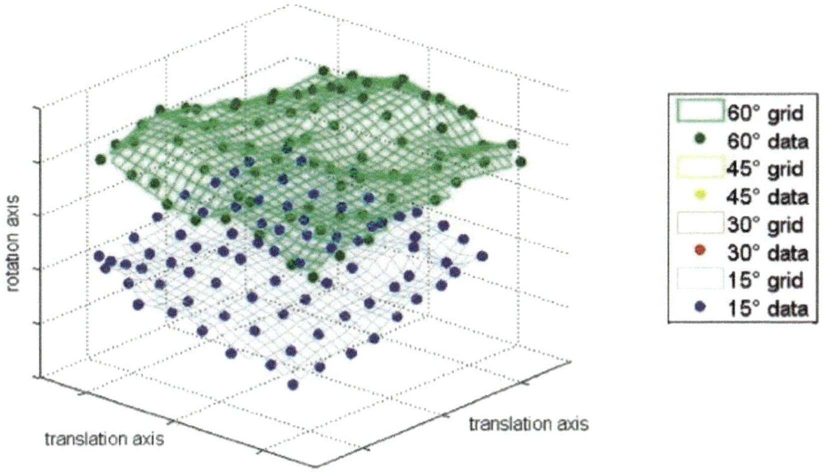

Fig. 3.7. Data is obtained by the camera with a variation of the three parameters (x, y, θ). For clarity, we represent only a subset of this data: points whose initial coordinates in the 2D world are of type (i, j, α) where i and j are integers and $\alpha = 15$ or 60. The resulting embedding shows as expected the appearance of a "millefeuilles", where the vertical axis perfectly relates to the rotation angle and the horizontal axis to x and y.

of predictors $\bar{Q}_{ij}$ for action sequences. The reason for the unreliability is the existence of a fundamental undetermination for sensory data evolution: in the case of pictures this comes from the fact that incoming pixels on the border of patches cannot be fully predicted from the content of the patch before the move. [72] solves the question of large-scale predictions using an implicit cyclic hypothesis (i.e. by assuming the signal to be periodic). Our work suggests a new way to tackle (or actually obviate) this problem.

The idea of the Land survey lemma is implicit in previous work of ours [74], and was already considered as a way to estimate the group structure of compensable transformations for an agent-environment system, i.e. transformation of the whole system leaving the agent's sensory inputs invariant. The most important difference with this work is that the notion of symmetry group for the agent's sensorimotor interactions presented here is more general and realistic than that of compensable transformations. To temper this improvement, it must be recalled however that we did not provide in this more general context an algorithmic way to distinguish between the different notions of locomotion depicted in section 3.4.1: it is only because we considered an example with no possible change in relative configuration that the transformation based metric at the basis of the algorithm is invariant to the symmetry group.

Theory of Invariants in Artificial Vision

The importance of the notion of invariance, and especially invariance to a symmetry group, is a standard of artificial vision. Indeed, the problem of recognizing objects, whose shape undergoes a geometric viewing transformation, often arises in machine vision tasks. For instance research done on structure from motion, i.e. reconstruction of three dimensional information from sequences of two dimensional projections, makes extensive use of the mathematical theory of invariants [75]. Much work has also been devoted to the development of multiscale geometric representations of shape which are invariant to a number of the typical viewing transformations in vision [76]. The main difference between these approaches and ours, apart from taking the perspective of cognitive science rather than that of artificial vision, is the kind of objects manipulated. [75] abstract from images points of interest that are represented as pair of viewfield coordinates, and [76] manipulate objects such as planar curves. We try to address the issue of sensory processing in the raw format in which sensory data are provided to the agent, without using an interpretation (e.g. in terms of geometric points or curves) of sensory data (e.g. of pictures). Actually, our aim precisely is ultimately to come to an interpretation of sensory data.

Algorithms for Uninterpreted Sensors and Effectors

Much work has been devoted to the development of algorithms for embodied agents that do not assume programmer's knowledge of the agent body. For instance in the domain of map learning, [77] present methods to learn increasingly complex interfaces to control in space a robot whose sensorimotor apparatus and environment are initially unknown. In the domain of developmental robotics, more precisely of the aquisition of a body schema, [78] take inspiration from earlier work to implement an algorithm aiming to learn its body map by measuring statistical dependence between its various sensors.

The main difference between these approaches and ours is that the metrics used do not measure the same aspects of sensorimotor interactions: even though spatial proximity of sensors presumably implies some form of statistical dependence between sensor outputs, the metric induced by the symmetry group of system dynamics and the metric induced by sensors' statistical dependence [79] *a priori* measure two different things. It would be interesting to further investigate the link between the two, but it seems at first glance that a systematic link can only be expected between sensors' statistical dependence and *topological* sensor relationships, and not between sensors' statistical dependence and sensors' spatial *metrical* relationships. As for side-information metrics, it must be stressed that there is no reason to be exclusive about the use of only one metric: depending on the task and objectives, different metrics may be appropriate and therefore the approaches are complementary rather than competing.

There is an additional difference between the previous approaches and ours: we deliberately take a holistic approach with respect to sensory inputs, i.e. we do not make the assumption that individual coordinates of sensory data are meaningful[13] since this actually is an assumption about the format of the sensory code. Consider for instance the case of color vision. The relevance of tackling sensory processing from the viewpoint of each different kind of photoreceptor is not clear: color information is an intrinsically multidimensional object. Apart from possible changes in subsequent computational abilities, the basis used by the agent's sensory system to represent this color information should not affect sensory abilities of the agent.

3.6 Conclusion

Many technical issues remain to be solved so as to implement the algorithm robustly enough and efficiently enough for real robotic systems. It is no surprise that the generality of the approach implies a large increase in computational load as compared to methods optimized for specific devices and tasks. This is an artifical version of the altricial-precocial trade-off that biological organisms have to face: adaptability versus time constraints in learning and processing [80]. In the simple examples provided in section 3.5.2, dealing with $c \times c$ patches and a actions requires estimating on the order of $O(a^2 \cdot c^4)$ parameters (for instance to learn the $\bar{Q}_{ij}$ matrices), which roughly requires handling a learning dataset on the order of $O(a^2 \cdot c^2)$ patches. Operations of linear algebra used to perform the estimation of predictors will then require on the order of $O(a^2 \cdot c^6)$ elementary operations, based on the fact that the complexity of matrix multiplication and inversion is at least cubic in the dimension of matrix entries [81]. Such a high price however results from the requirement of designing an adaptive system with *centralized* computational abilities, i.e. possibly implemented in a Turing machine. As for biological systems, the computational notion of complexity is not so clear, and one must be cautious when trying to apply similar frameworks to understand the abilities of biological systems and to conceive artificial agents mimicking such abilities.

It must also be stressed that an important aspect of the current approach is to constrain the choice of a metric over sensory inputs corresponding *to the same relative configuration of the agent-environment system*, without so far addressing the point of a constraint *across* different relative configurations. This is a critical point that calls for more theoretical work and is presumably linked with the notion of objects, as a change of relative configuration occurs in particular as a result of displacements of objects in the environment relative to each other and relative to the agent.

Beside the previous warnings, many interesting developments can be envisioned. In addition to displacements, a large variety of transformations on

[13] While such an assumption is made when sensors are implicitly *defined* as coordinates of sensory inputs.

sensory inputs could be considered. In the case of a visual system, blurring, lighting, or focus are target examples, provided that the linear model 3.7 holds. The structure of transformations with respect to sequences of actions will then be more complex than that of a displacement group, yielding new criteria to distinguish between actions. The coordination of several metrics, as mentioned in the overview of related work, is another possible line of development. In particular this could be used to measure distances between sensory inputs that cannot in any way result from agent actions: this would provide a necessary complement to our work.

In conclusion, we have suggested that dimension reduction could be a first step towards an unsupervised interfacing between an agent's sensory system and its cognitive function, therefore a first step towards an unsupervised self-insertion of the agent in space. We started by giving a survey of standard approaches to dimensionality reduction, and noted their (unjustified) reliance on an *a priori* metric on sensory inputs. We then proposed a general mathematical framework to describe an embodied agent and showed that this framework suggested the use of specific metrics grounded on sensorimotor interactions. We finally described an algorithm aiming at learning such metrics, and put this algorithm to use on some simple examples.

References

1. O'Regan, J.K., Noë, A.: A sensorimotor account of vision and visual consciousness. Behavioral and Brain Sciences 24(5), 939–973 (2001)
2. Lungarella, M., Sporns, O.: Mapping information flow in sensorimotor networks. PLoS Computational Biology 2(10), 1301–1312 (2006)
3. Bajcsy, R.: Active perception. IEEE Proceedings 76(8), 996–1006 (1988)
4. Aloimonos, J., Weiss, I., Bandyopadhyay, A.: Active vision. International Journal of Computer Vision 1(4), 333–356 (1988)
5. Ballard, D.: Animate vision. Artificial Intelligence 48(1), 1–27 (1991)
6. Eklundh, J., Pahlavan, K.: Head, eye and head-eye system. In: SPIE Applications of AI X: Machine Vision and Robotics, Orlando, Fla. (1992)
7. Findlay, I., Gilchrist, J.M.: Active vision: The psychology of looking and seeing. Oxford University Press, Oxford (2003)
8. Philipona, D., O'Regan, J.K.: Color naming, unique hues, and hue cancellation predicted from singularities in reflection properties. Visual Neuroscience 23(3-4), 331–339 (2006), `http://cognitivesystems.org/cosybook/chap3.asp#Philipona06`
9. Sloman, A.: Image interpretation: The way ahead? In: Braddick, O., Sleigh, A. (eds.) Physical and Biological Processing of Images, Proceedings of an international symposium organised by The Rank Prize Funds, London, pp. 380–401. Springer, Heidelberg (1982)
10. Kaelbling, L.P., Littman, M.L., Cassandra, A.: Planning and acting in partially observable stochastic keywords. Artificial Intelligence 101, 99–134 (1998)

11. Beeson, P., MacMahon, M., Modayil, J., Provost, J., Savelli, F., Kuipers, B.: Exploiting local perceptual models for topological mapping. In: Proc. IJCAI Workshop on Reasoning with Uncertainty in Robotics (RUR 2003), Acapulco, Mexico, pp. 15–22 (2003)
12. Reiter, R.: Knowledge in action: logical foundations for specifying and implementing dynamical systems. MIT Press, Cambridge (2001)
13. Kuipers, B.: The spatial semantic hierarchy. Artificial Intelligence 119, 191–233 (2000)
14. von Helmholtz, H.: The Facts in Perception. In: Selected Writings of Hermann Helmholtz, pp. 115–147. Wesleyan University Press (1878)
15. Poincaré, H.: La Science et l'Hypothese, Flammarion (1902)
16. Piaget, J.: The origins of intelligence in children. W.W. Norton & Company, Inc., New York (1936)
17. Gibson, J.: The senses considered as perceptual systems. Houghton Mifflin, Boston (1966)
18. Chalmers, D.: The conscious mind: In search of a fundamental theory. Oxford University Press, New York (1996)
19. Zeki, S., Bartels, A.: The autonomy of the visual systems and the modularity of conscious vision. Philosophical Transactions of the Royal Society B: Biological Sciences 353(1377), 1911–1914 (1998)
20. Rees, G., Kreiman, G., Koch, C.: Neural correlates of consciousness in humans. Nature Reviews Neuroscience 3, 261–270 (2002)
21. Crick, F., Koch, C.: A framework for consciousness. Nature Neuroscience 6(2), 119–126 (2003)
22. Tenenbaum, J.B., de Silva, V., Langford, J.C.: A global geometric framework for nonlinear dimensionality reduction. Science 290, 2319–2323 (2000)
23. Reinsel, G.C., Velu, R.P.: Multivariate reduced-rank regression: Theory and applications. Lecture Notes in Statistics. Springer-Verlag Telos (1998)
24. Tipping, M., Bishop, C.: Probabilistic principal component analysis. Tech. rep., Neural Computing Research Group, Aston University, Birmingham, UK (1997)
25. Artac, M., Jogan, M., Leonardis, A.: Incremental pca for on-line visual learning and recognition. In: Kasturi, C.S.R., Laurendeau, D. (eds.) Proceedings 16th international conference on pattern recognition, pp. 781–784 (2002)
26. Voegtlin, T.: Recursive principal components analysis. Neural Networks 18(8), 1051–1063 (2005)
27. Friedman, J.H., Tukey, J.W.: A projection pursuit algorithm for exploratory data analysis. IEEE Transactions on Computers 23, 881–890 (1974)
28. Friedman, J.H.: Exploratory projection pursuit. Journal of the American Statistical Association 82, 249–266 (1987)
29. Collins, M., Dasgupta, S., Schapire, R.: A generalization of principal components analysis to the exponential family. In: Dietterich, T.G., Becker, S., Ghahramani, Z. (eds.) Advances in neural information processing systems, vol. 14. MIT Press, Cambridge (2002)
30. Kambhatla, N., Leen, T.: Dimension Reduction by Local Principal Component Analysis. Neural Computation 9(7), 1493–1516 (1997)
31. Verbeek, J.J., Vlassis, N., Kröse, B.: Coordinating mixtures of probabilistic Principal Component Analyzers. Tech. rep., Computer Science Institute, University of Amsterdam, The Netherlands, iAS-UVA-02-01 (February 2002)

32. Vijayakumar, S., Schaal, S.: Local dimensionality reduction for locally weighted learning. In: IEEE International Symposium on Computational Intelligence in Robotics and Automation, pp. 220–225 (1997)
33. Scholkopf, B., Smola, A., Muller, K.: Nonlinear Component Analysis as a Kernel Eigenvalue Problem. Neural Computation 10(5), 1299–1319 (1998)
34. Chang, K., Ghosh, J.: A unified model for probabilistic principal surfaces. IEEE transactions on pattern analysis and machine intelligence 23, 22–41 (2001)
35. Fletcher, P., Lu, C., Joshi, S.: Statistics of Shape via Principal Geodesic Analysis on Lie Groups. In: IEEE Computer Society Conference on Computer Vision and Pattern Recognition, vol. 1. IEEE Computer Society, Los Alamitos (1999/2003)
36. Torgerson, W.: Multidimensional scaling: I. Theory and Method. Psychometrika 17, 401–419 (1952)
37. Shepard, R.: The analysis of proximities: Multidimensional scaling with an unknown distance function. II. Psychometrika 27(3), 219–246 (1962)
38. Kruskal, J.B.: Multidimensional scaling by optimizing goodness of fit to a nonmetric hypothesis. Psychometrika 29, 1–27 (1964)
39. Sammon, J.: A nonlinear mapping for data structure analysis. IEEE Transactions on Computers 18(5), 401–409 (1969)
40. Belkin, M., Niyogi, P.: Laplacian Eigenmaps for Dimensionality Reduction and Data Representation. Neural Computation 15(6), 1373–1396 (2003)
41. Kohonen, T.: Self-organizating maps. Springer, Heidelberg (1997)
42. Demartines, P., Hérault, J.: Cca: Curvilinear component analysis. In: 15th workshop GRETSI (1995)
43. Martinetz, T., Berkovich, S., Schulten, K.: Neural-gas' network for vector quantization and its application totime-series prediction. IEEE Transactions on Neural Networks 4(4), 558–569 (1993)
44. Fritzke, B.: A growing neural gas network learns topologies. In: Tesauro, D.S.T.G., Leen, T.K. (eds.) Advances in neural information processing systems, vol. 7, pp. 625–632. MIT Press, Cambridge (1995)
45. Chigirev, D., Bialek, W.: Optimal manifold representation of data: an information theoretic approach. In: Advances in Neural Information Processing Systems 16
46. Roweis, S.T., Saul, L.K.: Nonlinear dimensionality reduction by locally linear embedding. Science 290, 2323–2326 (2000)
47. Lee, J.A., Lendasse, A., Verleysen, M.: Curvilinear distance analysis versus isomap. In: European symposium on artificial neural networks, pp. 185–192 (2002)
48. Balasubramanian, M., Schwartz, E., Tenenbaum, J., de Silva, V., Langford, J.: The Isomap Algorithm and Topological Stability. Science 295(5552), 7 (2002)
49. Lee, J., Verleysen, M.: Nonlinear dimensionality reduction of data manifolds with essential loops. Neurocomputing 67, 29–53 (2005)
50. Broomhead, D.S., Kirby, M.: A new approach to dimensionality reduction: Theory and algorithms. SIAM Journal of Applied Mathematics 60(6), 2114–2142 (2000)
51. de Silva, V., Tenenbaum, J.: Global Versus Local Methods in Nonlinear Dimensionality Reduction. In: Advances in Neural Information Processing Systems, pp. 721–728 (2003)
52. Pless, R.: Differential structure in non-linear image embedding functions. In: Proceedings of the 2004 Conference on Computer Vision and Pattern Recognition Workshop (CVPRW 2004), vol. 1, pp. 10–17. IEEE Computer Society, Washington (2004)

53. Kuipers, B.: Consciousness: Drinking from the firehose of experience. In: Proceedings of the National Conference on Artificial Intelligence, vol. 20, pp. 1298–1305 (2005)
54. Van der Schaft, A., Schumacher, H.: An introduction to hybrid dynamical systems. LNCIS
55. Haken, H.: Synergetic computers and cognition: a top-down approach to neural nets. Springer, New York (1991)
56. Thom, R.: Structural stability and morphogenesis. Benjamin-Addison Wesley
57. Isidori, A.: Nonlinear Control Systems. Springer Communications and Control Engineering Series (1995)
58. Thoroughman, K., Shadmehr, R.: Learning of action through adaptive combination of motor primitives. Nature 407, 742–747 (2000)
59. Zinkevich, M., Balch, T.: Symmetry in markov decision processes and its implications for single agent and multiagent learning. In: Eighteenth International Conference on Machine Learning, pp. 632–640 (2001)
60. Kolar, I., Slovak, J., Michor, P.W.: Natural operations in differential geometry. Springer, Heidelberg (1993)
61. Montgomery, R.: Optimal control of deformable bodies and its relation to gauge theory. In: Ratiu, T. (ed.) The geometry of hamiltonian systems, vol. 22. MSRI Publications, Springer (1991)
62. Marsden, J.E., Ostrowski, J.: Symmetries in motion:geometric foundations of motion control. Nonlinear Sci. Today
63. Kelly, S.D., Murray, R.M.: Geometric phases and robotic locomotion, Tech. rep., California institute of technology (1994)
64. Mason, R., Burdick, J.W.: Propulsion and control of deformable bodies in an ideal fluid. In: Proceedings IEEE International Conference on Robotics and Automation, vol. 1, pp. 773–780 (1999)
65. Shapere, A., Wilczek, F.: Geometry of self-propulsion at low Reynolds number. Journal of Fluid Mechanics Digital Archive 198, 557–585 (2006)
66. Murray, R.M., Sastry, S.S., Zexiang, L.: A mathematical introduction to robotic manipulation. CRC Press, Inc., Boca Raton (1994)
67. Goodwine, B., Burdick, J.: Trajectory generation for kinematic legged robots. In: Proc. IEEE Int. Conf. Robotics and Automation, vol. 3, pp. 2689–2696 (1997)
68. Bump, D.: Lie groups. Springer, Heidelberg (2004)
69. Souvenir, R., Pless, R.: Isomap and Nonparametric Models of Image Deformation. In: IEEE Workshop on Motion and Video Computing, vol. 2, pp. 195–200 (2005)
70. Peltonen, J., Klami, A., Kaski, S.: Improved learning of riemannian metrics for exploratory data analysis. Neural Networks 17, 1087–1100 (2004)
71. Xing, E., Ng, A., Jordan, M., Russell, S.: Distance Metric Learning with Application to Clustering with Side-Information. In: Advances in Neural Information Processing Systems, vol. 15, pp. 521–528 (2003)
72. Rao, R., Ruderman, D.: Learning Lie groups for invariant visual perception. In: Proceedings of the 1998 Conference on Advances in Neural Information Processing Systems II, pp. 810–816 (1999)
73. Bengio, Y., Delalleau, O., Le Roux, N., Paiement, J.-F., Vincent, P., Ouimet, M.: Spectral dimensionality reduction. Cirano working papers (2004)
74. Philipona, D., O'Regan, J.K., Nadal, J.-P., Coenen, O.-M.: Perception of the structure of the physical world using unknown sensors and effectors. In: Advances in Neural Information Processing Systems, vol. 16, pp. 945–952 (2004)

75. Bazin, P., Boutin, M.: Structure from Motion: A New Look from the Point of View of Invariant Theory. SIAM Journal on Applied Mathematics 64(4), 1156–1174 (2004)
76. Olver, J.P., Sapiro, G., Tannenbaum, A.: Differential invariant signatures and flows in computer vision:a symmetry group approach, Tech. rep., University of Minnesota and MIT (1993)
77. Pierce, D., Kuipers, B.: Map learning with uninterpreted sensors and effectors. Artificial Intelligence 92, 169–229 (1997)
78. Hafner, V., Kaplan, F.: Interpersonal maps and the body correspondence problem. In: Demiris, Y., Dautenhahn, K., Nehaniv, C. (eds.) Proceedings of the third international symposium on imitation in animals and artifacts, Herderfordshire, UK, pp. 48–53 (2005)
79. Crutchfield, J.P.: Information and its metric, Nonlinear Structures in Physical Systems - Pattern formation. Chaos and Waves, 119–130 (1990)
80. Sloman, A., Chappell, J.: The Altricial-Precocial Spectrum for Robots. In: International Joint Conference on Artificial Intelligence, vol. 19, pp. 1187–1192 (2005)
81. Giorgi, P., Jeannerod, C., Villard, G.: On the Complexity of Polynomial Matrix Computations. In: ISSAC 2003, Philadelphia, USA, pp. 135–142. ACM Press, New York (2003)

4

Categorical Perception

Mario Fritz[1], Mykhaylo Andriluka[1], Sanja Fidler[2], Michael Stark[1], Aleš Leonardis[2], and Bernt Schiele[1]

[1] Technische Universität Darmstadt, Darmstadt, Germany
lastname@cs.tu-darmstadt.de
[2] VICOS Lab, University of Ljubljana, Slovenia
firstname.lastname@fri.uni-lj.si

The ability to recognize and categorize entities in its environment is a vital competence of any cognitive system. Reasoning about the current state of the world, assessing consequences of possible actions, as well as planning future episodes requires a concept of the roles that objects and places may possibly play. For example, objects afford to be used in specific ways, and places are usually devoted to certain activities. The ability to represent and infer these roles, or, more generally, categories, from sensory observations of the world, is an important constituent of a cognitive system's perceptual processing (Section 1.3 elaborates on this with a very visual example).

In the CoSy project, a substantial amount of work has been conducted on the advancement of methods that recognize and categorize objects and places by using different modalities, namely, vision, language, and laser range data. Our progress contributes to our effort to build systems that evolve through interaction with its environment in an ultimately live-long learning process.

While this chapter describes our contribution to modeling, learning and representing of visual categories, Chapter 7 shows how to combine the visual information with other modalities in a multi-modal learning process (e.g. speech/language as detailed in Chapter 8). Finally, Chapter 9 and 10 shows how we integrated these concepts in a autonomous systems to understand the implications of our progress in categorization on an interactive evolving system.

4.1 Introduction

Recently there has been significant progress in visual class recognition ranging from visual category modeling (e.g. [1]), learning (e.g. [2]), robustness (e.g. [3]) and scalability to more classes (e.g. [4]). Interestingly, there is a surprising diversity in the proposed approaches. E.g. the employed features representations range from local to global and object representations range from generative

H.I. Christensen et al. (Eds.): Cognitive Systems, COSMOS 8, pp. 131–164.
springerlink.com

to discriminant models. In the following we start with a discussion of the most important design and paradigm choices highlighting the implied merits and draw-backs. This leads to the conclusion, that no single paradigm alone is powerful enough to deal with the implicit challenges of object categorization. Instead we argue in this chapter that the object categorization problem requires a wide variety of approaches. This chapter therefore summarizes various approaches that have been explored during the CoSy project. For each of these approaches we will discuss the respective strengths and weakness and will discuss how these approaches advance the state of the art in object categorization.

Fig. 4.1. Illustration of some challenges encountered in visual categorization. We observe high intra-class variation and low inter-class variation. Observed effects occur at local or global scale with respect to members of the same category or contrasted to other categories.

Figure 4.1 illustrates one of the fundamental challenges in visual categorization: Many object classes (such as dogs) have high intra-class variability but may share similar features and appearance across different classes (e.g. dogs, cows and horses). Such similarities then lead to low inter-class variation. This challenge taken together with other challenges related to object categorization (robustness, viewpoint changes, partial occlusion, light conditions, etc.) have lead to a wide variety of approaches. In the following we discuss three of the most prominent design choices in more detail, namely generative vs discriminative modeling, local vs. global object features, and the importance of the learning paradigm.

Generative vs. Discriminative Paradigms

There is a fundamental difference whether an approach models all that is common to all category members (intra-class variation as illustrated in Figure 4.1 by the dogs) or all that distinguishes themselves with respect to others (inter-class variation as illustrated in Figure 4.1 by the other animals). Technically speaking – in the generative paradigm – the object model represents all properties or object instances of an object class. In contrast – in the discriminative paradigm – the model retreats to a simpler mapping from the observed properties to the predicted category label (discriminant or conditional learning). Both views have a multitude of benefits and drawbacks. As the generative model of an object class is typically learned independently of other object classes the generative paradigm lends itself to parallelism and incremental learning. Generative methods also enable principled ways of handling missing data (e.g. due to occlusion) while inference is typically computationally expensive. Discriminant models on the other hand are designed to capture the difference between different categories which often leads to higher precision. Typically, this comes at the price of a large training set including a substantial amount of background data.

Local vs. Global Object Representation Paradigms

Another important design choice for object category representation is the appropriate level of locality used in the feature representation. Approaches range from local interest point-based approaches (like the popular SIFT-feature over edge-segments proposed by [5]) to fully global representations (like the global HOG-representation proposed by [6]). On the one hand, local features can be made robust to scale and rotation changes and therefore enable robustness to partial occlusion. On the other hand, they require sophisticated models to describe the geometric and topological structure of object classes. In contrast, global feature descriptors inherently describe the geometric and topological structure by describing the global appearance of an object. On the downside, however, they do not expose the desired robustness to partial occlusion and it becomes more difficult to generalize across-instance.

Different Learning Paradigms

Over the years various approaches have been proposed for the recognition of object categories often based on models learned directly from image data. The approaches, however, vary greatly in the amount of supervision provided for the training data. The types of annotation varies from pixel-level segmentations, (e.g. [7]) over bounding-box annotations (e.g. [8]) and image level annotation (e.g. [2, 9]) to unsupervised methods (e.g. [10, 11, 12]), which do not even require the information which category is presented in which image.While approaches using more supervision tend to require less training data, there is a clear desire to use less supervision typically at the price to use more unlabeled training data.

Central Role of Representation

As motivated in Section 1.4.2, representation plays a central role in a cognitive system. The way categories are stored, organized and related to each other determines the capabilities to learn, evolve and therefore to interact of the overall system. In Chapter 7 on multi-modal learning and in particular in our integration efforts in Chapter 9 and 10, we'll detail how representation affects the capabilities of the overall system.

The above discussion leads to the conclusion that no single combination of these paradigms is powerful and versatile enough for the general problem of object categorization. It is commonly believed that relying on a single paradigm of addressing the category phenomenon won't do justice to the overwhelming diversity encountered in categorial perception. Throughout this chapter we therefore promote learning-based approaches, that are capable to adapt to the particularities of each categories. In this sense we have explored – during the course of the CoSy-project – a variety of different object representation integrating and advancing the state of the art for different of the above mentioned paradigms.

We perceive the *representation of visual categories* as a central element of each vision system, as it influences scalability in the number of categories, capability to evolve over time, addressable categories and it is the starting point of categorial reference in the interplay with other components in an integrated system. The following therefore presents the different approaches explored both w.r.t. its underlying representations as well as w.r.t. the above mentioned paradigms. In particular, we pursue this idea to progress towards scalable representations in Section 4.2.1, representations for affordance-base categorization in Section 4.2.2, representation by generative decompositions in Section 4.3 and representation of dynamics in Section 4.4. To conclude this introduction, we outline how these 4 directions progress over the previous state-of-the-art in categorial perception.

4.1.1 Towards Hierarchical Scalable Representations

As discussed above one of the important design choices is the locality of the feature representations. The first approach described in section 4.2.1 aims to bridge the gap between purely local to more global representations by automatically learning a hierarchy from local to more global representations. Such a hierarchical representations are believed to make approaches more scalable and allow for successive evidence aggregation. The proposed hierarchical representation is learned bootom-up and in an unsupervised fashion leading from local to more global representations.

Also computational considerations suggest that matching should be performed hierarchically, in order to gradually and coherently limit the otherwise computationally prohibited search space [13, 14, 15, 16, 17, 18, 19, 20]. We propose a novel approach to representing object categories within an *indexable, hierarchical compositional framework*.

4.1.2 Towards Representations for Affordance-Based Categorization

A second approach (Section 4.2.2) explores the possibility to represent objects related to their visual affordances. In this case we explore and analyze a variety of local region descriptors in a generative object model that is trained in a supervised fashion.

Most local region descriptors, such as the popular SIFT descriptor, explicitly encode local appearance, and have hence shown impressive results on objects with sufficient local appearance statistics, such as cars and motorbikes [2, 21]. However, for many object categories that might be important for interactions between a cognitive agent and its environment, local appearance might not be the most useful cue. Tools, cutlery, things in office and kitchen environments, and other man-made artifacts seem to require features that capture the respective shape and geometry rather than local appearance. In particular, categorizing those objects according to supported functions and affordances demands an explicit representation of geometric properties.

4.1.3 Representations and Discovery of Object Classes by Generative Decompositions

The third approach combines different paradigms ranging from unsupervised learning of generative object models to supervised learning for discriminant object detection. Object representations for categorization tasks should be applicable to a wide range of objects, scaleable to handle large numbers of object classes, and at the same time learnable from a few training samples. While such a scalable representation is still illusive today, it has been argued that such a representation should have at least the following properties: it should enable sharing of features [22], it should combine generative models with discriminative models [23, 24] and it should combine both local and global as well as appearance- and shape-based features [3]. Additionally, we argue that such object representations should be applicable both for unsupervised learning (e.g. visual object discovery) as well as supervised training (e.g. object detection). We achieve such a representation by learning a generative decomposition of localized oriented gradients that exploit the co-occurence statistic in the presented data. The obtain components range from local to global and lend themselves to unsupervised learning learning as well as supervised state-of-the-art detection.

4.1.4 Representations of Object Dynamics

An important type of object classes is articulated such as people and animals. Therefore our fourth and last approach presented in this chapter considers the challenging problem of modeling the dynamics of articulated objects. For this we explore a generative object model that represents a person as the topology of individual body parts. Furthermore a generative model is learned to represent the dynamics of the human walking cycle. The aim is to improve detection and tracking in cluttered scenes using a monocular, potentially moving camera. Probably the most fundamental difficulty in detection and tracking many people in cluttered scenes is that many people will be partially and also fully occluded for longer periods of times. Consequently, both the detection of people in individual frames as well as the data-association between people detections in different frames are highly challenging and ambiguous. The developed motion model lends to a more detailed interpretation of the observed sequences.

4.2 Low-Level Features and Hierarchical Representation Learning

4.2.1 Towards Scalable Representations for Visual Categorization

Motivation

Hierarchical representations can derive and organize the features at multiple levels that build on top of each other by exploiting the sharability of features among more complex compositions or objects themselves [25, 26, 27, 17]. This endows them with high computational efficiency and expressive power that surpasses the capabilities of flat representations. A number of hierarchical recognition systems have been proposed and confirmed the success of such representations in object categorization tasks [28, 29, 27, 16, 25, 30, 21, 31, 32, 33, 26]. However, an *automatic* construction of the visual hierarchy that would scale well with the number of image categories (that are in nature in the order of tens of thousands) is still an open issue.

We propose a novel approach to representing object categories within an *indexable, hierarchical compositional framework*. We develop a *bottom-up, statistical approach* that makes use of simple atomic features, i.e. oriented edges, to gradually learn more complex contour compositions. The learned library of features, i.e. *parts*, is organized with accordance of principles of efficient indexing which ensures that local retrieval of models during the online image processing stage will run in a roughly constant time despite the exponential increase in size along the hierarchical layers. The learned compositions can then be combined into objects with minimal human supervision, whereby the hierarchical sharability of features and the efficient indexability constraints could be a step towards scalable representations of object categories.

We substantiate our choice of design principles proposed to devise a plausible representation of object categories with respect to the shape modality:

Hierarchical Compositionality. Compositionality refers to hierarchical representations defined in terms of parts and their spatial relations where the hierarchically constructed entities are built from a relatively small number of lower-level constituents [34]. Computational benefits of compositionality in terms of storage, processing demands and the exponential expressive power have long been emphasized in the computer vision literature [18, 16, 35, 20, 36]. Since each hierarchical unit is shared among many more complex higher layer compositions, the computational cost is highly reduced compared to searching for each complex interpretation in isolation. In addition, while the receptive field sizes increase with the level of hierarchy, formed compositions are designed to respond to only smaller spatial subsets in their receptive fields. This offers much higher robustness and faster processing over the traditional neural networks approaches, which mainly integrate over the complete receptive fields both spatially as well as all feature values in the layer below.

A part of the biological evidence could potentially support such a line of architecture [37, 38]. Additionally, attempts have been made to map the mathematical theory onto the neuronal structure of the visual cortex [34].

Computational Plausibility. Hierarchical representations do not necessarily imply computational efficiency. Our main motivation for building a hierarchical visual representation is to enable fast *indexing and matching* of image features against hierarchically organized stored prototypes in order to avoid the computationally prohibitive linear search in the number of higher-level compositions and objects themselves [36, 13].

Statistics Driven Learning. Features and their higher level combinations should be learned in an unsupervised manner (at least in the first stages of the hierarchy) in order to avoid hand-labelling of massive image data as well as to capture the regularities within the visual data as effectively and compactly as possible [39, 40, 41, 19, 20]. Once the *visual building blocks* are learned, learning of novel objects can proceed mainly in the higher hierarchical layers and can thus operate fast and with no or minimal human supervision.

Robust and Repeatable Detection. To achieve robustness against noise and clutter, the features comprising the individual hierarchical layers should be manifested as models to enable a robust verification of the presence of their underlying components [17, 13, 18]. Models should incorporate loose geometric relations to achieve the spatial binding of features [41, 15], yet encode enough flexibility to ensure repeatability and gain discrimination gradually – through composition within the hierarchy.

Learning a Hierarchy of Parts (LHOP)

The Compositional Library of Parts

Let $\mathcal{L}_n$ denote the n-th Layer of the hierarchical library. Parts within the hierarchy are defined recursively in the following way. Each part in $\mathcal{L}_n$ codes spatial relations between its constituent subparts from a layer below. Formally, each composite part in $\mathcal{L}_n$ is characterized by a *central subpart* and a list of remaining subparts with their positions relative to the center:

$$\mathcal{P}_\ell^n = \big(\mathcal{P}_{central}^{n-1}, \{(\mathcal{P}_j^{n-1}, \boldsymbol{\mu}_j, \Sigma_j)\}_j\big),$$

where $\boldsymbol{\mu}_j = (x_j, y_j)$ denotes the relative position of subpart $\mathcal{P}_j^{n-1}$, while Σ_j denotes the variance of its position around (x_j, y_j).

The hierarchy starts with a fixed $\mathcal{L}_1$ composed of a set of arbitrary filters. Here we choose a set of Gabor filters that best respond to oriented edges. An example of a $\mathcal{L}_3$ composition is depicted in Figure 4.2.

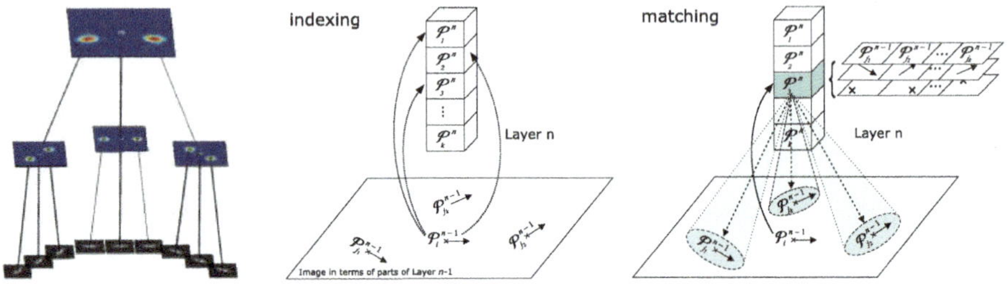

Fig. 4.2. Left: Example of a $\mathcal{L}_3$ composition. Middle: Indexing – evoking higher level composite hypotheses. Right: Matching – verification of a composite library part.

Hierarchical Processing: Indexing and Matching Scheme

Each image is first processed with a fixed Layer 1, the procedure that differs from the hierarchical processing further on. The hierarchical processing steps are all general in its traversal from one layer to the next, thus described in its general form.

Processing with Layer 1. For any given image, the process starts by describing the image in terms of discrete points denoting local oriented edges. This is done *on every scale* – each rescaled version of the original image (a Gaussian pyramid with two scales per octave) is processed separately. First, each image in the pyramid is filtered by 11×11 Gabor filters. By extracting local maxima of the Gabor energy function that are above a low threshold, an image (on each scale) is transformed into a list of $\mathcal{L}_1$ parts; $\{\pi_k^1\}_i$. In general, let π_k^n stand for a *realization* of the $\mathcal{L}_n$ part $\mathcal{P}^n$ with a corresponding location at which it was detected in an image; $\pi_k^n = \{\mathcal{P}^n, x_k, y_k\}$ (here k denotes the successive number of the found part).

Hierarchical Processing. In order to find a higher level image interpretation, the local neighborhoods around the detected $\mathcal{L}_{n-1}$-parts are compared against the composite, $\mathcal{L}_n$-parts stored in the hierarchical library. Each part $\pi_k^{n-1} = (\mathcal{P}_k^{n-1}, x_k, y_k)$ in the image under consideration is subjected to the *indexing and matching procedure* – efficient local search for higher level compositions.

The part $\mathcal{P}_k^{n-1}$ encoded in π_k^{n-1} plays the role of the central part in only a subset of all compositions at layer $\mathcal{L}_n$ of the library. This list is an internal part of the library and can be accessed in constant time during the online processing of images – the process referred to as *indexing*. The *matching* step demands comparing the local spatial neighborhood of π_k^{n-1} against the allowable (retrieved in the indexing step) prototypical compositions within the hierarchical library. Matching of one such composition demands checking for the presence of all subparts pertaining to the composition at hand at their relative locations, (x, y), and positioned within the allowed variances, Σ, with respect to the position of the central part π_k^{n-1}. The indexing and matching procedure is schematically depicted in Figure 4.2.

Unsupervised Learning of Part Compositions

The basic idea behind the learning procedure is to extract statistically salient compositions that encode spatial relations between the constituent parts from the layer below. Each modeled relation between components allows also for some displacement (variance) in spatial position.

The learning algorithm is in principle general – proceeding in the same manner when building each additional layer. It will thus be described in its general form.

The learning process consists of three stages, namely, (1) the local inhibition performed around each image feature (part), followed by (2) the statistical updating of the so-called *spatial maps* that capture pairwise geometric relations between parts, and finally, (3) learning the higher order compositions by tracking co-occurrence of spatial pairs. We must emphasize that each final composition can have a varying number of subcomponents (the number can be anything from 2 and larger).

Learning is performed by gathering statistics over a large body of natural image data processed up to the last (learned) layer in the hierarchical library, e.g. $\mathcal{L}_{n-1}$. Each image is thus represented by a list of parts with corresponding locations, $\{\pi_k^{n-1}\}_k$. A small local neighborhood around each π_k^{n-1} will be inspected in a two-stage process. The first, most crucial step aims to reduce the unnecessary redundancy coded in neighboring parts, referred to as *local inhibition*. Since each π_k^{n-1} is an $(n-1)$-th order composition, it is in fact a set union of a subset of $\mathcal{L}_1$ image parts. Within the inhibition step we remove all neighboring parts around π_k^{n-1} that have a large set intersection with respect to the $\mathcal{L}_1$ image parts. This step removes all features that code a large portion of edge structure already coded by π_k^{n-1}. In the next step, learning is performed by tracking frequent co-occurrences of part types and their relative locations.

The learning process commends by forming a set of all allowable pairs of part identities. The list is accompanied by a set of empty matrices, where the dimensions correspond to the spatial extent of the local neighborhoods. The prepared set thus contains information of type: $\mathcal{C}_{k,j}^n := (\mathcal{P}_k^{n-1}, \mathcal{P}_j^{n-1}, \mathcal{V}_{k,j})$, where $\mathcal{V}_{k,j}$ represents a local spatial voting space for the corresponding combination of pairs of parts.

Structure of small neighborhoods in terms of part locations is inspected around each part, π_k^{n-1}. The philosophy of local receptive field processing is the following: the location of each part $\pi_j^{n-1} = (\mathcal{P}_j^{n-1}, x_j, y_j)$ within the neighborhood of and relative to π_k^{n-1} will update the voting space $\mathcal{V}_{k,j}$ in $\mathcal{C}_{k,j}$ accordingly:

$$x := cx + x_j - x_k, \quad y := cy + y_j - y_k$$
$$\mathcal{V}_{k,j}(x, y) = \mathcal{V}_{k,j}(x, y) + 1,$$

where (cx, cy) denotes the center of the spatial map $\mathcal{V}_{k,j}$.

After all images are processed, we detect voting peaks in the learned spatial maps $\mathcal{V}_{k,j}$, and for each peak, a spatial surrounding area is formed – modeled by a Gaussian distribution, $(\boldsymbol{\mu}_j, \Sigma_j)$.

In the final step, the local image neighborhoods are checked once again by projecting the learned spatial pairs and repeating the learning process by increasing the number of subparts modeled in a composition (the reader is referred to [20] for details) or by tracking the most frequent co-occurrences of the projected spatial pairs.

The final selection of composite parts follows the indexibility constraint, i.e., each part of the lower, $(n-1)$-th Layer, must not index into too many higher layer compositions. Thus the compositions acquired in the learning procedure are sorted according to their decreasing probabilities and only a number of statistically most salient compositions consequently define the next layer. We set the upper bound to the order of $10 - 20$ times the number of parts in the previous, $(n-1)$-th Layer, meaning that on average each part in $\mathcal{L}_{n-1}$ indexes into 10 to 20 composite parts in $\mathcal{L}_n$. The thresholds used are chosen to comply with the available computational resources and affect only the number of finally selected parts and therefore the efficiency of the representation.

Category-Specific Higher Layers

Learning the lower-layer sharable parts in a category-independent way can only get so far - the overall statistical significance drops, while the number of parts reaches its critical value for learning. Thus, learning of higher layers proceeds only on a subset of parts - the ones that are the most repeatable in a specific category. Specifically, the learning of higher layers is performed in images of individual categories, whereby the final categorical layer then combines the most repeatable parts through the object center to form the representation of a category.

Experimental Results

We applied our method to a collection of 1500 images containing a number of diverse categories (cars, faces, mugs, dogs, etc.). The learned compositional hierarchy consisted of 160 parts on Layer 2 and 553 Layer 3 parts (a few examples from both layers are depicted in Fig. 4.3). The complete learning process took approximately 5 hours on one core of an Intel Core-2 CPU 2.4 Ghz computer.

To put the proposed hierarchical concept in relation to other hierarchical approaches as well as other categorization methods, which focus primarily on shape information, the approach was tested on the Caltech 101 database [42]. The Caltech 101 dataset contains images of 101 different object categories with the additional background category. The number of images varies from 31 to 800 per category, with the average image size of roughly 300×300 pixels.

Each image was processed on 3 different scales, spaced apart by $\sqrt{2}$. The average processing times per image per layer obtained with our C++ implementation are reported in Table 4.1. Most of the processing time is spent filtering an image with 6 Gabor filters ($\mathcal{L}_1$), which has not been optimized for performance. The features were combined with a linear SVM for multi-class classification. For this experiment we used 15 images for training and 15 images for testing, disjunct from the training set. The results, averaged over 8 random splits, are reported in Table 4.1 with classification rates of other hierarchical approaches shown for comparison.

Classification was also tested by varying the number of training examples. For testing, 50 examples were used for categories where this was possible and less otherwise. The classification rate was normalized accordingly. In all cases, the result was averaged over 8 random splits. The results are presented and compared with other categorization methods in Table 4.1. We must emphasize that the proposed model focuses on *shape* information *only*.

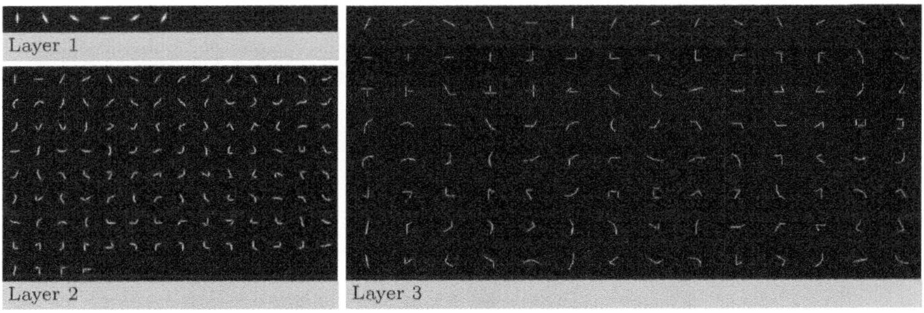

Fig. 4.3. $\mathcal{L}_1$ (fixed), and learned $\mathcal{L}_2$ and $\mathcal{L}_3$ parts (only a subset is shown) used in the Caltech 101 experiments

Discussion

We presented a novel approach to building a representation of object categories. The method learns a hierarchy of flexible compositions in an unsupervised manner in lower, category-independent layers, while requiring minimal supervision to learn higher, categorical layers.

Furthermore, the design of parts is incremental, where new categories can be continuously added to the hierarchy. Since the hierarchy is built as an efficient indexing machine, the system can computationally handle an exponentially increasing number of parts with each additional layer. The results

Table 4.1. Left top and right: Average classification rate (in percentage) on Caltech 101. Left bottom: Average processing times per layer per image.

	$N_{train} = 15$	$N_{train} = 30$
Serre et al. [29]	44	/
Mutch et al. [43]	51	56
Ranzato et al. [33]	/	54
Ommer et al. [26]	/	61.3
Wolf et al. [44]	51.18	/
our method	**60.5**	**66.5**

	Processing time per 300 × 300 image
Layer 1	1.6 s
Layer 2	0.54 s
Layer 3	0.66 s

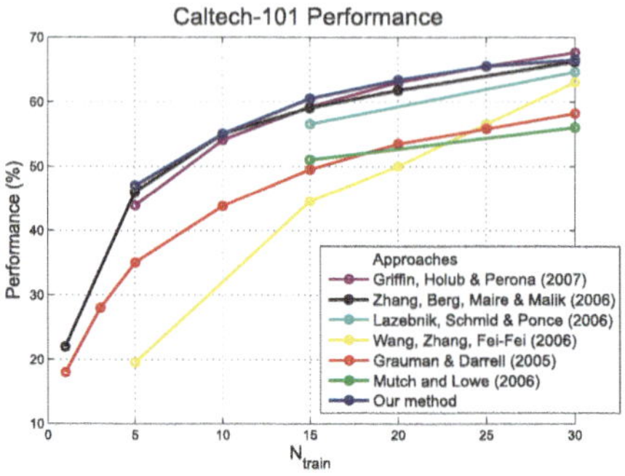

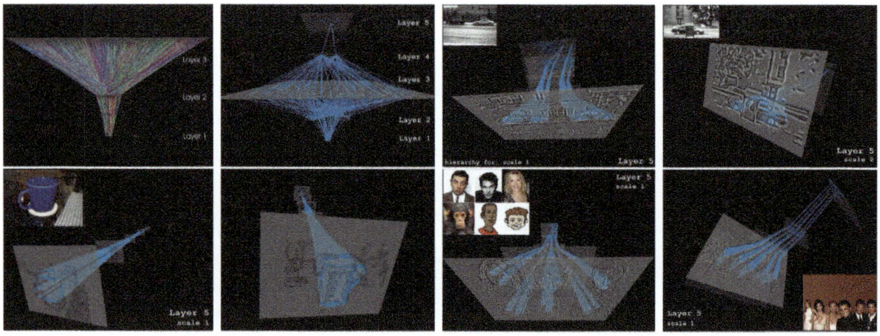

Fig. 4.4. Top left two images: learned 3–layer hierarchy for the Caltech experiment; learned hierarchy for faces with compositional links shown. Examples of detections of categories cars, cups, and faces, where the first three layers in the library are common to all three categories.

(a)
(b)
(c)

Fig. 4.5. Example images from the *TUD-Shape* ((a), (b)), and *TUD-Shape2* (c) data sets, depicting the categories *cup*, *fork*, *hammer*, *knife*, *mug*, *pan*, *pliers*, *pot*, *saucepan*, and *scissors*.

show that only a small number of higher layer parts are needed to represent individual categories, thus the proposed scheme would potentially allow for an efficient representation of a large number of visual categories.

4.2.2 Representations for Functional and Affordance-Based Categorization

Motivation

Recent work in object categorization often uses interest operators in connection with local region descriptions to sparsely represent objects in images. Typically, local regions are sampled around interest points (*e.g.*, from Harris-Laplace, Hessian-Laplace, or Salient Regions detectors) at characteristic scales, and then described by local region descriptors. The resulting sparse representation is then vector-quantized using a codebook of visual words, and either used directly for recognition (bag-of-words approaches [45]), or enriched by information about the spatial layout of feature occurences.

While these local feature-based representations have been successfully used for the recognition of object classes with sufficient local appearance statistics, they have rarely been employed in the context of functional or affordance-based categorization of objects, where, intuitively, an object's shape and geometry should be represented explicitly. Historically, the recognition of geometric objects such as cups and tables has been an important focus of object recognition [46]. In recent work however, it is largely underrepresented with some notable exceptions [47, 48]. We therefore compare in [78], on a novel data collection of 10 geometric object classes[1] (see Figure 4.5), various shape-based features with appearance-based descriptors, such as SIFT. The analysis includes a direct comparison of feature statistics as well as results within standard recognition frameworks, incorporating varying degrees of spatial information.

Appearance-Based *vs.* Shape-Based Features

We briefly introduce the features and interest point detectors used in our comparison. The shape-related features are k-Adjacent Segments [47], Geometric Blur [50], and Shape Context [51]; the appearance-based region descriptors are SIFT [5] and GLOH [52]. As interest point detectors, we employ Harris-Laplace [53], Hessian-Laplace [54], and Salient Regions [55].

k-*Adjacent Segments (*k-*AS)*

k-Adjacent Segments have been proposed as an extension to *contour segment networks*, a graph-based method for template-matching hand-drawings to image databases [56]. [47] demonstrates how k-AS can be incorporated into a

[1] `http://www.mis.informatik.tu-darmstadt.de/data`

general object recognition framework. We extract k-AS features using the original implementation[2]. First, *edgels* are detected via the Berkeley natural boundary detector [57]. Second, neighboring edgels are chained, and further linked to form L-, T- and higher-order junctions. Last, edgel-chains are replaced by straight line approximations (*contour segments*), and joined into a global *contour segment network* for the image. A k-AS descriptor then describes the geometric layout of a group of k adjacent segments in that network w.r.t. relative position, orientation, and length. The dimensionality of k-AS features is $n = 4 * k - 2$. For the typical choices of $k \in \{2, 3\}$, n is 6 or 10, rendering k-AS a comparably low dimensional descriptor.

Local Region Descriptors

Geometric Blur *(GB)*. We use the original implementation[3] of the Geometric Blur [50] region descriptor from [58]. Geometric Blur first extracts $c = 4$ channels of oriented edge energy [59] to obtain a sparse signal S. In S, the region centered at interest point location x_0 is blurred with a spatially-varying Gaussian kernel G_d to obtain the Geometric Blur $B_{x_0}(x) = S * G_{\alpha x+\beta}(x_0 - x)$. $B_{x_0}(x)$ is then sub-sampled over all channels at n distinct locations in a circular grid. The final descriptor is the concatenation of all $c \times n$ samples. Throughout all experiments, we use the standard values for $\alpha = 0.5$, $\beta = 1$ and $n = 51$, resulting in a descriptor of length 204.

Shape Context *(SC)*. Shape Context [51] is originally based on edge information. For a given interest point location, it accumulates the relative locations of nearby edge points in a coarse log-polar histogram. We compute a histogram containing 9 spatial bins over 4 edge orientation channels. Bin size increases w.r.t. distance from the interest point center. Note that this is similar in spirit to spatially varying blur, but results in a smaller descriptor (length 36).

SIFT. The Scale Invariant Feature Transform [5] descriptor is a 3D histogram over local gradient locations and orientations, weighted by gradient magnitude. It uses 4×4 location and 8 orientation bins, *i.e.*, 128 in total.

GLOH. Gradient Location Orientation Histograms [52] is an extension of the SIFT descriptor. It uses 17 bins for location and 16 bins for orientation in a histogram over a log-polar location grid, and reduces descriptor dimensionality to 128 by PCA. We use the implementation[4] of [52] for SC, SIFT and GLOH. All descriptors are made invariant to in-plane rotation by aligning the region to the dominant gradient direction before descriptor computation.

[2] `http://www.vision.ee.ethz.ch/~ferrari/release-kas.tgz`
[3] `http://www.cs.berkeley.edu/~aberg/demos/gb_demo.tar.gz`
[4] `http://www.robots.ox.ac.uk/~vgg/research/affine/descriptors.html`

Interest Point Detectors

We compute local region descriptors based on detections of the following interest point detectors: **Harris-Laplace** *(HarLap)* [54] is an extension to Harris corners [60]. It selects corners at locations where a Laplacian attains an extremum in scale-space. The **Hessian-Laplace** *(HesLap)* [54] detector responds to blob-like structures. It searches for local maxima of the Hessian determinant, and selects a characteristic scale via the Laplacian as for Harris-Laplace. The **Salient Regions** *(SalReg)* detector [55] identifies local image regions that are non-predictable across scales by measuring entropy over local intensity histograms. We use publicly available implementations for HarLap/HesLap[5], and SalReg[6].

Feature Evaluation

We evaluate the combined performance of feature detectors and descriptors at three different levels. First, we compute statistics over clusterings of local feature descriptors (codebooks). Second, we represent objects by means of occurrence statistics over codebook matches, and analyze classification performance in a Bayesian framework. Third, we investigate the impact of gradually adding location information to that object representation, by jointly boosting localized histograms of codebook matches over all object categories.

Cluster Precision

We follow the argumentation of [61] and base our evaluation on a mid-level representation of image features common to many computer vision techniques. In particular, we analyze the statistics of clusterings of feature descriptors. In order to quantify how well a clustering of feature descriptors reflects the separation of object classes, we introduce a refinement to *cluster precision* [61]. Intuitively, we want to measure to what extent features of a given class are grouped together by clustering. High scores should be obtained by big clusters with features from many instances of a single object class, and low scores by small clusters with few features, but from multiple classes. In a second experiment, we evaluate the generalization capabilities of a given feature type over an unseen test set. Features are extracted from an independent set of images, and matched against a codebook built from training images. For each test feature belonging to class a, we determine the precision P_{C_a} over all *matched* clusters with respect to class a. The average precision over all classes is called *matching precision*.

Results. We measure cluster precision for 9 different compression ratios ($\#Features/\#Clusters$) over codebooks generated from 200 training images

[5] `http://www.robots.ox.ac.uk/~vgg/research/affine/detectors.html`

[6] `http://www.robots.ox.ac.uk/~timork/salscale.html`

(see Figures 4.6 (a) to (c)). We observe that cluster precision changes substantially if we vary detectors for a given descriptor, while it remains relatively stable over varying descriptors for a given detector. Appearance-based (SIFT, GLOH) and shape-based descriptors (GB, SC) do not show great differences in cluster precision. Both perform comparably well. *k*-AS are worst, but still comparable to DoG-SIFT. We report that cluster precision results transfer to large extents to matching precision over previously unseen data, but omit the plots for brevity.

Naïve Bayes

The second level of our evaluation represents objects in terms of occurrence statistics (counts of nearest-neighbor matches) over codebook entries and trains a multi-class-classifier on a training set of such representations. We use an analogous approach to *Multinomial Naïve Bayes* [62] for text classification, and model the posterior distribution of an object class, given occurrence statistics over a codebook, as a multinomial distribution.

Results. We measure classification accuracy over fixed numbers of clusters from $n = 50$ to 1600, increasing by powers of 2 (see Figures 4.6 (d) to (f)). Dependent on the detector, $n = 1600$ corresponds to compression ratios of 9 (HarLap), 52 (HesLap), 21 (SalReg), 5 (DoG), 5 (*2*-AS) and 16 (*3*-AS). For each feature type, we train a Naïve Bayes classifier on a training set of 200 images (20 per category), and test on an independent test set of 100 images (10 per category). The differentiation between descriptors with fixed detectors is more pronounced for Naïve Bayes than for cluster precision. In particular, appearance-based features lead on average. GB and SC perform on a comparable level to SIFT and GLOH, but only for individual detectors (SalReg for GB, HesLap for SC). *k*-AS exhibit relatively weak discriminative power for Naïve Bayes classification. SC offers a good compromise between strong (GB) and weak (*k*-AS) discrimination.

Localized Bag-of-Words

We measure the impact of adding location information in terms of classification accuracy in a Joint Boosting [63] framework. The object representation on the third level of our evaluation is based on histograms of feature occurrences over a codebook, and inspired by [56]. We divide a rectangular image region into a grid of cells. For each cell, a local histogram over soft-matched codebook entries is computed, and concatenated to form the object representation. A Joint Boosting algorithm is trained from object representations of a set of training images using a fixed number of boosting rounds, and tested against an independent set of test images. We use *decision stumps* [63] over histogram bins as weak classifiers for boosting. By varying the number of grid cells g, we regulate the tradeoff between rich feature statistics (small g) and more accurate localization (large g).

Results. We measure classification accuracy as for Naïve Bayes for varying numbers of grid cells $g \in \{1, 4, 9\}$. We assume known bounding boxes for training *and* test, and use them to anchor histogram grids. We fix the number of clusters to $n = 200$, and obtain histograms of length $g \times n \in \{200, 800, 1800\}$. While for *TUD-Shape*, boosting over localized bag-of-words lifts the discriminant power of all feature types to a comparable level for $g = 9$, shape-based features win on *TUD-Shape2* (see Figures 4.6 (g) to (i)): HesLap-GB and rotation invariant *2*-AS are best (42% respectively 44% accuracy). The best performing SIFT and GLOH combinations obtain 33% (HesLap-SIFT) and 31% (HesLap-GLOH). HarLap-SC obtains 35%.

Discussion

Local shape- and appearance-based features do not show great differences in terms of feature statistics over our shape-based data set. On average, the choice of detector is more important on average than the choice of descriptor. Hessian-Laplace with SIFT and GLOH is best on average. Shape-based

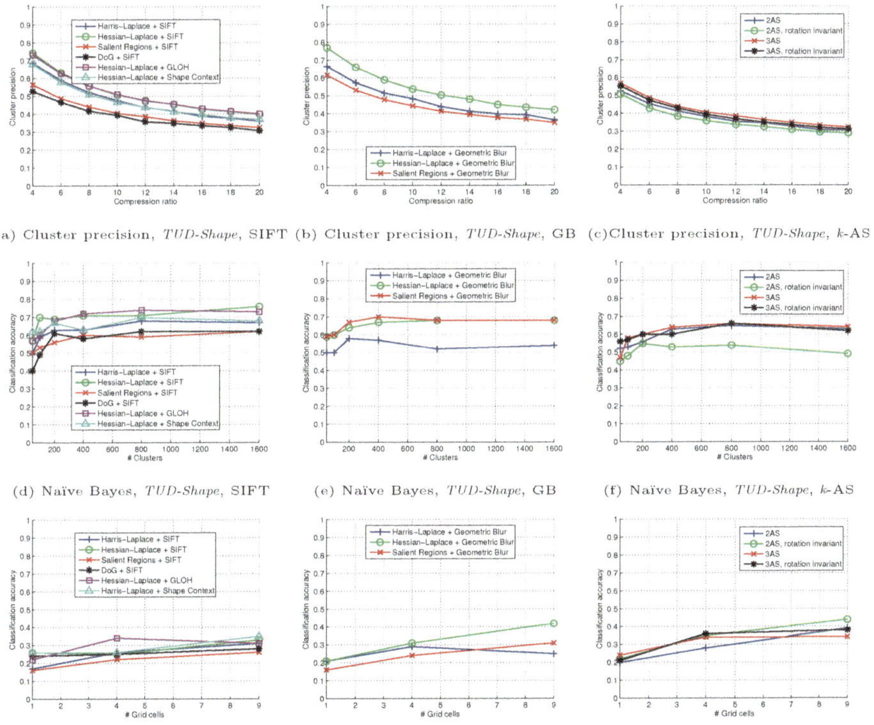

(a) Cluster precision, *TUD-Shape*, SIFT (b) Cluster precision, *TUD-Shape*, GB (c)Cluster precision, *TUD-Shape*, *k*-AS

(d) Naïve Bayes, *TUD-Shape*, SIFT (e) Naïve Bayes, *TUD-Shape*, GB (f) Naïve Bayes, *TUD-Shape*, *k*-AS

(g) Localized BOW, *TUD-Shape2*, SIFT (h) Localized BOW, *TUD-Shape2*, GB (i) Localized BOW, *TUD-Shape2*, *k*-AS

Fig. 4.6. *First row:* cluster precision, *second row:* naïve Bayes classification accuracy, *third row:* localized BOW classification accuracy for SIFT, GB, and *k*-AS. SIFT-plots include the best GLOH and SC curves.

features (Geometric Blur, k-Adjacent Segments) perform mostly worse than appearance-based ones for classification based on simple occurrence statistics. In particular, k-AS capture generic local shape properties rather than discriminant information for an object category. They hence benefit more from added location information than appearance-based features, and can even overtake appearance-based features on shape-based data. We make use of this result in Section 7.5, where we demonstrate how to incorporate k-AS features into a framework for the detection of functional object categories based on learned affordance cues.

4.3 Mid-Level Representation and Detection

4.3.1 Towards Adaptive Representations

The main focus of this section is therefore a new object representation that aims to combine the above mentioned properties to make a step towards more flexible and adaptive object representations applicable to a wide range of objects and suited both for unsupervised as well as supervised learning. Therefore, we propose a novel approach that allows to learn a low-dimensional representation of object classes by building a generative decomposition of objects. These learned decompositions of objects contain both local appearance features as well as global silhouette features shared across object classes. This generative model of objects is directly applicable to unsupervised learning tasks such as visual object class discovery. Second, we combine the low-dimensional and generative decomposition of objects with a discriminative learning framework to enable supervised training and competitive object class detection. Third, we provide empirical evidence that shows the properties of the approach (local vs. global features, feature sharing, unsupervised vs. supervised learning) and compares the approach with the state-of-the-art. Interestingly, the approach outperforms both unsupervised techniques as well as supervised techniques on various tasks on common databases.

4.3.2 Learning of Generative Decompositions

In this section we describe our approach to decomposition of multiple visual categories by combining dense gradient representations and topic models. Starting from the image, we first present our data representation. Then we describe how we apply the topic model to this representation and provide visualizations and insights for the obtained model as well as a quantitative evaluation on an unsupervised learning task. Inspired by [6], we compute gradients on each color channel of the images and use the maximum response to obtain a grid of histograms that overlays the image. Each histogram in the grid has 9 orientation bins equally spaced from $0°$ to $180°$ to represent the unsigned gradient orientation. An example of such an encoding is visualized

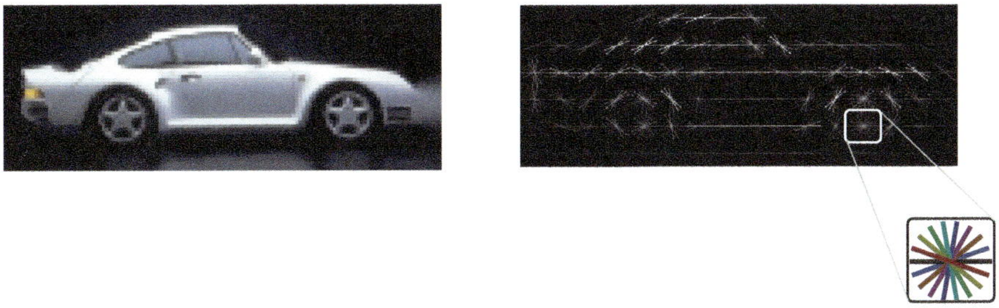

Fig. 4.7. Dense gradient histogram representation. The loupe shows the 9 possible edge orientation of the histogram bins that are interpreted as words.

in Figure 4.7. In each cell, the 9 possible edge orientations associated with the orientation bins are displayed by short lines. The grayscale value encodes the accumulated gradient magnitude in each bin. The size of the cells in the grid is 8×8 pixels.

As the following topic models operate on discrete word counts, we normalize the histograms to have a constant sum of discrete entries. We decided not to compute a redundant coding like the blocks in the HOG descriptor [6] as we believe that the introduced non-linearities by local normalization would hinder the fitting of the probabilistic model.

Topic Models

To define a generative process for our data representation, we employ probabilistic topic models [64, 65, 66] which were originally motivated in the context of text analysis. As it is common habit we adopt the terminology of this domain. In the following, a document d refers to a sequence of words $(w_1, w_2, \dots, w_{N_d})$, where each w_i is one word occurrence. The underlying idea of these models is to regard each document as a mixture of topics. This means that each word w_i of the total N_d words in document d is generated by first sampling a topic z_i from a multinomial topic distribution $P(z)$ and then sampling a word from a multinomial topic-word distribution $P(w|z)$. Therefore the word probabilities for the combined model are:

$$P(w_i) = \sum_{j=1}^{T} P(w_i|z_i = j)P(z_i = j) \tag{4.1}$$

where T is the number of topics and $P(w_i|z_i = j)$ as well as $P(z_i = j)$ are unobserved . According to the notation of [66], we will abbreviate

$\theta^{(d)}$: topic distribution $P(z)$ for document d
$\phi^{(j)}$: topic-word distribution $P(w_i|z = j)$ for topic j

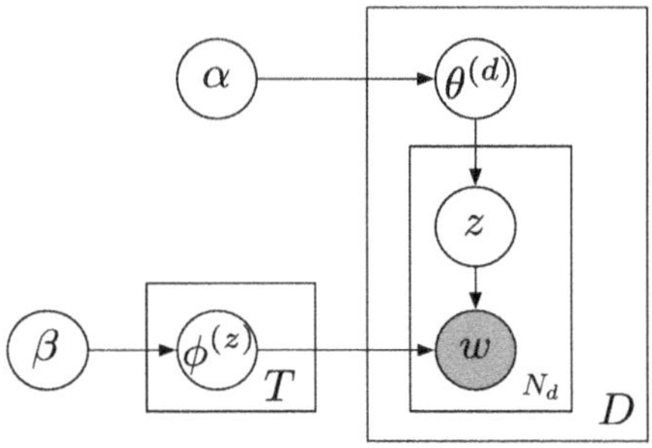

Fig. 4.8. LDA model as formulated by [66]

The particular topic models differ on the one hand in which additional hyperparameters/priors they introduce and on the other hand in how inference and parameter estimation is performed. We will discuss the *Latent Dirichlet Allocation* model [65] in some more detail focusing on the version presented in [66] that uses Gibbs sampling for inference and estimation. The graphical representation of this model is depicted in Figure 4.8. It visualizes the process that generates a total of D documents d, where each document has N_d words. Above we already described how each word w_i of a particular document is generated. In the full model, there are 2 additional hyperparameters, α and β, which place symmetric dirichlet priors on the topic distribution of each document $\theta^{(d)}$ and the topic-word distributions $\phi^{(j)}$ respectively. As the setting for α and β is common to all documents, these act as forces that impose global tendencies on these distributions. Intuitively, the prior α for the topic distribution θ favors co-activation (sharing) of multiple topics for each document for values larger than 1, whereas smaller values result in sparser topic distribution - ultimately having single topics explaining whole documents (clustering). Consequently, the sparseness of the topic-word distribution $\phi^{(j)}$ is affected by this choice. The second parameter β, has a direct smoothing effect on the topic distributions.

For more details on the models, inference and estimation, we refer to [65] and [67]. The idea behind the employed Gibbs sampling procedure is that all topic assignments z_i are initialized (typically randomly) and then iteratively updated in a random order. To perform such a single update, a topic is drawn from the conditional distribution $P(z_i|\Omega \setminus z_i)$ and assigned to z_i, where $\Omega \setminus z_i$ denotes all observed and unobserved variables but z_i. This is repeated for a fixed number of iterations.

To extend our findings to the detection task that we are aiming for in Section 4.3.4, we extract our representation on the multi-category, multi-view PASCAL'06 dataset [68], in order to obtain a decomposition that is shared across categories.

In the first row of Figure 4.9 13 of 100 topic distributions are visualized that were trained on the bounding box annotations of the training and validation data of the PASCAL'06 challenge. The rows below display the examples

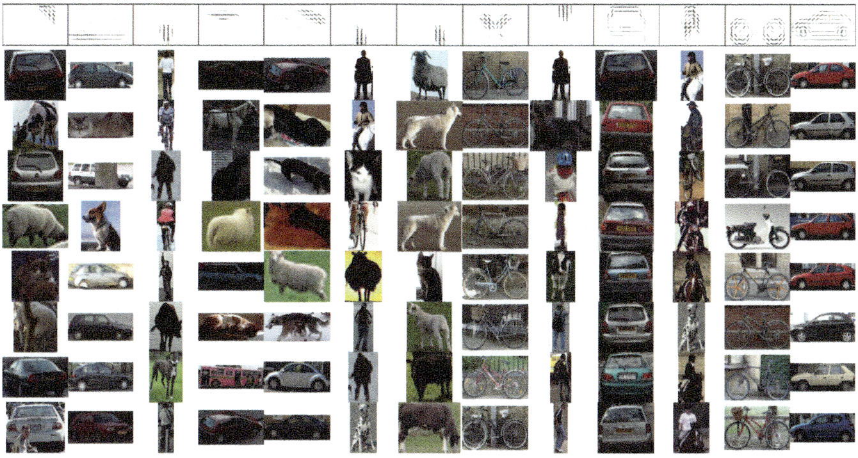

Fig. 4.9. First row: example topics that were learned by the proposed approach across categories and viewpoints for the 10 classes of the PASCAL'06 data. Below first row: training images that activated the topic above most. The topics model local structures, line segments as well as silhouette-like structures. The topics are distinctive enough to separate several category members and even view-points. On the other hand they are general enough to be shared across categories and viewpoints.

that activated this particular topic most. We observe that the topics capture different levels of objects, ranging from global silhouettes (car rear in column 10 and side view in column 13) over localized parts (legs in column 3, bicycle frame in column 8 and bicycle wheels in column 12) to line segments and corners (corner in column 1 and line segments in column 2 and 4) . The model discovers distinctive parts that even separate several examples of different categories and their viewpoints although no such information was available to the system during training. Importantly, we can see that other topics like those that got activated on legs are shared across several categories, which is a desirable property of a compact decomposition in order to be scalable [22].

4.3.3 Generative/Discriminative Hybrid Model for Detection

Based on these promising results, we describes a complete system for supervised multi-category detection that leverages the learned representation.

Combining Generative and Discriminant Models

Recently, the combinations of generative approaches with discriminative ones have shown to be very effective [23, 24]. The idea is that generative models can easily incorporate prior information to support learning from small samples, have increased robustness to noise and generally have more principled ways of dealing with missing data. Discriminative models on the other hand have shown to give superior performance for well posed learning tasks and a sufficient number of training examples. We also follow this idea and complement the generative topic model by a discriminative SVM classifier with an RBF kernel [69]. In particular we train an SVM to discriminate between the topic distributions $\theta^{(d)}$ which are inferred for images containing the category of interest and others that do not contain these. By doing so, we seek to profit from the above mentioned benefits of the generative model combined with the discriminative classifier.

Sliding Window Approach to Detection

As proposed in [6] a sliding window approach can be done efficiently in this setting if the sliding window is always shifted by exactly one cell in x or y direction. In this case, the gradient histograms of the cell grid are computed once and for each sliding window the relevant sub grid is used.

4.3.4 Results on Visual Category Detection

We evaluate our approach on the competition 3 of the PASCAL challenge 2006 [68] that poses a challenging detection task as 10 visual categories are to be detected from multiple viewpoints over a large scale range. Many images in training and test suffer from degradations due to occlusion, bad lighting conditions and show high variability in object appearance and background clutter.

We outperform all other competitors in the 3 categories bicycle, bus and car by improving the state-of-the-art [68] on this dataset by 5.75%, 9.14% and 5.67% in average precision respectively. In particular we surpass the fully global approach [6] that our method was motived by. Compared to [70] we improve on bicycles and bus only by 0.65% and 0.93%, but again significantly on cars with 8.87%. However, in contrast to [70] we do not use the viewpoint annotations to train our approach. For the other categories, we perform about average, but also showed some inferior results on the highly articulated categories. We are currently investigating means to make the approach less rigid and carry over the good results from the first 3 categories to the other ones.

For a more detailed analysis in different settings that show the merits of this approach supervised as well as unsupervised setting in comparison to local, global and shape-based approaches we refer to [71].

4.3.5 Discussion

This section described an approach to learn generative decomposition of visual categories from data that yield and effective representation for unsupervised as well as supervised categorization tasks. We argue that the flexibility and adaptivity we obtain through this learning-based approach results in

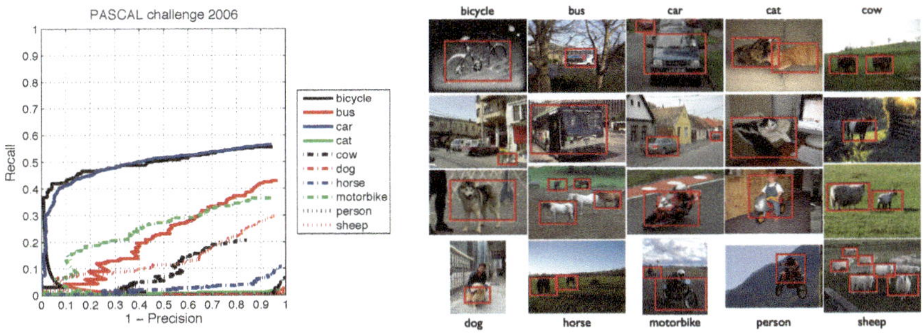

Fig. 4.10. Results on the PASCAL VOC challenge 2006. Precision-Recall curves and example detections.

the observed performance improvements over traditional, hand-crafted feature representations.

Additionally the learned decompositions add a certain structure to the observed objects and puts object into relation to each other. This is highly interesting to obtain object part reference and share knowledge across the observed object instances. In this sense, our model offers a high-level of introspection, that can be leveraged in tutor-based learning setting.

To overcome the limitations due to the rigid nature of the underlying coding, we currently investigate how to augment the approach with a deformation model to add more flexibility. However, it is yet unclear how to obtain a beneficial tradeoff between invariance and discriptiveness that results in an improved performance of the system. Progress in this direction would enable the system to generalize across more complex deviations in shape and appearance.

4.4 High-Level Representations and Dynamic Models

An important type of object classes is articulated such as people and animals. We study the issues associated to modeling of these articulated objects classes by addressing the task of tracking people in cluttered environments. To achieve reliable extraction of people-tracks as well as data-association across long periods of occlusion, the proposed approach combines recent advances in people detection with the power of dynamical models for tracking. Rather than to simply determine the position and scale of a person as is common for state-of-the-art people detectors [6, 72], we also extract the position and articulation of the limbs. This allows us to use a more powerful dynamical model that extends people detection to the problem of reliably extracting people-tracklets – people detections consistent over a small number of frames. In particular, we use a hierarchical Gaussian process latent variable model (hGPLVM) [73] to model the dynamics of the individual limbs. As we will show in the experiments this enables us to detect people more reliably than it would be possible from single frames alone. We combine this with a hidden Markov model (HMM) that allows to extend the people-tracklets, which cover only a small number of frames at a time, to possibly longer people-tracks. These people-tracks identify individuals over longer sequences of consecutive frames when that is appropriate, such as between major occlusion events. Tracking people over even longer periods of time is then achieved by associating people-tracks across potentially long periods of occlusion using both the dynamical model and an extracted appearance model, which allows identifying specific individuals throughout the sequence.

4.4.1 Appearance Model for Single-Frame Detection and Pose Estimation

Following the general idea of part-based object detection, we formulate the problem of locating an object from a specific class in a test image as search of the modes of the posterior probability distribution $p(L|E)$ of the object configuration L given the image evidence E [74].

In our model, the configuration is described as $L = \{\mathbf{x}^o, \mathbf{x}^1, \ldots, \mathbf{x}^N\}$, where $\mathbf{x}^o$ is the position of the object center and its scale, and $\mathbf{x}^i$ is the position and scale of part i. The image evidence, which here is defined as a set of local features observed in the test image, will be denoted as $E = \{\mathbf{e}_k^{app}, \mathbf{e}_k^{pos} | k = 1, \ldots, K\}$, where $\mathbf{e}_k^{app}$ is an appearance descriptor, and $\mathbf{e}_k^{pos}$ is the position and scale of the local image feature with index k. We will denote the combination of position, scale, and appearance of a local feature as $\mathbf{e}_k = (\mathbf{e}_k^{app}, \mathbf{e}_k^{pos})$.

Assuming that positions of object parts are independent of each other given person's position $\mathbf{x}^o$ and articulation a, it follows that

$$p(L|a, E) \approx p(\mathbf{x}^o) \prod_i p(\mathbf{x}^i|a, E) p(\mathbf{x}^i|\mathbf{x}^o, a). \tag{4.2}$$

If we assume that a particular image feature $\mathbf{e}_k$ belongs to part i of an object instance in the image with probability α, then it holds that

$$p(\mathbf{x}^i|a, E) = c_0 + c_1 \sum_{\mathbf{e}_k} p(\mathbf{x}^i|a, \mathbf{e}_k) + O(\alpha^2), \tag{4.3}$$

where c_0 and c_1 depend only on the image features E [75]. If α is sufficiently small, which is true for street scenes in which a particular person usually represents only a small portion of the image, we obtain

$$p(L|a, E) \approx \prod_i p(\mathbf{x}^i|\mathbf{x}^o, a) \left[\beta + \sum_{\mathbf{e}_k} p(\mathbf{x}^i|a, \mathbf{e}_k) \right], \tag{4.4}$$

where β can be seen as a regularizer for the evidence obtained from the individual image features, and we have additionally assumed uniform $p(\mathbf{x}^o)$.

As is common in models based on local feature representations, we introduce an object-specific *codebook* denoted as $\mathcal{C} = \{\mathbf{c}_j | j = 1, \ldots, J\}$. The part

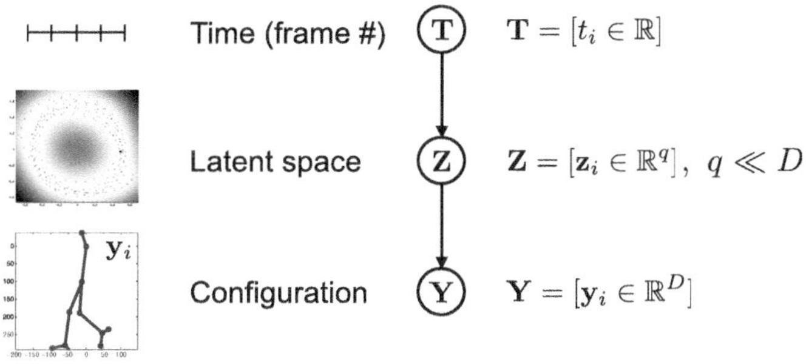

Fig. 4.11. Representation of articulations in the latent space

posterior with respect to a single image feature is computed by marginalization over the codebook entries:

$$p(\mathbf{x}^i|a, \mathbf{e}_k) = \sum_{\mathbf{c}_j} p(\mathbf{x}^i|a, \mathbf{c}_j, \mathbf{e}_k^{pos}) p(\mathbf{c}_j|\mathbf{e}_k^{app}). \tag{4.5}$$

$p(\mathbf{c}_j|\mathbf{e}_k^{app})$ is discrete distribution over codebooks based on a Gaussian similarity measure, and $p(\mathbf{x}^i|a, \mathbf{c}_j, \mathbf{e}_k^{pos})$ is learned from training data.

4.4.2 Representing the Dynamics of the Human Walking Cycles with Latent Variable Model

In order to integrate the evidence from several subsequence frames, it is necessary to incorporate prior knowledge about plausible human motions into the recognition framework. Such prior knowledge can be expressed in the form of prior distribution on the sequences of possible configurations of the person. Modelling such prior directly is difficult due to the high dimensionality of the space of pose sequences. Instead, several authors [76, 77, 78] have argued and shown that a low-dimensional representation is sufficient to approximate the pose dynamics. In the following we describe a Gaussian process latent variable model (GPLVM) which we use to obtain such a low-dimensional representation and discuss how it can be applied to reliable people detections in image sequences.

Let $\mathbf{Y} = [\mathbf{y}_1, \mathbf{y}_2, \ldots, \mathbf{y}_m]^T$ be a sequence of D-dimensional observations (here describing the relative joint angles of body limbs). GPLVMs model the D-dimensional observation space as the output of D Gaussian processes with an input space of dimensionality q, where $q < D$. Each observation $\mathbf{y}_i$ is associated with a q-dimensional latent point $\mathbf{z}_i$. The likelihood of the observation sequence $\mathbf{Y}$ given the latent sequence $\mathbf{Z} = [\mathbf{z}_1, \mathbf{z}_2, \ldots, \mathbf{z}_m]^T$ and model parameters θ is given by [79]:

$$p(\mathbf{Y}|\mathbf{Z},\theta) = \prod_{i=1}^{D} \mathcal{N}(\mathbf{Y}_{:,i}|0,\mathbf{K_z}), \tag{4.6}$$

where $\mathbf{Y}_{:,i}$ is the vector of values of feature i across all observations, and $\mathbf{K_z}$ is the covariance matrix with elements given by a covariance function $k(\mathbf{z}_i, \mathbf{z}_j)$. In this paper we use a squared exponential covariance function augmented by Gaussian white noise.

For a given $\mathbf{Y}$ we can find the positions of the latent points $\mathbf{Z}$ along with the model parameters θ by maximizing their likelihood from Eq. (4.6).

In addition to the low-dimensional latent representation of the data, GPLVMs provide a probabilistic mapping from the latent space to the observation space. One possibility to define a dynamic model in the latent space is to place a suitable prior on the elements of $\mathbf{Z}$. Such a prior can be given by a Gaussian process with time as input variable [73]. The structure of this prior is shown on the Fig. 4.11. Given the sequence of points in time, $\mathbf{T} = [t_1, t_2, \ldots, t_m]^T$ at which the observations $\mathbf{Y}$ were made, the prior over $\mathbf{Z}$ is given by

$$p(\mathbf{Z}|\mathbf{T}) = \prod_{i=1}^{q} \mathcal{N}(\mathbf{Z}_{:,i}|0,\mathbf{K_T}) \tag{4.7}$$

where $\mathbf{K_T}$ is the covariance matrix of the time points. The covariance function in the time space can again be taken as squared exponential, which ensures smoothness of the trajectories.

We now combine this prior with the likelihood from Eq. (4.6), and maximize *w.r.t.* $\mathbf{Z}$ and θ. Fig. 4.11 shows the 2 dimensional latent space obtained by applying this model to 11 walking sequences of different subjects, each containing one complete walking cycle. Walking cycles in each sequence are manually aligned so that we can interpret the frame number in each sequence as phase of the walking cycle. This hierarchical approach to GPLVM dynamics has several advantages over the auto-regressive prior proposed in [77]. In particular, it allows us to evaluate the likelihood of a sequence of poses, even if the poses occurred at unequally spaced time intervals. This arises, *e.g.*, when the subject was occluded or not detected for several frames. Additionally, for a given pose the model allows us to hypothesize both successive and previous poses, which we use to produce good initial hypotheses for the whole image sequence from a few good detections.

4.4.3 Robust Detection and Tracking of People in Image Sequences

The person detector which we have described in Sec. 4.4.1 provides hypotheses for position, scale, and body articulation in single frames based on the detection of individual body parts or limbs. To further improve the detection performance in image sequences it is desirable to incorporate temporal consistency among subsequent frames. In contrast to frequently used motion

Fig. 4.12. Detection and tracking of multiple people on "TUD-Crossing" dataset

model based on position and velocity of the person, we propose to use a more expressive kinematic limb model thereby leveraging the articulated tracking literature (*e.g.* [80, 81, 82, 76, 83, 78]).

Given the image evidence $E = [E_1, E_2, \ldots, E_m]^T$ in a sequence of m subsequent frames, we would like to recover the positions $\mathbf{X}^{o*} = [\mathbf{x}_1^{o*}, \mathbf{x}_2^{o*}, \ldots, \mathbf{x}_m^{o*}]^T$ of the person as well as the configurations of the limbs in each frame $\mathbf{Y}^* = [\mathbf{y}_1^*, \mathbf{y}_2^*, \ldots, \mathbf{y}_m^*]^T$ with $\mathbf{y}_j^*$ denoting the recovered limb orientations in the j-th frame. Assuming independence of the detections in each frame, the posterior factorizes as:

$$\begin{aligned} p(\mathbf{Y}^*, \mathbf{X}^{o*}|E) &\propto p(\mathbf{Y}^*)p(\mathbf{X}^{o*})p(E|\mathbf{Y}^*, \mathbf{X}^{o*}) \\ &\propto p(\mathbf{Y}^*)p(\mathbf{X}^{o*}) \prod_{j=1}^{m} p(E_j|\mathbf{y}_j^*, \mathbf{x}_j^{o*}). \end{aligned} \quad (4.8)$$

$p(E_j|\mathbf{y}_j^*, \mathbf{x}_j^{o*})$ is the likelihood of the image evidence E_j, and is given by the detection model described in the previous section. $p(\mathbf{X}^{o*})$ corresponds to a prior of human body speed, which we model as a broad Gaussian and $p(\mathbf{Y}^*)$ is given by the Gaussian process latent variable model.

Given limb likelihoods and the hGPLVM prior, we can maximize Eq. (4.8) to find the best pose sequence. This is equivalent to jointly solving the inverse kinematics in each frame of the sequence under soft constraints given by limb likelihoods and is similar to [84], except that in our case hints about limb positions are provided by a person detector instead of being manually given by the user. If we denote the training observations, their latent representation and model parameters by $\mathcal{M} = [\mathbf{Y}, \mathbf{T}, \mathbf{Z}, \boldsymbol{\theta}]$, the probability of the unknown pose sequence $\mathbf{Y}^*$, its latent representation $\mathbf{Z}^*$, and the person positions $\mathbf{X}^{o*}$ is given by

$$p(\mathbf{Y}^*, \mathbf{X}^{o*}, \mathbf{Z}^* | \mathcal{M}, E, \mathbf{T}^*) \propto \\ p(E | \mathbf{Y}^*, \mathbf{X}^{o*}) p(\mathbf{Y}^* | \mathbf{Z}^*, \mathcal{M}) p(\mathbf{Z}^* | \mathbf{T}^*, \mathcal{M}) p(\mathbf{X}^{o*}). \tag{4.9}$$

The first term is the detection likelihood from single-frame detections (see Eq. (4.8)). The second term is given by

$$p(\mathbf{Y}^* | \mathbf{Z}^*, \mathcal{M}) = \prod_{i=1}^{D} p(\mathbf{Y}^*_{:,i} | \mathbf{Z}^*, \mathbf{Y}_{:,i}, \mathbf{Z}), \tag{4.10}$$

where $p(\mathbf{Y}^*_{:,i} | \mathbf{Z}^*, \mathbf{Y}_{:,i}, \mathbf{Z})$ is a Gaussian process prediction of the pose sequence given a sequence of latent positions. The third term is given by the dynamics prior on the latent space:

$$p(\mathbf{Z}^* | \mathbf{T}^*, \mathcal{M}) = \prod_{i=1}^{q} p(\mathbf{Z}^*_{:,i} | \mathbf{T}^*, \mathbf{Z}_{:,i}, \mathbf{T}). \tag{4.11}$$

In our formulation, detecting people in a series of m frames therefore corresponds to finding pose sequences $\mathbf{Y}^*$ and people positions $\mathbf{X}^{o*}$ that maximize Eq. (4.9). The gradients of the second and third terms in 4.9 can be computed analytically, while we use a finite difference approximation for gradient of $p(\mathbf{Y}^* | F)$. Subsequently the local maxima can be found using any of the standard non-linear optimization methods using positions and poses of the person estimated by the single-frame detector as initialization.

Given people hypothesis jointly computed using evidence accumulated over several subsequent frames we can compute longer people tracks by considering these hypothesis as states in the hidden Markov model and computing their optimal sequence with Viterbi algorithm (see [85] for details). Fig. 4.12 shows several frames from a street sequence with multiple people, in which our system can detect and track most of the people in spite of reoccuring parial and full occlusions.

4.5 Outlook and Discussion

We have argued that that no single combination of these paradigms is powerful and versatile enough for the general problem of object categorization. It is commonly believed that relying on a single paradigm of addressing the category phenomenon won't do justice to the overwhelming diversity encountered in categorial perception. In the previous sections we have motivated and demonstrated the merits of learning-based approaches that integrate different paradigms of representing visual categories.

A hierarchical approach has been presented that learns a representation that aggregates evidence on multiple levels. Local features have been evaluated to reach beyond purely visual categories. An approach to learning generative decompositions was developed to bridge across different representations paradigms from local to global. Finally, a even more high-level view on the data was proposed that manages to address and represented dynamics of objects like pedestrians in a learning based manner.

Despite the integration of different views and approaches to the phenomena of visual categories, we are aware of the limitations, that still have to be overcome. Although the shown approach manage to add flexibility and range over larger portion of the design space than previous approaches did, they still address particular limitations of previous approaches and yet have not integrated all our lessons learnt into one single approach.

A common theme in our approaches is to fit statistical models to the observed data, which use little to no supervision to reconstruct the unobserved hidden structure in the data. This leads to higher-level representations which we successfully use to solve more complex tasks such as multi-viewpoint modeling and temporal modeling. While hierarchical statistical models have been used for a long time, many of your results are founded on the more recent success of robustly learning such complex model with minimal intervention.

To sum up, we want to emphasize the importance these learning-based approaches have for moving towards a larger number of addressable categories. We contributed to achieving the desired adaptivity and flexibility, we believe to be a key ingredient for scalable vision system for object categorization. Although there are still unresolved computational as well as conceptional challenges, we are confident that the shown improvements on learning of hierarchies, features and dynamics will help us to reach this goal.

Ultimately, the way categories are stored, organized and related to each other determines the capabilities to learn, evolve and therefore to interact of the overall system. In Chapter 7 on multi-modal learning and in particular in our integration efforts in Chapter 9 and 10, we elaborate on how representation affects the capabilities of the overall system.

References

1. Burl, M.C., Perona, P.: Recognition of planar object classes. In: IEEE Conference on Computer Vision and Pattern Recognition (CVPR 1996), p. 223. IEEE Computer Society, San Francisco (1996)
2. Fergus, R., Perona, P., Zisserman, A.: Object class recognition by unsupervised scale-invariant learning. In: CVPR 2003, pp. 264–271 (2003)
3. Leibe, B., Seemann, E., Schiele, B.: Pedestrian detection in crowded scenes. In: IEEE Conference on Computer Vision and Pattern Recognition (CVPR 2005) [86], `http://cognitivesystems.org/cosybook/chap4.asp#leibe05cvpr`
4. Varma, M., Ray, D.: Learning the discriminative power-invariance trade-off. In: IEEE International Conference on Computer Vision (ICCV 2007). IEEE Computer Society, Rio de Janeiro (2007)
5. Lowe, D.G.: Distinctive image features from scale-invariant keypoints. IJCV 60(2), 91–110 (2004)
6. Dalal, N., Triggs, B.: Histograms of oriented gradients for human detection. In: IEEE International Conference on Computer Vision (ICCV 2005) [87]

7. Leibe, B., Leonardis, A., Schiele, B.: Robust object detection with interleaved categorization and segmentation. International Journal of Computer Vision (IJCV) 77(1-3), 259–289 (2008), http://cognitivesystems.org/cosybook/chap4.asp#Leibe05c
8. Viola, P., Jones, M.: Rapid object detection using a boosted cascade of simple features. In: IEEE Conference on Computer Vision and Pattern Recognition (CVPR 2001), pp. 511–518. IEEE Computer Society, Kauai (2001)
9. Winn, J.M., Jojic, N.: Locus: Learning object classes with unsupervised segmentation. In: IEEE International Conference on Computer Vision (ICCV 2005) [87], pp. 756–763
10. Weber, M., Welling, M., Perona, P.: Unsupervised learning of models for recognition. In: Vernon, D. (ed.) ECCV 2000. LNCS, vol. 1842, pp. 18–32. Springer, Heidelberg (2000)
11. Sivic, J., Russell, B.C., Efros, A.A., Zisserman, A., Freeman, W.T.: Discovering objects and their locations in images. In: IEEE International Conference on Computer Vision (ICCV 2005) [87]
12. Fergus, R., Fei-Fei, L., Perona, P., Zisserman, A.: Learning object categories from google's image search. In: IEEE International Conference on Computer Vision (ICCV 2005) [87]
13. Ettinger, G.J.: Hierarchical object recognition using libraries of parameterized model sub-parts, Tech. rep., MIT (1987)
14. Tsotsos, J.K.: Analyzing vision at the complexity level. Behavioral and Brain Sciences 13(3), 423–469 (1990)
15. Mel, B.W., Fiser, J.: Minimizing binding errors using learned conjunctive features. Neural Computation 12(4), 731–762 (2000)
16. Amit, Y., Geman, D.: A computational model for visual selection. Neural Comp. 11(7), 1691–1715 (1999)
17. Amit, Y.: 2d Object Detection and Recognition: Models, Algorithms and Networks. MIT Press, Cambridge (2002)
18. Geman, S., Potter, D., Chi, Z.: Composition systems. Quarterly of Applied Mathematics 60(4), 707–736 (2002)
19. Fidler, S., Berginc, G., Leonardis, A.: Hierarchical statistical learning of generic parts of object structure. In: CVPR, pp. 182–189 (2006), http://cognitivesystems.org/cosybook/chap4.asp#s:fidler06
20. Fidler, S., Leonardis, A.: Towards scalable representations of visual categories: Learning a hierarchy of parts. In: CVPR 2007 (2007), http://cognitivesystems.org/cosybook/chap4.asp#s:fidler07
21. Mikolajczyk, K., Leibe, B., Schiele, B.: Multiple object class detection with a generative model. In: IEEE Conference on Computer Vision and Pattern Recognition (CVPR 2006) [88], pp. 26–36, http://cognitivesystems.org/cosybook/chap4.asp#Mikolajczyk06c
22. Torralba, A., Murphy, K.P., Freeman, W.T.: Sharing visual features for multiclass and multiview object detection. IEEE Trans. Pattern Analysis and Machine Intelligence 29(5)
23. Jaakkola, T.S., Haussler, D.: Exploiting generative models in discriminative classifiers. In: Proceedings of the 1998 conference on Advances in neural information processing systems II, pp. 487–493. MIT Press, Cambridge (1999)

24. Fritz, M., Leibe, B., Caputo, B., Schiele, B.: Integrating representative and discriminant models for object category detection. In: IEEE International Conference on Computer Vision (ICCV 2005) [87], `http://cognitivesystems.org/cosybook/chap4.asp#Fritz05`
25. Sudderth, E., Torralba, A., Freeman, W., Willsky, A.: Learning hierarchical models of scenes, objects, and parts. In: ICCV 2005, pp. 1331–1338 (2005)
26. Ommer, B., Buhmann, J.M.: Learning the compositional nature of visual objects. In: CVPR 2007 (2007)
27. Fleuret, F., Geman, D.: Coarse-to-fine face detection. IJCV 41(1/2), 85–107 (2001)
28. Fukushima, K., Miyake, S., Ito, T.: Neocognitron: a neural network model for a mechanism of visual pattern recognition. IEEE Systems, Man and Cybernetics 13(3), 826–834 (1983)
29. Serre, T., Wolf, L., Bileschi, S., Riesenhuber, M., Poggio, T.: Object recognition with cortex-like mechanisms. PAMI 29(3), 411–426 (2007)
30. Scalzo, F., Piater, J.H.: Statistical learning of visual feature hierarchies. In: Workshop on Learning, CVPR (2005)
31. Ullman, S., Epshtein, B.: Visual Classification by a Hierarchy of Extended Features. In: Ponce, J., Hebert, M., Schmid, C., Zisserman, A. (eds.) Toward Category-Level Object Recognition. LNCS, vol. 4170, pp. 321–344. Springer, Heidelberg (2006)
32. Agarwal, A., Triggs, B.: Hyperfeatures - multilevel local coding for visual recognition. In: Leonardis, A., Bischof, H., Pinz, A. (eds.) ECCV 2006. LNCS, vol. 3951, pp. 30–43. Springer, Heidelberg (2006)
33. Ranzato, M.A., Huang., F.-J., Boureau, Y.-L., LeCun, Y.: Unsupervised learning of invariant feature hierarchies with applications to object recognition. In: CVPR 2007 (2007)
34. Bienenstock, E., Geman, S.: Compositionality in neural systems. In: Arbib, M. (ed.) The Handbook of Brain Theory and Neural Networks, pp. 223–226. MIT Press, Cambridge (1995)
35. Zhu, S., Mumford, D.: Quest for a stochastic grammar of images. Foundations and Trends in Computer Graphics and Vision 2(4), 259–362 (2007)
36. Califano, A., Mohan, R.: Multidimensional indexing for recognizing visual shapes. Pattern Analysis and Machine Intelligence 16(4), 373–392 (1994)
37. Tsunoda, K., Yamane, Y., Nishizaki, M., Tanifuji, M.: Complex objects are represented in macaque inferotemporal cortex by the combination of feature columns. Nature Neuroscience (4), 832–838 (2001)
38. Brincat, S., Connor, C.: Dynamic shape synthesis in posterior inferotemporal cortex. Neuron 49(1), 17–24 (2006)
39. Barlow, H.B.: Cerebral cortex as a model builder. In: Rose, D., Dobson, V. (eds.) Models of the Visual Cortex, pp. 37–46. John Wiley, Chichester (1985)
40. Rolls, E.T., Deco, G.: Computational Neuroscience of Vision. Oxford Univ. Press, Oxford (2002)
41. Edelman, S., Intrator, N.: Towards structural systematicity in distributed, statically bound visual representations. Cognitive Science 27, 73–110 (2003)
42. Fei-Fei, L., Fergus, R., Perona, P.: Learning generative visual models from few training examples: an incremental bayesian approach tested on 101 object categories. In: IEEE CVPR 2004, Workshop on Generative-Model Based Vision (2004)

43. Mutch, J., Lowe, D.G.: Multiclass object recognition with sparse, localized features. In: CVPR 2006, pp. 11–18 (2006)
44. Wolf, L., Bileschi, S., Meyers, E.: Perception strategies in hierarchical vision systems. In: CVPR 2006, pp. 2153–2160 (2006)
45. Csurka, G., Dance, C., Fan, L., Willarnowski, J., Bray, C.: Visual categorization with bags of keypoints. In: SLCV (2004)
46. Mundy, J.L.: Object recognition in the geometric era: A retrospective. In: Ponce, J., Hebert, M., Schmid, C., Zisserman, A. (eds.) Toward Category-Level Object Recognition. LNCS, vol. 4170, pp. 3–28. Springer, Heidelberg (2006)
47. Ferrari, V., Fevrier, L., Jurie, F., Schmid, C.: Groups of adjacent contour segments for object detection, Rapport De Recherche Inria
48. Opelt, A., Pinz, A., Zisserman, A.: A boundary-fragment-model for object detection. In: Leonardis, A., Bischof, H., Pinz, A. (eds.) ECCV 2006, Part II. LNCS, vol. 3952, pp. 575–588. Springer, Heidelberg (2006)
49. Stark, M., Schiele, B.: How good are local features for classes of geometric objects. In: ICCV (2007), http://cognitivesystems.org/CoSyBook/chap4.asp#stark07iccv
50. Berg, A.C., Malik, J.: Geometric blur for template matching. In: CVPR, pp. 607–614 (2001)
51. Belongie, S., Malik, J., Puzicha, J.: Shape context: A new descriptor for shape matching and object recognition. In: NIPS, pp. 831–837 (2000)
52. Mikolajczyk, K., Schmid, C.: A performance evaluation of local descriptors. PAMI 27(10), 1615–1630 (2005)
53. Mikolajczyk, K., Schmid, C.: Scale & affine invariant interest point detectors. IJCV 60(1), 63–86 (2004)
54. Mikolajczyk, K., Tuytelaars, T., Schmid, C., Zisserman, A., Matas, J., Schaffalitzky, F., Kadir, T., Gool, L.J.V.: A comparison of affine region detectors. IJCV 65(1-2), 43–72 (2005)
55. Kadir, T., Zisserman, A., Brady, M.: An affine invariant salient region detector. In: Pajdla, T., Matas, J(G.) (eds.) ECCV 2004. LNCS, vol. 3021, pp. 228–241. Springer, Heidelberg (2004)
56. Ferrari, V., Tuytelaars, T., Gool, L.J.V.: Object detection by contour segment networks. In: Leonardis, A., Bischof, H., Pinz, A. (eds.) ECCV 2006. LNCS, vol. 3953, pp. 14–28. Springer, Heidelberg (2006)
57. Martin, D.R., Fowlkes, C., Malik, J.: Learning to detect natural image boundaries using local brightness, color, and texture cues. PAMI 26(5), 530–549 (2004)
58. Berg, A.C., Berg, T.L., Malik, J.: Shape matching and object recognition using low distortion correspondences. In: CVPR 2005 (2005)
59. Morrone, M., Burr, D.: Feature detection in human vision: a phase dependent energy model. Proc. Royal Soc. London Bulletin, 221–245 (1988)
60. Harris, C., Stephens, M.J.: A combined corner and edge detector. In: Alvey Conference, pp. 147–152 (1988)
61. Mikolajczyk, K., Leibe, B., Schiele, B.: Local features for object class recognition. In: ICCV [87], pp. 1792–1799, http://cognitivesystems.org/cosybook/chap4.asp#conf/iccv/MikolajczykLS05
62. McCallum, A., Nigam, K.: A comparison of event models for naive bayes text classification. In: AAAI, Workshop on Learning for Text Categorization (1998)
63. Torralba, A., Murphy, K., Freeman, W.: Sharing features: efficient boosting procedures for multiclass object detection. In: CVPR (2004)

64. Hofmann, T.: Unsupervised learning by probabilistic latent semantic analysis. Machine Learning
65. Blei, D.M., Ng, A.Y., Jordan, M.I.: Latent dirichlet allocation. Journal of Machine Learning Research
66. Griffiths, T.L., Steyvers, M.: Finding scientific topics. In: PNAS USA
67. Steyvers, M., Griffiths, T.L.: Probabilistic topic models. In: Handbook of Latent Semantic Analysis. Lawrence Erlbaum Associates, Mahwah (2007)
68. Everingham, M., Zisserman, A., Williams, C.K.I., Van Gool, L.: The PASCAL Visual Object Classes Challenge 2006 (VOC 2006) (2006), http://www.pascal-network.org/challenges/VOC/voc2006/results.pdf
69. Chang, C.-C., Lin, C.-J.: LIBSVM: a library for support vector machines (2001) Software, http://www.csie.ntu.edu.tw/~cjlin/libsvm
70. Chum, O., Zisserman, A.: An exemplar model for learning object classes. In: IEEE Conference on Computer Vision and Pattern Recognition (CVPR 2007). IEEE Computer Society, Minneapolis (2007)
71. Fritz, M., Schiele, B.: Decomposition, discovery and detection of visual categories using topic models. In: IEEE Conference on Computer Vision and Pattern Recognition (CVPR 2008) [89] (to appear), http://cognitivesystems.org/cosybook/chap4.asp#fritz08cvpr
72. Leibe, B., Seemann, E., Schiele, B.: Pedestrian detection in crowded scenes. In: IEEE Conference on Computer Vision and Pattern Recognition (CVPR 2005) [86], http://cognitivesystems.org/cosybook/chap4.asp#leibe05cvpr
73. Lawrence, N.D., Moore, A.J.: Hierarchical Gaussian process latent variable models. In: ICML 2007 (2007)
74. Felzenszwalb, P.F., Huttenlocher, D.P.: Pictorial structures for object recognition. IJCV 61, 55–79 (2007)
75. Williams, C.K.I., Allan, M.: On a connection between object localization with a generative template of features and pose-space prediction methods. Tech. Rep. EDI-INF-RR-0719, University of Edinburgh (2006)
76. Urtasun, R., Fleet, D.J., Fua, P.: 3D people tracking with Gaussian process dynamical models. In: IEEE Conference on Computer Vision and Pattern Recognition (CVPR 2006) [88]
77. Wang, J.M., Fleet, D.J., Hertzmann, A.: Gaussian process dynamical models. In: NIPS 2005 (2005)
78. Sminchisescu, C., Kanaujia, A., Metaxas, D.N.: BM^3E: Discriminative density propagation for visual tracking. PAMI 29, 2030–2044 (2007)
79. Lawrence, N.D.: Probabilistic non-linear principal component analysis with Gaussian process latent variable models. JMLR 6, 1783–1816 (2005)
80. Deutscher, J., Reid, I.: Articulated body motion capture by stochastic search. IJCV 61, 185–205 (2005)
81. Demirdjian, D., Taycher, L., Shakhnarovich, G., Grauman, K., Darrell, T.: Avoiding the "streetlight effect": Tracking by exploring likelihood modes. In: IEEE International Conference on Computer Vision (ICCV 2005) [87]
82. Sigal, L., Black, M.J.: Measure locally, reason globally: Occlusion-sensitive articulated pose estimation. In: IEEE Conference on Computer Vision and Pattern Recognition (CVPR 2006) [88]
83. Ramanan, D., Forsyth, D.A., Zisserman, A.: Tracking people by learning their appearance. PAMI 29, 65–81 (2007)
84. Grochow, K., Martin, S.L., Hertzmann, A., Popovic, Z.: Style-based inverse kinematics. In: SIGGRAPH (2004)

85. Andriluka, M., Roth, S., Schiele, B.: People-tracking-by-detection and people-detection-by-tracking. In: IEEE Conference on Computer Vision and Pattern Recognition (CVPR 2008) [89] (to appear)
86. IEEE Computer Society, San Diego, CA, USA (2005)
87. IEEE Computer Society, Beijing, China (2005)
88. IEEE Computer Society, New York, NY, USA (2006)
89. IEEE Computer Society, Anchorage, Alaska, USA (2008) (to appear)

5

Semantic Modelling of Space

Andrzej Pronobis[1], Patric Jensfelt[1], Kristoffer Sjöö[1], Hendrik Zender[2], Geert-Jan M. Kruijff[2], Oscar Martinez Mozos[3], and Wolfram Burgard[3]

[1] Royal Institute of Technology (KTH), Centre for Autonomous Systems, Stockholm, Sweden
{pronobis, patric, krsj}@csc.kth.se
[2] DFKI GmbH, Saarbrücken, Germany
{zender, gj}@dfki.de
[3] Albert-Ludwigs-Universität Freiburg, Department of Computer Science, Freiburg, Germany
{omartine, burgard}@informatik.uni-freiburg.de

5.1 Introduction

A cornerstone for robotic assistants is their understanding of the space they are to be operating in: an environment built by people for people to live and work in. The research questions we are interested in in this chapter concern spatial understanding, and its connection to acting and interacting in indoor environments. Comparing the way robots typically perceive and represent the world with findings from cognitive psychology about how humans do it, it is evident that there is a large discrepancy. If robots are to understand humans and vice versa, robots need to make use of the same concepts to refer to things and phenomena as a person would do. Bridging the gap between human and robot spatial representations is thus of paramount importance.

A spatial knowledge representation for robotic assistants must address the issues of human-robot communication. However, it must also provide a basis for spatial reasoning and efficient planning. Finally, it must ensure safe and reliable navigation control. Only then can robots be deployed in semi-structured environments, such as offices, where they have to interact with humans in everyday situations.

In order to meet the aforementioned requirements, i.e. robust robot control and human-like conceptualization, in CoSy, we adopted a spatial representation that contains maps at different levels of abstraction. This stepwise abstraction from raw sensory input not only produces maps that are suitable for reliable robot navigation, but also yields a level of representation that is similar to a human conceptualization of spatial organization. Furthermore, this model provides a richer semantic view of an environment that permits the robot to do spatial categorization rather than only instantiation.

H.I. Christensen et al. (Eds.): Cognitive Systems, COSMOS 8, pp. 165–221.
springerlink.com

This approach is at the heart of the Explorer demonstrator (cf. Chapter 10), which is a mobile robot capable of creating a conceptual spatial map of an indoor environment. In the present chapter, we describe how we use multimodal sensory input provided by a laser range finder and a camera in order to build more and more abstract spatial representations.

5.1.1 Related Work

Research in spatial representations for mobile robots has yielded different multi-layered environment models. Vasudevan *et al.* [1] suggest a hierarchical probabilistic representation of space based on objects. The work by Galindo *et al.* [2] presents an approach containing two parallel hierarchies, spatial and conceptual, connected through anchoring. Inference about places is based on objects found in them. Furthermore, the *Hybrid Spatial Semantic Hierarchy* (HSSH), introduced by Beeson *et al.* [3], allows a mobile robot to describe the world using different representations, each with its own ontology.

Other different cognitively inspired approaches to robot navigation convey route descriptions from a technically naïve user to a mobile robot. These approaches need not necessarily rely on an exact global self-localization, but rather require the execution of a sequence of strictly local, well-defined behaviors in order to iteratively reach a target position. Kuipers [4] presents the *Spatial Semantic Hierarchy* (SSH). Alternatively, the *Route Graph* model is introduced by Krieg-Brückner *et al.* [5]. Both theories propose a cognitively inspired multi-layered representation of the *map in the head*, which is at the same time suitable for robot navigation.

Additionally, several approaches on mobile robotics extend metric maps of indoor environments with semantic information. The work by Diosi *et al.* [6] creates a metric map through a guided tour. The map is then segmented according to the labels given by the instructor. Martinez Mozos *et al.* [7] extract a topological semantic map from a metric one using supervised learning. Alternatively, Friedman *et al.* [8] use *Voronoi Random Fields* for extracting the topologies. Although these works use range measurements as main input data, other sensors have been used for similar tasks. Torralba *et al.* [9] use processed images to distinguish between different place categories in the environment. Pronobis *et al.* [10] also use vision to recognize the different places that form an indoor environment. Finally, the combination of different sensory modalities can improve the recognition, as shown in Rottmann *et al.* [11] and Pronobis *et al.* [12]. More detailed review of different approaches to place classification can be found in Section 5.8.

The multi-layered representation presented in this chapter differs from the previous work primarily in the level of integration achieved. First, each of the layers of the representation advances the state of the art in its corresponding area. Second, the advanced techniques are combined into a single, coherent model, representing the world at various levels of abstraction (e.g. metric, topological, semantic, conceptual) based on information coming from multiple sources (vision, range sensors, verbal cues etc.). In particular, the model integrates the approaches of [7] and [12] for the semantic classification of places with visual object search algorithms [13] and the metric mapping based on the M-Space representation [14]. Moreover, the representation is designed for human-robot interaction and the models generated using the aforementioned techniques are combined with a common-sense ontology of an indoor environment. This bridges the gap in spatial understanding between the robot and humans and allows to include information extracted from verbal cues into the representation.

5.1.2 Outline

The rest of this chapter is organized as follows. First, we highlight the background for the research on spatial representations for mobile robots (Section 5.2). Then, we provide an overview of our spatial model (Section 5.3) and describe each of the levels of representation in detail (Sections 5.4–5.6). Finally, we present the algorithms used to augment the representation with semantic object and place information (Sections 5.7 and 5.8, respectively) and report results of performed experiments (Section 5.9). We conclude the chapter with a brief summary in Section 5.10.

5.2 Background

An approach to endowing autonomous robots with a human-like conceptualization of space inherently needs to take into account research in sensor-based mapping and localization for robots as well as findings about human spatial cognition.

Research in cognitive psychology addresses the inherently qualitative nature of human spatial knowledge. In accordance with experimental studies, it is nowadays generally assumed that humans adopt a partially hierarchical representation of spatial organization [15, 16]. The basic units of such a qualitative spatial representation are topological regions [17], which correspond to more or less clearly bounded spatial areas. The borders may be defined physically, perceptually, or may be purely subjective to the human. It has been shown that even in natural environments without any clear physical or perceptual boundaries, humans decompose space into topological hierarchies by clustering salient landmarks [18].

Aside from the functionality of the cognitive map, another relevant question from cognitive science is how people categorize spatial structures. Categories determine how people can interact with, and linguistically refer to entities in the world. Basic-level categories represent the most appropriate name for a thing or an abstract concept. The basic-level category of a referent is assumed to provide enough information to establish equivalence with other members of the class, while distinguishing it from non-members [19, 20]. We draw from these notions when categorizing the spatial areas in the robot's conceptual map. We are specifically concerned with determining appropriate properties that allow us to meaningfully refer to spatial entities in a situated dialogue between the robot and its user.

5.3 Overview of the Spatial Model

This section gives an overview of the multi-layered conceptual spatial model adopted in the CoSy project. The model forms a basis for spatial understanding, reasoning, navigation, and human-robot interaction in the integrated robotic system. In this framework, the space is modeled at different levels of abstraction that range from low-level metric maps for robot localization and navigation to a conceptual layer that provides a human-like space decomposition and categorization. An illustration of the model and the main levels of representation is presented in Figure 5.1.

The lower layers of the model are derived from sensory input. These layers combine a metrical, line-based representation of spatial structure modeling occupied space, and a navigation graph of virtual markers modeling free space. Different methods are used to gradually construct more abstract representations. On higher levels, we regard topological regions and spatially situated objects as the primitive entities of spatial conceptualization. The robot must be able to assign human concepts to such spatial entities in order to meaningfully act in, and talk about, an environment. Many places in indoor environments are designed in a way that makes their structure, general appearance, and spatial layout afford specific actions; corridors and staircases are examples of this. Other places afford more complex actions provided by objects that are located there. For instance, the concept of a living room applies to rooms that are suited for resting. Having a rest, in turn, can be afforded by certain objects, such as couches or TV sets. The representation allows for combining cues provided by the basic geometrical shape, general visual appearance, perceived objects, and possibly situated dialogue to provide reliable semantic descriptions of space.

The rest of this section provides an overview of each of the layers of the spatial representation.

5.3.1 Metric Map

The lowest level of the spatial model is represented by a metric map. The map encodes spatial boundaries in the environment using lines as basic spatial primitives and supports self-localization of the robot. It is anchored to a metric world co-ordinate system, which is also used as a basis for the higher level representations. The positions of lines as well as of a robot on the metric map are established and maintained by a module for Simultaneous Localization

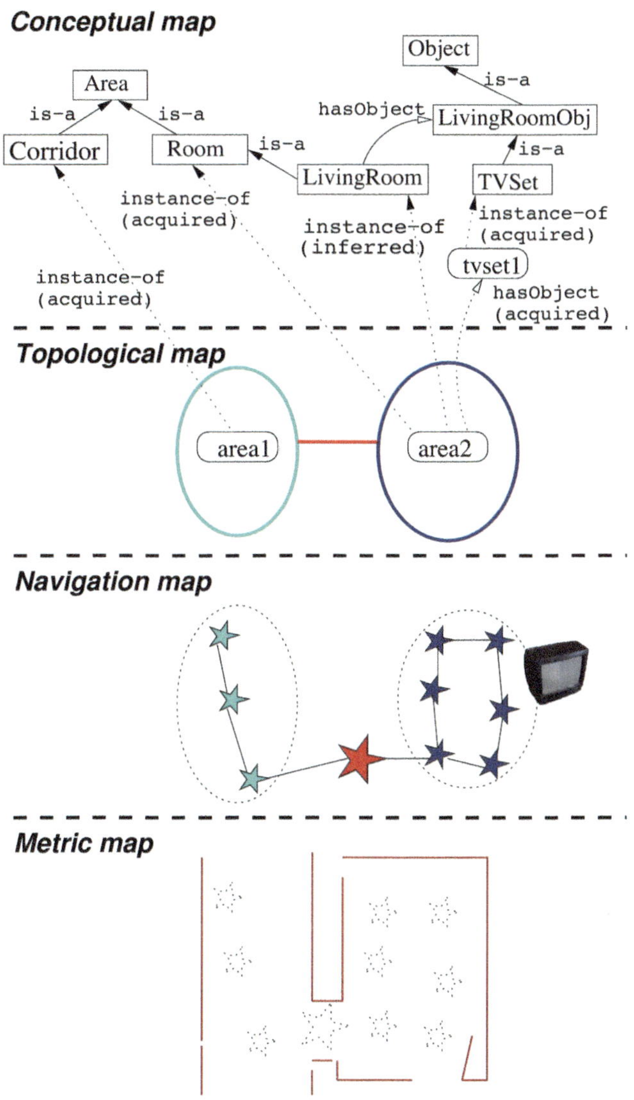

Fig. 5.1. The multi-layered structure of the conceptual spatial model

and Mapping (SLAM) [14]. Section 5.4 gives more details about the applied SLAM algorithm and other approaches that have been investigated.

In comparison to a representation based on an occupancy grid [21], the line-based map does not directly provide a description of the free space but rather of the surfaces in the environment that can be described by lines. However, since the global co-ordinate system of the metric map is purely internal to the robot and humans are not able to easily evaluate quantitative spatial descriptions, the metric map alone is not sufficient to support human-robot dialogues.

5.3.2 Navigation Map

The navigation map provides the next layer of representation, which establishes a model of free space and its connectivity, i.e. reachability. It is based on the notion of a roadmap of virtual free-space markers as described in [22, 23] and implemented as a graph of nodes that are anchored to the metric map. As the robot navigates through the environment, a marker or navigation node is dropped whenever the robot has traveled a certain distance from the closest existing node. Nodes are connected following the order in which they were visited. More information about the navigation graph can be found in Section 5.5.

It is also in the navigation graph that the spatial representation is augmented with semantic information about the environment. First, the semantic category of a place is extracted using a place classification algorithm [12] and stored in the nodes. Doors detected in the environment are represented as doorway nodes and added to the graph. Finally, objects detected by an object search component [13] are also stored on this level of the map. Section 5.7 and Section 5.8 present the algorithms used to detect objects and extract semantic place information from the geometry and appearance of the environment.

5.3.3 Topological Map

The navigation map provides a basis for further, topological abstractions. A topological map consisting of connected areas is built by segmenting the navigation graph into interconnected sets of nodes separated by recognized doors (doorway nodes). This layer of abstraction corresponds to human-like qualitative segmentation of an indoor space into distinct regions (e.g. rooms). On this level, semantic place information extracted and accumulated over entire regions is evaluated to determine appropriate semantic categories for areas in the topological graph. More information about this process can be found in Section 5.5.

5.3.4 Conceptual Map

On the highest level of abstraction, the system is endowed with a conceptual map. The conceptual map builds up a further interpretation of spatial organization. The topological areas together with their place categories form the basic spatial entities. A description logic-based reasoner is used to infer more fine-grained semantic information for the areas. The reasoner integrates knowledge about areas and observed objects with a common-sense ontology of an indoor environment. This ontology represents a taxonomy of areas and objects and the relations between objects and areas. Since there is a strong connection between typical objects found in an area and the semantic category of the area, this layer can also be used to constrain expectations about which objects are likely to be observed, given that the basic-level concept of an area is known (for example through a situated dialogue with a human user). Section 5.6 provides more details about the conceptual map.

5.4 Metric Mapping

This section gives details about the metric mapping algorithms that were investigated in CoSy and used to maintain the metric representation in the spatial model. Metric maps can be represented in many ways. The two most common approaches are based on occupancy grids [24, 25, 26, 27, 28, 29] and features [30, 31, 32, 33, 14]. Occupancy grids discretize the world into cells. Each grid cell holds a value representing the probability that the corresponding area in the environment is occupied. Feature-based maps on the other hand abstract the sensor data into a set of features. In a structured environment – of which most office environments are examples – lines, corners and edges are common features. The features can be parameterized by, for instance, their color, length, width, position, etc. One of the main advantages of this type of representation is that it requires very few assumptions about the world, whereas one has to settle on a set of features to parameterize the map beforehand.

Mapping (building a model of the environment) and localization (finding the position in the environment) are often treated as two separate problems. Maps were made assuming that the position is known and the position is calculated given a map. However, for an autonomous agent that explores an unknown environment these two task are intrinsically linked, and form a chicken-and-egg problem; to perform mapping one needs the position and to perform position one needs the map. This leads to Simultaneous Localization and Mapping (SLAM) which has been a thriving research area for more than a decade.

In this chapter we will focus on feature-based representations. A feature-based map can in general be written

$$\mathcal{M} = \{f_j \mid j = 1, \dots, M\}, \tag{5.1}$$

where f_j is a feature and M is the number of features in the map.

5.4.1 M-Space

A number of different types of features have been used for mapping. Depending on the type of application the model of the environment is 2D or 3D. For most indoor applications a 2D representation is used; navigation in cluttered environments often requires a 3D representation. When taking the step outdoors the world is less structured and it becomes increasingly likely that the ground is not flat which also calls for a 3D model.

To motivate the work with the so called M-Space representation, let us first consider how to represent a line segment. Such a line segment could for example offer a 2D abstraction of a wall in an indoor environment. Four parameters are needed to fully specify the line segment. On a real robot, the sensors may not be able to detect its end points; even if the end points are within the range of the sensors they are often hard to detect due to occlusions. This implies that a measurement typically only constrains the position of the sensor to a certain distance and relative angle with respect to the wall. In other words, all dimensions of a wall are typically not constrained by one single measurement.

There are a number of ways to represent a line segment. To name a few; slope and intersection ($y = k_y x + m_y$), end points, distance and direction (infinite line [34]), center point, length and orientation. All of these suffer one or more problems, some of which are addressed by the so called SP-model [35].

A characteristic property of SP-model is that each feature element has its own local reference frame. The frame of reference is chosen with the axes along the directions of symmetry. A line, for example, has the x-axis along the direction of the line. A plane will have a normal that coincides with the z-axis. The main advantage of using a local frame is that the description of the uncertainty can be made independent of the global position of the features. This avoids lever-arm effects that can result when for example using direction and orientation to represent a line. The local frames also help to make frame transformations and differentiations thereof more standardized. Another key concept in the SP-model is the so called *binding matrix*, B. The binding matrix is a row-selection matrix. The self-binding matrix selects the DOFs that are not part of the motion symmetry, i.e. the DOFs of a feature that are constrained and have probabilistic information attached to them. The binding matrices offer a machinery for making partial observations of a feature. This is useful, for example, when observing a single point on a line. A limitation with the SP-model is that one has to attach a frame to all features. For some types, such as lines, it is difficult to model the extent, e.g. the length, in a probabilistic way within the SP-model framework. In [36], the length of lines is estimated and modeled but it relies on always detecting both end points at the same time and making a *direct measurement* of the length. An *indirect measurement* is not possible as the origin of the reference frame cannot be observed, not being attached to anything observable, but just defined to be in the middle of the line.

The so called M-Space representation builds on the SP-model[1]. It also attaches a local frame to each feature element and allows for a generic treatment of many types of features. The measurement subspace, or M-space, is

[1] For a detailed description see [14] from which this text is derived.

an abstraction of the measured subspace of the feature space that reflects symmetries and constraints. The idea is that the features are parameterized to fully specify their location and extent (the feature space) but that they can be initialized in a subspace corresponding to the information provided by the sensors.

For example, when representing a line segment the extent is accommodated for in the representation even though only the distance to and the orientation of the line is known initially. We cannot represent the uncertainty with regard to changes in the coordinates along the length of the line by a Gaussian distribution. However, the uncertainty regarding changes perpendicular to the line and regarding the orientation can be approximated by a Gaussian. Let $\delta\mathbf{x}_p$ denote the M-space corresponding to a small change in feature coordinates $\delta\mathbf{x}_f$. Here the subscript, p, stands for small perturbations in the M-space. The actual values of the M-space coordinates, $\mathbf{x}_p$, are never needed or considered. It is only the changes to them that enter into the estimates. These changes are used to make adjustments to the feature coordinates $\mathbf{x}_f$. The uncertainty estimate is an estimate of the distribution of $\delta\mathbf{x}_p$ values around a mean of zero. The adjustments to the feature coordinates are made to maintain this zero-mean. No re-centering step like in the SP-model is required with this view of the uncertainty. The uncertainty is defined in a frame attached to the feature and can be projected into the global frame using the current global coordinates of the feature. The statistics are represented in an analytic way rather than in the strict geometric sense of the SP-model. In most cases, the differences are in the second order corrections to the covariances.

The relation between the feature space coordinates and the M-space coordinates is defined by a projection matrix, $B(\mathbf{x}_f)$, similar to the binding matrix in the SP-model. The projection matrix relates small changes $\delta\mathbf{x}_p$ to small changes $\delta\mathbf{x}_f$. An important difference to the binding matrix is that the projection matrix is a function of the individual feature and changes with time. The rather involved re-centering step in the SP-model is replaced by re-evaluating the projection matrices.

A common issue in feature-based SLAM is that one cannot initialize a feature after the first observation. A single observation typically does not contain enough information to do so reliably. Among the reasons behind this we find for example

- The entire feature is not detected at once.
  - In the case of a line segment, the end points might not have been detected if the line is partially occluded or long.
  - When using monocular vision, only the bearing to the feature can be initialized from a single image.
- Measurements are noisy. Even though a feature is fully observed it is good practice to get a second opinion from new measurement data to reject false measurements.

The M-space representation offers a solution to these problems by allowing the M-space dimensionality to change over time. Features are typically initialized with zero M-space dimensions and with time, as more information is gathered, more dimensions will be added. Consider mapping a wall as a line segment. The life cycle of the line segment might be

1. First detection: feature initialized with 0 M-space dimensions.
2. Re-observed N times: the wall's existence and quality are confirmed. The distance and orientation of the wall added to the M-space.
3. Start point detected: M-space dimensionality goes up to 3
4. End point detected: the feature reaches full dimensionality of 4.

The importance of the ability to let the dimensions of a feature grow over time is well illustrated by a horizontal line feature observed by a camera. A single image does not contain information to pinpoint the location of a feature. The assumption that the line feature is horizontal implies that a single observation will be enough to provide information about the relative orientation of the robot. That is, even if the robot moves parallel under the line and is unable to use triangulation to fix the position of the line in space the observations of the line can help reduce the angular uncertainty of the robot. This is useful in, for example, a corridor where the motion often is parallel to the linear structures found in the ceiling.

5.4.2 Single Camera Bearing Only SLAM

Range sensors such as laser scanners are still the most common sensors for systems that perform mapping and localization in settings where robustness is key e.g. in industrial applications. However, cameras are becoming increasingly interesting as the performance keeps increasing while the price keeps going down due to the large demand from the consumer market e.g. for mobile phones. One of the main advantages of a camera over a range sensor is that the information that it provides is so much richer and not limited to a few hundred distance measurements typically lying in a plane. The hard part is to get the information out from the image.

There is a rich literature on single camera bearing-only SLAM. Most of these use point features extracted from the image to define landmarks in the map [37, 38]. Given standard feature detectors such as SIFT [39], there can be several hundreds of features and thus potential landmarks per frame. A single SIFT descriptor is not discriminative enough in itself, especially in man-made environments where structures like corners give raise to many SIFT points with very similar descriptors. When used for object recognition [39] it is a combination of descriptors extracted from the object that provide the discriminative strength. This idea is used in the vSLAM approach [40] where the SIFT points are used to recognize places. In [41], we present a framework where a few stable (over time and in space) SIFT features are identified and used as landmarks (location features). The rest of the SIFT features are

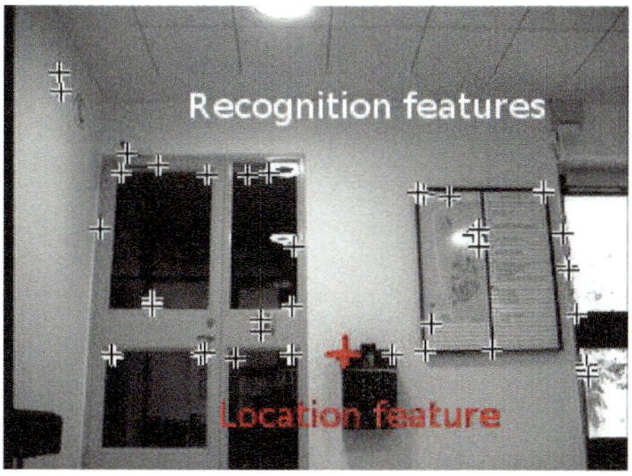

Fig. 5.2. A few stable features are identified and used to define the location of landmarks. The rest of the features are used to improve the matching/recognition of these.

used to strengthen the matching of features (recognition features). This is illustrated in Figure 5.2. When matching against a landmark, the matching is performed not only with the feature defining the position of the landmark but also the rest of the feature extracted from the two frames. This greatly improves the robustness of the matching.

5.4.3 Using Visual Attention for SLAM

Choosing useful landmarks which are easy to track, stable over several frames, and easily re-detectable when returning to a previously visited location is important in order to get a visual SLAM system working. Getting few but good, rather than many and bad landmarks reduces the issue of complexity. In [42, 43, 44], we suggest the application of a biologically motivated attention system [45] to find salient regions in images. Attention systems are designed to favor regions with a high uniqueness such as a red fire extinguisher on a white wall. Such regions are especially useful for visual SLAM because they are discriminative by definition and easy to track and re-detect. We show that salient regions have a considerably higher repeatability than Harris-Laplacians and SIFT key-points. Active gaze control is also used and has been shown to enhance the performance over using a statically mounted camera. The strategy to steer the camera consists of three behaviors: a *tracking* behavior identifies the most promising landmarks and prevents them from leaving the field of view. A *re-detection* behavior actively searches for expected landmarks to support loop-closing. Finally, an *exploration* behavior investigates regions with no landmarks, leading to a more uniform distribution of landmarks. The advantage of the active gaze control is to obtain more informative landmarks (e.g. with a better baseline), a faster loop closing, and a better distribution of landmarks in the environment. Figure 5.3 shows the difference in the robot pose uncertainty when driving the same trajectory with active camera control

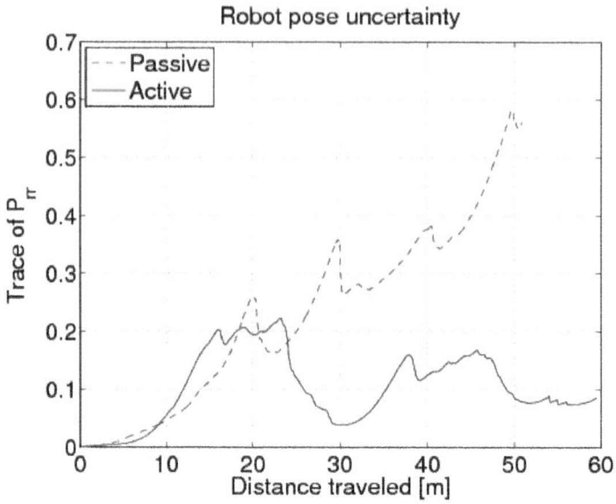

Fig. 5.3. A comparison of the robot pose uncertainty with camera control (active) or without (passive)

and without. This example illustrates that the active camera control allows the robot to reduce the uncertainty by seeing landmarks that it could otherwise not have seen. For more details please refer to [43, 44].

5.4.4 Visual Scans

One of the more popular and successful ways to do metric mapping is to use scan matching [46, 47, 29]. In a feature-based setting the scan can be considered to be a feature which is defined by the scan distances themselves. Building the map boils down to finding the position from which the scans were acquired in such a way that the laser scans align and for a consistent map.

In [48] we present an idea to use so called visual scans in much the same way as laser scans are used in scan matching. Using a stereo camera, a 3D point cloud is calculated by extracting and matching SIFT features in each image. This 3D point cloud forms the visual scan. Aside from the position, each point also has a descriptor saying something about the appearance which can be used for matching.

The map is defined by a number of reference scans. A reference scan is added when there is not enough overlap between the current visual scan and any of the other visual scans. The advantage of this representation is that it gives a very rich description of the environment (a dense point cloud which can have hundreds or even thousands of points) while the estimation problem only needs to deal with the parameters defining the position of the sensor (3 parameters in 2D and 6 in 3D, compared to $3N$ with N points treated independently). An example is shown in Figure 5.4.

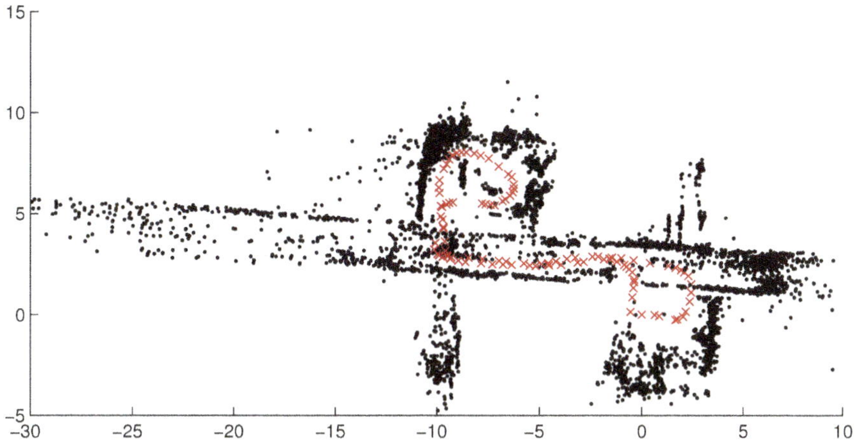

Fig. 5.4. A map with 126 visual reference scans. Together these scans contain 8333 points. This shows how the visual scans help define a dense representation at the same time as providing a low-dimensional estimation problem.

5.5 Navigation and Topological Maps

Above the bottom layer – the metric map – the spatial model contains the navigation and topological maps. As explained above, the metric map encodes boundaries in the environment and is used to ensure safe and reliable navigation and obstacle avoidance. In contrast, the navigation and topological layers encode more abstract information about the space accessible to the robot, particularly important from the functional point of view. This information is encoded in the form of graph-like structures in which the links represent connectivity between spatial entities at different spatial scales. The graph constituting the navigation map consists of nodes representing small unbounded free space regions in the environment. The topological representation, in its turn, models the indoor environment in terms of larger bounded areas connected by detected doors.

The representations play two main roles in the system. First, they discretize the continuous free space into a finite number of spatial units. These units are then used for tasks such as planning or interaction with the robot. For example, the high level task "go to the office", can be translated to the low level action "go to the closest navigation node attached to the topological node representing the office". If this position is occupied the robot can choose to go to the next navigation node. Discretization of space drastically reduces the number of combinations that have to be considered during the planning process.

The second important function of the navigation and topological representations is preserving additional information about the surroundings. Here,

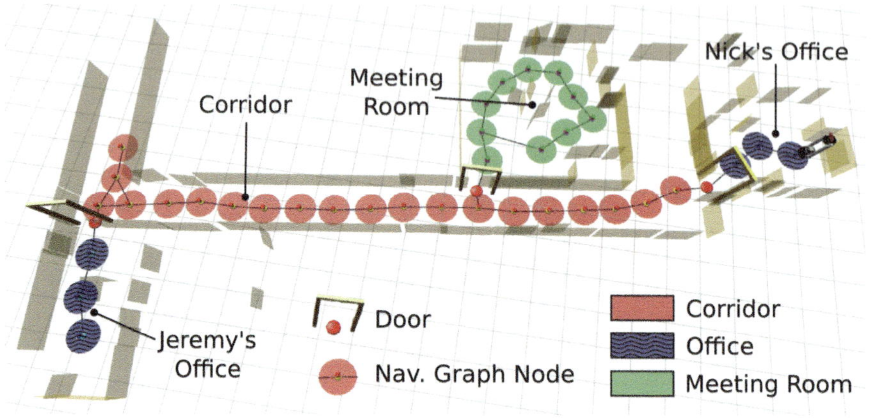

Fig. 5.5. An example of a navigation map overlayed on a metric map. The free space navigation nodes are represented by circles and are assigned to different topological areas based on the separation established by the doorway nodes. The colors of the nodes indicate the functional category of the areas as recognized by a place classification algorithm.

semantic information about places extracted from the sensory input is accumulated and stored together with navigation and topological nodes. Moreover, information about objects found in the environment is tied to nodes in the navigation graph. This is used to link the representations with the conceptual map and allows to refer to places in terms of their functional category or detected objects. As an example, the order "go to the TV", can be processed and translated into "got to the closest node from which you saw the TV".

The rest of this section focuses on the process of generating the navigation and topological representations based on the information encoded in the metric map and additional cues extracted from the sensory input.

5.5.1 Building the Navigation Graph

The navigation map provides the first discretization of the continuous space described by the model. It is represented in the form of a graph built as the robot explores the environment and is based on the notion of a roadmap of virtual free space markers [22, 23]. A free space navigation node is dropped whenever the robot has traveled a certain distance from the closest existing node (approximately 1 meter). Each node is anchored to the metric map and is assigned (x, y) co-ordinates. Nodes are connected following the order in which they were generated. This order is given by the trajectory that the robot follows during the map acquisition process. The final graph serves for planning and autonomous navigation in the already visited part of the environment. An example of a navigation graph overlayed on a metric map is presented in Figure 5.5. This simple representation proved to be very powerful during real-world experiments with the integrated system.

As the robot navigates through the environment, additional information about the surroundings collected on the way is assigned to the closest navigation nodes. Moreover, special doorway nodes are added to the navigation graph at the points where doors are detected in the environment (see Figure 5.5). As explained below, these nodes play an important role in building the spatial model.

5.5.2 Space Segmentation and Topological Graph

The structure of indoor environments allows for introducing larger scale and more abstract representations than the navigation graph. In such environments a room is an important concept. Different rooms can be associated with different owners and functionalities. Moreover, rooms are spatial entities commonly referred to in the natural language. The ability to segment space into rooms becomes crucial for an artificial mobile cognitive system. Therefore, as another layer, the spatial model builds a topological graph consisting of areas and links which represents rooms in the environment and their connectivity.

The structure of the topological graph is built based on the assumption that the transition between two areas happens through a door. This creates a human-like qualitative segmentation of an indoor space into distinct regions. A door detection algorithm is used to generate doorway nodes in the navigation graph whenever the robot passes through a narrow opening. The width of the opening is selected so that it matches typical doorways in the environment. Information about the door opening, such as width and orientation, is stored along with the detected position of the doorway in the doorway node. The doorway nodes are used to segment the navigation graph and assign navigation nodes to areas in the topological graph.

More complex door models such as those in [49, 50] can be used for more robust door detection. However, such models put additional constraints on how doors have to look to be recognized. The only assumption in the model described here is that the door is a narrow opening which the robot passes through. No assumptions are made regarding the door leaf (e.g. swinging or sliding) or special structure around the door. This can be beneficial for a robot that has to operate in different environments. An alternative would be to use a learning approach, such as in [51], where both visual features and the motion of the door are taken into account.

5.5.3 Adding Object Information

Objects and landmarks play an important role in understanding spatial structure. They are important cues that often determine the actions that can be performed in a particular area. Moreover, objects are nameable features commonly used in describing spatial locations. For this reason, visual search algorithms are used in the system to perform autonomous exploration and detection of objects typically found in indoor environments. Detailed information about the algorithms applied for this purpose can be found in Section 5.7.

The presented spatial model incorporates information about objects. This information is later used by the last conceptual layer. First, however, the objects must be tied to their spatial locations. This is the role of the object nodes which are connected with the navigation nodes of the navigation graph. The object nodes store information about the type of the recognized object and its metric location. The nodes are then linked to the closest navigation node.

5.5.4 Adding Semantic Place Information

Many places in an indoor environment can be characterized by semantic categories corresponding to their inherent functionality. Rooms constitute a good example as they can be categorized as offices, kitchens, meeting rooms etc. However, semantic descriptions can also be assigned to smaller regions such as a printer area in a corridor. The semantic place category is usually reflected in the objects located in that place, but also in the general appearance and geometrical layout.

One of the roles of the navigation nodes and topological areas in the spatial model is to store semantic information about the places to which they correspond. This information is used to link the lower layers modeling spatially allocated regions with the spatial concepts of the conceptual layer. A specialized component performing multi-modal place classification is used to extract semantic descriptions from the sensory input of a robot. Visual and laser range sensory data acquired in an environment are analyzed and compared to place models in order to produce beliefs about the place categories. In the simplest case, the component can be used to distinguish between two basic place categories: a corridor or a room (e.g. based on the clutteredness and geometric layout). Further specialization can be performed in the conceptual layer based on object information or situated dialogue. However, more specific place categories can also be recognized directly by the place classification system. More information about this process can be found in Section 5.8.

As will be shown through experiments in Section 5.9, the place classification system is able to classify a place with high accuracy given a single data sample (e.g. one image and laser scan) corresponding to only one viewpoint. However, in this case, the task is to provide a reliable and stable label for the whole region covered by a navigation node or a topological area. Since the sensors employed are not omnidirectional, it is necessary to accumulate and fuse the incoming information. However, the data that the robot gathers are not evenly spread over different viewpoints. On the contrary, in many cases the sensors receive a continuous stream of non-informative data (e.g. when the robot is parked close to a wall blocking the view). The system must be able to deal with such problems as temporary lack of informative cues, long-term occlusions or large variability affecting certain viewpoints.

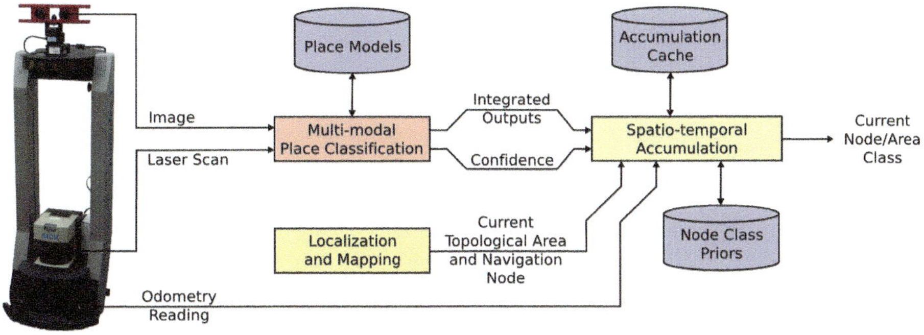

Fig. 5.6. Generating semantic labels for navigation nodes and topological areas using multi-modal place classification

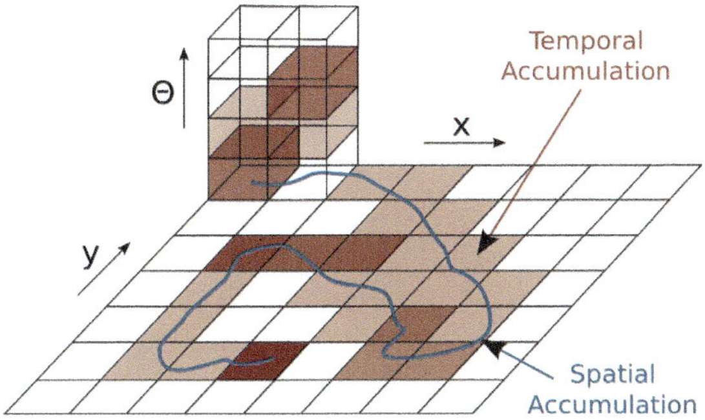

Fig. 5.7. Illustration of the spatio-temporal accumulation process. As the robot explores the environment, the beliefs collected on the way accumulate over time within the bin corresponding to the current pose (x, y, θ) and over space in different bins.

For this reason, the place information produced by the place classification algorithm is accumulated over time and space as presented in Figure 5.6. For each multi-sensory data sample, place classification provides a set of beliefs encoded as a vector of real-valued outputs (see Section 5.8 for details). The confidence of the final decision is also measured and provided by the classification component. The beliefs are fused using a confidence-based spatio-temporal accumulation algorithm. The algorithm relies on information about the current position on the navigation and topological map provided by the localization and mapping system. The spatio-temporal accumulation process is performed within the region covered by the current node and area. When the robot moves to a different node, the collected information is used to update the semantic label attached to the map and saved as a future prior. When the robot enters a location which was already explored, the previously stored beliefs are loaded and can be refined by further exploration.

The principle behind the confidence-based spatio-temporal accumulation algorithm is illustrated in Figure 5.7. As the robot explores the environment, it moves with a varying speed. The robot has information about its own movement provided by the wheel encoders (odometry). As errors accumulate over time, this information can only be used to estimate relative movement rather than absolute position. Although the accurate metric information could be used instead, odometry is sufficient for our application. The spatio-temporal accumulation process creates a sparse histogram along the robot pose trajectory described by the metric position (x, y) and heading (θ). The size of the histogram bins is adjusted so that each bin roughly corresponds to a single viewpoint. Then, as the robot moves, the beliefs about the current semantic category accumulate within the bins. An average of the outputs is calculated in a manner similar to the Discriminative Accumulation Scheme (DAS, [10]) used in the framework of cue integration. This is what we call the temporal accumulation. It prevents a single viewpoint from becoming dominant due to long-term observation. Since each viewpoint observed by the robot will correspond to a different bin, performing accumulation across the bins (this time spatially) allows for generating the final outputs to which each viewpoint contributes equally. In order to exclude most of the misclassifications before they get accumulated, the decisions are filtered based on the confidence value provided by the place classification component. Finally, the best hypothesis is calculated. It is assigned to the navigation node and topological area representing the spatial region over which the accumulation was performed.

5.6 Conceptual Map

The conceptual map provides the link between the low-level maps and the communication system used for situated human-robot dialogue by grounding linguistic expressions in representations of spatial entities, such as instances of rooms or objects. It is also in this layer that knowledge about the environment stemming from other modalities, such as object recognition and dialogue, is anchored to the metric and topological maps.

Based on the work by Zender [52], our system is endowed with a commonsense OWL ontology of an indoor environment (see Figure 5.8) that describes taxonomies (*is-a* relations) of room types and typical objects found therein through *has-a* relations. These conceptual taxonomies have been handcrafted and cannot be changed online. However, instances of the concepts are added to the ontology during run-time. Through fusion of *acquired* and *asserted* knowledge gathered in an interactive map acquisition process [53] and through the use of the *innate conceptual* knowledge, a reasoner can *infer* information about the world that is neither given verbally nor actively perceived. This way linguistic references to spatial areas can be generated.

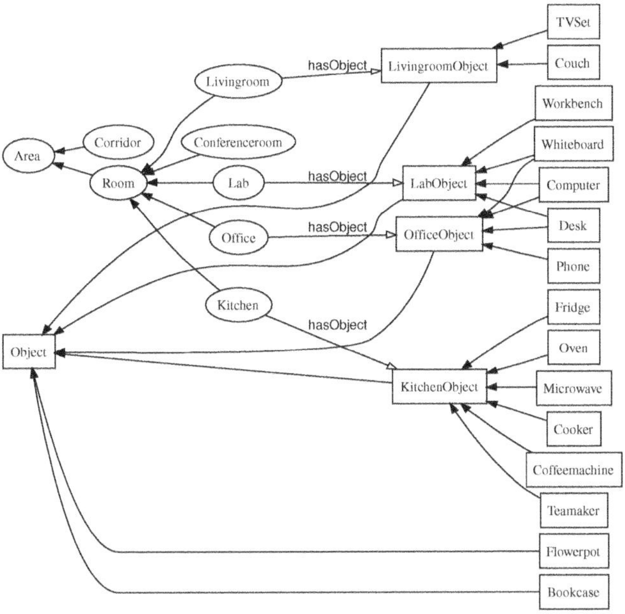

Fig. 5.8. Illustration of a part of the commonsense ontology of an indoor office environment. Solid arrows denote the taxonomical `is-a` relation.

Acquired Knowledge

While the robot moves around constructing the metric and topological maps, our system derives higher-level knowledge from the information in these layers. Each topological area, for instance, is represented in the conceptual map as an ontological instance of the type `Area`. Furthermore, as soon as reliable information about the semantic classification of an area is available, this is reflected in the conceptual map by assigning the area's instance a more specific type. Information about recognized objects stemming from the vision subsystem is also represented in the conceptual map. Whenever a new object in the environment is recognized, a new instance of the object's type, e.g. `Couch`, is added to the ontology. Moreover, the object's instance and the instance of the area where the object is located are related via the `hasObject` relation. This process is shown in Figure 5.1.

Asserted Knowledge

During a guided tour with the robot, the user typically names areas and certain objects that he or she believes to be relevant for the robot. Typical assertions in a guided tour include "You are in the corridor," or "This is the charging station." Any such assertion is stored in the conceptual map, either by specifying the type of the current area or by creating a new object instance of the asserted type and linking it to the area instance with the `hasObject` relation.

Innate Conceptual Knowledge

We have handcrafted an ontology (Figure 5.8) that models conceptual commonsense knowledge about an indoor office environment. On the top level of the conceptual taxonomy, there are the two base concepts `Area` and `Object`. `Area` can be further partitioned into `Room` or `Corridor`. The basic-level subconcepts of `Room` are characterized by the instances of `Object` that are found there, as represented by the `hasObject` relation.

Inferred Knowledge

Based on the knowledge representation in the ontology, our system uses a description logic-based reasoning software that allows us to move beyond a pure labeling of areas. Combining and evaluating acquired and asserted knowledge within the context of the innate conceptual ontology, the reasoner can infer more specific categories for known areas. For example, combining the acquired information that a given topological area is classified as a room and contains a couch with the innate conceptual knowledge given in our commonsense ontology, it can be inferred that this area can be categorized as being an instance of `LivingRoom`. Conversely, if an area is classified as a corridor and the user shows the robot a charging station in that area, no further inference can be drawn. The most specific category the area instantiates will still be `Corridor`.

Our method allows for multiple possible classifications of any area because the main purpose of the reasoning mechanisms in our system is to facilitate human-robot interaction. The way people refer to the same room can differ from situation to situation and from speaker to speaker, as reported by Topp *et al.* [54]. For example, what one speaker prefers to call the kitchen might be referred to as the recreation room by another person. Since our aim is to be able to resolve all such possible referring expressions, our method supports ambiguous classifications of areas.

5.7 Object Detection and Recognition

In this section, we discuss how the robot can use active vision for perceiving objects and landmarks in the environment. The process is active in that it is based on active search, primed by interpretations established at other levels of spatial representation. Active vision provides information about objects in the environment. It covers object recognition and determines object pose relative to the world coordinate system adopted by the metric layer in which all other representations are grounded.

The rest of this section gives details about the approach adopted in CoSy for object search and localization (Section 5.7.1) and presents results of an experiment evaluating different methods for object distance estimation (Section 5.7.2).

5.7.1 Object Search and Localization

In our early work [55], the search for objects was performed while exploring the environment to cover space or when guided by the user. This is not sufficient. The user is able to show the robot all objects, but this is a tedious task and does not have the right level of autonomy expected from an autonomous agent. When object search runs in parallel with exploration, it is driven by the laser scanner which has a 180° field of view compared to the camera having about a fourth of that. This means that the camera is not guaranteed to see all parts of the environment.

View planning as a research area is well established. The so called *art gallery problem* [56] is defined as finding a minimal set of viewpoints from which all the parts of the environment can be observed. This problem is akin to the problem of planning for finding objects in the environment. The main difference is that one also needs to take into account the limitations of the observer, i.e. the camera. One of the most important limitations comes from the finite resolution of the camera and the fact that objects have different sizes. Even if a small object is in the field of view it will not be detected if the camera is too far away. Similarly, a large object can typically not be detected if the camera is too close. A system taking these constraints into account is presented in [57, 13]. This system uses a combination of a visual attention mechanism in the form of the RFCH algorithm [58], camera zoom and SIFT recognition [39] for finding the objects. View planning is carried out by selecting views from the nodes in the navigation graph. Briefly, when searching for objects the system first analyzes the map of the environment and performs the view planning. The robot then visits each view point and performs the visual search. The visual attention mechanism tells the systems what parts of the image to investigate further and the system does so by zooming in, thus gathering more pixels from the potential object. When the object is close, SIFT recognition is used to verify the identity of the object. The distance to the object is estimated from visual cues. The distance is used to control the zooming and to estimate the position of the object.

This method relies on the visual attention system not to produce too many views to investigate further. In [59] a method for accumulating over time the available visual evidence for the presence of objects was investigated. This would allow the object recognition algorithm to run in parallel to the exploration and could also handle object detection/recognition algorithms that provide information that is too weak to act on immediately.

The input to the view planning is, besides the objects to search for, a grid representation of the world and a navigation graph. The grid resolution is 0.5m [13]. The grid cells represent possible object locations, and the plan is constructed such that each grid cell is observed from a distance appropriate

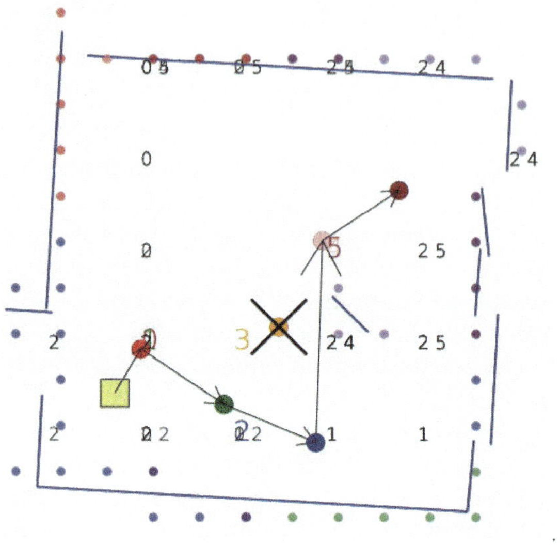

Fig. 5.9. An example of a planned path for object search. In this example, one of the nodes is not used in the plan.

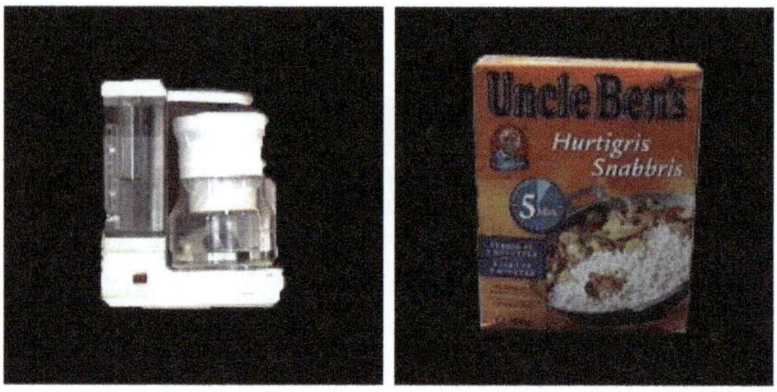

Fig. 5.10. Two of the training images (coffee machine and rice package) provided to the robot beforehand to learn the appearance of the objects

for each object. Figure 5.9 shows an example of a plan from the view planning. There may be several views associated with a node corresponding to different viewing directions.

To represent an object for recognition, the system uses one segmented image of each object in a close-up view. Two examples of training images are shown in Figure 5.10. To be able to plan for detecting the objects and to determine the distance to a detected object, the real world and image size of the object are also stored in the object database.

As a consequence of using a single view for each object, the system can only recognize the object from one side. Using a multi-view representation of the object is a natural extension to this work.

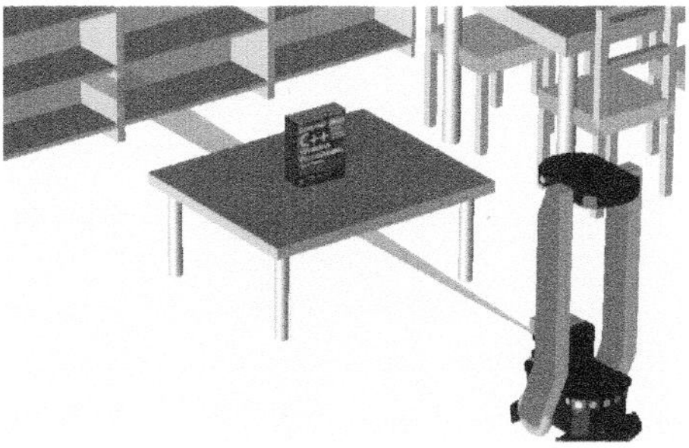

Fig. 5.11. Distance estimation provided by the laser may not be reliable: instead of the distance to the object on the table, the distance to the shelf is measured

In [60], the distance estimate used to determine zoom levels was based directly on the robot's laser sensor. However, the distance provided by the laser is often misleading, as Figure 5.11 shows: the laser sensor is placed about 30cm above the floor and if an object is not at that height, the estimate may be wrong. The approach works only for objects that are placed on the floor or are located close to walls (for example, in a bookshelf). If the distance estimate is wrong, the final zoom may either not be sufficient to make the object occupy enough of the image, or otherwise may be too large causing only a small part of the object to be seen. Furthermore, even if the object is recognized, its estimated position might be inaccurate. To address these issues, we have looked at two alternative ways for distance estimation.

Using the Vote Matrix

Using the RFCH vote matrix for distance estimation consists of measuring how many cells are part of the object and treating the area they occupy in the image as an approximation of the object's size. Here, cells are considered to be associated with a hypothesis if their degree of match is above the threshold and if there is an 8-connected path to the hypothesis with cells of monotonically increasing value. Only the strongest hypothesis and its associated 8-connected cells are taken into account, because it is likely to be the most reliable.

Given the object's actual size stored in the training database, the distance is then computed as:

$$D = \frac{W_{real} \dfrac{W_{im}}{2D_{vote}}}{\tan\left(\dfrac{\alpha}{2}\right)}$$

where D stands for the estimated distance (meters); W_{real}, for the real width of the object (meters); W_{im}, for the width in pixels of the camera image; D_{vote}, for the width in pixels of the bounding box of the cells associated with a hypothesis and α, the horizontal viewing angle. This procedure is fast and approximate, but sufficiently accurate to allow the object search algorithm to assign a valid zoom.

Using SIFT

SIFT produces a scale parameter for each key point extracted. For each matched pair of key points in the training and recognition image, the quotient of the keys' scale parameter gives an estimate of their relative apparent size and hence their distance, according to:

$$D = \frac{W_{real} \dfrac{W_{im}}{2W_{tr}} \dfrac{S_{tr}}{S_{real}}}{\tan\left(\dfrac{\alpha}{2}\right)}$$

where S_{tr} denotes the scale of the point extracted from the training image; S_{real}, the scale of the point extracted from the recognition image, and W_{tr}, the width of the object in the training image in pixels.

As mismatched key point pairs can produce incorrect scale parameters, the final estimate of the object distance is taken as the median of the distance estimates from all matches. Experiments indicate that an adequate estimate is obtained given 10 or more SIFT matches. With 4 matches or more a passable rough estimate is typically obtained (within about 30%). If there are fewer than 4 matches, the result is likely to be very poor (most likely based on some other structure than the object) and is not used.

The drawback of the above method is that extracting SIFT features from an image is computationally expensive, and using it to guide the zoom process may take too long to be feasible. Another problem is the number of SIFT features required to obtain a robust estimation; when the object is small in the image (i.e. resolved by few pixels), it is unlikely that enough matches will be available.

5.7.2 Object Distance Estimation

Figure 5.12 presents the results of distance estimation using RFCH and SIFT without magnification, performed on five different test objects. As expected, performance deteriorates for both methods at long range, due to the decreased size of the object in the image, and for RFCH also partly to the discretization of the vote cells.

It is notable that the values obtained through both methods tend towards the low end. The reason for this are mainly outliers, erroneously assigned values of 0.5–1m, caused by large background structures being mistaken for a close-up object. Compared to RFCH, SIFT exhibits a far more accurate and dependable estimate at short range. However, its quality rapidly deteriorates

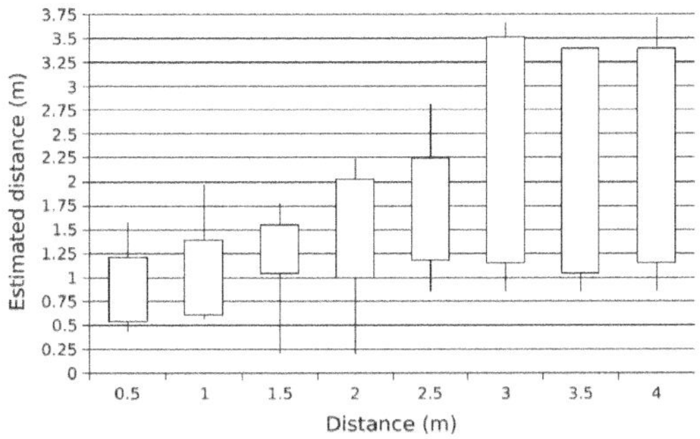

(a) RFCH distance estimates

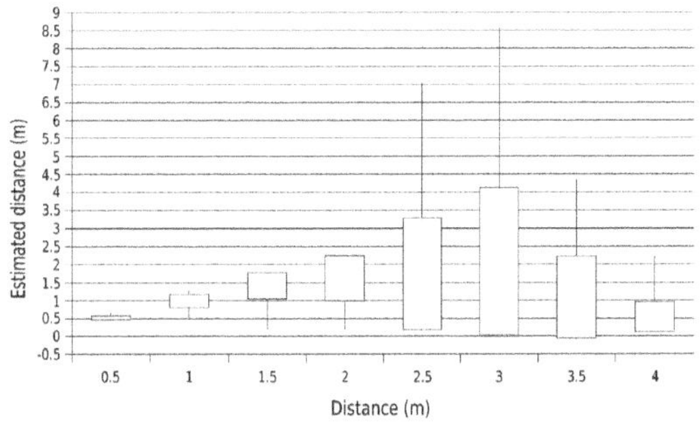

(b) SIFT distance estimates

Fig. 5.12. Distance estimation results; all objects. Top image RFCH, bottom SIFT. Boxes signify one standard deviation about the average for each distance; lines signify the most extreme values.

at longer distances, as can be seen by inspecting the average value of the estimates beyond 2.5m in Figure 12(b). This is because a certain level of detail is needed to extract SIFT keys. In contrast, RFCH, though most reliable at medium ranges (as demonstrated by the standard deviations in Figure 12(a)), retains the ability at long range to provide very rough approximations, generally adequate for the purpose of selecting a zoom level for the next step. For the final distance estimate, it should be pointed out that SIFT is used – but the magnification of the image will correspond to shifting the diagram in Figure 12(b) into the 0.5m–1m region where the method is most effective.

Figure 5.13 highlights the differences between RFCH and SIFT in distance estimation. Here, for each test image, the absolute error of the distance estimate is compared between the two methods and the percentage of cases where each of the methods gives better estimate is plotted. The graph shows that RFCH becomes more reliable at 2m range or above.

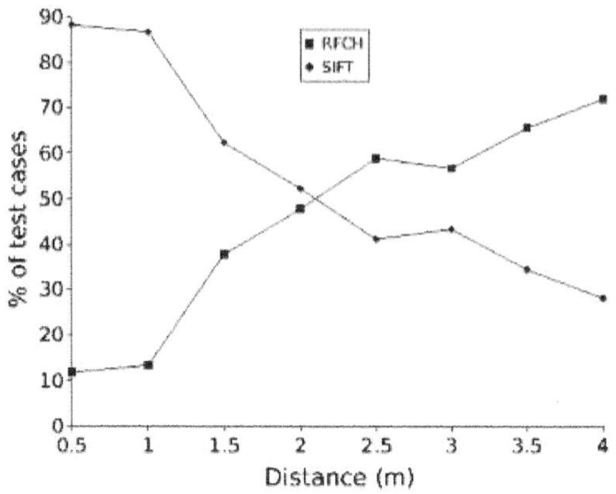

Fig. 5.13. Proportion of instances in which RFCH and SIFT provide the best estimate

5.8 Place Classification

This section presents a multi-modal place classification algorithm able to identify places and recognize semantic place categories. The method effectively utilizes information from different robotic sensors by fusing multiple visual cues and laser range data [12]. The presented approach was used for real-time semantic labeling of the spatial entities represented by the conceptual spatial model.

Place classification, as considered in this section, can be described as a supervised pattern recognition problem of assigning a region in an environment to one of predefined place classes based on multi-modal sensory input and a set of place models. First, the place models are build from a collection of labeled data samples acquired in places belonging to the modeled classes. The models store intrinsic visual and geometric properties of the classes. Then, the algorithm is presented with data samples acquired in one of the same places or in a novel place belonging to one of the same categories, possibly under different conditions and after some time (where the time range goes from some minutes to several months). The goal is to classify correctly as much of the sensory data samples as possible.

The ability to classify places based on their visual and geometric properties is an important competence for a mobile cognitive agent in two fundamental scenarios. First, place classification can be used to recognize previously visited places. In this scenario, place classification becomes a key element of topological and hybrid localization systems, providing them with means for global localization and loop closing [61, 62, 12]. Second, place classification can be used to assign novel places to semantic categories, and thus augment space representations with semantic information [9, 7, 63]. In both cases, this is a challenging classification problem due to large variability and dynamics of real-world environments. First, viewpoint variations cause the sensors to capture different aspects of the same place, which often can only be learned if enough training data are provided. Moreover, real-world environments are usually dynamic and their appearance changes over time. The recognition system must be robust to variations introduced by changing illumination (e.g. during sunny days and at night) and due to human activity (people might appear in the images, objects and furniture can be relocated).

Place classification is a widely researched topic. Purely geometric solutions based on laser range data have proven to be successful for certain tasks [7, 64, 61]. However, the limitations of such solutions inspired many researchers to turn towards vision which nowadays is becoming tractable in real-time applications. The proposed methods employed either perspective [9, 65, 66, 10, 62] or omnidirectional cameras [67, 68, 69, 70]. The main differences between the approaches relate to the way the scene is perceived. Several approaches employ local features, computed from distinct parts of an image [66, 69, 70]. Other use global features, derived from the whole image [67, 68, 9]. Recently, several authors observed that robustness and efficiency can be improved by combining information provided by different visual cues [10, 62] or different sensors, such as a camera and a laser range finder [11, 71, 12].

The algorithm presented here is able to perform robust place classification under different types of variations that occur in indoor environments over a span of time of several months. The method relies on robust descriptors [72, 39, 7] and discriminative classifiers [73, 74] known for their superior generalization abilities. The reliability is further improved by integrating multiple cues and modalities. The system uses different types of visual information provided by global and local image descriptors and geometric cues derived from laser range scans. The cues are combined using a high-level cue integration scheme that learns how to optimally weight each cue [12]. The system is able to measure its own level of confidence and fuse information over time and space in order to provide a reliable decision. Finally, in case of dynamic environments, where the long-term variability cannot be handled by the generalization abilities of the algorithm, the internal representation can be incrementally updated to maintain a stable performance as proposed in [75].

The rest of this section motivates the choice of modalities (Section 5.8.1), provides an overview of the architecture of the place classification system (Section 5.8.2) and gives details about the algorithms used to extract and classify the geometric and visual cues (Section 5.8.3 and Section 5.8.4). Then, the method used for cue integration is described in Section 5.8.5. The section concludes with a discussion on the need for adaptive models for place classification in dynamic environments (Section 5.8.6). All the algorithms were experimentally evaluated on robotic platforms operating in realistic environments. The experiments and obtained results are presented in Section Section 5.9.

5.8.1 Multiple Cues and Modalities for Place Classification

Nowadays, robots are usually equipped with several sensors, typically a laser range finder and a camera (or cameras), providing both geometrical and visual information about the environment. The ability to effectively integrate multiple cues, possibly extracted from multiple sensory modalities, becomes an important feature of a place classification system. First of all, as each sensor usually captures a different aspect of the environment, using multiple cues allows for obtaining more descriptive representation. A laser range scanner can be a valuable source of geometrical information, while vision is necessary if a robot requires a notion of human-like appearance-based concepts. Good descriptive and discriminative abilities along with robustness are the two crucial features of a place classification system with a great influence on its overall performance. The visual sensor is an irreplaceable source of distinctive information about a place. However, this information tends to be noisy and difficult to analyze due to the susceptibility to variations introduced by changing illumination and everyday activities in the environment. At the same time, laser range finders provide much more stable and robust geometric cues. These cues, however, are unable to uniquely represent the properties of different places. This leads to the problem of perceptual aliasing [76]. Clearly, each modality has its own characteristics. Interestingly, the weaknesses of one often correspond to the strengths of the other.

It is important to note that even alternative interpretations of the information obtained by the same sensor can be valuable. In this work we concentrate on two different types of visual cues based on global and local image features. Global features are derived from the whole image and thus can capture general properties of the whole scene. In contrast, local features are computed locally, from distinct parts of an image. This makes them much more robust to occlusions and viewpoint variations, but requires a costly matching process in order to find feature correspondences.

The different properties of the cues result in different performance and error patterns on the place classification task. This is illustrated in Figure 5.14 which shows distributions of errors made by three single-cue place classification algorithms for five different place classes (see Section 5.9.1 for details). It is apparent that each of the cues makes errors according to a different pattern. The cue integration scheme should exploit this fact in order to increase the overall performance. The experimental results reported in Section 5.9.2 show that the performance of a place classification system can indeed be boosted by combining the stability of geometrical solutions with the versatility of different visual cues.

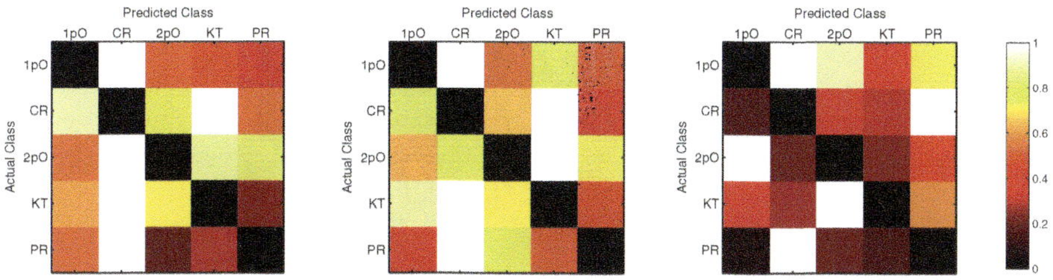

(a) Global visual cues (b) Local visual cues (c) Laser range cues

Fig. 5.14. Distributions of errors made by three single-cue place classification algorithms for five different place classes (1pO - one person office, CR - corridor, 2pO - two persons office, KT - kitchen, PR - printer area). Bright colors indicate the classes most often confused with the actual class. The diagonal elements were removed.

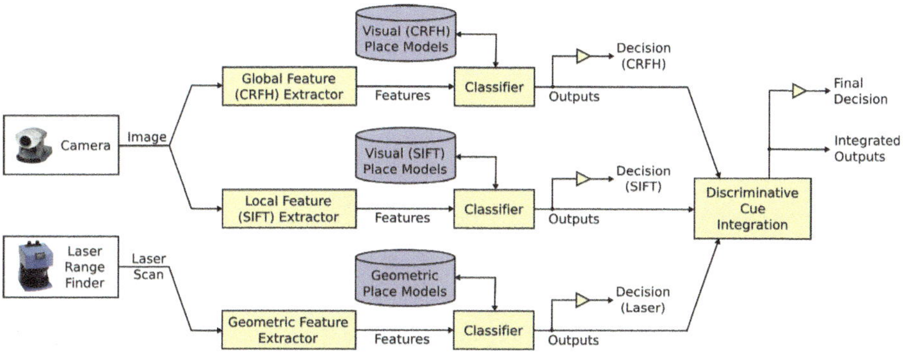

Fig. 5.15. Architecture of the multi-modal place classification system

5.8.2 Architecture of the Place Classification System

The architecture of the place classification system described in this section is illustrated in Figure 5.15. The system relies on two visual cues corresponding to two different types of image features (local based on the SIFT descriptor [39] and global based on the Composed Receptive Field Histograms [72]) as well as simple geometrical cues extracted from laser range scans [7]. The cues are processed independently. For each cue, there is a separate path in the system which consists of two main building blocks: a feature extractor and a classifier. Each classifier produces a set of outputs indicating its soft decision for all place classes. These outputs can be used directly to obtain the final decision separately for each cue. In cases when several cues are available, the single-cue outputs are combined using a high-level discriminative accumulation scheme producing integrated outputs from which the final decision is derived. As was described in Section 5.5.4, the integrated outputs can be accumulated over time and space if the system is used on a mobile platform. Since each of the

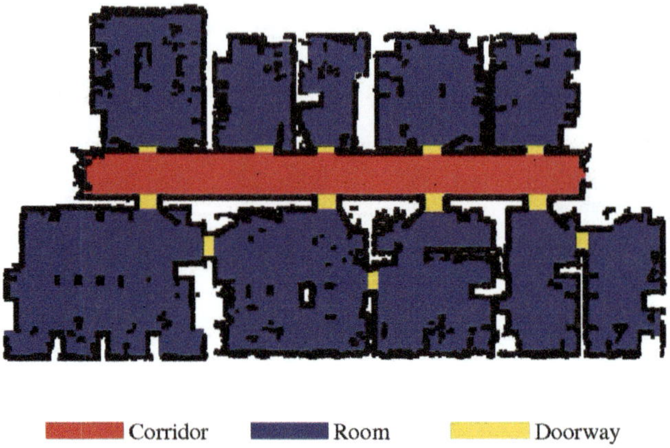

Fig. 5.16. An occupancy grid map built on the ground floor of the building 52 at the University of Freiburg. Some natural divisions can be extracted from this map e.g. corresponding to rooms, doorways and a corridor.

cues is treated independently, the system can decide to acquire and process additional information only when necessary e.g. only in difficult cases. This scheme is referred to as Confidence-based Cue Integration [10].

5.8.3 Laser-Based Place Classification

This section presents our approach to place classification based on geometric features extracted from laser range data. Many places in indoor environments can be distinguished due to their different structure. This structure can be unique for an instance of a place, but can also be characteristic for a whole semantic place category. For example, the bounding box of a corridor is usually longer than that of rooms or hallways. At the same time, rooms are typically smaller than hallways, and also more cluttered than corridors and hallways. As an example, Figure 5.16 shows a typical hand-labeled division of an environment into three categories of places.

As illustrated in Figure 5.15, the place classification algorithm first extracts a set of simple geometrical features from the scan acquired by the range sensor. Figure 5.17 shows an example of a scan taken by a mobile robot in a corridor. Each feature is represented by a numerical value computed from the beams of the scan or from a polygon representing the covered area. Single features alone are not sufficient for reliable places classification. Here, the AdaBoost algorithm is used to boost the simple features into a strong classifier. As is shown in Section 5.9.1, different classification algorithms, such as Support Vector Machines [73], can also be used to successfully derive a place class from the geometrical features. A brief description of the features and the AdaBoost algorithm is given below.

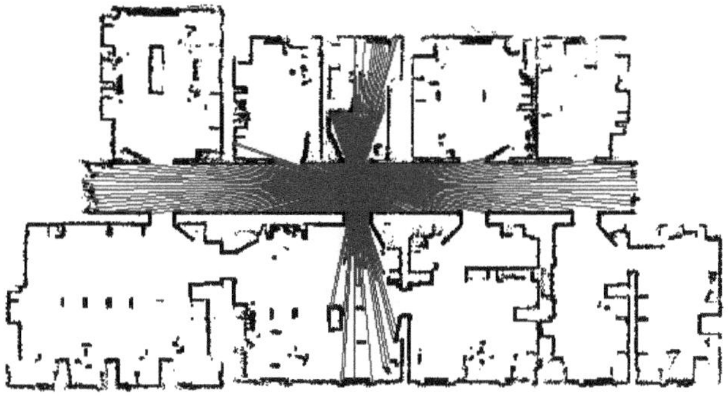

Fig. 5.17. A range scan covering the complete 360° field of view acquired by a mobile robot in a corridor

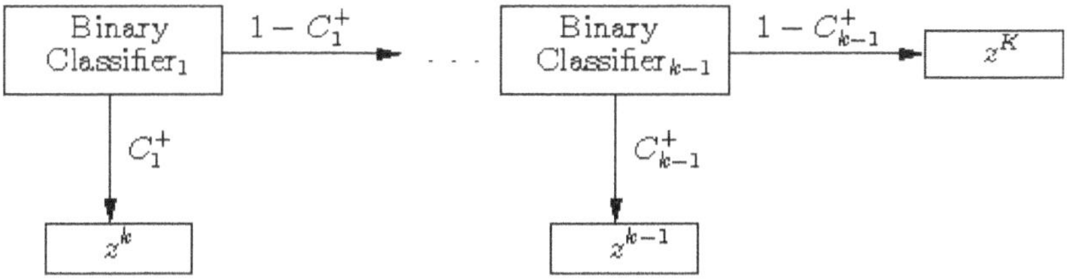

Fig. 5.18. A decision list built for K classes using binary classifiers. The output of each binary classifier is the probability z^k that the classified example belongs to the k-th class.

Classification Using AdaBoost

The AdaBoost algorithm, introduced in [74], is one of the most popular boosting algorithms. This algorithm takes as an input a training set of positive and negative examples. On each round, AdaBoost calls a weak learning algorithm repeatedly to select a weak hypothesis. The key idea is to maintain a weight distribution over the training examples. This distribution indicates the importance of the examples at the beginning of the training process and later is controlled by the algorithm. Below, a modified version of the original algorithm is described which outputs a confidence value for each positive and negative classification [77].

The original AdaBoost algorithm was designed for binary classification problems. However, to label places in the environment, we need the ability to handle multiple classes. One way to construct a multi-class classifier is to arrange several binary classifiers into a decision list. Each element of such a list represents one binary classifier which determines if an example belongs to one specific class. In addition, each binary classifier outputs a confidence value C_k^+ for a positive classification of its class k. Figure 5.18 illustrates the structure of the probabilistic decision list.

In the decision list, each test example is fed into the first binary classifier, which outputs a confidence value C_1^+ for a positive classification. Then the example is passed to the next binary classifier. This process is repeated until the last element in the list. The complete output of the decision list is represented by a histogram z. In this histogram, the k-th bin stores the probability that the classified location belongs to the k-th class according to the sequence of classifiers in the decision list. This probability can be computed as follows:

$$z^k = C_k^+ \prod_{j=1}^{k-1} (1 - C_j^+), \tag{5.2}$$

where the confidence value for the last K-th bin is equal to 1. In the multi-cue framework, these probability values are used as the outputs which are integrated by the cue integration function (see Figure 5.15).

One important question in the context of a sequential classifiers is the order in which the individual binary classifiers are arranged. A good strategy is to order the classifiers in increasing order according to their training error rate. Compared to the optimal order, the classifier generated by this heuristic performed only 1.3% worse on average for an application with several classes demonstrated in [78]. In several cases, the sequence generated by this heuristic turned out to be the optimal one.

Simple Features from Sensor Range Data

Let's assume that the mobile robot is equipped with a range sensor covering the 360° field of view. Each laser observation $z = \{b_0, ..., b_{M-1}\}$ contains a set of beams b_i. Each beam b_i consists of a tuple (α_i, d_i) where α_i is the angle of the beam relative to the robot and d_i is the length of the beam. Each training example for the AdaBoost algorithm is just one observation z and its classification y. Thus, the set of training examples is given as

$$E = \{(z_i, y_i) \mid y_i \in Y = \{\text{Room}, \text{Corridor}, \ldots\}\}. \tag{5.3}$$

In this approach, each laser observation is represented by a set of simple geometric features expressed using single real values. All features are rotationally invariant to make the classification dependent only on the (x, y)-position of the robot and not of its orientation. Most of the features are standard geometrical characteristics often used in shape analysis and pattern recognition. We define a feature f as a function that takes as argument one observation and returns a real value: $f : Z \rightarrow \Re$, where Z is the set of all possible observations. Figure 5.19 shows graphically some of these features used. The complete list of features, together with their mathematical definition, can be found in [77].

Common configurations on real mobile robots have only one laser scanner covering the 180° in front of the robot. In these cases the values corresponding to the rear laser scan can be set to zero. A more advanced solution is

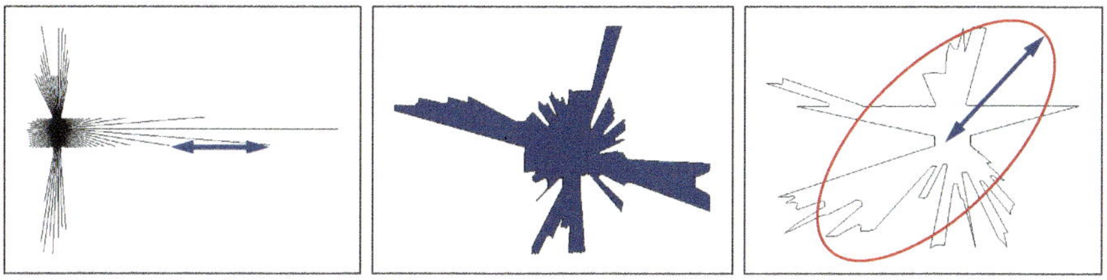

Fig. 5.19. Examples of features generated from laser range data, namely the average distance between two consecutive beams, the perimeter of the area covered by a scan, and the mayor axis of the ellipse that approximates the polygon described by the scan. Here, the laser beams cover a 360° field of view.

to maintain a local map around the robot. This local map can be updated during the movements of the robot, and then used to simulate the rear laser beams [77]. This approximation have shown good results in several indoor experiments [7, 63].

5.8.4 Vision-Based Place Classification

This section describes the visual place classification algorithms proposed in [12] that constitute the two paths of the vision-based channel in the multi-modal system presented in Figure 5.15. Each of the paths is built around a Support Vector Machine (SVM) classifier [79] and a different type of visual feature, global or local, extracted from the same image frame (see Section 5.8.1 for the distinction between the feature types). The global features are represented using a rich global descriptor, Composed Receptive Field Histograms (CRFH, [72]). The local features are based on the Scale Invariant Feature Transform (SIFT, [39]). Both have already been proved successful in the domain of vision-based localization [10, 37, 66].

The rest of the section describes the feature extraction algorithms and sketches the theory behind SVMs which will also form a basis for the cue integration scheme presented in Section 5.8.5.

Global Visual Features: Composed Receptive Field Histograms

CRFH is a multi-dimensional statistical representation of the occurrence of responses of several image descriptors applied to the image. This idea is illustrated in Figure 5.20. Each dimension corresponds to one descriptor and the cells of the histogram count the pixels sharing similar responses of all descriptors. This approach allows to capture various properties of the image as well as relations that occur between them. We tested a wide variety of combinations of image descriptors with several scale levels. On the basis of an evaluation of performance and computational cost, we build the histograms from either first order or second order Gaussian derivative filters applied to the illumination channel at two scales. This resulted in either 4- or 6-dimensional histograms.

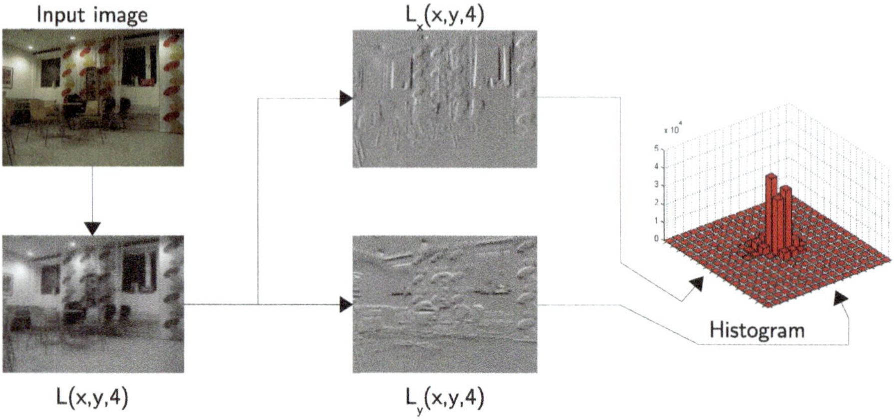

Fig. 5.20. The process of generating multi-dimensional receptive field histograms shown on the example of the first-order derivatives computed at the same scale $t = 4$ from the illumination channel

Multi-dimensional histograms can grow extremely fast if the number of dimensions grows. However, most of the cells are usually empty [72]. Storing only those that are non-zero allows for reducing the amount of required memory and performing operations such as histogram accumulation and comparison efficiently.

In case of SVMs, special care must be taken in choosing an appropriate kernel function which acts as a similarity measure between the feature vectors. In this work, the χ^2 kernel [80] was used for the CRFH descriptors. The χ^2 kernel belongs to the family of exponential kernels, and is given by

$$K(\boldsymbol{x}, \boldsymbol{y}) = \exp\left\{-\gamma\chi^2(\boldsymbol{x}, \boldsymbol{y})\right\}, \qquad \chi^2(\boldsymbol{x}, \boldsymbol{y}) = \sum_i \frac{||x_i - y_i||^2}{||x_i + y_i||}. \tag{5.4}$$

Local Features: Scale Invariant Feature Transform

The process of local feature extraction consists of two stages: *interest point detection* and *description*. The interest point detector identifies a set of characteristic points in the image that could be re-detected even in spite of various transformations. The role of the descriptor is to extract robust features from the local patches located at the detected points. Here, we used the scale, rotation, and affine invariant interest point detector based on the difference-of-Gaussians (DoG) operator [81] and the SIFT descriptor [39]. Figure 5.21 presents local patches located at the interest points detected in three typical images acquired in an indoor environment.

In case of local features, the similarity between two images is measured by solving the correspondence problem. Thus, in order to couple the local

Fig. 5.21. Local patches at the interest points detected in three typical images acquired in an indoor environment. The size of the patches illustrate the scale at which the points were detected.

descriptors with the SVMs, the match kernel proposed in [82] was used. The match kernel is given by

$$K(\boldsymbol{L}_h, \boldsymbol{L}_k) = \frac{1}{n_h} \sum_{j_h=1}^{n_h} \max_{j_k=1,\ldots,n_k} \left\{ K_l(\boldsymbol{L}_h^{j_h}, \boldsymbol{L}_k^{j_k}) \right\}, \tag{5.5}$$

where $\boldsymbol{L}_h, \boldsymbol{L}_k$ are local feature sets and $\boldsymbol{L}_h^{j_h}, \boldsymbol{L}_k^{j_k}$ are two single local features. The sum is always calculated over the smaller set of local features and only some fixed amount of best matches is considered in order to exclude outliers. The local feature similarity kernel K_l can be any Mercer kernel. Here, the RBF kernel based on the Euclidean distance was used for the SIFT features:

$$K_l(\boldsymbol{L}_h^{j_h}, \boldsymbol{L}_k^{j_k}) = \exp\left\{ -\gamma ||\boldsymbol{L}_h^{j_h} - \boldsymbol{L}_k^{j_k}||^2 \right\}. \tag{5.6}$$

Support Vector Machines

Support Vector Machines are a binary discriminative classifier known for their superior generalization abilities. Consider the problem of separating the set of labeled training data $(\boldsymbol{x}_1, y_1), (\boldsymbol{x}_2, y_2), \ldots, (\boldsymbol{x}_n, y_n)$ into two classes, where $\boldsymbol{x}_i \in \Re^N$ is a feature vector and $y_i \in \{-1, +1\}$ its class label. Assuming that the two classes can be separated by a hyperplane in some Hilbert space $\mathcal{H}$, then the optimal separating hyperplane is the one which has maximum distance to the closest points in the training set resulting in a discriminant function

$$f(\boldsymbol{x}) = \sum_{i=1}^{n} \alpha_i y_i K(\boldsymbol{x}_i, \boldsymbol{x}) + b. \tag{5.7}$$

The classification result is then given by the sign of $f(\boldsymbol{x})$. The values of α_i and b are found by solving a constrained minimization problem, which can be done efficiently using the SMO algorithm [83]. Most of the α_i's take the value of zero; those $\boldsymbol{x}_i$ with nonzero α_i are the "support vectors". In cases where the two classes are non-separable, the optimization is formulated in such a way that the classification error is minimized and the final solution remains identical. The mapping between the input space and the usually high dimensional feature space $\mathcal{H}$ is done using kernels $K(\boldsymbol{x}_i, \boldsymbol{x})$.

The extension of SVM to multi-class problems can be done in several ways. Three different approaches were used in this work:

1. *Standard one-against-all (OaA) strategy.* If M is the number of classes, M SVMs are trained, each separating a single class from all other classes. The decision is then based on the distance of the classified sample to each hyperplane, and the sample is assigned to the class corresponding to the hyperplane for which the distance is largest.
2. *Modified one-against-all strategy.* In [10], a modified version of the OaA principle was proposed. The authors suggested to use distances to precomputed average distances of training samples to the hyperplanes (separately for each of the classes), instead of the distances to the hyperplanes directly. In this case, the sample is assigned to the class corresponding to the hyperplane for which the distance is smallest. Experiments presented in this paper and in [10] show that in many applications this approach outperforms the standard OaA technique.
3. *One-against-one (OaO) strategy.* In this case, $M(M-1)/2$ two-class machines are trained for each pair of classes. The final decision can then be taken in different ways, based on the $M(M-1)/2$ outputs. A popular choice is to consider as output of each classifier the class label and count votes for each class; the test image is then assigned to the class that received more votes.

In each of the aforementioned cases, the classified sample is processed by a set of binary classifiers. Each of these classifiers produces a value of the discriminant function as defined by Eq. (5.7). In the multi-cue framework, these values are used as the outputs which are integrated by the cue integration function (see Figure 5.15).

Support Vector Machines do not provide any out-of-the-box solution for estimating the confidence of the decision; however, it is possible to derive confidence information and hypotheses ranking from the distances between the samples and the hyperplanes. In order to estimate the confidence of the decision provided by the single-cue place classification algorithms and both confidence estimates and the hypotheses ranking for the final decision of the multi-modal place classification system, we used the distance-based confidence estimation method proposed in [10].

5.8.5 Discriminative Cue Integration

This section describes the SVM-based Discriminative Accumulation Scheme (SVM-DAS) algorithm [12]. The algorithm is used to integrate cues from one or multiple modalities in the place classification system presented in Figure 5.15.

Various cue integration methods have been proposed in the robotics and machine learning community [71, 84, 10, 85, 86, 87]. These approaches can be described according to various criteria. For instance, [88] suggest to classify them into two main groups, *weak coupling* and *strong coupling*. Assuming that each cue is used as input of a different classifier, weak coupling is when the output of two or more independent classifiers are combined. Strong coupling is instead when the output of one classifier is affected by the output of another classifier, so that their outputs are no longer independent. Another possible classification is into *low level* and *high level* integration methods, where the emphasis is on the level at which integration happens. We call *low level integration methods* those algorithms where cues are combined together at the feature level, and then used as input to a single classifier [87, 71]. Another strategy is to keep the cues separated and to integrate the outputs of individual classifiers, each trained on a different cue [85, 84, 12]. We call such algorithms *high level integration methods*, of which voting is the most popular [89]. These techniques are more robust with respect to noisy cues or sensory channels. Moreover, they allow to divide the learning problem into several smaller sub-problems. Additionally, not all cues need always be used and the algorithm can decide on the number of cues that should be extracted for each particular classification task [10].

SVM-DAS is a technique performing weak coupling, high level, non-linear cue integration. For each cue, the method requires training a separate classifier which provides a set of outputs encoding the relation of the classified sample to the place models for the particular cue. The integration is performed by feeding the outputs to a Support Vector Machine. Compared to previous high-level discriminative accumulation methods [84, 10], SVM-DAS gives several advantages. First, it accumulates cues with a more complex, possibly non-linear function, by using the SVM framework and kernels. Such approach makes it possible to integrate outputs of different classifiers such as SVM and AdaBoost. Moreover, it learns the weights for each cue very efficiently from the training data, therefore making it possible to accumulate large numbers of cues without computational problems. At the same time, SVM-DAS preserves the important property of the previous methods to perform correct classification even when each of the single cues gives misleading information.

Suppose, there are P cues and therefore, P single-cue classifiers. Each classifies a single cue $T_p(\boldsymbol{I})$, where $p = 1 \ldots P$, extracted from the sensory input $\boldsymbol{I}$. Then, each classifier produces a set of outputs $\{O_h^p(T_p(\boldsymbol{I}))\}_{h=1}^{H_p}$, where H_p defines the number of outputs for the p-th cue. The outputs are used as an input to an SVM, and the parameters of the integration function are learned during the optimization process, for instance using the SMO algorithm [83] (see Section 5.8.4 for a brief overview of the theory behind SVMs). This gives raise to the following integration function of SVM-DAS:

$$O_g^{\Sigma P}(\boldsymbol{I}) = \sum_{i=1}^{n} \alpha_i^g y_i K(\boldsymbol{O}_i, \boldsymbol{O}) + b^g, \; g = 1, \ldots, G,$$

where K is the kernel function and $\boldsymbol{O}$ is a vector containing all the outputs for all cues:

$$\boldsymbol{O} = \left[\{O_h^1(T_1(\boldsymbol{I}))\}_{h=1}^{H_1}, \ldots, \{O_h^P(T_P(\boldsymbol{I}))\}_{h=1}^{H_P}\right].$$

The parameters y_i, α_i^g, b^g and the support vectors $\boldsymbol{O}_i$ are inferred from the training data either directly or efficiently during the optimization process. The number of the final outputs G and the way of obtaining the final decision depends on the multi-class extension used with SVM-DAS. We tested the one-against-all extension, for which $G = M$, and the one-against-one extension, for which $G = M(M-1)/2$, where M is the number of classes. In both cases, we observed a very similar performance.

In case of SVM-DAS, the nonlinearity is given by the choice of the kernel function, thus in the case of the linear kernel the method is linear. For the experiments reported in this section, we used the non-linear RBF (Gaussian) kernel given by

$$K(\boldsymbol{x}, \boldsymbol{y}) = \exp\left\{-\gamma||\boldsymbol{x} - \boldsymbol{y}||^2\right\}. \tag{5.8}$$

5.8.6 Adaptive Place Classification

In most cases, the place classification systems are trained off-line or once they are trained the representation remains static. However, in the real, dynamic world, learning cannot be a single act. It is simply not possible to create a static model which could explain all the variability observed over time. Continuous information acquisition and exchange, coupled with an ongoing learning process, is necessary to provide the system with a valid world representation and preserve stable performance. In artificial autonomous agents constrained by limited resources, continuous learning must be performed in an incremental fashion. It is not feasible to rebuild the internal model from scratch every time new information arrives; neither is it possible to store all the previously acquired data for that purpose. The model must be updated and the updating process must have certain properties. First, the knowledge representation must remain compact and free from redundancy to fit into the limited memory and maintain a fixed computational complexity. Second, the model cannot grow forever even though new information is constantly arriving. The updating process should be able to gradually filter out unnecessary information.

Here, we focus on the scenario in which incremental learning is applied to place models in order to provide adaptability to different types of variations observed in real-world environments. As the experiments described in Section 5.9.2 show, the multi-modal place classification system presented in this chapter is able to cope with illumination and pose changes as well as short-term dynamic variations. Moreover, since it relies on multiple sensors, it can deliver satisfactory results despite dynamic variations that occurred during the period of around 6 months. Still, this variability is clearly affecting the performance of the system. As it is not possible to predict *a priori* how the environment is going to change, the only possible long-term strategy is to update the representation over time, learning incrementally from the new data recorded during use.

To experimentally verify the usefulness of adaptive models for place classification, we implemented and tested the memory-controlled approximate incremental extension of the SVM algorithm proposed in [90]. Approximate techniques [91, 92, 90] seem to be better suited for our problem because, at each incremental step, they discard non-informative training vectors, thus reducing the memory requirements. Other methods, such as [93, 94], instead require storing in memory all the training data. The basic principle behind the memory-controlled method is to combine the fixed-partition incremental extension [92] with an algorithm for controlling the memory growth [95]. Every time a new batch of data becomes available for the learning algorithm, the knowledge stored in previously built model in the form of support vectors is combined with the incoming data and used to train a new model. Then, a support vector reduction algorithm is applied to the model, which filters out redundant information by eliminating those vectors that can be expressed by a linear combination of the others. This permits keeping the model compact and provides the algorithm with forgetting capabilities. For more details, the reader is referred to [90, 75]. The results of the experimental evaluation of the method on place classification data are presented in Section 5.9.3.

5.9 Experiments with Place Classification

This section describes several series of experiments we conducted to evaluate the performance of the place classification algorithms presented in Section 5.8 on both uni-modal and multi-modal data. First, we performed experiments with single-cue place models to verify their properties and test their robustness to different types of variations e.g. introduced by illumination changes and long-term human activity (Section 5.9.1). Then, the evaluation was repeated for systems based on different combinations of cues and modalities to see if the robustness can be improved by cue integration (Section 5.9.2). In the next experiment, we took a different approach and tried to tackle long term variability by using adaptive place models (Section 5.9.3). Finally, we combined multi-modal place classification with a localization and mapping component implementing the first three layers of the spatial model and run an experiment where the task was to build a representation of a novel indoor environment (Section 5.9.4).

5.9.1 Single-Cue Place Classification

This section reports results of two experiments performed using single-cue place classification systems. The aim of the first experiment was to test the

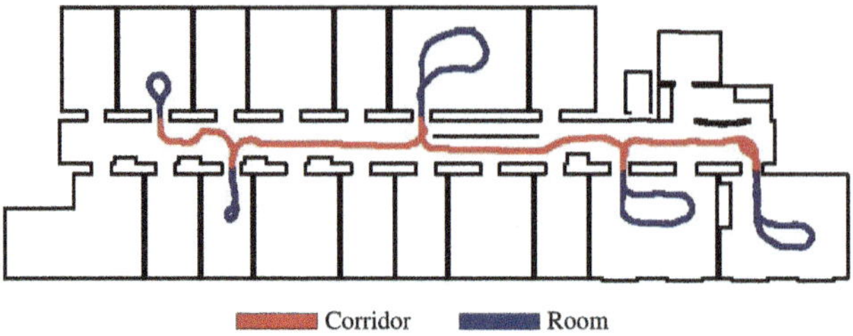

Fig. 5.22. Trajectory followed by the robot during acquisition of the training data for the room vs. corridor classification experiment. Labels were attached to the data based on the position of the robot and are marked on the plot using different colors.

ability of a system relying purely on laser range data to perform classification of places in a typical office environment into two classes: a corridor and a room. The second experiment aimed at evaluating robustness of various cues on the place classification task despite substantial variability that occurred in a realistic indoor environment over the period of several months.

Semantic Place Classification Using Laser Range Data

As explained in Section 5.5.4, providing even basic semantic descriptions, such as a room or a corridor, for regions of space can enhance functionality of a mobile cognitive agent operating in an indoor environment and interacting with a user. In such a scenario, the robot is often facing the user which affects the information captured using the laser range sensor. In order to provide reliable classification during these experiments, we used the approach based on the simple geometric features and the AdaBoost classifier presented in Section 5.8.3. We simulated the rear-view laser scanner by ray-tracing in the local obstacle map. Then, the simulated and the real scans were used together as a 360° laser range finder.

In order to test the method, we used data acquired along trajectories of the robot being driven through rooms and corridors found on two different floors of the CAS/CVAP laboratory at the Royal Institute of Technology in Stockholm, Sweden. To train the classifier, we used the scans acquired on the 6th floor along the trajectory shown in Figure 5.22. The robot was then moved to the 7th floor of the same building, which contains a similar structure. On this floor, we classified two different trajectories established in opposite directions. The classification rates for all the poses of the robot during its movement ranged from 93.18% to 96.8%. A more extensive set of experiments using these approach is shown in [77].

Single-Cue Place Classification Under Large Variability

In this experiment, we tested the robustness of four different single-cue place classification algorithms to different types of variations, such as those introduced by changing illumination or human activity over a long period of time [12]. We evaluated performance of two SVM models trained on global visual features (CRFH, Section 5.8.4) and local visual features (SIFT, Section 5.8.4) as well as SVM and AdaBoost models trained on the laser range cues (here referred to as L-AB and L-SVM, Section 5.8.3). The design of these experiments was partially based on findings from our previous work on visual place classification [65]. A video presenting the setup, experimental procedure and visualization of the results for the original experiments described in [65] can be found in [96].

The evaluation was performed on the IDOL2 database [97, 75]. The database comprises 24 labeled sequences of images at the resolution of 320x240 pixels synchronized with laser scans and odometry data acquired using two mobile robot platforms (PeopleBot and PowerBot) over a time span of 6 months. The acquisition was performed in a five room subsection of a larger office environment, selected in such way that each of the five rooms represented a different functional area: a one-person office (1pO), a two-persons office (2pO), a kitchen (KT), a corridor (CR), and a printer area (PR). Example pictures showing interiors of the rooms are presented in Figure 5.23. The appearance of the rooms was captured under three different illumination conditions: in cloudy weather, in sunny weather, and at night. The robots were manually driven through each of the five rooms while continuously acquiring images and laser scans at a rate of 5fps. The acquisition process was conducted in two phases. Two sequences were acquired using each robot for each type of illumination conditions over the time span of more than two weeks, and another two sequences for each setting were recorded 6 months later. Thus, the sequences captured variability introduced not only by illumination but also natural activities in the environment (presence/absence of people, furniture/objects relocated etc.). It is important to note that, even for sequences acquired within a short time span under similar illumination conditions, variations still exist from everyday activities and viewpoint differences during acquisition. The captured variability is illustrated in Figure 5.23. More detailed information about the database can be found in [98].

We conducted two sets of experiments for each cue on 12 data sequences from the IDOL2 database acquired with the PowerBot (additional experiments can be found in [12]). For each single experiment, we trained the models on one sequence and tested on another. The first set consisted of 12 experiments, performed on different combinations of training and test data acquired closely in time and under similar illumination conditions. Then, we increased the complexity of the problem and performed experiments on 24 pairs of training and test sets, obtained 6 months from each other and under different illumination settings. As a measure of performance we used the percentage of properly classified samples (classification rate) calculated separately for each of the rooms and then averaged with equal weights independently of the number of samples acquired in each room.

(a) Variations introduced by illumination

(b) Variations observed over time

(c) Remaining rooms (at night)

Fig. 5.23. Examples of pictures taken from the IDOL2 database showing the interiors of the rooms, variations observed over time and caused by activity in the environment as well as introduced by changing illumination

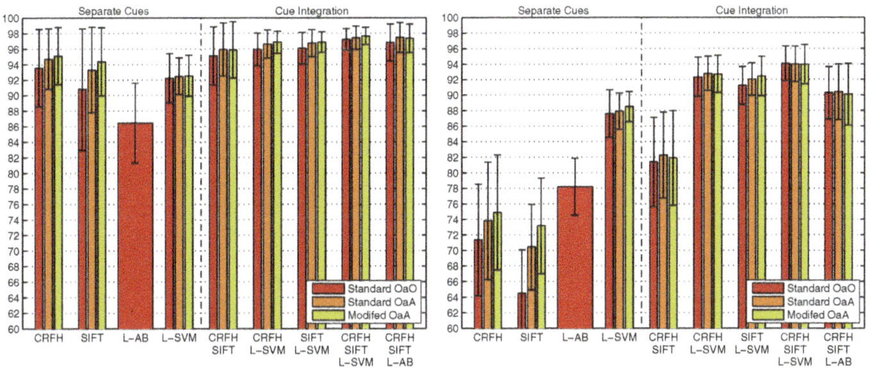

(a) Stable illumination, close in time (b) Varying illumination, distant in time

Fig. 5.24. Results of the experiments evaluating performance of four single-cue place classification systems and systems based on several combinations of multiple cues

In each experiment, we evaluated the performance of all four types of models: CRFH, SIFT, L-AB, and L-SVM. For SVM, we tried the three multi-class extensions described in Section 5.8.4. The results are presented in Figure 5.24a,b (the first four bar groups). First, the results for the three different multi-class extensions are in agreement with [10] - for single cues, the modified one-against-all algorithm gives the best performance independently of the modality on which the classifier was trained. Second, we see that under stable conditions, the vision-based methods outperform the systems based on laser range cues (95.1% for CRFH and 92.5% for L-SVM). It is also apparent that the variations that occurred over the long period of time pose a challenge for both modalities. In this case, vision also suffers from the large variations in illumination which do not affect the geometric cues. Furthermore, we can see that there is a significant difference in performance between the two laser-based solutions in favor of the SVM-based method.

A detailed analysis of the distribution of errors made by all the SVM-based models can be found in Figure 5.14 and Section 5.8.1. The fact that there are large discrepancies between the error patterns indicates that effective cue integration might result in increased performance.

5.9.2 Combining Multiple Cues and Modalities

The experiments described in this section were designed to evaluate performance of the SVM-DAS cue integration scheme and multi-cue place classification system presented in Section 5.8 and [12]. Since SVM-DAS performs high level cue integration, separate models must be trained for each of the combined cues. In this case, we used the models obtained during the single-cue experiments presented in the previous section. Moreover, we used the same

Table 5.1. Average percentages (with standard deviations) of test samples for which all cues had to be used in order to retain the maximal recognition rate

Cues (**Primary cue**)	Percentage of test samples
CRFH + SIFT	29.5±22.1
CRFH + **L-SVM**	32.7±20.3
SIFT + **L-SVM**	33.3±22.4
SIFT + CRFH + **L-SVM**	40.8±21.9

experimental setup, so that the results can be easily compared. A detailed description of all the experiments performed can be found in [12].

We tested the integration method with several combinations of different cues and modalities. The results are reported in Figure 5.24a,b (the last 5 bar groups). First, we combined the two visual cues. We see that the robustness of a purely visual recognition system can be greatly improved by integrating different types of cues, in this case local and global. This can be observed especially for the experiment where the algorithms had to tackle the largest variability. Despite that, the error distributions in Figure 5.14 indicate that we should expect largest gain when different modalities are combined. As we can see from Figure 5.24 this is indeed the case. By combining one visual cue and one laser range cue (e.g. CRFH + L-SVM), we exploit the descriptive power of vision in case of stable illumination conditions and the invariance of geometrical features to the visual noise. Moreover, if the computational cost is not an issue, the performance can be further improved by using both visual cues instead of just one. To test the ability of SVM-DAS to integrate outputs of different classifiers, we combined the SVM models trained on visual cues with AdaBoost model based on geometrical features (L-AB). The method obtained a large improvement in comparison to each of the individual cues. For instance, the recognition rate increased by 12.2% on average in the most difficult case.

Although it is clear that the performance can be significantly improved by using multiple cues, each of the cues introduces additional computational cost. This cost can be significantly reduced by taking the approach presented in [10] which combines confidence estimation methods with high level cue integration. Since, in most cases, decisions based on only one cue are correct, the system could decide to use additional sources of information only when necessary i.e. when the decision based on a single cue is not confident enough. Table 5.1 presents the results of applying the method to the experiments presented in this section. We see that, in general, the decision can be based on the fastest cue (marked with bold font in Table 5.1) and the maximal performance can be retained despite using additional cues only in approximately 35% of cases. Additional cues will be used more often when the variability is large, and rarely for less difficult cases.

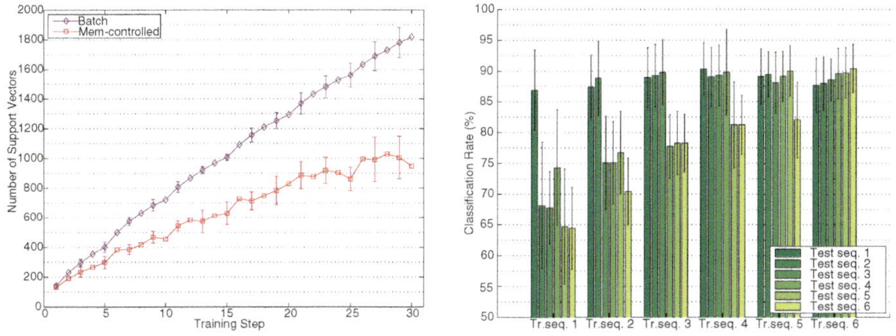

(a) Support vectors (b) Performance of memory-controlled

Fig. 5.25. Average results of the experiments with adaptive place classification: the number of support vectors stored in the model after each step and the classification rates obtained by testing the models after every fifth step with all the available test sets. The training and test sets marked with the same indices were acquired under similar conditions.

5.9.3 Adaptive Place Classification

This section gives an overview of the experimental evaluation of an adaptive place classification model presented in [75]. The experiments were based on the IDOL2 database described in Section 5.9.1 and focused on the ability of the algorithm to adapt to long-term variations. We used the memory-controlled incremental SVM algorithm for training the place models and the visual global features (CRFH) to represent the sensory data. Preliminary experiments showed that the behavior of the algorithm was very similar for the local features.

We considered a case where the algorithm needed to incrementally gain robustness to variations introduced by changing illumination and human activities, while at the same time using its adaptation ability to handle long-time changes in the environment. We first trained the system on three image sequences from the database acquired at roughly the same time but under different illumination conditions. Then, we repeated the same training procedure on sequences acquired 6 months later. In order to increase the number of incremental steps and differentiate the amount of new information introduced by each set of data, each sequence was again divided into five subsequences. Thus, in total, there were 30 incremental steps. Since the IDOL2 database consists of pairs of sequences acquired under roughly similar conditions, each training sequence has a corresponding one which could be used for testing. As a measure of performance we used the percentage of properly classified samples (classification rate) averaged over all the rooms.

The experiment was repeated 12 times for different orderings of training sequences and we compared the results of the incremental method to the batch SVM algorithm. Figure 5.25a shows the average amounts of support vectors stored in the models at each incremental step for both methods. Figure 5.25b reports the classification rate measured every fifth step (every time the system completes learning a whole sequence) for the incremental technique. In order to emphasize the need for adaptation as well as to visualize how the learning process affects the performance on the past test data, the figure shows recognition rates for all testing sets used throughout the experiment. By observing the rates for a classifier trained on the first sequence only, we see that the system achieves best performance on a test set acquired under similar conditions. The classification rate is significantly lower for other test sets especially for images acquired 6 months later, even under similar illumination conditions. At the same time, the performance greatly improves when incremental learning is performed on new batches of data. The classification rate decreases for the old test sets; at the same time, the size of the model tends to stabilize and the incremental model is much more compact than the one produced by the batch method. The results presented provide clear evidence of the capability of the discriminative methods to perform incremental learning for vision-based place classification, and their adaptability to variations in the environment.

5.9.4 Semantic Labeling of Space

We performed a real-time experiment to test the multi-modal place classification system together with other components implementing the multi-layered spatial model on a mobile robot platform. The experiment was performed during working hours in a typical office environment. Following the findings of the off-line experiments described in Section 5.9.2, we built the multi-modal place classification system based on visual and laser range cues integrated using SVM-DAS. For efficiency reasons, we used only global features (CRFH) for the vision channel. The system was implemented in the CAST (The CoSy Architecture Schema Toolkit, see Chapter 2) framework and run on a standard 2.5GHz dual-core laptop. The whole experiment was videotaped and a video presenting the setup, experimental procedure and visualization of the results can be found in [99].

The experiment was performed in the building of the School of Computer Science at the University of Birmingham, United Kingdom. The interior of the building consists of several office environments located on three floors. For our experiments, we selected three semantic categories of rooms that could be found in the building: a corridor, an office and a meeting room. In order to train the system, and build place models for these three classes, we first performed acquisition of training data in different parts of the building. To build the model of an office, we acquired data in three different offices: Aaron's office (1st floor), Robert's office (1st floor) and Richard's office (ground floor). To create a representation of the corridor class, we recorded data in 2 corridors, one on the ground floor and one on the 1st floor. The acquisition was performed at night. Finally, to train the model of a meeting room, we used an instance on the 2nd floor. The meeting room belonged to the part of the environment

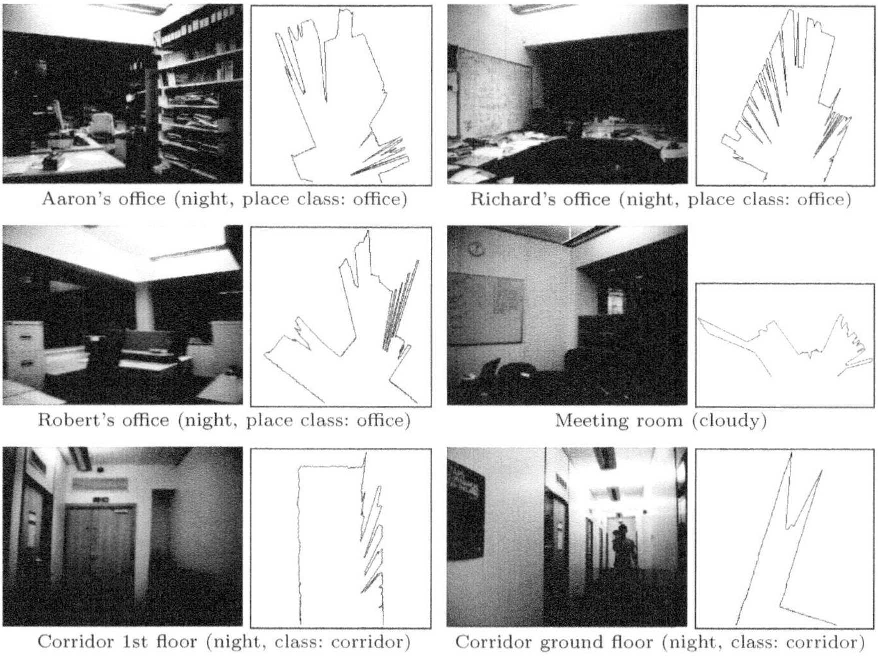

(a) Samples from the data sequences used to train the models of place classes

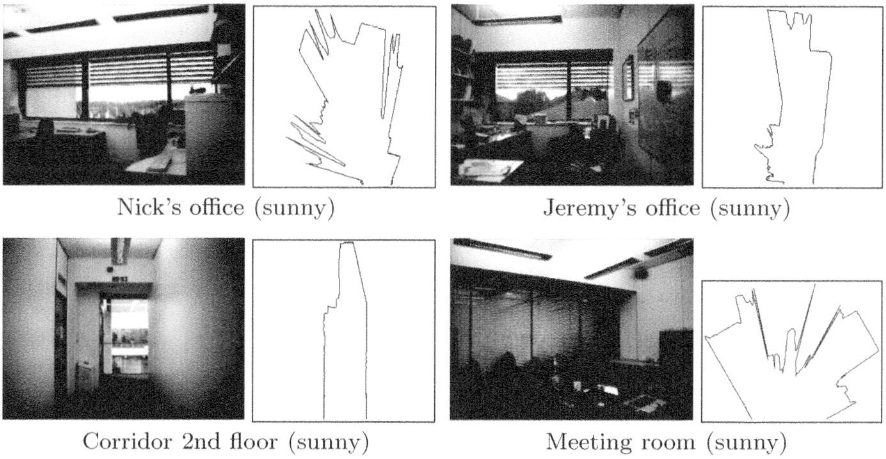

(b) Samples acquired during the test run

Fig. 5.26. Examples of images and laser scans (synchronized) taken from the data sequences used for training the models of place classes (a) and acquired during the test run (b) in each of the rooms considered during the semantic labeling experiment. The figure illustrates the within-category variations for corridors and offices as well as other types of variability observed for each place class.

where later we conducted the final test. The robot was manually driven around each room and data samples were recorded at the rate of 5 fps. In case of the meeting room, the $1st$ floor corridor as well as Aaron's and Richard's offices, the acquisition was repeated twice. Examples of images and laser scans acquired in each of the rooms can be found in Figure 5.26a.

We trained the place models separately for each modality on a dataset created from one data sequence recorded in each of the rooms. Since one of the advantages of SVM-DAS is the ability to infer the integration function from the training data, after training the models, we trained the integration scheme. We used the additional data sequences acquired in some of the rooms and trained SVM-DAS on the outputs of the uni-modal models tested on these data.

Three days after the training data were collected, we performed a real-time experiment in the lab on the 2nd floor in the same building. The experiment was conducted during the day during sunny weather. The part of the environment that was explored by the robot consisted of 2 offices (Nick's office and Jeremy's office), a corridor and a meeting room. The interiors of the rooms and the influence of illumination can be seen in the images in Figure 5.26b. An automatically generated map of the environment is presented in Figure 5.5.

During the experiment, the task of the robot was to build a multi-layered spatial representation of the environment and semantically label the navigation graph nodes and areas. The only knowledge given to the robot before the experiment consisted of the models of the three place classes: "office", "corridor" and "meeting room". The robot started in Nick's office, and was manually driven through the corridor to Jeremy's office. Then, it was taken to the meeting room where the autonomous exploration mode was turned on. The robot used a frontier-based algorithm based on [100]. After the meeting room was explored, the robot was manually driven back to the Nick's office where the experiment finished. The semantic labeling process was running on-line and the place classification was performed approximately at the rate of 5 times per second. The final semantic map build during the run is shown in Figure 5.5. We can see that the system correctly labeled all the areas in the environment.

The fact that the data were stored allowed for detailed performance analysis of the place classification system, similar to the one presented in Section 5.9.2. The results are displayed in Figure 5.27. When we look at the overall classification rate for all the data samples in the test sequence, we see that vision significantly outperformed laser in this experiment (66% vs. 84%). Still, the performance of the system was boosted by additional 8% compared to vision alone when the two modalities were integrated. The gain is even more apparent if we look at the detailed results for each of the classes (the first three charts in Figure 5.27). We see that the modalities achieved different performance, but also different error patterns, for each class. Clearly, the system based on laser range data is a very good corridor detector. On the other hand, vision was able to distinguish between the offices and the meeting room almost perfectly.

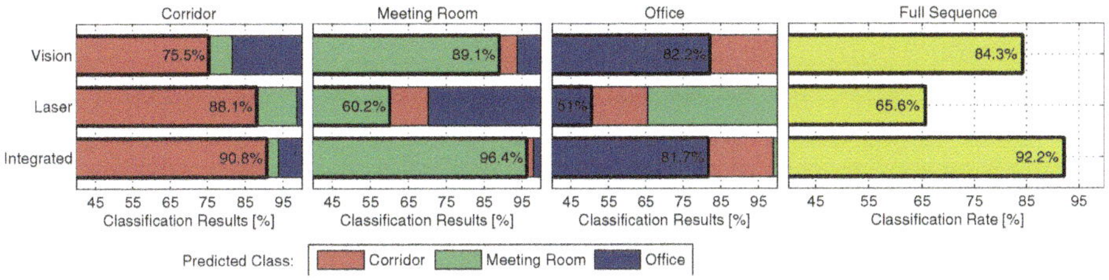

Fig. 5.27. Place classification results obtained on the dataset recorded during the test run. The first three bar charts show the results separately for each place class: "corridor", "meeting room" and "office". The charts show the percentage of the samples that were properly classified (most left bars marked with thick lines), but also how the misclassifications were distributed. The chart on the right presents the percentage of properly classified samples during the whole run.

Finally, the integrated system always achieved the performance of the more reliable modality and for two out of three classes outperformed the uni-modal systems.

5.10 Summary

We set out to create a spatial representation that would help to bridge the gap between how humans and robots represent space to facilitate interaction and support spatial reasoning. In part supported by findings in cognitive psychology and also inspired by such work as by Kuipers [4], we proposed a layered spatial model.

At the lowest level, our representation consists of a metric map that supports navigation and localization. This chapter presented a number of different approaches to how this metric map can be represented and implemented. In the integrated system, the so called M-Space feature representation [14] was used with laser range data. Much of the CoSy research on metric mapping concentrated on investigating methods for a vision-only strategy. This is also where most of the contributions to science in the area of metric mapping are found [48, 41, 14, 44]. However, since the metric map is the foundation of the spatial model and is fundamental for proper functioning of the entire system, reliability had to take priority. The framework used has however been tested in vision-only setups as described in [41, 14].

The navigation map was designed to provide a way of representing the free space in the model. As it provides coarse discretization of space it limits the state space of the path planning tasks and is also extremely useful for storing semantic information. While not a new idea, the navigation graph has been shown in this research to be a powerful representation that supports tasks beyond pure path planning for which it was originally designed [23]. One of the avenues for future work is to investigate how information about the appearance of a place (e.g. detected objects, visual features extracted from a scene or metric place descriptors) can be used to not only introduce semantics into the model, but also support localization. A model that captures the graphical nature of the navigation layer and contains place descriptors and coarse metric information seems like a good candidate for such a joint representation.

The navigation and topological maps allow to segment space into topological regions and associate semantic place information with those regions. A purely geometric method was investigated for categorizing places into rooms and corridors [7]. The experiment showed that the method is able to generate models valid even across different environments. In parallel, research was conducted on vision-based place classification. Extensive experiments demonstrated that places can be recognized and categorized reliably even using a perspective camera with limited field of view and in presence of different types of visual variations [10]. These two novel strands of work were integrated into a joint, multi-modal framework in [12]. This framework was used for semantic place categorization within the integrated system.

The conceptual map corresponds to the highest level of abstraction in the model and provides the link between the spatial model and the communication system used for situated human-robot dialogue. It grounds linguistic expressions in representations of spatial entities, such as instances of rooms or objects. The conceptual also allowed us to derive new knowledge from partial knowledge and a common sense ontology.

Each of the layers in the spatial model plays an important role in the system, providing a basis for different pieces of its functionality. Each layer also advances the state of the art in its corresponding area. As a whole, the model constitutes a versatile, but also coherent spatial representation. Compared to the work by Kuipers [4] our work uses a supervised paradigm and is focused on human-robot interaction. In fact, as will be explained in more detail in Chapter 8, the way we acquire the spatial model is in itself an example of human-robot interaction. This allows knowledge to be exchanged between robot and human during the mapping process which paves the way for a shared representation of space.

As clearly demonstrated in Section 5.9, the integration of information from many different cues and sensory modalities helps to improve the performance and comprehension of space. In a similar way, the layered spatial model provides the means for integrating information across different levels of abstractions. Chapter 10 will explain how the rest of the system interacts with the spatial model in the context of the Explorer scenario system.

References

1. Vasudevan, S., Gachter, S., Berger, M., Siegwart, R.: Cognitive maps for mobile robots an object based approach. In: Proceedings of the IROS 2006 Workshop: From Sensors to Human Spatial Concepts, Beijing, China (2006)
2. Galindo, C., Saffiotti, A., Coradeschi, S., Buschka, P., Fernández-Madrigal, J., González, J.: Multi-hierarchical semantic maps for mobile robotics. In: Proceedings of the International Conference on Intelligent Robots and Systems (IROS 2005), Edmonton, Alberta, Canada (2005)
3. Beeson, P., MacMahon, M., Modayil, J., Murarka, A., Kuipers, B., Stankiewicz, B.: Integrating multiple representations of spatial knowledge for mapping, navigation, and communication. In: Interaction Challenges for Intelligent Assistants, AAAI Spring Symposium, Stanford, CA, USA (2007)
4. Kuipers, B.: The Spatial Semantic Hierarchy. Artificial Intelligence 119, 191–233 (2000)
5. Krieg-Brückner, B., Röfer, T., Carmesin, H.-O., Müller, R.: A taxonomy of spatial knowledge for navigation and its application to the Bremen autonomous wheelchair. In: Freksa, C., Habel, C., Wender, K.F. (eds.) Spatial Cognition 1998. LNCS (LNAI), vol. 1404, pp. 373–397. Springer, Heidelberg (1998)
6. Diosi, A., Taylor, G., Kleeman, L.: Interactive SLAM using laser and advanced sonar. In: Proceedings of the International Conference on Robotics and Automation (ICRA 2005), Barcelona, Spain (2005)
7. Mozos, O.M., Triebel, R., Jensfelt, P., Rottmann, A., Burgard, W.: Supervised semantic labeling of places using information extracted from laser and vision sensor data. Robotics and Autonomous Systems Journal 55(5), 391–402 (2007), `http://cognitivesystems.org/CoSyBook/chap5.asp#MartinezMozos07a`
8. Friedman, S., Pasula, H., Fox, D.: Voronoi random fields: Extracting the topological structure of indoor environments via place labeling. In: Proceedings of the International Joint Conference on Artificial Intelligence (IJCAI 2007), Hyderabad, India (2007)
9. Torralba, A., Murphy, K.P., Freeman, W.T., Rubin, M.A.: Context-based vision system for place and object recognition. In: Proceedings of the International Conference on Computer Vision, ICCV 2003 (2003)
10. Pronobis, A., Caputo, B.: Confidence-based cue integration for visual place recognition. In: Proceedings of the IEEE/RSJ International Conference on Intelligent Robots and Systems (IROS 2007), San Diego, CA, USA (2007), `http://cognitivesystems.org/CoSyBook/chap5.asp#pronobis07iros`
11. Rottmann, A., Mozos, O.M., Stachniss, C., Burgard, W.: Place classification of indoor environments with mobile robots using boosting. In: Proceedings of the National Conference on Artificial Intelligence, Pittsburgh, PA, USA, pp. 1306–1311 (2005), `http://cognitivesystems.org/CoSyBook/chap5.asp#rottmann05aaai`
12. Pronobis, A., Martínez Mozos, O., Caputo, B.: SVM-based discriminative accumulation scheme for place recognition. In: Proceedings of the IEEE International Conference on Robotics and Automation (ICRA 2008), Pasadena, CA, USA (2008), `http://cognitivesystems.org/CoSyBook/chap5.asp#pronobis08icra`

13. López, D.G., Sjö, K., Paul, C., Jensfelt, P.: Hybrid laser and vision based object search and localization. In: Proceedings of the International Conference on Robotics and Automation, ICRA 2008 (2008), http://cognitivesystems.org/CoSyBook/chap5.asp#Galvez08a
14. Folkesson, J., Jensfelt, P., Christensen, H.: The m-space feature representation for slam. IEEE Transactions on Robotics 23(5), 1024–1035 (2007), http://cognitivesystems.org/CoSyBook/chap5.asp#Folkesson07a
15. Stevens, A., Coupe, P.: Distortions in judged spatial relations. Cognitive Psychology 10, 422–437 (1978)
16. McNamara, T.: Mental representations of spatial relations. Cognitive Psychology 18, 87–121 (1986)
17. Cohn, A.G., Hazarika, S.M.: Qualitative spatial representation and reasoning: An overview. Fundamenta Informaticae 46, 1–29 (2001)
18. Hirtle, S.C., Jonides, J.: Evidence for hierarchies in cognitive maps. Memory and Cognition 13, 208–217 (1985)
19. Brown, R.: How shall a thing be called? Psychological Review 65(1), 14–21 (1958)
20. Rosch, E.: Principles of categorization. In: Rosch, E., Lloyd, B. (eds.) Cognition and Categorization, pp. 27–48. Lawrence Erlbaum Associates, Hillsdale (1978)
21. Moravec, H.P.: Sensor fusion in certainty grids for mobile robots. AI Magazine 9, 61–74 (1988)
22. Latombe, J.C.: Robot Motion Planning. Academic Publishers, Boston (1991)
23. Newman, P., Leonard, J., Tardós, J., Neira, J.: Explore and return: Experimental validation of real-time concurrent mapping and localization. In: Proceedings of the International Conference on Robotics and Automation (ICRA 2002), Washington, D.C., USA, pp. 1802–1809 (2002)
24. Moravec, H., Elfes, A.: High resolution maps from wide angle sonar. In: Proceedings of the International Conference on Robotics and Automation (ICRA 1985), pp. 116–121 (1985)
25. Schiele, B., Crowley, J.L.: A comparison of position estimation techniques using occupancy grids. In: Proceedings of the International Conference on Robotics and Automation (ICRA 1994), vol. 2, pp. 1628–1634 (1994)
26. Burgard, W., Fox, D., Henning, D., Schmidt, T.: Estimating the absolute position of a mobile robot using position probability grids. In: Proceedings of the National Conference on Artificial Intelligence (AAAI 1996), Portland, Oregon, USA, pp. 896–901 (1996)
27. Duckett, T., Nehmzow, U.: Mobile robot self-localisation using occupancy histograms and a mixture of gaussian location hypotheses. Robotics and Autonomous Systems 34(2–3), 119–130 (2001)
28. Hähnel, D., Schulz, D., Burgard, W.: Map building with mobile robots in populated environments. In: Proceedings of the International Conference on Intelligent Robots and Systems (IROS 2002), vol. 1, pp. 496–501 (2002)
29. Hähnel, D., Burgard, W., Fox, D., Thrun, S.: An efficient FastSLAM algorithm for generating maps of large-scale cyclic environments from raw laser range measurements. In: Proceedings of the International Conference on Intelligent Robots and Systems (IROS 2003), vol. 1, pp. 206–211 (2003)
30. Leonard, J.J., Durran-Whyte, H.F., Cox, I.J.: Dynamic map building for an autonomous mobile robot. The International Journal of Robotics Research 11(4), 286–298 (1992)

31. Arras, K., Vestli, S.: Hybrid, high-precision localisation for the mail distributing mobile robot system mops. In: Proceedings of the International Conference on Robotics and Automation (ICRA 1998), pp. 3129–3134 (1998)
32. Dissanayake, M.G., Newman, P., Clark, S., Durrant-Whyte, H., Corba, M.: A solution to the simultaneous localization and map building (SLAM) problem. IEEE Transactions on Robotics and Automation 17(3), 229–241 (2001)
33. Frese, U., Schrder, L.: Closing a million-landmarks loop. In: Proceedings of the International Conference on Intelligent Robots and Systems (IROS 2006), Beijing, China (2006)
34. Arras, K., Tomatis, N., Siegwart, R.: Multisensor on-the-fly localization using laser and vision. In: Proceedings of the International Conference on Intelligent Robots and Systems (IROS 2000), Takamatsu, Japan, pp. 462–476 (2000)
35. Tardós, J.: Representing partial and uncertain sensorial information using the theory of symmetries. In: Proceedings of the International Conference on Robotics and Automation (ICRA 1992), vol. 2, pp. 1799–1804 (1992)
36. Castellanos, J.A., Tardós, J.D.: Mobile Robot Localization and Map Building: A Multisensor Fusion Approach. Kluwer Academic Publishers, Dordrecht (1999)
37. Se, S., Lowe, D.G., Little, J.: Vision-based mobile robot localization and mapping using scale-invariant features. In: Proceedings of the International Conference on Robotics and Automation (ICRA 2001), Seoul, Korea (2001)
38. Elinas, P., Sim, R., Little, J.J.: σSLAM: Slam: Stereo vision slam using the rao-blackwellised particle filter and a novel mixture proposal distribution. In: Proceedings of the International Conference on Robotics and Automation (ICRA 2006), Orlando, FL (2006)
39. Lowe, D.G.: Distinctive image features from scale-invariant keypoints. International Journal of Computer Vision 60(2), 91–110 (2004)
40. Goncalves, L., di Bernardo, E., Benson, D., Svedman, M., Ostrovski, J., Karlsson, N., Pirjanian, P.: A visual front-end for simultaneous localization and mapping. In: Proceedings of the International Conference on Robotics and Automation (ICRA 2005), Barcelona, Spain, pp. 44–49 (2005)
41. Jensfelt, P., Kragic, D., Folkesson, J., Björkman, M.: A framework for vision based bearing only 3D SLAM. In: Proceedings of the International Conference on Robotics and Automation (ICRA 2006), Orlando, FL (2006)
42. Frintrop, S., Jensfelt, P., Christensen, H.I.: Attentional landmark selection for visual slam. In: Proceedings of the International Conference on Intelligent Robots and Systems (IROS 2006), Beijing, China (2006)
43. Frintrop, S., Jensfelt, P.: Active gaze control for attentional visual SLAM. In: Proceedings of the International Conference on Robotics and Automation, ICRA 2008 (2008), `http://cognitivesystems.org/CoSyBook/chap5.asp#Frintrop08a`
44. Frintrop, S., Jensfelt, P.: Attentional landmarks and active gaze control for visual SLAM. IEEE Transactions on Robotics, special Issue on Visual SLAM 24 (2008), `http://cognitivesystems.org/CoSyBook/chap5.asp#Frintrop08b`
45. Frintrop, S.: Vocus: A visual attention system for object detection and goal-directed search. Ph.D. thesis, University of Bonn (July 2005)
46. Lu, F., Milios, E.: Robot pose estimation in unknown environments by matching 2d range scans. In: Proceedings of the Conference on Computer Vision and Pattern Recognition (CVPR 1994), pp. 935–938 (1994)

47. Gutmann, J.-S., Konolige, K.: Incremental mapping of large cyclic environments. In: Proceedings of the International Symposium on Computational Intelligence in Robotics and Automation (CIRA 1999), pp. 318–325 (1999)
48. Bertolli, F., Jensfelt, P., Christensen, H.I.: Slam using visual scan-matching with distinguishable 3D points. In: Proceedings of the International Conference on Intelligent Robots and Systems, IROS 2006 (2006), http://cognitivesystems.org/CoSyBook/chap5.asp#Bertolli06a
49. Jensfelt, P.: Approaches to mobile robot localization in indoor environments. Ph.D. thesis, Signal, Sensors and Systems (S3), Royal Institute of Technology, SE-100 44 Stockholm, Sweden (2001)
50. Tapus, A., Ramel, G., Dobler, L., Siegwart, R.: Topology learning and recognition using bayesian programming for mobile robot navigation. In: Proceedings of the International Conference on Intelligent Robots and Systems (IROS 2004), Sendai, Japan, pp. 3139–3144 (2004)
51. Anguelov, D., Koller, D., Parker, E., Thrun, S.: Detecting and modeling doors with mobile robots. In: Proceedings of the International Conference on Robotics and Automation, ICRA 2004 (2004)
52. Zender, H.: Learning spatial organization through situated dialogue, Master's thesis, Dept. of Computational Linguistics, Saarland University, Saarbruecken, Germany (2006)
53. Topp, E.A., Christensen, H.I.: Tracking for following and passing persons. In: Proceedings of the International Conference on Intelligent Robots and Systems (IROS 2005), Edmonton, Alberta, Canada (2005)
54. Topp, E.A., Hüttenrauch, H., Christensen, H., Severinson Eklundh, K.: Bringing together human and robotic environment representations – a pilot study. In: Proceedings of the International Conference on Intelligent Robots and Systems (IROS 2006), Beijing, China (2006)
55. Kruijff, G.-J.M., Zender, H., Jensfelt, P., Christensen, H.I.: Situated dialogue and spatial organization: What, where... and why? International Journal of Advanced Robotic Systems, Special Issue on Human-Robot Interaction 4(1), 125–138 (2007), http://cognitivesystems.org/CoSyBook/chap5.asp#kruijff07jars
56. Lee, D.T., Lin, A.K.: Computational complexity of art gallery problems. IEEE Transactions on Information Theory 32(2), 276–282 (1986)
57. López, D.G.: Combining object recognition and metric mapping for spatial modeling with mobile robots. Master's thesis, Royal Institute of Technology (July 2007)
58. Ekvall, S., Kragic, D.: Receptive field cooccurrence histograms for object detection. In: Proceedings of the International Conference on Robotics and Automation, IROS 2005 (2005)
59. Sjö, K., Paul, C.: Object localization using bearing only visual detection. In: Proceedings of the 10th International Conference on Intelligent Autonomous Systems (2008), http://cognitivesystems.org/CoSyBook/chap5.asp#Sjoe08b
60. Ekvall, S., Kragic, D., Jensfelt, P.: Object detection and mapping for service robot tasks. Robotica: International Journal of Information, Education and Research in Robotics and Artificial Intelligence 25(2), 175–187 (2007)
61. Brunskill, E., Kollar, T., Roy, N.: Topological mapping using spectral clustering and classification. In: Proceedings of the International Conference on Intelligent Robots and Systems (IROS 2007), San Diego (2007)

62. Siagian, C., Itti, L.: Biologically-inspired robotics vision monte-carlo localization in the outdoor environment. In: Proceedings of the International Conference on Intelligent Robots and Systems (IROS 2007), San Diego, CA, USA (2007)
63. Zender, H., Mozos, Ó.M., Jensfelt, P., Kruijff, G.-J.M., Burgard, W.: Conceptual spatial representations for indoor mobile robots. Robotics and Autonomous Systems 56(6), 493–502 (2008), http://cognitivesystems.org/CoSyBook/chap5.asp#zender08ras_fs2hsc
64. Buschka, P., Saffiotti, A.: A virtual sensor for room detection. In: Proceedings of the International Conference on Intelligent Robots and Systems, IROS 2002 (2002)
65. Pronobis, A., Caputo, B., Jensfelt, P., Christensen, H.I.: A discriminative approach to robust visual place recognition. In: Proceedings of the IEEE/RSJ International Conference on Intelligent Robots and Systems (IROS 2006), Beijing, China (2006), http://cognitivesystems.org/CoSyBook/chap5.asp#pronobis06iros
66. Filliat, D.: A visual bag of words method for interactive qualitative localization and mapping. In: Proceedings of the International Conference on Robotics and Automation (ICRA 2007), Roma, Italy (2007)
67. Ulrich, I., Nourbakhsh, I.: Appearance-based place recognition for topological localization. In: Proceedings of the International Conference on Robotics and Automation (ICRA 2000), San Francisco, CA, USA (2000)
68. Blaer, P., Allen, P.: Topological mobile robot localization using fast vision techniques. In: Proceedings of the International Conference on Robotics and Automation (ICRA 2002), Washington, DC, USA (2002)
69. Murillo, A.C., Guerrero, J.J., Sagues, C.: Surf features for efficient robot localization with omnidirectional images. In: Proceedings of the the International Conference on Robotics and Automation (ICRA 2007), Roma, Italy (2007)
70. Valgren, C., Lilienthal, A.J.: Incremental spectral clustering and seasons: Appearance-based localization in outdoor environments. In: Proceedings of the International Conference on Robotics and Automation (ICRA 2008), Pasadena, CA, USA (2008)
71. Tapus, A., Siegwart, R.: Incremental robot mapping with fingerprints of places. In: Proceedings of the International Conference on Intelligent Robots and Systems (IROS 2005), Edmonton, Alberta, Canada (2005)
72. Linde, O., Lindeberg, T.: Object recognition using composed receptive field histograms of higher dimensionality. In: Proceedings of the International Conference on Pattern Recognition (ICPR 2004), Cambridge, UK (2004)
73. Vapnik, V.: Statistical Learning Theory. Wiley and Son, New York (1998)
74. Freund, Y., Schapire, R.E.: A decision-theoretic generalization of on-line learning and an application to boosting. In: Vitányi, P.M.B. (ed.) EuroCOLT 1995. LNCS, vol. 904, pp. 23–37. Springer, Heidelberg (1995)
75. Luo, J., Pronobis, A., Caputo, B., Jensfelt, P.: Incremental learning for place recognition in dynamic environments. In: Proceedings of the IEEE/RSJ International Conference on Intelligent Robots and Systems (IROS 2007), San Diego, CA, USA (2007), http://cognitivesystems.org/CoSyBook/chap5.asp#luo07iros
76. Kuipers, B., Beeson, P.: Bootstrap learning for place recognition. In: Proceedings of the 18th National Conference on Artificial Intelligence, AAAI 2002 (2002)

77. Mozos, O.M.: Semantic place labeling with mobile robots, Ph.D. thesis, University of Freiburg, Freiburg, Germany (July 2008)
78. Rottmann, A.: Bild- und laserbasierte klassifikation von umgebungen mit mobilen robotern, Master's thesis, University of Freiburg, Department of Computer Science (2005) (in German)
79. Cristianini, N., Taylor, J.S.: An Introduction to SVMs and Other Kernel-based Learning Methods. Cambridge University Press, Cambridge (2000)
80. Chapelle, O., Haffner, P., Vapnik, V.: SVMs for histogram-based image classification. Transactions on Neural Networks 10(5)
81. Rothganger, F., Lazebnik, S., Schmid, C., Ponce, J.: 3D object modeling and recognition using local affine-invariant image descriptors and multi-view spatial constraints. International Journal of Computer Vision 66(3)
82. Wallraven, C., Caputo, B., Graf, A.: Recognition with local features: the kernel recipe. In: Proceedings of the International Conference on Computer Vision, ICCV 2003 (2003)
83. Platt, J.C.: Fast training of support vector machines using sequential minimal optimization. In: Schölkopf, B., Burges, C., Smola, A. (eds.) Advances in Kernel Methods: Support Vector Learning, pp. 185–208. MIT Press, Cambridge (1999)
84. Nilsback, M.E., Caputo, B.: Cue integration through discriminative accumulation. In: IEEE Computer Society Conference on Computer Vision and Pattern Recognition, CVPR 2004 (2004)
85. Poggio, T., Torre, V., Koch, C.: Computational vision and regularization theory. Nature 317
86. Triesch, J., Eckes, C.: Object recognition with multiple feature types. In: Proceedings of the International Conference on Artificial Neural Networks, ICANN 1998 (1998)
87. Matas, J., Marik, R., Kittler, J.: On representation and matching of multicoloured objects. In: Proceedings of the International Conference on Computer Vision, ICCV 1995 (1995)
88. Clark, J., Yuille, A.: Data fusion for sensory information processing systems. Kluwer Academic Publishers, Dordrecht (1990)
89. Duda, R., Hart, P., Stork, D.: Pattern Classification, 2nd edn. Wiley, Chichester (2001)
90. Pronobis, A., Caputo, B.: The more you learn, the less you store: Memory-controlled incremental SVM, IDIAP-RR 51, IDIAP (2006), `http://cognitivesystems.org/CoSyBook/chap5.asp#pronobis06idiap`
91. Domeniconi, C., Gunopulos, D.: Incremental support vector machine construction. In: Proceedings of the International Conference on Data Mining, ICDM 2001 (2001)
92. Syed, N.A., Liu, H., Sung, K.K.: Incremental learning with support vector machines. In: Proceedings of the International Joint Conferences on Artificial Intelligence, IJCAI 1999 (1999)
93. Cauwenberghs, G., Poggio, T.: Incremental and decremental support vector machine learning. In: Advances in Neural Information Processing Systems, NIPS 2000 (2000)
94. Orabona, F., Castellini, C., Caputo, B., Luo, J., Sandini, G.: Indoor place recognition using online independent support vector machines. In: 18th British Machine Vision Conference (BMVC 2007), Warwick, UK (2007)
95. Downs, T., Gates, K.E., Masters, A.: Exact simplification of support vector solutions. Journal of Machine Learning Research 2

96. Pronobis, A., Caputo, B., Jensfelt, P., Christensen, H.I.: A discriminative approach to robust visual place recognition (2006), http://cognitivesystems.org/cosybook/videos.asp#robVisPR
97. The KTH-IDOL2 database, http://cognitivesystems.org/cosybook/datasets.asp#idol
98. Luo, J., Pronobis, A., Caputo, B., Jensfelt, P.: The KTH-IDOL2 database. Tech. Rep. CVAP304, Kungliga Tekniska Hoegskolan, CVAP/CAS (October 2006), http://cognitivesystems.org/CoSyBook/chap5.asp#luo06kth
99. Pronobis, A.: Multi-modal semantic labeling of space (2008), http://cognitivesystems.org/cosybook/videos.asp#MMsemLab
100. Yamauchi, B.: A frontier-based approach for autonomous exploration. In: Proceedings of the International Symposium on Computational Intelligence in Robotics and Automation, Monterey, CA, pp. 146–151 (1997)

6

Planning and Failure Detection

Michael Brenner[1], Christian Plagemann[1], Bernhard Nebel[1],
Wolfram Burgard[1], and Nick Hawes[2]

[1] Albert-Ludwigs-Universität Freiburg, Department of Computer Science, Freiburg, Germany
{brenner,plagem,nebel,burgard}@informatik.uni-freiburg.de

[2] Intelligent Robotics Laboratory, School of Computer Science, University of Birmingham, Birmingham, UK
nah@cs.bham.ac.uk

6.1 Introduction

The capacity for planful behavior is one of the major characteristics of an intelligent agent. When acting in realistic environments, however, reasoning about how to achieve one's goals is complicated significantly, both by the limited perceptions of the agent and the high dynamics of the environment, especially when other intelligent agents are present. Fortunately, when acting continuously in such an environment, agents can actively try to reduce their uncertainties, for example by deliberative exploration, cooperation with others, and monitoring of failures.

In this chapter, we will present representations and techniques developed in the CoSy project for continual planning, acting, and failure detection in dynamic domains with multiple human and artificial agents. These methods intelligently interleave planning, communicating, acting and execution monitoring, thereby avoiding consideration of all possible contingencies and detecting failures as early as possible.

We have used these methods successfully in our implemented robot systems. The robots thus created are able to pursue long-term goals intelligently and autonomously, to react to failures and exogenous changes dynamically and to proactively interact with each others and with humans. See Chapters 9 and 10 for more detailed descriptions of continual planning in the robot scenarios.

Continual Planning

Practical cognitive agents, e.g. robots acting in the real world, are faced with environments that make planning particularly hard: usually, an agent has only very limited knowledge about the current state of the world, mainly because its sensing capabilities are limited. Worse, other agents (including "nature" itself)

H.I. Christensen et al. (Eds.): Cognitive Systems, COSMOS 8, pp. 223–264.
springerlink.com

may change the world without the agent noticing, such that its former knowledge about the world may even become incorrect. In such highly dynamic, partially observable, multiagent environments it is practically impossible to take all possible contingencies into account in advance; the corresponding AI Planning problems, *e.g.* conformant, contingent or probabilistic planning, are highly intractable.

Fortunately, in many environments an alternative planning approach is possible: Instead of considering many possible futures in advance, an agent may execute parts of its plan in order to gather additional information, thereby reducing the number of possible contingencies that it has to take into account for the remaining planning. This technique of interleaving planning, plan execution and execution monitoring is called Continual Planning (CP)[1]. CP is often advocated as a practical approach to planning in dynamic or incompletely known domains, but has not been investigated in much detail in the literature. In particular, the following questions must be answered by any CP approach: If the agent does not have a complete plan when it starts acting, how can it nevertheless behave in a goal-directed manner? How can the agent decide which parts of the problems solving process it should postpone? Do CP agents *plan* their later replanning and if so, how? What is the role of knowledge-gathering actions in CP? In this chapter, we present a new principled approach to CP that tries to answer these questions.

Our approach is based on the idea of *active knowledge-gathering*: instead of planning for possible contingencies, agents try to learn more about the state of the world directly. In order to enable agents to reason about how they can gather additional knowledge it is necessary to explicitly model the agents' beliefs as well as their sensing capabilities as part of their formal planning domain. Agents can then *plan* how to extend their knowledge.

For realistic planning problems it is infeasible to branch over the possible outcomes of sensing actions and to generate a plan for all possible contingencies in advance. Instead, we want to emulate human situated problem solving, and enable the planner to *postpone* decisions to the moment where the actual perception has been made. In other words, we want to "hide" potential conditional subplans until the agent has enough information to plan concretely. For this purpose, we introduce the concept of *assertions*. Assertions are a special kind of action specified in the planning domain. In contrast to normal actions assertions cannot be directly executed. Instead, the domain designer asserts that the effects of the assertion can be achieved by a subplan if its preconditions are satisfied. In contrast to the similar concept of *methods* in Hierarchical Tasks Networks (HTNs), no decompositions need to be specified for assertions; this is done by the planner itself in in later planning phases.

[1] The term *continuous* planning is also often used to describe this form of planning. We prefer the term Continual Planning, since it refers to a repeatedly interrupted and partially postponed planning process rather than a permanently ongoing one. See the survey by desJardins et al. for additional reasons [1] .

Failure Detection and Monitoring

For robotic systems deployed to realistic environments, it cannot be assumed that the sensors and actuators are perfectly reliable in the sense that their input always corresponds to the expected input and that there is no malfunction of the sensors or actuators. Such situations, however, cannot always be dealt with on the abstract level of the planning system—which may realize that an action has failed but which typically has too little insight into the lower-level processes to be able to infer the *cause* of the failure in order to react appropriately. For this reason, we have studied the failure detection and monitoring problem also on the lower system level, closer to perception and actuation. The monitoring system on this level, which is described in Sec. 6.4, provides (symbolic) status information about processes and low-level tasks to the planner.

We follow the probabilistic state estimation approach to failure detection and describe the system to be monitored as a dynamic Bayesian network [2]. Particle filtering [3] is then applied in conjunction with learned proposal distributions for inference. As we show in experiments with a real robot, learning proposal distributions for failure modes from previous experience or in simulation has the potential of drastically increasing the robustness of the detector while at the same time reducing the number of samples required.

As a visual motivation for the general approach and as a test bed for evaluation, we consider (a) localizing an autonomous, wheeled robot relative to a known environment, (b) detecting collisions with (unseen, movable) obstacles online and (c) estimating their physical properties (e.g., location and weight) such that they can be dealt with safely. Our system does not require additional hardware such as inertial sensors or bumper rings.

Continual Collaborative Planning

Dealing with limited knowledge, execution failures and other uncontrollable changes is generally important for planning in dynamic environments, but becomes even more important in multiagent environments. Since other agents are the major source of dynamics in such environments, uncertainty can be greatly reduced by monitoring their behaviour and, in particular, by communicating with them. However, CP in multiagent environments raises additional questions: How does the concept of interleaving planning, execution and monitoring generalise to the multiagent cases of collaborative planning and synchronised execution? Which role does communication play there? Are there fundamental differences between action planning and communication planning? This chapter sheds some light on these questions by proposing the concept of Collaborative Continual Planning (CCP). We show how CCP agents can behave more successfully than non-collaborative agents in dynamic environments.

Since Continual Planning approaches can only be tested in environments where agents can actually interleave planning, execution and sensing, we have developed MAPSIM, a software environment that can automatically generate multiagent simulations from planning domains. In other words, MAPSIM interprets a planning domain description as an executable model of the environment in which agents can perceive, plan, act and engage in dialogues for task-orientated collaboration. While MAPSIM was developed to investigate the approach to continual planning presented her, it can also be used for evaluating agents that use different (continual) planning methods.

Since CCP switches between planning, execution and communication as necessary, it is particularly suited to domains where communicative and physical actions must be interleaved in order to act successfully. In such domains, the continual update of the agents' goals during CCP forms the basis of our approach to mixed-initiative *situated dialogue*. Indeed, in the CoSy robot demonstrators (see Chapters 9 and 10) CCP is used for high-level pragmatic reasoning in Human-Robot Interaction, i.e. for reasoning about the role communicative actions play in a plan to achieve the overall, non-communicative goals of the agent. Thus, CCP complements the specific techniques for situated dialogue that will be presented in Chapter 8.

6.2 The Multiagent Planning Language MAPL

Planning in dynamic multiagent (MA) environments means reasoning about the environment, about (mutual) beliefs, perceptual capabilities and the possible physical and communicative actions of oneself and of others. All of these elements can be modeled in the multiagent planning language MAPL that we have developed. In this section, we introduce MAPL informally and discuss its suitability for building cognitive systems; formal definitions can be found elsewhere [4].

MAPL is a multiagent variant of PDDL, the de facto standard language for classical planning [5]. One important extension in MAPL is the use of multi-valued state variables (MVSVs) instead of propositions. For example, a state variable *color(ball)* would have exactly one of its possible *domain* values *red, yellow,* or *blue* compared to the three semantically unrelated propositions *(color ball red), (color ball yellow), (color ball blue)*, all or none of which could be true in a given STRIPS state. MVSVs have been used successfully in classical planning in recent years [6], but they also provide distinctive benefits when used for multiagent planning. In particular, we can use MVSVs to model *knowledge* and *ignorance* of agents: If no value is known for a state variable it is *unknown* (contrast this with the closed world assumption of classical planning where what is not known to be true is *false*). This concept can also extended to beliefs about other agents' beliefs and mutual beliefs which are modeled by so-called **belief state variables**.

In our robot systems (see Chapters 9 and 10), MAPL state descriptions are used to maintain a representation of the robot's knowledge in a form that is suitable for long-term and cross-component planning and execution. The

Binding subarchitecture (cf. Chapter 2) fuses information into a representation that uses multi-valued features, too, and thus can easily translated into (and back from) MAPL.

MAPL **actions** are similar to those of PDDL. In MAPL, every action has a **controlling agent** who will execute the action and, in particular, controls *when* this will happen. Agents are fully autonomous when executing actions, *i.e.* there is no external synchronization or scheduling component. As a consequence an action will only be executed if, in addition to its preconditions being satisfied, the controlling agent *knows* that they hold. Implicitly, all MAPL actions are extended with such **knowledge preconditions**. Similarly, there are implicit **commitment preconditions**, intuitively describing the fact that another agent will only execute actions if he has agreed to do so.

A MAPL domain can define three different ways to affect the beliefs of agents (necessary, *e.g.* in order to satisfy knowledge preconditions): Sensing, copresence (joint sensing) and communication. All three are MAPL actions that have knowledge effects. **Sensor models** describe the circumstances in which the current value of a state variable can be perceived. **Copresence models** are multiagent sensor models that induce mutual belief about the perceived state variable. Informally, agents are copresent when they are in a common situation where they do not only perceive the same things but also each other. Individual and joint sensing are important for multiagent systems because they help avoiding *communication*: An agent does not need to ask for what he senses himself, and he does not need to verbalize what he knows to be perceived by the other agents as well. Communicative acts come in two forms: as **declarative statements**, *i.e.* actions that (similarly to sensory actions) can change the belief state of another agent in specific circumstances, or as **requests** which correspond to *questions* and *commands*. Requests do not have to be modeled explicitly in the MAPL domain, but automatically generated during CCP; this is discussed in the next section.

Similarly to other planning languages, MAPL (in its current form) describes actions in a purely qualitative way: no costs or probabilities are associated with them. This permits the use of performant planning techniques—however, it prohibits the representation of additional, quantitative information that might be available. This is of particular importance for sensor models where most qualitative action effects are unknown until execution, while some quantitative information might be available and helpful for selecting the right operators, e.g. the reliability of a sensory action or its expected information gain. For a specific model of planning for visual processing with quantitative information, see the POMDP formalisation and planning approach presented in Chapter 2.

MAPL **goals** correspond to PDDL goal formulae. However, MAPL has two additional goal-like constructs: **Temporary subgoals** (TSGs) are mandatory, but not necessarily permanent goals, *i.e.* they must be satisfied by the plan at some point, but may be violated in the final state. **Assertions**, on the other hand, describe *optional* "landmarks", *i.e.* TSGs that may helpful in achieving specific effects in later phases of the continual planning processes, which cannot be fully planned for yet because of missing information [7, 4]. Assertions will be described in more detail in Section 6.3.

MAPL plans differ from PDDL plans in being only *partially ordered.* This is inevitable since we assume that there is no central executive which could guarantee a totally ordered execution. We use the term **asynchronous plans** since MAPL plans also allow for *concurrent* occurrence of actions. An asynchronous plan that guarantees that the implied knowledge preconditions will be satisfied during execution (*e.g.* by explicitly naming the perceptions to be made and speech acts to be used) is called **self-synchronizing plan** because it "explains" how the agents can coordinate their behavior during execution.

6.3 Continual Planning

The term "Continual Planner" is often used simply to describe a system that replans in light of state changes which have rendered its previous plan invalid, *i.e.* a systes that does execution monitoring and replanning. While these are indeed important aspects of any continual planning approach, many such systems will only start acting when they have found a *complete* plan. In other words, while the execution process can be interrupted, planning is monolithic.

In this chapter, we argue for an extended notion of continual planning where agents can *suspend* the planning process and start acting. For example, in the Explorer scenario (see Chapter 10) it is crucial that the robot can physically move to another room before it can make detailed plans about how to achieve its given task there.

But when and why should agents suspend planning? How can they act purposefully when they have not finished planning yet? And when will they resume planning again? We will answer these questions in this section. The key idea of our approach is that agents will deliberately *postpone* those parts of their planning process that concern currently indeterminable contingencies. However, postponing subproblem resolution will only be allowed if instead they engage in *active information gathering.* This ensures that, as soon as the additional knowledge is available, the planning process can be resumed and the plan can be further refined.

6.3.1 Assertions

In Continual Planning, we want to allow the planning agent to deliberately postpone parts of its planning to later phases in the plan-execution-monitoring cycle when it has gathered more information. In other words, we want to "hide" a conditional subplan until the agent has enough information to resolve the contingency. For this purpose, we introduce the concept of *assertions.*

Assertions are virtual MAPL actions that a planner can reason about and include in a plan as usual, but that can never be executed. The name "assertion" comes from the role these actions play for continual planning: they *assert* that, once their precondition will have been achieved, their effects will be achievable, too – although not by executing the assertion, but by means of a new plan. Therefore, when during plan execution an assertion actually becomes executable, it is *expanded*, *i.e.* a new planning phase is triggered. In this planning phase the planner can make use of the information gained during the last execution phase and can replace the assertion with concrete, executable actions. In fact, a planner need not even wait to expand an assertion until it is executable. It is sufficient that a specific subset of preconditions of the assertion is satisfied in the current state to inform the planner that replanning is now possible. These special preconditions are called the *replanning conditions*. Formally we define assertions as follows:

Definition 1. *An* ***assertion*** *is an event e with a distinguished, non-empty set of preconditions $repl(e) \subseteq pre(e)$, called the* replanning conditions. *Preconditions $p \notin repl(e)$ are called the* ordinary preconditions *of e. If $repl(e) = \emptyset$, then e is an* ordinary action. *We denote the set of assertions with $\mathcal{E}_{ass} \subseteq \mathcal{E}$.*

When the replanning condition of an assertion has been satisfied, it can be *expanded*, *i.e.* it can be replaced by a subplan that achieves the asserted effects.

Definition 2. *An assertion e is* ***expandable*** *in a state s if $repl(e) \subseteq s$. For a plan P, an assertion $e \in P$ is said to be* ***permanently expandable*** *in a state s if e is expandable in s and there is no event $e' \in P$ with $e' \prec e$ that threatens any replanning condition $c \in repl(e)$.*

Usually, assertions that become expandable during the execution of a plan P will also be permanently expandable. However, sometimes a replan condition is only satisfied temporarily. For example, an agent may know that another agent will soon destroy it. In such a case, it will not make sense yet to devise a concrete plan for the assertion. Therefore, the semantics of plans with assertions (or *assertional plans*) will rest on permanently expandable assertions.

The power of assertions for Continual Planning results from the following change in the semantics of plans:

Definition 3. *The state res(s,P) resulting from the* ***symbolic execution*** *of an an assertional plan P in the current state s is defined as follows:*

If any assertion $e \in P$ is permanently expandable in s, then $res(s, P)$ is undefined. Otherwise, $res(s, P)$ is defined as $res(s, P_{TO})$ where P_{TO} is an arbitrary total order of P.

If $res(s, P)$ is defined for some state s and plan P, we say that P is ***valid*** *in s.*

This definition describes how assertions will trigger a new planning phase: as soon as they are permanently expandable, they will render an otherwise valid plan obsolete, thereby forcing the continual planner to switch back into planning mode.

Def. 3 makes an important distinction that is not present in other planning formalisms: it distinguishes between the current state s and the projected future states $res(s, P)$. In particular, replanning conditions work as normal action preconditions as long as long as they do not hold in s. When, however, the plan has been executed until the point where an assertion is actually permanently expandable in s, the semantics of assertions changes and they will trigger replanning.

This unusual semantics induces an interesting interplay between planning and execution: Assertions will first enable the planner to find a plan and start its execution. Later, however, the same assertions will render the plan obsolete again and force the planner to switch back into planning mode. This process will be specified in detail by Alg. 2 in next section.

Note that all directly executable actions in a valid plan P, *i.e.* those events $e \in P$ whose preconditions are satisfied in the current state s, can never be assertions (otherwise their replanning conditions would be satisfied, too, and the plan would no longer be valid). This is important, because it means that when an agent selects an action to execute among the ones currently enabled in the plan, it can never be an assertion. Thus, assertions are completely *virtual* actions that only appear in plans, but never in execution.

In summary, assertions are virtual actions that can be used to enable planning for active, *goal-directed information gathering.* Instead of branching on possible perceptions, assertions help to deliberately postpone contingency resolution until the missing information is available. Thus, assertions not only abstract from subproblems in a plan (in a way similar to hierarchical task networks; cf. Sec. 6.10), but additionally lead to a *temporal* decomposition of the planning *process.* The next section shows, how this is accomplished in a planning algorithm.

6.3.2 Assertional Planning

Basically, Continual Planning consists of three interleaved phases:

1. (Re-)planning in order to determine how to achieve the current goals from the current state
2. Plan execution
3. Perception of self-induced or exogenous changes to the world

A continual planning agent that tightly integrates these phases is described in Alg. 1. Execution and monitoring of expected changes are particularly interdependent. Therefore Alg. 1 has only two major subprocesses, which are detailed in Algs. 2 and 3. In Section 6.7 we will further extend this model with *collaborative* capabilities.

Algorithm 1. Continual Planning Agent(S, G)

$P = \emptyset$
while $S \not\supseteq G$ do
  $P =$ MonitoringAndReplanning(S, G, P)
  if $P = \emptyset$ then
    return "cannot achieve goal G"
  $(S, P) =$ ExecutionAndStateEstimation(S, P)
return "goal reached"

Alg. 2 shows how a new planning phase is triggered: Step (1) describes *plan monitoring*, *i.e.* checking whether a changed current state or an expandable assertion makes the current plan invalid. If this is the case, Alg. 2 first determines the prefix of the asynchronous plan that is still executable (step 2), then extends it with a newly planned suffix. Note that since a multiagent plan is only partially ordered, the "obsolete suffix" consists only of those actions that *causally* rely on preconditions that have become unachievable. Of course, simple replanning is also possible if *plan stability* is not an issue. However, especially in collaborative settings where agents need to reach agreement over their plans, breaking plan stability is costly since it may lead to *asynchronous backtracking*, *i.e.* plan revision recursively concerning other agents [8].

Algorithm 2. MonitoringAndReplanning(S, G, P)

if $res(S, P) \not\supseteq G$ or any assertion $e \in P$ is permanently expandable then (1)
  $E = \mathcal{E} \setminus \{e \in \mathcal{E}_{ass} \mid repl(e) \subseteq S\}$
  $P' =$ RemoveObsoleteSuffix(P) (2)
  $P'' =$ Planner($E, res(S, P'), G$) (3)
  $P =$ concat(P', P'') (4)
return P

In order to exploit the power of state-of-the-art planning systems, step (3) of Alg. 2 uses calls to an unspecified classical planner Planner (supporting the STRIPS subset of PDDL) as a subroutine. In order to allow this modularisation and still be compliant with the semantics of plans with assertions, the algorithm *removes* all expandable assertions from the set of possible events before calling the Planner subroutine. However, it may be more convenient to modify the planner (as we have done in our implementation) so that it assures itself that expandable assertions are not included in a plan. Since most modern non-temporal planners produce totally-ordered plans, the generated must be converted to asynchronous plans. We use a straightforward algorithm for doing this that adds new actions to an asynchronous plans as early as possible but making sure that no conflict arise. Cf. Bäckström's work for a discussion of several variants of this approach [9].

Algorithm 3. EXECUTIONANDSTATEESTIMATION(S, P)

$e =$ *choose* a first level event from P
$S' = app(S, e)$ (1)
$exp =$ EXPECTEDPERCEPTIONS(S', $\mathcal{E}_{sense}$) (2)
if $agt(e) =$ self then
  EXECUTE(e) (3)
$perc =$ GETSENSORDATA() (4)
if $perc \supseteq exp$ then (5)
  remove e from P
$S =$ FUSE(S', $perc$) (6)
return (S,P)

Alg. 3 describes the execution phase of Continual Planning: First, it non-deterministically chooses an event e on the initial level of the plan, *i.e.* one whose preconditions are satisfied in the current state. Note that, due to Def. 3, e can *not* be an assertion. Steps (1-2) first compute the state and perceptions expected as a result of executing event e. If the planning agent is also the executing agent, he will execute e (step 3), otherwise the agent will do nothing but try to *observe* e.

Steps (4-5) try to match the expected results with the real perceptions after executing e. This is also the case if the planning agent was not the executing agent of e, but wants to determine if another agent has executed e. Note that by not removing e from the plan until its occurrence is perceived the CP algorithm enters a waiting loop: Alg. 3 will be executed repeatedly until Alg. 2 detects that external events have made P invalid. Step (6) tries to estimate the next state by fusing old knowledge, new perceptions and expected changes (that might not have been observable, but predicted by e). Basically, FUSE applies those expected effects that do not contradict *perc*. Contradiction is defined in terms of mutually exclusive state variable assignments.

```
(:action move_A
 :agent (?a - agent)
 :parameters (?to - location)
 :variables (?from - location ?d - door)
 :replan (KIF ?a (doorstate ?d))
 :precondition (and (pos ?a : ?from)
      (entrance ?d ?from) (entrance ?d ?to))
 :effect (pos ?a : ?to))
```

Fig. 6.1. Assertion for planning movements between rooms that ignore the state of doors until perceived during execution

An Example

Consider a scenario in which a robot has to explore an apartment, *i.e.* it must enter all rooms at least once. The robot has a map of the apartment, but does not know which of the doors between adjacent rooms are open. It can sense the state of all doors in a room as soon as it enters it (this is specified by a MAPL sensor model). A *contingent* plan for this scenario modelled as a *tree* where each possible perception during plan execution spawns a new plan branch could result in a number of branches exponential in the number of doors in the worst case (depending on the layout of the map) [10, 11]. This is due to the fact that, while a contingent planning algorithm will *reason* about future sensing, it cannot actively try to reduce uncertainty by actually *executing* sensing actions. In contrast, this kind of *active information gathering* is encouraged by the assertion `move_a` shown in Fig. 6.1. Note that this assertion requires only that the robot knows *whether* a door is open or not, *i.e.* `(KIF ?a (doorstate ?d))` or, formally, $doorstate(?d)^{?a}_{KIF} = \top$. However, in order to satisfy this condition, the robot will have to plan a `sense-door` sensing action which, when executed, will provide the robot with the information really needed, *i.e.* the real value of $doorstate(?d)^{?a}$, which it can then use in the next planning phase to concretise its plan for exploring the appartment.

6.4 Probabilistic Monitoring of Dynamic Processes

In this section, we introduce our approach to monitoring low-level processes in the robotic system. This provides the basis for robustly verifying and refuting assertions made by the planning system and it provides (symbolic) status information about processes and low-level tasks to the planner allowing it to react appropriately.

In our approach, the system to be monitored is modelled as a dynamic Bayesian network (DBN) [2] and its state is estimated sequentially using sampling-based estimators in conjunction with learned sampling functions. As the main contribution, which goes beyond the application to the process monitoring task, we present a novel way of improving the generation of state samples in particle filters. This step has to be performed frequently and it has a major influence on the efficiency and robustness of the filter. Specifically, we show how Gaussian processes (GPs), that is, a state-of-the-art Bayesian approach to regression and classification, can be used to incorporate prior knowledge into the state estimation process.

DBNs are widely used to model the dynamic behavior of artifical and natural systems. In a DBN, the variables of interest are indexed by time and related to one another using conditional probability distributions (CPDs), which in general span multiple time frames to describe the dynamics. The result is a (typically sparse) network of local dependencies for which efficient learning and inference mechanisms have been developed. A particularly flexible and, thus, often-used inference technique are the so called particle filters (PFs). They represent the state of the system by a finite set of weighted samples and perform inference by updating the sample set iteratively using the DBN structure, the conditional models, and the set of observed (thus constant) variables.

Our primary goal is to improve the generation of state samples in the particle filter algorithm [3], which is a step that has to be performed frequently and that has a major influence on the efficiency and robustness of the filter. In particular, we consider data-driven sampling, i.e. the generation of state hypotheses at locations that are most promising given the latest sensor measurements. Our major contribution is, that the sampling policy is learned from past experience, using Gaussian process (GP) regression for continuous state variables and classification for discrete ones. Given the GPs ability to also estimate the uncertainty of its predictions, we can derive a sound way of integrating these *informed proposals* into the particle filter framework. As a result, the estimation process is more robust since the sampling density is higher in the important parts of the state space while it is also more efficient because less samples are "wasted" in unlikely regions.

As a running example and test-bed for our approach, we consider the problem of online failure detection and recovery for a mobile robot. Concretely, we face the task of (a) localizing an autonomous, wheeled robot relative to an environment, (b) to detect collisions with (unseen and movable) obstacles online and (c) to estimate their physical parameters such that they can be dealt with safely.

In the following, we first formalize the sequential state estimation problem and describe how to solve it using particle filters. We then introduce the concept of Gaussian process proposals and specifically focus on the mobile robot localization and collision detection problem outlined above.

6.4.1 Sequential State Estimation

One of the fundamental problems in robotics and engineering is to estimate the state of a system given a sequence of acquired sensor measurements and action commands that have been executed. In order to perform this task sequentially, i.e. in an iterative way whenever new information is available, the system is typically modeled using a dynamic Bayesian network. Assuming a discretization of time—either by modeling iterations of equal duration or by taking the finite set of measurement and action events—, there is a finite set of time-indexed variables that has to be modeled. The DBN factorizes the joint posterior over these variables to a (typically sparse) network of dependent variables and their corresponding conditional probability distributions. Here, these 'local models' can be limited to a certain time frame or they can span several to model the behavior of the system over time.

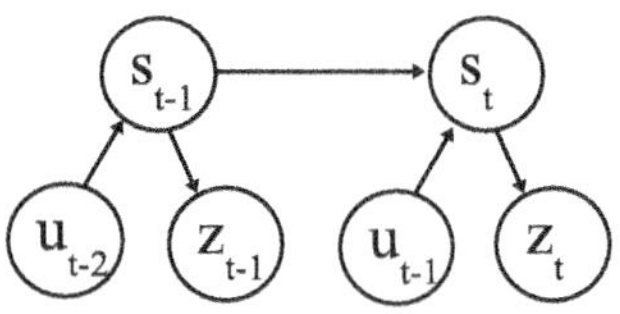

Fig. 6.2. DBN for the mobile robot domain

Mobile Robot Localization

In the case of mobile robot localization, for instance, the (dynamic) state of the system is the three-dimensional pose $s_t = x_t = (x_t, y_t, \theta_t)$ of the robot relative to its environment. Additional variables include the sensor measurements $z_{1:t}$ as well as the action commands $u_{0:t-1}$ that have been issued—w.l.o.g. assuming time-synchronous state, action, and measurement variables for ease of notation. A typical DBN for such a system is composed of a *prior* distribution over poses $p(s_0)$, a *transition model* $p(s_{t+1}|s_t, u_t)$ and an *observation model* $p(z_t|s_t)$. Figure 6.2 shows how these CPDs are connected to yield the typical DBN for state estimation problems. It can be read from the graph, that each variable is independent from all others given its direct predecessors (see [2] for a general discussion of graphical models). Given the modeled independence assumptions, the state posterior can be formulated recursively as

$$p(s_t \,|\, z_{1:t}, u_{0:t-1}) = \int p(s_t \mid s_{t-1}, z_t, u_{t-1})\, p(s_{t-1} \mid z_{0:t-1}, u_{0:t-2})\, ds_{t-1} \qquad (6.1)$$

$$\propto \underbrace{p(z_t \mid s_t)}_{\text{obs. model}} \int \underbrace{p(s_t \mid s_{t-1}, u_{t-1})}_{\text{transition model}}\, \underbrace{p(s_{t-1} \mid z_{0:t-1}, u_{0:t-2})}_{\text{recursive term}}\, ds_{t-1}\,.$$

In order to achieve a tractable approximation of this equation, one has to make additional assumptions about the distributions involved. Important classes of such approximations include

Discrete filters (DF) where the state space is discretized—typically using a regular grid—which turns the integral in the equation above into a sum over possible predecessor states,

Kalman filters (KF) for which the distributions over states $p(s_t|\ldots)$ and the observation model are assumed Gaussian and the transition model is linear, the extended Kalman filter (EKF) for which the assumption of linearity is relaxed, and

Particle filters (PF) which represent the distributions over states by a weighted set of state hypotheses which are manipulated individually according to the transition and observation models.

Particle filters are flexible in terms of distributions that can be represented, they are easy to implement, and they can convert heuristics into provably correct algorithms through the concept of *proposal distributions* (see Chapter 5 in [2]).

6.4.2 Particle Filters for Nonparametric Bayesian Filtering

Particle filters seek to approximate the integral in Eq. (6.1) using Monte Carlo integration. Concretely, they represent distributions over states $p(s_t|\dots)$ by sets of weighted samples $\mathcal{X}_t = \{\langle s_t^{[i]}, w_t^{[i]}\rangle\}_{i=1}^n$. Starting from a prior distribution $\mathcal{X}_0 \sim p(s_0)$, the particle set is updated sequentially by executing the following steps in each iteration:

1. *Sampling* (or *motion update*): The next generation of particles $\{s_t^{[i]}\}$ is obtained from the generation $\{s_{t-1}^{[i]}\}$ by sampling from a proposal distribution π.
2. *Importance Weighting* (or *observation update*): Importance weights $w_t^{[i]}$ are assigned to the individual particles according to

$$w_t^{[i]} = \frac{p(s_{1:t}^{[i]} \mid z_{1:t}, u_{1:t})}{\pi(s_{1:t}^{[i]} \mid z_{1:t}, u_{1:t})} \propto \frac{p(z_t \mid s_t^{[i]})p(s_t^{[i]} \mid s_{t-1}^{[i]}, u_t)}{\pi(s_t \mid s_{1:t-1}^{[i]}, z_{1:t}, u_{1:t})} \cdot w_{t-1}^{[i]} \tag{6.2}$$

   The weights account for the fact that the proposal distribution π is in general not equal to the target distribution of successor states. It can be shown that for all proposal functions π with $p(x) > 0 \Rightarrow \pi(x) > 0$ and the weight update (6.2), the PF algorithm is guaranteed to approximate the correct posterior [3].
3. *Resampling*: Particles are drawn with replacement proportional to their importance weight.

In the special case of $\pi \equiv p(s_t|s_{t-1}, u_t)$, that is if sampling is performed according to the transition model, the relationship between Eq. (6.1) and the filtering procedure can directly be seen and the recursive weight update (6.2) simplify to a multiplication of all particle weights with the respective observation likelihoods $p(z_t|s_t^{[i]})$. For a formal derivation of the particle filtering principle, see for example [12] or [3]. Note that resampling exchanges 'likelihoods' by 'frequencies' without altering the represented distribution. In the limit of infinitively many and densely-sampled particles, the representation of a distribution by non-constant importance weights is equivalent to the one where all weights are equal, but the particle density varies accordingly. Thus, the purpose of the resampling step is to distribute more particles to the probable areas of the state space and less to the improbable ones.

The robustness and efficiency of particle filters strongly depends on the proposal distribution π that is used to sample the new state hypotheses in the sampling step. If the proposal distribution differs too much from the true posterior, there is a high risk of filter divergence. On the other hand, if π is too flat (e.g., uniform in the extreme case) many particles are wasted.

In the following section, we extend the mobile robot DBN to also include a discrete failure mode variable and additional continuous variables for the failure's parameters. As this increases the dimensionality of the state space and also adds highly unlikely transitions (failures occur only rarely), the transition

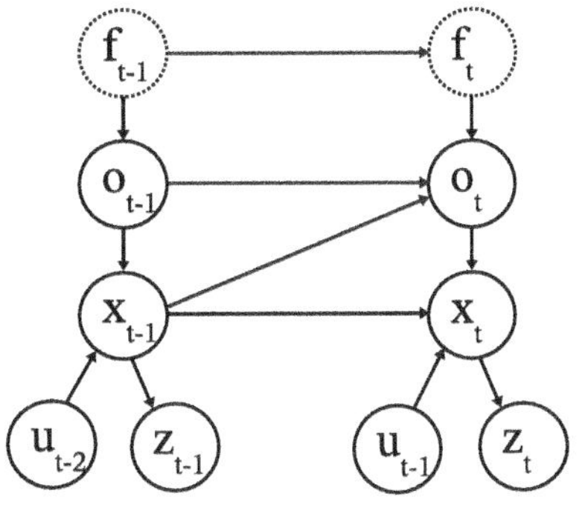

Fig. 6.3. Extended DBN to also include the discrete failure mode f and the continuous failure parameters o. With $s_t = (f_t, o_t, x_t)$, we denote the entire state of the system to highlight the relation to the standard model depicted in Fig. 6.2.

model becomes a sub-optimal choice as proposal distribution. For this reason, we then introduce an *informed proposal* distribution, which utilizes information about the failure mode contained in the most recent sensor measurement to yield better pose estimates.

6.4.3 Modeling the Influence of Failures Using Hybrid DBNs

In realistic environments there typically exist external influences that can cause drastic changes to the behavior of a dynamic system. Consider for example a mobile robot colliding with an undetected obstacle while executing a motion command. Such situations can be modeled by extending the DBN by a discrete failure mode variable f_t and additionally continuous failure-dependent state variables o_t (see Fig. 6.3). With $s_t = (f_t, o_t, x_t)$, we denote the entire state of the system to highlight the relation to the standard model depicted in Fig. 6.2. We assume the commonly used constant failure rate model (see [13]), in which failure events do not depend on the other state variables, but rather occur according to an individually specified prior distribution. Since the observations are independent of the failure state (f_t, o_t) given the pose x_t of the robot, the observation model remains $p(z_t \mid s_t)$. The state transition model can be factorized as $p(x_t, f_t, o_t \mid x_{t-1}, f_{t-1}, o_{t-1}, u_{t-1}) =$

$$\underbrace{p(f_t \mid f_{t-1})}_{\text{failure event model}} \cdot \underbrace{p(o_t \mid f_t, x_{t-1}, o_{t-1})}_{\text{failure parameter model}} \cdot \underbrace{p(x_t \mid o_t, x_{t-1}, u_{t-1})}_{\text{transition model}} \quad . \tag{6.3}$$

Note that this is a direct translation of the graphical model depicted in Fig. 6.3 and that we denote with $s_t = (f_t, o_t, x_t)$ the state of the system including failure variables and the pose of the robot.

The constant failure rate model states that failure events are distributed exponentially depending on a failure rate parameter λ, i.e.

$$p(f_t) = 1 - e^{-\lambda\cdot(t-\tilde{t})} \ , \tag{6.4}$$

where $\tilde{t}$ denotes the time of the last failure event. For such a model, the mean time between failures becomes $MTBF = \frac{1}{\lambda}$. For realistic failure rates λ, this model results in extremely low failure probabilities per filter iteration. Assume, for example, a mean time of 30 minutes between collisions of a service robot with unseen obstacles. This implies $\lambda = \frac{1}{1800s} = 0.000\overline{5}$ and with a filter frequency of $\delta t = 0.1$ seconds yields a failure probability of $p(f_t \mid \neg f_{t-1}) \approx 0.000056$ within one iteration. For such a small value, just one of 20,000 particles would be sampled to a failure mode on average, if the transition model was used directly as proposal distribution for this variables. Thus, one would either need an extremely large particle set or would risk that failures remain undetected. This problem is amplified by the fact that not only the discrete failure mode has to be sampled, but also the unknown continuous failure parameters. Since in general, there is no prior knowledge about the parameters of randomly occurring failures, we assume a uniform distribution

$$p(o_t \mid f_t, x_{t-1}, o_{t-1}) = \mathcal{U}_{[o_{min}, o_{max}]}(o_t) \tag{6.5}$$

over a certain interval. Note that this model applies only to the case where the system transitions into a failure state. The evolution of failure parameters within a failure state is typically governed by a much more peaked distribution similar to the motion model of the robot. In Section 6.6, we describe our model for the evolution of collision parameters based on rigid body dynamics. This model is able to track collisions sufficiently accurate, if the initial collision parameters have been estimated well enough. To achieve this, we introduce an *informed proposal* distribution in the next section, which utilizes information about the failure mode contained in the most recent sensor measurement.

6.5 Gaussian Processes Proposals for Failure Events

If system models containing states with extremely low a-priori probabilities are to be estimated using particle filters, it is obvious that the state transition model should not be employed as the proposal distribution directly (see the numerical example in the previous section). To address the problem of low sampling probabilities for important parts of the state space, [14] introduced the risk sensitive particle filter (RSPF) that incorporates a learned risk function to force the filter into less likely but important states. While this approach ensures a reasonable amount of samples in the important failure modes, it cannot adapt to the specific situation the robot is in when the sampling decision is made. In contrast, we propose to use *learned* proposal distributions that provide informed guesses about what the states with higher a-posterior probabilities are.

6.5.1 Data-Driven Proposal Distributions

In sequential importance sampling [15], an arbitrary proposal distribution can be used to sample the relevant areas of the state space as long as (a) all possible states have a non-zero possibility of being chosen and (b) the importance weights of the particles are adjusted appropriately. Proposal distributions that depend on the most recent sensor measurements or on features extracted by separate algorithms are typically denoted as *data-driven proposals* or *detector-mediated proposals* [16]. Such proposals aim at approximating the optimal proposal $p(s_t \mid s_{t-1}, z_t, u_{t-1})$ which includes the most current sensor measurement z_t. It can be shown [2] that such a *fully informed proposal* minimizes the variance of the importance weights making it the optimal choice.

Assuming that the proposal $\pi(s_t \mid s_{t-1}, z_t, u_{t-1})$ in the failure detection scenario also factorizes according to our transition model (6.6), i.e., $\pi(x_t, f_t, o_t \mid x_{t-1}, f_{t-1}, o_{t-1}, u_{t-1}) =$

$$\underbrace{\pi_f(f_t \mid f_{t-1})}_{\text{failure event proposal}} \cdot \underbrace{\pi_o(o_t \mid f_t, x_{t-1}, o_{t-1})}_{\text{failure parameter proposal}} \cdot \underbrace{\pi(x_t \mid o_t, x_{t-1}, u_{t-1})}_{\text{state proposal}}, \tag{6.6}$$

then the weight for particle i at time t becomes

$$\begin{aligned} w_t^{[i]} &= \frac{p(s_{1:t}^{[i]} \mid z_{1:t})}{\pi(s_{1:t}^{[i]} \mid z_{1:t})} = \frac{p(z_t \mid s_{1:t}^{[i]}, z_{1:t-1})\, p(s_{1:t}^{[i]} \mid z_{1:t-1})}{\underbrace{p(z_t \mid z_{1:t-1})}_{=:1/\eta}\, \pi(s_{1:t}^{[i]} \mid z_{1:t})} \\ &= \eta \cdot \frac{p(z_t \mid s_t^{[i]})\, p(s_t^{[i]} \mid s_{t-1}^{[i]})}{\pi(s_t^{[i]} \mid s_{1:t-1}^{[i]}, z_{1:t})} \cdot \underbrace{\frac{p(s_{1:t-1}^{[i]} \mid z_{1:t-1})}{\pi(s_{1:t-1}^{[i]} \mid z_{1:t-1})}}_{=w_{t-1}^{[i]}} \\ &= \eta \cdot w_{t-1}^{[i]} \cdot p(z_t \mid s_t^{[i]}) \cdot \frac{p(x_t^{[i]} \mid o_t^{[i]}, x_{t-1}^{[i]})}{\pi(x_t^{[i]} \mid s_{1:t-1}^{[i]}, z_{1:t}^{[i]})} \cdot \\ &\quad \frac{p(f_t^{[i]} \mid f_{t-1}^{[i]})}{\pi_f(f_t^{[i]} \mid s_{1:t-1}^{[i]}, z_{1:t}^{[i]})} \cdot \frac{p(o_t^{[i]} \mid f_t^{[i]}, x_{t-1}^{[i]}, o_{t-1}^{[i]})}{\pi_o(o_t^{[i]} \mid s_{1:t-1}^{[i]}, z_{1:t}^{[i]})} . \end{aligned} \tag{6.7}$$

Here, we left out the control variables u for brevity. The normalizing factor η is constant for all particles i. π_f denotes the proposal for the failure event and π_o the one for the failure parameters.

We now propose to learn π_f and π_o from data. Formally, the task is to learn a mapping from a feature vector F_t extracted from $\langle s_{1:t-1}^{[i]}, z_{1:t}^{[i]} \rangle$ to the variables f_t and o_t. Any feature vector F_t as well as any learned models π_f and π_o can be used as a proposal as long as the assumptions

- $\pi_f(f \mid F_t) \neq 0$ for all f with $p(f \mid s_{1:t}, z_{1:t-1}) \neq 0$
- $\pi_o(o \mid f_t, F_t) \neq 0$ for all o with $p(o \mid s_{1:t}, z_{1:t-1}) \neq 0$.

hold, which means that all possible failure states are assigned a non-zero probability of being chosen. Visually speaking, Equation 6.7 states that after each filter iteration, the particle weights have to be multiplied with the current observation likelihood $p(z_t \mid s_t^{[i]})$ and with two correction terms for the two learned proposal distributions. To calculate these correction terms for a specific sample $s_t^{[i]}$, we divide the probabilities defined in Equations 6.4 and 6.5 by the likelihoods according to which the state variables $f_t^{[i]}$ and $o_t^{[i]}$ have been drawn from π_f and π_o. Another precondition for the learned proposals therefore is the availability of likelihoods for sampled values. As we will see in the next section, Gaussian processes always ensure the non-zero probability assumptions, can easily be sampled from, and provide the transition probabilities needed for the weight correction described above.

6.5.2 Learning Sampling Models from Data

A major benefit of using Gaussian processes for learning and representing π_f and π_o is that by the availability of predictive uncertainties, the learned proposal distributions can directly be used in the sampling-based estimation scheme. The estimated state variables can be sampled directly from the Gaussian Processes. In the classification as well as the regression case.

In our concrete scenario, we found that the discrepancy between the current motion estimate (velocity and heading) w.r.t. an additional source (e.g. a laser scan-matcher or an inertial sensor) provides a good feature for learning proposals for the hidden failure states. Concretely, we take the extracted features F_t as input vectors $\mathbf{x}$ and failure parameters o_t as targets y. Given a training set of such quantities and a properly chosen covariance function k, we can compute the predictive distribution $\pi_o(o_t \mid f_t, F_t) := \mathcal{N}(\mu, \sigma^2)$ as detailed, for example, in [17]. This (normal) distribution naturally meets the requirements for proposal distributions named in the previous section. It can be sampled from directly, it supplies the likelihoods of sampled values, it has an infinite support, and it therefore only assigns non-zero likelihoods.

An important aspect that has been left out so far is the choice of covariance function and how its parameters can be set. The covariance function plays an important role in the Gaussian process framework as it represents the prior knowledge about the underlying function g. By changing its form and parameters, one can control the generalization behavior and smoothness of the predictor. A common choice of covariance function (see [18]) that is also used in this work is

$$k(\mathbf{x}, \mathbf{x}') := a_0 + v_0 \cdot \exp\left(-\frac{1}{2}\sum_{d=1}^{m} w_d(x_d - x'_d)^2\right)$$

with a constant component and a nonlinear, stationary term, which depends only on the distance between input points. The parameters of the covariance

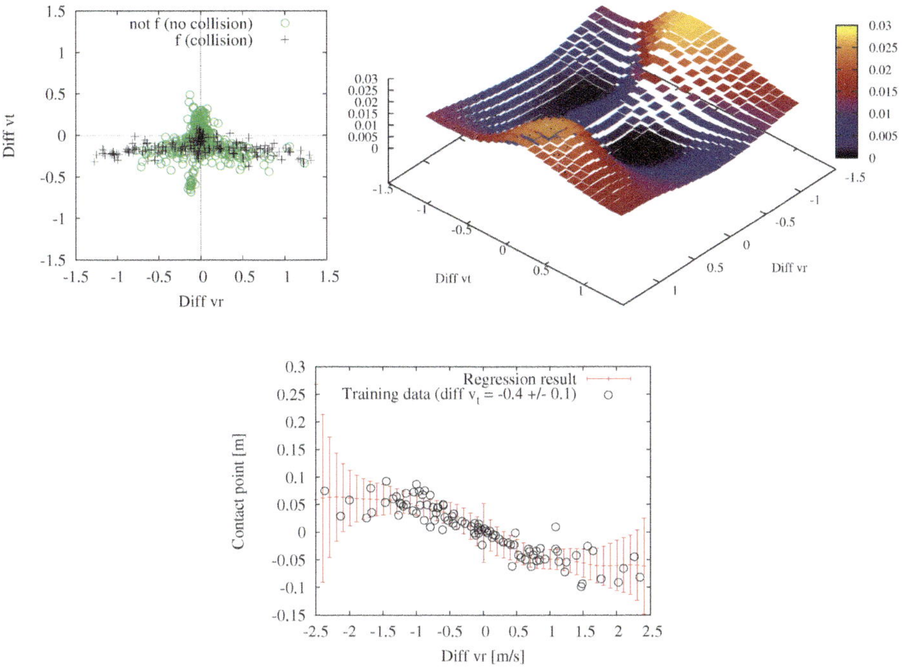

Fig. 6.4. The collision event training set, in which two feature values are mapped to class instances (top left), the learned class probabilities $\pi_f(F_t)$ using Gaussian process classification (top right). The learned regression model for the collision parameters visualized by a cut through the two-dimensional pdf that maps velocity deviations to contact points for a collision (bottom).

function are called hyperparameters of the Gaussian process. They can either be fixed by maximizing the likelihood of the given data points or, for fully Bayesian treatment, can be integrated over using parameter-specific prior distributions. For our experiments reported in Section 6.6, we employed the latter strategy with Gamma priors on the hyperparameters. We set these priors to favor smooth regression functions to avoid overfitting to the training data. The analytically intractable integration over the hyperparameters is approximated using Markov Chain sampling and the prediction results are cached on a fine-grained grid.

Binary classification problems can consistently be modeled in this framework by including for every binary target t_i a real-valued latent variable l_i, such that

$$p(t_i = 1) = \frac{1}{1 + e^{-l_i}} , \tag{6.8}$$

which is known as the logistic model that links class probabilities to real values, see [19]. The latent variables can now be given a Gaussian process prior as in the regression setting and predictions of class probabilities can be performed by predicting the corresponding latent variables and evaluating Equation 6.8. For the failure detection problem, we again use feature vectors F_t as inputs and binary failure labels f_t as targets. The predicted class probabilities for new features then directly define the failure event proposal $\pi_f(f_t \mid F_t)$.

6.5.3 Predicting Collision Events and Parameters

As discussed in Section 6.4.3, typical failure rates result in extremely low failure event probabilities. We therefore propose to train a Gaussian process classifier π_f for predicting failure events online. The learning task can be simplified by not taking the raw sensor measurements and state variables as inputs to the learner directly, but choosing lower dimensional features of these instead. In our failure detection setting, where we aim at detecting collisions with unseen obstacles

In this section, we apply the sequential failure detection approach described above to the problem of collision detection for mobile robots under noisy sensor measurements and without additional hardware like bumpers or inertial sensors. For learning the collision event proposal and the proposal for the contact point of the robot with the obstacle, we achieve excellent results with simple features based on the rotational and the translational velocity of the robot. Thus, we use as input to the Gaussian process classifier the two-dimensional feature vector $F_t = (\Delta v_t, \Delta v_r)$, where Δv_t is the difference between the translational velocity estimated by the particle filter and the one estimated by local laser scan matching. Furthermore, Δv_r is the difference of the rotational velocities respectively. A training set of 500 automatically labeled trajectories was generated by simulating random collisions with different obstacles using the 3D simulator Gazebo [20]. The top left diagram in Fig. 6.4 shows the gathered data points, the middle diagram in the same figure shows the learned class probabilities depending on the two velocity differences described above. As can be seen from the left diagram, Δv_t is negative for nearly all "collision" data points, which corresponds to the fact that the robot is slowed down when a collision occurs. The data points for "no collision" are spread widely and do not separate well from the "collision" data points due to noisy sensor measurements and imperfect labeling of the collision events. This makes the classification problem a hard one. It should be stressed, that we use this classifier as a proposal distribution for collisions rather than as a collision detector directly, because the features are too ambiguous to allow for perfect instantaneous classification. Experiments with a real robot (see Section 6.6) showed that this yields high detection rates with a low number of false alarms.

Given a collision event, the continuous collision parameters o have to be estimated to simulate the effects on the system and to continue the tracking process. Since the task is not to fully track the pushed obstacle over time, a simple model that abstracts from the obstacle's geometry and exact pose has proven sufficient to describe the effects on the robot. A collision with an unseen obstacle is represented by the obstacle mass m and the contact point c on the front of the robot. Therefore, the collision parameters are $o = (m, c)$. We learn the proposal distribution $\pi_o(o_t|F_t)$ for the parameters o_t using the same velocity-based features and simulated training set as described above and the Gaussian process regression technique. The lower diagram of Fig. 6.4 depicts a cut through the learned 2-dimensional distribution for the collision parameter c, which is the contact point of the obstacle on the front of the robot. The point of contact is measured in meters from the center of the robot's front to the right. It can be seen from the diagram that unexpected clockwise rotations ($\Delta v_r < 0$) of the robot are mapped to positive values for the contact point, which corresponds to a collision on the righthand side of the robot.

6.6 Implementation of Sensor-Level Monitoring

Our failure detection system described in the has previous two sections has been implemented on a real robot and was tested in an office environment. Before presenting experimental results, we describe the motion model we implemented for our localization system. The most widely used motion model for mobile robots is based on the wheel encoder measurements (see [12]). This information, rather than the actual control command, is taken as control input u_{t-1}, which under normal circumstances results in accurate predictions of the performed movement. Under the influence of failures like collisions or wheel slip, however, the motion of the wheels is not consistent with the whole robot's motion any more. A more appropriate model for such situations that still is efficient enough to be evaluated online is based on simple rigid body dynamics (see [21]). We model the robot as a rigid body in the (two-dimensional) plane, represented by a set of constant values and the variable state vector $x_t = (pos_x, pos_y, pos_\theta, vel_t, vel_r)$ which includes the translational velocity vel_t and the rotational velocity vel_r. In each filter iteration, the wheel thrusts are calculated from the actual velocity command that was sent to the motors. From this, the next state vector is computed by numerical simulation using the physical relationships between forces, acceleration, and speed. Due to space limitations, we refer to [22] for details about rigid body physics. With this model, collisions with another rigid object at a given point of contact can be simulated using the same type of physical abstraction, namely computing the impulse, the resulting forces, and ultimately the influence on the robot's state vector. At the same time, this model describes how the point of contact between the robot and the obstacle changes over time and therefore defines the transition model for failure parameters of Equation 6.6. From our experience, this physical model achieves the best balance between accuracy and efficiency. Simpler models fail to handle important test cases while more complex models have too many free parameters to be evaluated in real time.

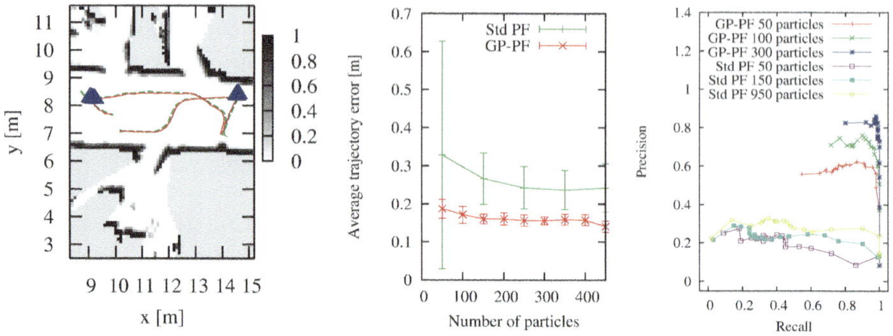

Fig. 6.5. Results of our experimental evaluation with a real robot colliding with undetected obstacles. Left: Two correctly detected collisions (triangles), the estimated trajectory (solid line), and the ground truth (dotted). Middle: Average deviation of the estimated trajectory from the ground truth in meters for a varying number of particles. Right: Estimation details around a correctly detected collision including the manually labeled true collision event. One filter iteration corresponds to 0.1 seconds.

6.6.1 Evaluation

To quantitatively evaluate the usefulness of our approach to failure detection, we compared it to a particle filter that implements the same process model with the standard uninformed proposals described in Section 6.4. The parameters of the standard filter were optimized for best tracking performance and failure detection rates to ensure comparability. We recorded data by manually steering the robot through the environment and arranged for two collisions, one with boxes of milk and the other one with a box of lemonade bottles. Both obstacles were placed at arbitrary positions and the obstacle heights were too low for the laser sensor to detect them. The left plot in Fig. 6.5 depicts a typical test run, in which our system successfully tracked the pose of the robot and detected the two collisions. On the recorded data set, we tested our improved particle filter with Gaussian process proposals (GP-PF) as well as the standard particle filter (Std PF) for different parameter settings. Each filter was executed 50 times for each parameter setting.

Figure 6.6 gives the failure detection performance of the different filters. The detection rate is defined as the number of correctly identified failures relative to the full number. The false positives rate is the amount of false alarms relative to the number of detections. The ground truth collision events were manually entered and a collision was counted as correctly detected, when the marginal failure likelihood exceeded a threshold Θ after a maximum of six filter iterations (0.6 seconds) after the true failure event. The threshold Θ was optimized independently for each filter to ensure unbiased comparison.

The middle diagram in Fig. 6.5 gives the average deviation of the tracked poses of the robot compared to the ground truth trajectory. The ground truth

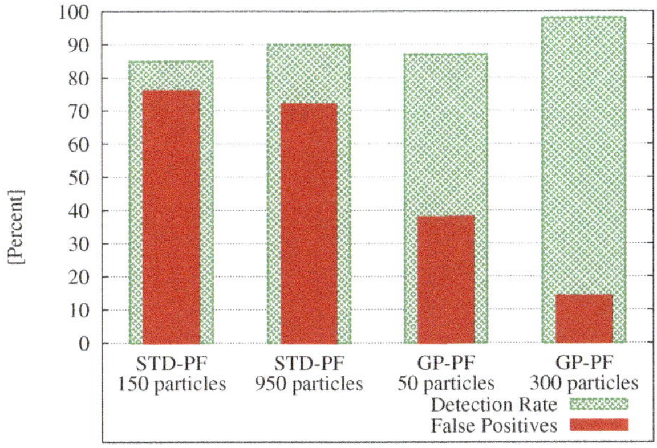

Fig. 6.6. Detection results with the optimized standard particle filter (STD-PF) and our approach that uses Gaussian process proposals (GP-PF)

trajectory was computed using a scan matcher. The results visualized in this diagram show that our system stays around ten centimeters closer to the true trajectory and produces less variance in these estimates than the standard approach. This is mainly due to the fact that the failure parameter (here, the point of contact with the obstacle) is estimated more accurately. To give an impression about the accuracy with which our filter estimates the point of contact and thereby the path of the robot, one failure event is depicted in detail in the diagram of Figure 6.7. It can be seen, that the estimated failure likelihood increases shortly after the labeled failure event and that the heading angle of the robot is correctly estimated.

The detection rates as well as the tracking results show that learned Gaussian process proposals can indeed increase the reliability and efficiency of online state estimation approaches. The time requirements for the improved particle filter are around 10% to 15% higher than for the standard implementation without Gaussian process proposals. Nevertheless, the implemented system with 200 particles still processes one minute of recorded data in less than 23 seconds on a PC with a 2800 MHz CPU.

6.7 Continual Collaborative Planning

In the previous sections, we have considered the dynamics of the world outside the agent mainly as a source of unforeseeable, uncontrollable events that require careful monitoring, failure diagnosis and, possibly, replanning. However, quite often the dynamics of an environment is due to other agents. In this case, an agent may, instead of just trying to react to their actions, try to actively interact with these agents to learn more about the current world state

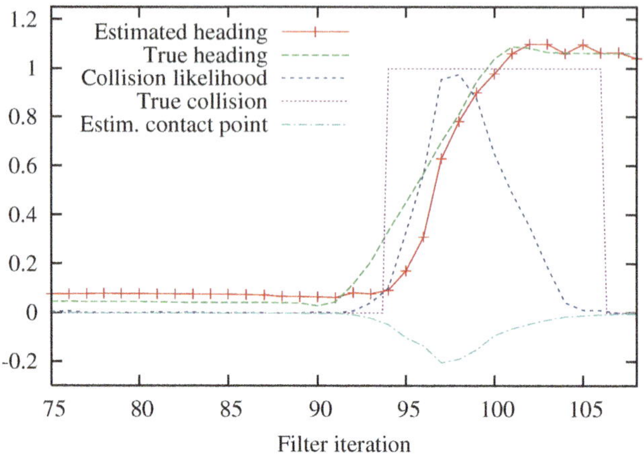

Fig. 6.7. Measurements and filter estimates during a typical collision event. Between iteration 90 and 95, the robot hits an undetected obstacle.

or even negotiate future plans with them. Such *proactive* multiagent planning is similar in spirit to the techniques for active infomation gathering presented in the earlier sections of this chapter.

For Continual *Collaborative* Planning (CCP) we must incorporate the Continual Planning algorithm (Algs. 2 and 3) into a *distributed* algorithm that allows communication between agents. The basic CCP algorithm is shown in Alg. 4. As is customary in Distributed Systems, the current state of the algorithm depends on the messages received from other agents [23].

As long as no message is sent or received the basic Continual Planning algorithm is executed as discussed above. However, during his individual CP process the agent may now also ask others to achieve subgoals. This arises naturally whenever an agent devises a plan that includes actions by another agent. Instead of executing such an action himself, the planning agent must send an appropriate *request* to the other agent. The planning agent will keep a *whitelist* for requests that have been accepted and a *blacklist* for those which have been denied.

To incorporate the generation of speech acts into continual planning, we slightly extend Alg. 3 to Alg. 5. The only visible changes occur in steps 4-5 in which requests are determined and communicated. Note, however, that already step (3), *i.e.* the execution of a planned action, may refer to a communicative action. For example, MAPL domains generally define *tell-val* actions, *i.e.* actions for informing another agent about a state variable value. If an agent receives such a speech act (step 4 of Alg. 4) it adds this information to its knowledge base, *i.e.* CCP assumes the sincerity of other agents.

The specific request is computed by the function SELECTBESTREQUEST in steps 4 of Alg. 5. This is a black-box function whose realisation may differ

Algorithm 4. CCP $\text{AGENT}(S, G)$

$P = \emptyset$

Received no message:
  if S satisfies G do
    return "goal reached"
  $P = \text{MONITORINGANDREPLANNING}(S, G, P)$
  if $P = \emptyset$ then
    return "cannot achieve goal G"
  $(S, P) = \text{EXECUTIONANDSTATEESTIMATION}(S, P)$

Received request(e) **from** agent a:
  $sg = \text{TRANSLATEREQUESTTOGOAL}(e)$ (1)
  $P = \text{MONITORINGANDREPLANNING}(S, sg, \emptyset)$ (2)
  if $P = \emptyset$ then
    send "cannot execute request e" **to** a
  else
    send "accept request e" **to** a
    add sg to G as temporary subgoal (3)

Received (tell-val vx) **from** agent a:
  add $v = x$ to S (4)

Received "cannot execute request e" **from** agent a:
  add e to *blacklist* (5)

Received "accept request e" **from** agent a:
  add e to *whitelist* (6)

Algorithm 5. $\text{EXECUTIONANDSTATEESTIMATION}(S, P)$

e = choose a first level event from P
$S' = app(S, e)$ (1)
$exp = \text{EXPECTEDPERCEPTIONS}(S', \mathcal{E}_{sense})$ (2)
if $agt(e)$ = self then
  $\text{EXECUTE}(e)$ (3)
else
  $e' = \text{SELECTBESTREQUEST}(e, P, blacklist, whitelist)$ (4)
  send request(e') **to** $agt(e)$ (5)
$perc = \text{GETSENSORDATA}()$ (6)
if $perc \supseteq exp$ then (7)
  remove e from P
$S = \text{FUSE}(S', perc)$ (8)
return (S,P)

between CCP implementations and applications. SELECTBESTREQUEST can take into account the *whitelist* and *blacklist* to determine requests that have not already been accepted or refused or which would be *subsumed* by such earlier requests.

SELECTBESTREQUEST also takes into account not only the event in question itself, but also its context in the agent's plan: If the plan contains several actions by another agent, *i.e.* a whole subplan, it is often best not to request execution of the actions individually, but to ask for the achievement of its end result. This will give the executing agent the liberty to find the best way to combine execution of the request with its individual goals. In other situations or applications, in contrast, it might be crucial for the requesting agent that its original plan is executed as precisely as possible.

Since the decision about the most appropriate request is highly dependent on the domain of application and, in particular, the form of cooperativity among agents, the general CCP framework does not commit to a specific algorithm for SELECTBESTREQUEST. Our standard implementation uses a simple algorithm that determines the maximal asynchronous subplan of P that uses only *one* specific agent. Then SELECTBESTREQUEST will request all actions on the final level of this plan. However, we are currently investigating variants that determine where it is reasonable to "outsource" larger subproblems to *groups* of other agents.

CCP Process Flow

CCP agents are first and foremost individual CP agents. Collaboration among agents arises only when one CP agent includes actions by another agent into his plan, *i.e.* when it wants to be helped. Whether collaboration is generally desirable or only a last resort largely depends on the domain of application: sometimes collaboration can raise a plan's utility dramatically, *e.g.* because of concurrent execution, but often the need for synchronisation and negotiation outweighs these advantages. We abstract from this issue here and assume an appropriate cost model that PLANNER can exploit exists for each planning domain, *e.g.* one that assigns additional costs to actions intended to be performed by other agents.

When an agent receives a request to perform an action, it first translates this request to a new subgoal (step 1 of Alg. 4). If the request corresponds directly to a grounded MAPL action, the subgoal is determined by the *effects* of this action. However, we want to allow for other forms of requests, too; in particular we want to be able to model requests posed by *human* users, *e.g.* during Human-Robot-Interaction. The reason for translating such requests to goals is twofold: What matters to the requesting agent is usually not the exact action, but the end result achieved, *i.e.* the achievement of a goal or of a precondition for a subsequent action of the requesting agent.[2] Additionally,

[2] This is also where *assertions* play a central role again. By referring to an assertion in a request, an agent a can ask another agent b for some behaviour even without knowing what b knows about the current situation, *i.e.* a does not need to reason about the abstraction level of b's ensuing continual planning process. Instead, b translates the request to a goal corresponding to the effects of the assertion and will then use the requested assertion, another one or none at all, depending on its own degree of knowledge.

agents may want to use *referring expressions* in their requests, because they cannot make an *unique name assumption* for objects they talk about. This is of particular importance when interaction happens in natural language, *e.g.* in Human-Robot-Interaction. Referring expressions can be modelled as constraints on the goal state to achieve [24]. The translation process has been described in our previous work [24]. In a nutshell, it maps a request to the effect of the corresponding operator scheme. If referring expressions are used, a more complex formula describing the constraints on the operator parameters is produced.

In the basic form of CCP described here, agents will check whether they can achieve the new subgoal (step 2 of Alg. 4) and, if possible, are always willing to adopt it (step 3). If the request is conflicting with their previous goals, they will reject it. The requesting agent will then add this request to a *blacklist* which is taken into consideration when requests are chosen again (step 4 of Alg. 5).

In practice often a more complex strategy of subgoal acceptance will be necessary. For instance, we have experimented with fixed and dynamic hierarchies among agents where higher-ranking agents will not accept requests from lower-ranking ones. Evaluating the resulting CCP variants is an important topic for future work. Our aim in this book, however, is to introduce CCP as a general framework for collaborative planning; we therefore restrict our attention to the setting with maximal cooperativity among agents.

Temporary Subgoals

A request is adopted as a *temporary* subgoal (step 3 of Alg. 4), *i.e.* the agent commits to achieving it at some point during the CCP run, but it need not necessarily stay true later on. Temporary subgoals (TSG) are not first-class constructs in MAPL; instead we exploit the fact that new goals and actions can be added to the planning domain on-the-fly during CP. Essentially, for each TSG sg a unique constant c_{sg} is generated and *(achieved* c_{sg}*)* is added to G as a conjunct. Then an artificial action *(achieve* c_{sg}*)* is created whose precondition is sg. Since only *(achieve* c_{sg}*)* can achieve *(achieved* c_{sg}*)* this enforces the planner to satisfy the precondition, *i.e.* sg. However, once *(achieved* c_{sg}*)* has been made true sg can become false again without preventing G from becoming satisfied. Thus sg is only enforced to be temporarily true.

Requests and TSGs add another important element to CCP: just as the beliefs of an agent are continually revised during the CP process, the use of TSGs lead to continual *goal revision.* Although Alg. 4 only *adds* new TSGs to G, we can in practice remove TSGs that are already satisfied from G. During CCP, agents thus try to achieve continually expanding and shrinking goal sets.

6.8 MAPSIM

Plan execution and plan monitoring are essential parts of Continual Planning algorithms. Therefore continual planning must, in contrast to classical planning, be implemented and tested in some sort of "reality". Providing such a reality, *e.g.* a simulation, is not a trivial task. Being able to provide realities in a domain-independent way, such that continual planning can be evaluated for arbitrary planning domains, is even harder. This might be a reason for the comparatively few systematic approaches to continual planning in the literature (cf. Sec. 6.10).

For evaluating CCP across varied MAPL domains and scenarios, we have developed a simulation testbed for multiagent planning, called MAPSIM. MAPSIM is a *simulation generator* that automatically transforms MAPL domains into multiagent simulations. MAPSIM parses and analyses a MAPL domain and turns it into perception, action, and communication models for CCP agents. During the simulation, MAPSIM maintains and updates the global world state and it uses the sensor models to compute individual and joint perceptions of agents. In other words, MAPSIM interprets the planning domain as an *executable model* of the environment. Thus, MAPSIM allows designers of DCP algorithms to evaluate their approaches on various domains with minimal effort.

Agents interact with MAPSIM by sending *commands* that directly derived from MAPL actions selected for execution during CCP. The simulator then executes the action, *i.e.* it checks the preconditions and applies effects as specified in the MAPL domain. If the controlling agent of a command is not identical to the agent who sent it to the simulator this is interpreted as a *request* which, of course, is not directly executed but passed on to the corresponding agent. MAPSIM also accepts some specific commands for acknowledging subgoal acceptance and subgoal achievement. When a command corresponds to a speech act, it is "heard" by the addressees, *i.e.* its effects are added to the beliefs of the addressees. In short, MAPSIM allows agents to directly implement Alg. 4 without having to care about the internals of of the "world" they are acting in. However, MAPSIM does not enforce the use of CCP. Agents can use arbitrary deliberative or reactive methods to determine their behaviour and their reactions to requests. We believe that this can make MAPSIM a valuable evaluation tool even when a (distributed) continual planning algorithm is used that differs significantly from CCP.

Interaction is very common during CCP runs in MAPSIM: agents plan and execute speech acts in order to gather information, negotiate requests and ensure self-synchronised plan execution by informing each other about state changes. Thus, CCP can be regarded not only as multiagent planning approach, but also as a model of collaborative, situated *dialogue*. Fig. 6.8 shows parts of one such interaction in a scenario where one agent, Anne, wants another, R2D2, to bring her coffee. Since it is hard to study dialogue behaviour in different variants of CCP based only on the logging format shown

MAPSIM run starts. There are 2 agents: Anne and R2D2.
(1) Anne: request R2D2 'give R2D2 coffee Anne'.
(2) R2D2: accept_request 'give R2D2 coffee Anne'.
(3) R2D2: request Anne 'tell_val Anne R2D2 pos(coffee)'.
(4) Anne: execute 'tell_val Anne R2D2 pos(coffee)'.
(5) R2D2: ack_achieved 'tell_val Anne R2D2 pos(coffee)'.
(6) ...

Fig. 6.8. A MAPSIM interaction in the *Household* domain

in Fig. 6.8, we have extended MAPSIM with a verbalisation module, called the *reporter*. The reporter observes all physical and communicative events in the simulation and verbalises them in English. All dialogues shown in the remainder of the chapter are unaltered outputs of this module. Fig. 6.9 shows the beginning of the MAPSIM run of Fig. 6.8 using the *reporter*.

MAPSIM run starts. There are 2 agents: Anne and R2D2.
(1) Anne: "Please bring me the coffee, R2D2."
(2) R2D2: "Okay."
(3) R2D2: "Where is the coffee, Anne?"
(4) Anne: "The coffee is in the kitchen."
(5) R2D2: "Thanks, Anne."
(6) ...

Fig. 6.9. Verbalisation of the interaction of Fig. 6.8 by the MAPSIM *reporter* agent

The reporter is a simple template engine that first determines an appropriate pattern depending on the command type currently executed, then recursively replaces templates with concrete arguments until a template-free sentence is generated. Base values for arguments are generated directly from analysing the MAPL domain. For example, operator names are assumed to directly correspond to verbs. Standard templates can be overridden by domain-specific patterns, but, surprisingly, this is often not even necessary to generate fairly natural English phrases. While, compared to "real" natural-language processing systems, this is a simplistic approach with obvious limits, the minimal effort needed to give MAPSIM basic linguistic capabilities is a noteworthy indication of the similarity between the MAPL representation and language.

Currently, dialogues are only *verbalised* by the reporter; the "real" communication between agents is performed as direct update of mental states as described by the MAPL speech acts. See Sec. 7.4.6 for a discussion of how agents can be enabled to communicate completely in a natural language, thereby enabling human-in-the-loop collaboration and dialogues.

MAPSIM is implemented in Python. The generic planner currently used by the CCP agents is a modified version of Axioms-FF [25] that prevents the use of assertions enabled in the current state, as required by definition3 and allows to express belief introspection with derived predicates. To enable the use of a classical PDDL planner like FF, MAPL is first compiled to PDDL: induced state variables are explicitly generated; speech acts that use state variables as parameters in MAPL are translated to a specific PDDL action for each such state variable; derived predicates for introspection of belief state variables are added to the PDDL domain; and MVSVs are translated to PDDL propositions.

6.9 Situated Dialogue as Continual Collaborative Planning

In this section, we will explain the use of CCP for scenarios that interleave planning, acting and sensing as well as action and interaction between multiple agents, *i.e.* for scenarios where agents engage in *situated dialogue* [26, 27]. In this chapter, we use examples from MAPSIM to discuss planning of interactions. However, the same approach has been used for interaction planning in our integrated robot systems (see Chapters 9 and 10). There, CCP is integrated with a specialised component for situated dialogue processing, presented in Chapter 8. CCP is used for high-level pragmatic reasoning, i.e. for planning how and why to use communication for achieving a (possibly non-communicative) goal. Linguistic situation-appropriate realisation of the planned "verbal behaviour", as well as the situation-aware interpretation of speech acts by other agents, are handled by the specialised dialogue component.

Figs. 6.10, 6.11, and 6.12 show situated dialogues between MAPSIM agents in different MAPL domains, generated using the CCP algorithm and automatically verbalised by the MAPSIM *reporter*. This shows how, due to the domain independence of both the CCP algorithm and the MAPSIM implementation, it is easily possible to evaluate different CCP variants and dialogue strategies across different domains and scenarios.

It is important to realize that none of the sample runs shows the execution of a single multiagent plan, but a *series* of plans, devised, partly executed and revised several times according to Alg. 4.

In Fig. 6.10 the necessity for collaboration stems from the fact that only MrChips can move to the kitchen to get coffee, but only MrData can open the kitchen door. At the beginning, both agents have different goals: MrData wants to have coffee and MrChips wants to be at MrData's service. Basically, this is a master-slave scenario; however, since both agents have individual goals, incidentally it is MrChips that takes the initiative. MrChips' original goal is very simple: he wants to achieve *(has-goal MrChips)*. This goal is directly achieved by step 1 and 2 of the dialogue. However, step 2 already is

MAPSIM run starts. There are 2 agents: MrChips and MrData.
(1) MrChips: "What can I do for you, MrData?"
(2) MrData: "Please bring me the coffee, MrChips!"
(3) MrChips: "Where is the coffee, MrData?"
(4) MrData: "The coffee is in the kitchen, MrChips!"
(5) MrChips: "Please open the kitchen door, MrData!"
(6) MrData opens the kitchen door.
(7) MrChips moves to the kitchen.
(8) MrChips takes the coffee.
(9) MrChips moves to the livingroom.
(10) MrChips: "I have the coffee, MrData!"
(11) MrChips: "Please take the coffee, MrData!"
(12) MrData takes the coffee.
(13) MrData: "Thanks for the coffee, MrChips!"
MAPSIM terminates successfully.

Fig. 6.10. Mixed-initiative dialogue between two artificial agents in MAPSIM (*Household* domain)

a request by MrData that provides MrChips with a new temporary subgoal that he will try to achieve during the rest of the dialogue.

To see why and how MrData generated this request, consider MrData's individual planning process: knowing that the coffee is in the kitchen, he can generate a plan that involves MrChips going to the kitchen and bringing back the coffee. Alg. 4 uses the function SELECTBESTREQUEST to determine the appropriate subgoal that MrData will request MrChips to achieve. Our standard implementation uses a simple algorithm FINDINDIVIDUALSUBPLAN (omitted from this chapter) that determines the largest subplan of P that uses only *one* specific agent. Then SELECTBESTREQUEST chooses an action on the final level of this plan as the best request. In our example, MrChips can directly ask for the last action in his plan, namely MrData giving him the coffee.

The strategy of requesting long-term effects rather than immediately possible actions is also exemplified in Fig. 6.11. Its main advantage is that it abstracts from the level of detail on which the individual agents can plan (due to the differences in their knowledge). In Fig. 6.11, the "Boss" agent may not know whether the preconditions for the "put" action are already satisfied, *i.e.* his own plan may have included knowledge-gathering actions and assertions. However, by asking for the expected result, he gives the robot the opportunity to find its own appropriate solution.

In the *household* scenario, MrChips adopts the new goal to provide MrData with coffee (according to step 3 of Alg. 4). However, since he does not know where the coffee is his next plan must resort to a fairly abstract assertion, *(fetch-A MrChips coffee)*, that completely hides the position of the target object as well as the possible complex planning necessary to reach that position.

(1)	Boss: "Please put obj1 on obj2, Robot."
(2)	Robot picks up obj1.
(3)	Robot puts obj1 on obj2.

Fig. 6.11. Long-term request in the *object manipulation* domain

Although very abstract, the assertion guides MrChips' planning by means of its *replanning condition (KIF MrChips (pos coffee))* which, in words, says that in order to fetch the coffee he must at least find out where it is. This leads to the following multiagen plan by MrChips. Here, MrChips takes the initiative again by asking MrData about where the coffee is:

MrChips: request MrData 'tell_val MrData MrChips pos(coffee)'
MrData: execute 'tell_val MrData MrChips pos(coffee)'
MrChips: execute 'fetch-A MrChips coffee'
MrChips: execute 'give MrChips MrData coffee'

Alg. 2 declares MrChips' plan as valid, so Alg. 3 executes the first action: MrChips requests MrData to tell him the position of the coffee. The domain-specific verbalisation template maps a request for telling the value of state variable *pos* to the wh-question "where?" as shown in step 3 of the Fig. 6.10.

Note that in the further course of the dialogue both agents switch seamlessly between communicative and physical actions. However, unnecessary verbalisation of action effects is avoided, because both agents also reason about their mutual perceptions. For example, MrData does not verbalise having opened the door, because in his plan he can apply a sensor model for MrChips, thereby deducing that MrChips will perceive the opening of the door himself. In this manner, CCP both enforces knowledge preconditions, but also avoids unnecessary communication about them. In many applications, however, communication is necessary for *grounding* the collaborative process or simply acknowledging understanding [28]. Fig. 6.12 shows an example (from a Search and Rescue planning domain) where CCP was run with enforced acknowledgements upon subgoal adoption and achievement. Especially in those Collaborative Planning environments that include both artificial and human agents such acknowledgements are crucial.

(1)	Ambulance: "Please extinguish house1, Firebrigade."
(2)	Firebrigade: "Okay."
(3)	Firebrigade refills watertank.
(4)	Firebrigade extinguishes house1.
(5)	Ambulance: "Thanks for extinguishing house1, Firebrigade."

Fig. 6.12. Acknowledgements for subgoal adoption and achievement

6.10 Related Work

This work integrates ideas from several subfields of AI, in particular Classical and Distributed Planning, Multiagent Systems, Epistemic Logic, and Reasoning about Actions and Change.

Planning in dynamic, incompletely known domains can be modelled as a *conformant*, *contingent* or *probabilistic* planning task. All of These approaches compute conditional plans or policies for possible contingencies such that the agent can react adequately when faced with them. Unfortunately, this increased flexibility comes at the cost of being computationally much harder than classical planning [29, 30]. Thus, these approaches scale badly in dynamic multiagent environments with large numbers of unobservable features and exogenous events. Therefore, Continual Planning is often advocated as a practical approach to planning in such environments [31]. In practice, this often amounts to not more than repeatedly switching between planning and execution. Previous work that more tightly integrates planning, monitoring, execution and information gathering includes [32, 33, 34, 35, 36]. Similarly to our work, these approaches explicitly model knowledge and knowledge-gathering actions. Instead of the concept of assertions which enables us to postpone parts of the planning process, yet reason about its outcomes, these approaches use runtime variables to represent unknown sensing results. Runtime values can be used as action parameters in the remainder of plan and thus allow for reasoning about unknown future knowledge, although this reasoning is heavily limited because nothing is known about the variable beside the fact that has been sensed. Because of the limitations of runtime variables, MAPL does not yet support them. Instead, our use of MVSVs enables a planner to non-deterministically *guess* one of the possible value of the MVSVs domain if this is desired by the domain modeller.

Our asynchronous plans are similar to Boutilier and Brafman's [37] multi-actuator plans. They model interacting effects of concurrent actions by specific kinds of conditional effects of the individual agents. A plan must provide simultaneity constraints ensuring that the interaction really takes place as planned. The authors assume that an external synchronisation mechanism will ensure that during execution the constraints are met by the agents. Cox and Durfee's [38] and Clement's [39] coordination algorithms provide such mechanisms. Our approach, however, rests on the assumption that executing agents are truly autonomous and there is no external instance to synchronise them. Therefore it must allow agents to synchronise plan execution on their own. This is achieved by explicitly including the knowledge and perceptions necessary for coordinated execution into the plans of agents. Since only plans that include these information are valid multiagent plans, the planning algorithm itself can (and is forced to) ensure synchronised execution. It is worth studying, however, how special-purpose coordination algorithms like those of Cox or Clement could be integrated with our Continual Planning approach, thereby potentially simplifying the actual planning process.

Planning with sensing actions has often been described in the planning literature [40, 41, 42, 10, 43]. To our knowledge, none of these models extends planning for sensing to the concept of *copresence* [44]. Copresence has been much discussed in literature on pragmatics, e.g. by Clark and colleagues [45]. To our knowledge, ours is the first work that allows agents to explicitly reason about copresence in order to self-coordinate multiagent plan execution without explicit communication.

The explicit inclusion of beliefs and mutual beliefs in our planning approach follows BDI models of multiagent planning, *e.g.* the SharedPlans model of Grosz and Kraus [46] that describes the role of (mutual) beliefs as necessary conditions for planful MA behaviour. Our formalism and implementation does not cover all aspects of these models (yet); in particular, we do not model *intentions* explicitly (yet). However, by explicitly modelling perception and copresence, our approach complements existing BDI approaches to MA plans, since it can explain how knowledge conditions for joint behaviour can be achieved during plan execution.

The need for Distributed Continual Planning (DCP), *i.e.* Continual Planning in multiagent settings, was pronounced clearly by desJardins and colleagues in [1]. Most work within this field is based on hierarchical representations of multiagent plans [47, 48, 49, 50, 51]. Indeed, the expansion of assertions is similar to the decomposition of HTN schemata [52, 53, 54]. In our approach, however, the abstraction hierarchy need not be explicitly given by the domain designer, but is resolved by the planner itself. Also the purpose of the abstraction is different from HTN planning: while HTN decompositions embody knowledge about how to solve subtasks, assertions essentially represent a way to postpone parts of the planning process. Thus Continual Planning with assertions produces a *series* of *non-hierarchical* plans, whereas HTN produces one abstraction hierarchy. Note also that while it has been proposed in textbooks that HTN planners may leave parts of the plan hierarchy unexpanded until a plan has been partially executed, we are not aware of work that describes how exactly an HTN planner should make such decisions.

Our CCP approach to *dialogue* is close in spirit to existing frameworks for collaborative dialogue, in particular the one of Lochbaum [55]. In contrast to other approaches to dialogue planning that use formal state descriptions mainly for the specification and validation of the computational approaches, *e.g.* [28, 56, 57], we directly reason on the formal logical representations of the agents' beliefs (*i.e.* the MAPL states). In this respect, our work mostly resembles Sadek's approach to dialogue planning [58]. Due to our use of a general-purpose planning method, other approaches can deal with more elaborate linguistic phenomena. However, CCP seems to be able to explain pragmatic aspects of a dialogue better (or at least more explicitly) than other approaches, because of the inherently causal reasoning underlying all dialogue created. In particular, CCP is suited for *situated* dialogue planning, because the generated dialogues directly depend on the agent's knowledge about the current situation and its physical actions in the world.

Common approaches to *failure detection* are model-based, that is, they involve reasoning about potential failures based on an explicit system model describing the structural and behavioral properties of the system. This has been approached (a) from within the *AI community* using symbolic reasoning with a focus on large systems with many interacting components and (b) from the *control theory community* concentrating on fewer components with complex dynamics and higher noise levels. Most of the established approaches found in the literature have originally been designed for discrete, static, and noise-free domains. More recently, they have typically been extended to also handle continuous variables—if just by discretizing the space—and to respect the dynamic and uncertain nature of the processes. It is, nevertheless, common agreement that open questions in this area include the handling of noise, diagnosis of hybrid systems, and model building without excessive human engineering.

The close coupling between a system and its environment makes it hard in general to detect abnormal behavior using instantaneous statistical tests only without tracking possible failure modes over time [59]. This is especially the case for mobile systems, such as service robots. By modeling faults as special states in a (typically hybrid) state space model, any state estimation technique can—in principle—be used for estimating their posterior likelihoods given a sequence of observations. Approaches found most often in the FDI literature include multi-hypothesis tracking [60, 61], single model tracking [62], or one step look-ahead particle filtering [63]. In particle filter based approaches to fault diagnosis, the system is typically modeled by a non-linear Markov jump process [64] or a dynamic mixture of linear processes [63]. Verma et al. [65] introduce the variable resolution particle filter for failure detection. Their approach is similar to ours in that they build an abstraction hierarchy of system models. They model the robot as a hybrid system with discrete modes of operation and corresponding continuous state variables. They assume that the probabilities of failure modes can be estimated from the state tracking performance of their associated system models. As described in Sec. 6.4, we also approach the problem using sequential state estimation in hybrid models and—in contrast to previous work—apply Gaussian process learning for sampling the failure mode more efficiently and robustly based on prior experience.

Approaches that deal with the time efficiency of particle filters include [66] in which real-time constraints are considered for single system models or techniques in which a Rao-Blackwellized particle filter is used to coordinate multiple models for tracking moving objects [67]. Related work addressing the efficiency of particle filtering also includes the regularized particle filter [68] and the Parzen particle filter [69]. Early references to the general idea include [70], in which the authors propose to exchange the delta-Dirac kernels of a particle filters for Gaussian kernels in order to estimate fixed system parameters and dynamic state variables jointly.

6.11 Conclusion

Acting deliberatively is hard in realistic dynamic environments that cognitive agents can only partially observe and influence. For all practical purposes, deliberation must be combined with reactive behaviour that takes newly perceived changes into account quickly. Often, however, dynamic environments demand an even more *proactive* behaviour: agents must actively try to gather new or reconfirm outdated information before they can determine appropriate solutions to their problems. To that end, they must be able to reason about how and when to extend or update their knowledge, and then plan their own cycle of planning, acting and replanning.

In this chapter, we have investigated the interplay between planning, acting and behaviour monitoring in dynamic multiagent environments. We have introduced several theoretical and practical tools that can be used for building practical cognitive agents for such environments:

The Multiagent Planning Language MAPL. Intelligent agents in dynamic multiagent environments must be able to explicitly reason about their own and others' sensory and communicative capabilities, their beliefs and mutual beliefs, and about the necessary conditions for joint behaviour. Most formal languages for AI Planning, notably the de facto standard PDDL, do not permit this. We have therefore developed the Multiagent Planning Language MAPL that extends PDDL and thus also enables extending existing planning methods for multiagent planning. MAPL plans can freely interleave physical action, sensing and communication, and thus forms the basis for our CP and CCP algorithms. In particular, MAPL plans guarantee that during execution all agents are provided (by perception or communication) with the necessary knowledge to autonomously, *i.e.* without a central scheduler or synchronisation component, execute their parts of a MA plan.

Continual Planning with Assertions. We have developed a new algorithmic framework for continual planning, i.e. integrated planning, plan execution, and plan monitoring. Previous approaches to continual planning have mostly regarded it as just a case of replanning in light of new information. However, such a purely reactive approach is insufficient in dynamic environments where knowledge is limited from the start. Such environments demand more *proactive* behaviour: agents must actively try to gather new (or reconfirm outdated) information before they can determine appropriate solutions to their problems. To that end, they must be able to reason about how and when to extend or update their knowledge, and then plan their own cycle of planning, acting and replanning. In CoSy, we have introduced a principled approach for this kind of continual planning. Based on the novel concept of *assertions* agents can decide autonomously which parts of of planning problem they can already solve in detail and for which they must gather additional information.

Failure Detection and Robust Monitoring. To be able to check assertions and to reason about the outcome of actions, the robot requires the ability to monitor lower-level dynamic processes. We developed a novel approach to process-monitoring based on particle filtering that enables one to use prior knowledge about failure modes learned from experience or in simulation. We derived a general framework for integrating learned failure models into algorithms for sequential state estimation. The system was implemented, tested, and demonstrated on the task of online collision detection for a mobile robot.

Continual Collaborative Planning. Other agents are the major source of exogenous change in most dynamic environments. However, their presence is not only a source of uncertainty, but can also be exploited: by interacting with others, an agent can reduce its uncertainty about the present and constrain possible contingencies in the future effectively. Again, sharing information and negotiating plans with others is a continual process that is interleaved with physical action (e.g. to enable communication in the first place) and sensing (e.g. to establish a context for situated references). Therefore, we have extended our continual planning approach to Continual Collaborative Planning (CCP), which integrates planning, acting, sensing and communication. During CCP, agents will not only continually update their beliefs, but also revise their *goals* as a result of negotiations with other agents. This leads to particularly dynamic collaborative behaviour.

The Multiagent Planning Simulator MAPSIM. Continual planning approaches can only be tested in environments where agents can actually interleave planning, execution and sensing. The CoSy demonstration scenarios provide realistic settings for such an evalutation; however, we also wanted to provide a generic way to evaluate CCP and other distributed continual planning approaches across a wide range of multiagent planning domains and problems. We have therefore developed MAPSIM, a software environment that can automatically generate MA simulations from MAPL domains. In other words, MAPSIM interprets a formal MAPL domain as an executable model of the environment in which agents can perceive, plan, act and engage in negotiations for task-orientated collaboration.

Situated Dialogue as Continual Collaborative Planning. When several agents are situated in a common environment they usually interact physically as well as verbally. Verbal interaction in such environments, i.e. situated dialogue, both reflects the past and influences the future physical behaviour of the agents. Thus situated dialogue can often be regarded as continual collaborative planning. Interestingly, the role of *communication* in CCP is twofold: A dialogue move can be part of the collaborative *planning* process; however, it is also the *execution* of a communicative action and, just like the execution of a physical action, it changes the "world" in ways that may lead to previously unforeseen changes in plans and, consequently, additional interactions. Since goals and plans of agents are continually revised, dialogues generated by CCP naturally include mixed-initiative subdialogues and interleaved physical and communicative actions, as exemplified in the CoSy demonstrator systems.

Continual Planning in a Complex Cognitive Architecture. Large cognitive systems, e.g. the CoSy demonstrators that are built on the CAST architecture and include many substantially different subarchitectures, do not only act in dynamic environments, but can themselves be considered complex multiagent systems. We have therefore applied CCP not only for planning interactions with humans or other robots, but also for planning *internal* processes flows of a cognitive system. Each of the "agents", i.e. the subarchitectures, in such a system can provide a plethora of information. One important realisation made while integrating our continual planner into the CAST architecture was that the abundance of information potentially available to the planner is obstructive to its success. Instead, what is needed is a *demand-driven* way to provide information to the planner. Again, this need turns out to be a need that can be satisfied by a continual planning approach: by including *less* information in the state initially given to the planner, but extending it with meta-level and self-referential knowledge (e.g. knowledge about which subarchitecture may have or may provide additional knowledge), the continual planner can determine on its own which information it additionally requires for solving a problem. This may go as far as the planner requiring more detailed information about the planning *ontology* or the planning *operators*, all *during* the continual planning process.

Summary. During the CoSy project, we have developed a new framework for proactive continual planning and execution monitoring. The framework has been successfully used in simulation as well as in complex robot systems (*c.f.* Chapters 9 and 10) for a variety of tasks ranging from monitoring hardware failures to engaging in human-robot dialogues. We believe that our approach of encouraging information gathering on all levels of a cognitive architecture is a significant step towards building agents that show intelligent proactive, yet robust behaviour, *i.e.* agents that can rightfully be called autonomous.

References

1. DesJardins, M., Durfee, E., Ortiz, J.C., Wolverton, M.: A survey of research in distributed, continual planning. The AI Magazine
2. Kevin, M.: Dynamic bayesian networks: Representation, inference and learning, Ph.D. thesis, UC Berkeley (2002)
3. Doucet, A., de Freitas, N., Gordan, N. (eds.): Sequential Monte-Carlo Methods in Practice. Springer, Heidelberg (2001)

4. Brenner, M., Nebel, B.: Continual planning and acting in dynamic multiagent environments. Journal of Autonomous Agents and Multiagent Systems (Accepted for publication), http://www.cognitivesystems.org/cosybook/chap6.asp#Brenner/etal:2008
5. Fox, M., Long, D.: PDDL 2.1: an extension to PDDL for expressing temporal planning domains. JAIR
6. Helmert, M.: The Fast Downward planning system. Journal of Artificial Intelligence Research 26, 191–246 (2006)
7. Brenner, M., Nebel, B.: Continual planning and acting in dynamic multiagent environments. In: PCAR 2006: Proceedings of the 2006 international symposium on Practical cognitive agents and robots, pp. 15–26. ACM, New York (2006), http://www.cognitivesystems.org/cosybook/chap6.asp#Brenner/etal:2006a, http://doi.acm.org/10.1145/1232425.1232431
8. Yokoo, M., Hirayama, K.: Algorithms for distributed constraint satisfaction: a review. Autonomous Agents and Multi-Agent Systems 3(2)
9. Bäckström, C.: Computational aspects of reordering plans. JAIR 9, 99–137 (1998)
10. Petrick, R., Bacchus, F.: A knowledge-based approach to planning with incomplete information and sensing. In: Proc. AIPS 2002 (2002)
11. Hoffmann, J., Brafman, R.: Contingent planning via heuristic forward search with implicit belief states. In: Biundo, S., Myers, K.L., Rajan, K. (eds.) ICAPS, pp. 71–80. AAAI, Menlo Park (2005)
12. Thrun, S., Burgard, W., Fox, D.: Probabilistic Robotics. MIT Press, Cambridge (2005)
13. Ng, B., Pfeffer, A., Dearden, R.: Continuous time particle filtering. In: Proceedings of the 19th IJCAI, Edinburgh (2005)
14. Thrun, S., Langford, J., Verma, V.: Risk sensitive particle filters. In: NIPS (2001)
15. Doucet, A.: On sequential simulation-based methods for bayesian filtering, Tech. rep., Signal Processing Group, Departement of Engeneering, University of Cambridge (1998)
16. Khan, Z., Balch, T.R., Dellaert, F.: An mcmc-based particle filter for tracking multiple interacting targets. In: Pajdla, T., Matas, J(G.) (eds.) ECCV 2004. LNCS, vol. 3024, pp. 279–290. Springer, Heidelberg (2004)
17. Rasmussen, C.E., Williams, C.K.: Gaussian Processes for Machine Learning. The MIT Press, Cambridge (2006)
18. MacKay, D.J.C.: Introduction to Gaussian processes. In: Bishop, C.M. (ed.) Neural Networks and Machine Learning. NATO ASI Series, pp. 133–166. Kluwer Academic Press, Dordrecht (1998)
19. Neal, R.: Monte carlo implementation of gaussian process models for bayesian regression and classification, Tech. rep., Dept. of Computer Science, University of Toronto (1997)
20. Koenig, N., Howard, A.: Design and use paradigms for gazebo, an open-source multi-robot simulator. technical report. Tech. rep., USC Center for Robotics and Embedded Systems, CRES-04-002 (2004)
21. Plagemann, C., Stachniss, C., Burgard, W.: Efficient failure detection for mobile robots using mixed-abstraction particle filters. In: Christiensen, H. (ed.) European Robotics Symposium 2006. STAR Springer tracts in advanced robotics, vol. 22, pp. 93–107. Springer, Heidelberg (2006), http://www.cognitivesystems.org/cosybook/chap6.asp#Plagemann/etal:2006

22. Witkin, A., Baraff, D.: An introduction to physically based modeling. In: SIGGRAPH 1997 Course Notes (1997)
23. Lynch, N.: Distributed Algorithms. Morgan Kaufmann, San Francisco (1996)
24. Brenner, M.: Situation-aware interpretation, planning and execution of user commands by autonomous robots. In: Proceedings of IEEE RO-MAN 2007 (2007), http://www.cognitivesystems.org/cosybook/chap6.asp#Brenner:2007a
25. Thiebaux, S., Hoffmann, J., Nebel, B.: In defense of axioms in PDDL. In: Proc. IJCAI (2003)
26. Kruijff, G., Brenner, M.: Modelling spatio-temporal comprehension in situated human-robot dialogue as reasoning about intentions and plans. In: Proceedings of the Symposium on Intentions in Intelligent Systems, AAAI Spring Symposium Series 2007, Stanford University, Palo Alto, CA (2007), http://www.cognitivesystems.org/cosybook/chap6.asp#Kruijff/Brenner:2007
27. Brenner, M., Kruijff-Korbayová, I.: A continual multiagent planning approach to situated dialogue. In: Proceedings of the 12th Workshop on the Semantics and Pragmatics of Dialogue (Semdial), London, UK (2008), http://www.cognitivesystems.org/cosybook/chap6.asp#Brenner/etal:2008a
28. Traum, D.: A computational model of grounding in natural language conversation, Ph.D. thesis, Univ. of Rochester (1994)
29. Littman, M., Goldsmith, J., Mundhenk, M.: The computational complexity of probabilistic planning. JAIR
30. Rintanen, J.: Constructing conditional plans by a theorem-prover. JAIR 10, 323–352 (1999)
31. Russell, S., Norvig, P.: Artificial Intelligence: A Modern Approach, 2nd edn. Prentice-Hall, Englewood Cliffs (2003)
32. Ambros-Ingerson, J.A., Steel, S.: Integrating planning, execution and monitoring. In: Proc. AAAI 1988, Saint Paul, MI, pp. 83–88 (1988)
33. Etzioni, O., Hanks, S., Weld, D., Draper, D., Lesh, N., Williamson, M.: An approach to planning with incomplete information. In: Proc. KR 1992, pp. 115–125 (1992)
34. Knoblock, C.A.: Planning, executing, sensing, and replanning for information gathering. In: Mellish, C. (ed.) Proceedings of the Fourteenth International Joint Conference on Artificial Intelligence, pp. 1686–1693. Morgan Kaufmann, San Francisco (1995)
35. Etzioni, O., Golden, K., Weld, D.S.: Sound and efficient closed-world reasoning for planning. Artificial Intelligence 89(1-2), 113–148 (1997)
36. Golden, K.: Leap before you look: Information gathering in the puccini planner. In: Proc. AIPS 1998, pp. 70–77 (1998)
37. Boutilier, C., Brafman, R.: Partial order planning with concurrent interacting actions. JAIR
38. Cox, J.S., Durfee, E.H.: An efficient algorithm for multiagent plan coordination. In: Proc. AAMAS 2005 (2005)
39. Clement, B., Barrett, A.: Continual coordination through shared activities. In: Proc. AAMAS 2003 (2003)
40. Levesque, H.J.: What is planning in the presence of sensing? In: Proc. AAAI 1996, pp. 1139–1146. MIT Press, Cambridge (1996)
41. Golden, K., Weld, D.: Representing sensing actions: The middle ground revisited. In: Proc. KR 1996 (1996)

42. Weld, D.S., Anderson, C.R., Smith, D.E.: Extending graphplan to handle uncertainty and sensing actions. In: AAAI/IAAI, pp. 897–904 (1998)
43. Petrick, R.P.A., Bacchus, F.: Extending the knowledge-based approach to planning with incomplete information and sensing. In: Proc. ICAPS 2004, pp. 2–11 (2004)
44. Lewis, D.: Convention. A Philosophical Study. Harvard University Press, Cambridge (1969)
45. Clark, H.H., Marshall, C.R.: Definite reference and mutual knowledge. In: Elements of discourse understanding. Cambridge University Press, Cambridge (1981)
46. Grosz, B.J., Kraus, S.: Collaborative plans for complex group action. Artificial Intelligence 86
47. Durfee, E., Montgomery, T.: Coordination as distributed search in hierarchical behavior space. IEEE Transactions on Systems, Man, and Cybernetics
48. Myers, K.L.: Cpef: A continuous planning and execution framework. The AI Magazine 20(4)
49. Durfee, E.H.: Distributed continual planning for unmanned ground vehicle teams. AI Magazine 20(4), 55–61 (1999)
50. DesJardins, M., Wolverton, M.: Coordinating a distributed planning system. The AI Magazine 20(4)
51. Clement, B., Durfee, E.: Top-down search for coordinating the hierarchical plans of multiple agents. In: Proc. AGENTS 1999 (1999)
52. Yang, Q.: Intelligent Planning: A decomposition and abstraction based approach. Springer, Heidelberg (1997)
53. Erol, K., Hendler, J., Nau, D.: Complexity results for hierarchical task-network planning. Annals of Mathematics and Artificial Intelligence 18, 69–93 (1996)
54. Nau, D., Cao, Y., Lotem, A., Munoz-Avila, H.: SHOP: Simple hierarchical ordered planner. In: Dean, T. (ed.) Proceedings of the 16th International Joint Conference on Artificial Intelligence (IJCAI 1999). Morgan Kaufmann, Stockholm (1999)
55. Lochbaum, K.E.: A collaborative planning model of intentional structure. Computational Linguistics
56. Rich, C., Sidner, C.L.: Collagen: A collaboration manager for software interface agents. User Modeling and User-Adapted Interaction Special Issue on Computational Models for Mixed Initiative Interaction
57. Blaylock, N., Allen, J., Ferguson, G.: Managing communicative intentions with collaborative problem solving. In: Current and New Directions in Dialogue. Kluwer, Dordrecht (2003)
58. Sadek, M.D.: Dialogue acts are rational plans. In: Proceedings of the ESCA/ETR Workshop on Multi-Modal Dialogue, Maratea, Italy (1991)
59. Dearden, R., Clancy, D.: Particle filters for real-time fault detection in planetary rovers. In: Proceedings of the Thirteenth International Workshop on Principles of Diagnosis, pp. 1–6 (2002)
60. Hofbaur, M.W., Williams, B.C.: Hybrid estimation of complex systems. IEEE Transactions on Systems, Man, and Cybernetics, Part B 34(5), 2178–2191 (2004)
61. Mehra, R., Rago, C., Seereeram, S.: Autonomous failure detection, identification and fault-tolerant estima tion with aerospace applications. In: IEEE Aerospace Conference, Aspen, CO, Proceedings, CO; UNITED STATES, vol. 2 (1998)

62. Washington, R.: On-board real-time state and fault identification for rovers. In: Proceedings of the IEEE International Conference on Robotics and Automation (ICRA), pp. 1175–1181 (2000)
63. de Freitas, N., Dearden, R., Hutter, F., Morales-Menendez, R., Mutch, J., Poole, D.: Diagnosis by a waiter and a mars explorer. In: Invited paper for Proceedings of the IEEE, special issue on sequential state estimation (2003)
64. Driessen, J., Boers, Y.: An efficient particle filter for nonlinear jump markov systems. IEEE Sem. Target Tracking: Algorithms and Applications, Sussex, UK (2004)
65. Verma, V., Thrun, S., Simmons, R.: Variable resolution particle filter. In: Proc of IJCAI (2003)
66. Kwok, C., Fox, D., Meila, M.: Real-time particle filters. In: Advances in Neural Information Processing Systems 15 (NIPS), pp. 1057–1064 (2002)
67. Kwok, C., Fox, D.: Map-based multiple model tracking of a moving object. In: Nardi, D., Riedmiller, M., Sammut, C., Santos-Victor, J. (eds.) RoboCup 2004. LNCS (LNAI), vol. 3276, pp. 18–33. Springer, Heidelberg (2005)
68. Musso, C., Oudjane, N., Le Gland, F.: Improving regularized particle filters. In: Doucet, A., de Freitas, N., Gordon, N. (eds.) Sequential Monte Carlo Methods in Practice, New York. Statistics for Engineering and Information Science, ch. 2, pp. 247–271. Springer, Heidelberg (2001)
69. Lehn-Schiøler, T., Erdogmus, D., Principe, J.C.: Parzen particle filters. In: ICASSP, vol. 5, pp. 781–784 (2004)
70. Liu, J., West, M.: Combined parameter and state estimation in simulation-based filtering. Sequential Monte Carlo Methods in Practice

7

Multi-modal Learning

Danijel Skočaj[1], Matej Kristan[1], Alen Vrečko[1], Aleš Leonardis[1], Mario Fritz[2], Michael Stark[2], Bernt Schiele[2], Somboon Hongeng[3], and Jeremy L. Wyatt[3]

[1] VICOS Lab, University of Ljubljana, Ljubljana, Slovenia
first.last@fri.uni-lj.si
[2] Technische Universität Darmstadt, Darmstadt, Germany
lastname@informatik.tu-darmstadt.de
[3] Intelligent Robotics Laboratory, School of Computer Science, University of Birmingham, Birmingham, UK
{jlw,sxh}@cs.bham.ac.uk

7.1 Introduction

The main topic of this chapter is learning, more specifically, multimodal learning.

In biological systems, learning occurs in various forms and at various developmental stages facilitating adaptation to the ever changing environment. Learning is also one of the most fundamental capabilities of an artificial cognitive system, thus significant efforts have been dedicated in CoSy to researching a variety of issues related to it.

Learning involves numerous deep problems that stretch far beyond straightforward applications of current statistical methods. In particular, learning should take place in a rich interaction of an artificial cognitive system with the environment or with a human (tutor) exploiting multiple modalities to arrive at new concepts and/or extend the current ontologies or merely to adapt to different circumstances.

One of the main topics which was addressed within the CoSy project was the selection of appropriate representations that allow for their efficient creation and modification as the new data is acquired, and the levels and types of supervision that guide the learning processes. In this context, the exploitation of multiple modalities is crucial as they provide means for robust and efficient learning that is not possible within a single modality. For example, the data of one modality may act as a means of supervision for learning within another modality, or, through a simple dialogue inherent ambiguities present in a visual signal can be resolved. Similarly, an artificial cognitive system can learn important insights into how the environment is structured, including causal relations, by observing activities or exercising certain actions.

H.I. Christensen et al. (Eds.): Cognitive Systems, COSMOS 8, pp. 265–309.
springerlink.com

Therefore, the learning we are addressing in this chapter involves information from multiple modalities. One modality helps another modality by providing additional information, i.e., one modality supervises another modality. In Sections 7.2 and 7.3, we address the multimodal learning in the context of interaction between the visual and communication modalities. The PlayMate system we have developed has the capability of relating the visual information and the information obtained through the communication subsystem. In this way, the learning of visual concepts can be supervised by language. In Sections 7.4 and 7.5, the pure visual information is interacting with the information produced by performing certain actions (pushing, grasping). Broadly speaking we can view visual information and the information obtained from actions as two separate sub-modalities. In this view, sections 7.4 and 7.5 address interaction between the visual sub-modality and the action sub-modality. Again, visual cues are learnt that could not have been learnt without having access to information from different origins (sub-modalities). Note that in the literature, the multimodal learning, as defined here, is also frequently referred to as cross-modal learning. Both terms emphasize the interaction between different modalities during the learning process and are often used interchangeably; so they are in this book.

In terms of the levels of supervision, a variety of different modes have been applied, ranging from fully supervised, weakly supervised, to unsupervised. As for the types of supervision they were achieved through dialogues, perception of affordance cues and exploratory strategies. The approaches presented in Sections 7.2 and 7.3 focus on analysing different levels of supervision, which is achieved through a dialogue with a human teacher. The integrated system we have developed facilitates such kind of research and provides means for applying different learning strategies in an interactive learning settings.

The remainder of the chapter contains four sections which address the issues mentioned above in different contexts.

Section 7.2 presents an interactive framework for continuous learning of visual concepts. The main goal that was set is to learn associations between automatically extracted visual features and words describing the scene (visual attributes and spatial relations) in a user friendly, natural, open-ended, and continuous manner. The system facilitates interactive learning of basic visual concepts in a dialogue with a human tutor. The learning can be performed with different levels of supervision. The developed framework also supports unlearning, which enables recovery from any errors accumulated in the learned models. The entire learning process is fully integrated with other parts of the system that provide information needed by the learner and use the information produced by the learner. The entire system thus provides visual input and enables verbal and non-verbal communication with a tutor facilitating continuous and interactive cross-modal learning.

In Section 7.3, crossmodal learning of visual categories is performed in a way that combines supervised and unsupervised training methods. While supervised methods tend to produce more accurate results, unsupervised methods are highly attractive due to their potential to use masses of unlabeled training data. The proposed novel method uses unsupervised training to obtain visual groupings of objects and a cross-modal learning scheme to overcome the inherent limitations of purely unsupervised training. The method uses a unified and scale-invariant object representation by Scale-Invariant Patterns (SIPs) that allows us to handle labeled as well as unlabeled information in a coherent way. One of the potential settings is to learn object category models from many unlabeled observations and a few dialogue interactions that can be ambiguous or even erroneous. Initial experiments demonstrate the ability of the system to learn meaningful generalizations across objects already from a few dialogue interactions.

Section 7.4 addresses supervised-learning of human actions and activities. A cognitive robot that interacts with humans needs not only to detect motion patterns, but also to put them in the context of understanding human intention. Towards this goal, a new representation of intentional actions has been proposed, which models the causal relations between hand motion, object states, and the effects of actions. It has been shown, by reproducing a key result from the literature on action learning in children, that the proposed representation has properties in common with action learning in humans. Human movement analysis has also been put in the context of activity understanding. This extension allows the robot to relate the concept of objects to the situations that the agent tends to perform certain kinds of actions, and also to learn about the reliability of a class of execution styles.

Section 7.5 addresses multimodal learning in the context of an embodied cognitive agent. In this study, the challenge of functional object categorization is approached from a completely different angle. Namely, the functional category representations are acquired by observing few prototypical human-object interactions rather than explicitly modeling physical object properties. Naturally, the set of functional categories that the system's local feature-based vision module is able to represent is restricted to those that are characterized by distinct visual features, e.g., the bent shape of a mug handle suggests how to grasp the mug in a specific way. Such distinct visual features are termed *affordance cues*, and the functional object category detection is based on the recognition of these cues. The results are reported for the detection of two functional object categories learned by the system, which demonstrate their generalization capabilities across and beyond basic level categories. In fact, it is shown that the system supports the interpretation of these categories as composite functional ones.

7.2 Continuous Learning Framework

7.2.1 Introduction

An important characteristic of a system that operates in a real-life environment is the ability to expand its current knowledge. The system has to create and extend concepts by observing the environment – and has to do so continuously, in a life-long manner. An integral part of such a *continuous learning* system is a method for incremental updating of previously learned representations, which has to fulfill several requirements: (i) the learning algorithm should be able to update the current representations (and create new ones if necessary), (ii) it should not require access to old (previously processed) original data, (iii) the representations should be kept compact; they should allow improving the information content while not requiring an increase of memory requirements with each observation, and (iv) the computational effort needed for a single update should not depend on the amount of previously observed data. Furthermore, models should allow for error-recovery (*unlearning*) in cases when erroneous information gets incorporated into them. The learning process should thus create, extend, update, delete, and modify models of real-world concepts in a continuous, life-long manner, while still keeping the representations of the environment compact and efficient.

In this subsection we present an algorithm that satisfies these requirements when learning associations between low-level visual features and higher-level concepts. In particular, we address the problem of continuous learning of visual properties (such as colour or shape) and spatial relations[1](such as 'to the left of' or 'far away'). The main goal is to find *associations* between *words* describing these concepts and simple *visual features* extracted from images.

We tackle this problem by using a continuous learning paradigm in a cross-modal interaction between the system and the tutor. This interaction plays a crucial role in the entire learning process, since the tutor provides very reliable information about the scenes in question. This information can also be inferred by the system itself, reducing the need for tutor supervision, however also increasing the risk of false updates and degradation of the current knowledge. In this section we introduce and analyse several *different learning modes* requiring different levels of tutor supervision.

In our setting, the inputs in the learning process are partial descriptions of the scene; descriptions of the objects and relationships between them. An object can be labeled with several concept labels (e.g., object A can be 'yellow' and 'round' and positioned 'in the middle' of the image and 'to the left of' object B). These descriptions serve as learning examples. There are no negative examples (e.g., the information that 'yellow' is not 'red' is not provided). Therefore, the algorithm should build *reconstructive* representations without relying on *discriminative* information, which would discriminate between different classes (i.e., concepts) [1].

Similar problems have often been addressed by researchers from various fields, from psychology, to computational linguistics, artificial intelligence, and computer as well as cognitive vision. Our problem is closely related to the symbol grounding problem [2], which has been very often addressed by a plethora

[1] In the current implementation we only address a simplified geometrical aspect of spatial relations, without considering functional aspect and context.

of researchers [3, 4, 5, 6, 7, 8, 9]. Our work is closely related to that of Roy [5, 6], in that our framework focuses on learning qualitative linguistic descriptions of visual object properties and scene descriptions. His system in [5] was designed to learn word forms and visual attributes from speech and video recordings, and subsequently, he extended this work for generating spoken descriptions of scenes [6]. Several papers have also been published addressing online learning, particularly with regard to object recognition [10, 11, 12]. In our system, however, we utilize, in a unified framework, continuous online learning of qualitative object properties and spatial relations in a setting with no negative examples where every sample can be labelled with multiple concept labels. Furthermore, our system facilitates unlearning and error-recovery as well. The system is completely integrated in a robot that on one hand provides useful multimodal information for the learning subsystem, and on the second hand, uses the knowledge and information provided by the learning subsystem to accomplish given tasks.

Fig. 7.1 shows a sample dialogue between the tutor and the system, which indicates the process of learning of object visual properties (colour, size, and shape). The tutor is describing the scene and verifying the state of the robot's knowledge by posing questions. The robot also asks the tutor when presented with an object whose visual properties it does not recognise. This example shows how the system can extend its knowledge in a dialogue with the tutor. This is one of our motivating scripts for the PlayMate scenario.

'H: This is a blue, large, and square object.'
'R: OK.'
...
'H: What do you see?'
'R: I don't know this object.'
'H: It is green, small, and triangular.'
'R: OK.'
...
'H: What do you see?'
'R: I see a small and square object.'
'H: Yes, that's correct.'
'R: OK.'
...
'H: What do you see?'
'R:I see a green object. I think it is also large.'
'H: Yes, that's correct.'
'R: OK.'
'H: It is also rectangular.'
'R: OK.'
...
'R: I don't know this object. What is it like?'
'H: It is blue, small, and rectangular.'
'R: OK.'
...
'R: Is this object large?'
'H: Yes, that's correct.'
'R: OK.'
'H: It is also yellow and circular.'
'R: OK.'
...
'R: Is this object red?'
'H: No, it is not red.'
'R: OK.'
'H: It is yellow.'
'R: OK.'
...
'H: What do you see?'
'R: I see a red, small, and triangular object.'

Fig. 7.1. An example dialogue between the tutor (H) and the system (R)

The integration aspect of the PlayMate is described in Chapter 9, while in this section we focus on the problem of learning. We first present different modes of learning requiring different levels of human interaction and then introduce the main incremental learning algorithm and the underlaying techniques for managing reconstructive representations. We then present the evaluation of the proposed methods. Finally, we summarize and discuss some open issues.

7.2.2 Different Modes of Learning

The interaction between a tutor and an artificial cognitive system plays an important role in a continuous learning framework. In this subsection we define several learning modes which alter the behaviour of the system and require different levels of tutor involvement.

When implementing a continuous learning mechanism, two main issues have to be addressed. Firstly, the representation, which is used for modeling the observed world, has to allow for updates when presented with newly acquired information, considering the requirements described in Section 7.2.1. Secondly, a crucial issue is the quality of the updating, which highly depends on the correctness of the interpretation of the current visual input. With this in mind, several learning strategies can be used, ranging from completely supervised to completely unsupervised. Here we discuss three such strategies:

- **Tutor-driven approach (*TD*).** The correct interpretation of the visual input is always correctly given by the tutor.
- **Tutor-supervised approach (*TS*).** The system tries to interpret the visual input. If it succeeds to do this reliably, it updates the current model, otherwise asks the tutor for the correct interpretation.
- **Exploratory approach (*EX*).** The system updates the model with the automatically obtained interpretation of the visual input. No intervention from the tutor is provided.

We further divide *tutor-supervised learning* into two sub-approaches:

- **Conservative approach (*TSc*).** The system asks the tutor for the correct interpretation of the visual input whenever it is not completely sure that its interpretation is correct.
- **Liberal approach (*TSl*).** The system relies on its recognition capabilities and asks the tutor only when its recognition is very unreliable.

Similarly, we also allow for **conservative** and **liberal exploratory** sub-approaches (***EXc***, ***EXl***).

It is obvious that the system is supposed to have a certain level of self-understanding; it should be able to estimate whether its current knowledge suffices to interpret the current scene, or it should ask a tutor for help. Therefore, it should have a recognition capability, i.e., the ability to interpret the visual input to some extent. And even more importantly, the system should be able to evaluate the reliability of this recognition process.

To formalise the above descriptions, let us assume that the recognition algorithm is able to provide information about the reliability of the recognition by giving one of the following five answers when asked to confirm the interpretation of the visual scene (e.g., the question may be: "Is this object circular?"): *'yes' (YES)*, *'probably yes' (PY)*, *'probably no' (PN)*, *'no' (NO)*, and *'don't know' (DK)*. Table 7.1 presents actions that are taken after an answer is obtained from the recognition process. The system can either *ask* the tutor for the correct interpretation of the scene (or the tutor provides it without being asked), *update* the model with its interpretation, or do nothing. As it is stated in Table 7.1, the system can communicate with the tutor all of the time (*TD* learning), often (*TSc*), occasionally (*TSl*) or even never (*EX* learning). This communication is only initiated by the tutor in the tutor-driven approach, while in other approaches the dialogue and/or the learning process is initiated by the system itself.

Table 7.1. Update table

	YES	PY	PN	NO	DK
TD	ask	ask	ask	ask	ask
TSc	update	ask	ask	/	ask
TSl	update	update	/	/	ask
EXc	update	/	/	/	/
EXl	update	update	/	/	/

To speed up the initial phase of the learning process and to enable development of consistent basic concepts, one could start with mainly tutor-driven learning with many user interactions. These concepts would then be used for updating with limited help from the user in tutor-driven or even exploratory manner.

7.2.3 Learning Algorithm

There are two major parts of the interactive learning framework discussed in this section. The first part is the *update* algorithm, which continually updates the representations of the visual concepts. The second is the *recognition* algorithm, which, using the representations, can produce quantitative answers to the queries from the user.

Our system does not have a priori access to the negative examples of the learnt concepts. Furthermore, we allow multiple concept labels for each input instance, which generally prohibits the update algorithm to exploit the discriminative information to improve the concepts. For those reasons, the update algorithm builds and maintains *reconstructive* representations of the observed visual features in a form of generative models. Each visual concept is associated with a visual feature which best models the observed visual data according to two criteria: *consistency* and *specificity*. The algorithm automatically determines which of the extracted visual features are *consistent* over all observations of the same visual concept; at the same time, the features also have to be *specific* for that particular concept.

The learning algorithm thus selects from a set of one-dimensional features[2] the feature whose values are most consistent and specific over all of the observed images representing the same visual concept (e.g., all images of large objects, or circular objects, or pairs of objects far apart etc.). Note that this process is performed incrementally, considering the current image (or a very recent set of images) and only the learnt representations of the concepts without accessing the previously processed images. For the purposes of establishing the concept-feature associations, the distribution each feature is initially modelled by a simple generative model for each concept separately. In our implementations, a single Gaussian is used for such simple generative model. These models (Gaussians) are then continually updated as new observations arrive. After each observation, the algorithm associates each concept with a single feature. Once an association with a particular feature is established, an additional, more detailed generative model is created for that feature. It is initialized by the simple model of that feature and then updated to a more accurate and complex one as new observations arrive. In our implementations we use kernel density estimates (KDE) for the detailed generative models, which will be described in more detail in the next subsection. After more and more observations arrive, it may turn out that some other feature agrees better with a given concept than the one to which it was initially associated. In that event, the concept-feature association is reset to a new feature and the generative model for the concept is reinitialized. The algorithm fulfills the consistency and specificity criteria by associating a concept with a feature whose generative model for that concept is most distinctive from the generative models of the other concepts. To measure the distinctiveness we require a measure of distance between the generative models. In our implementations we used the Hellinger distance [13] since it is a metric and returns the distance value on a constrained interval between zero and one, which makes it convenient for interpretation how much two distributions are similar (in contrast to Kullback-Liebler divergence). Therefore an analytic Hellinger distance was applied to evaluate distance between single Gaussians (simple generative models) and a numerical generalization of the Hellinger distance for the more detailed generative models (see, eg., [14]).

In the recognition stage, the algorithm first extracts feature vectors from the visual data and evaluates it against every concept (i.e., its generative model). For each concept, the algorithm returns a value between zero and one, which reflects the confidence that the observation agrees with that concept. Since our generative models, as we will see later, are in fact probability density functions (pdf), the confidence is evaluated as the integral over the feature values whose probability is smaller than that of the observed value. This result is then quantized and transformed into a form which is suitable for

[2] These features may be, for example, the median hue value of an object, area of segmented region, coordinates of the object center, distance between two objects, etc.

communication with the human operator, or into a form which is appropriate for any other particular mode of continuous learning presented in the previous subsection.

In online algorithms like the one presented in this section, the noise in the input data has a detrimental effect on the learnt representations. In an online semi-autonomous and exploratory framework discussed here, this problem is even more pronounced. If, for example, the recognition algorithm fails at some point to correctly interpret the visual scene and overestimates the reliability of its recognition, the generative models of the concepts tend to degrade and the performance of the system will typically decrease severely. However, since the system interacts with a human user, the user can help the system to recover from the errors through interaction, by, e.g., indicating to a system that its belief about a certain concept is wrong. This unlearning is a crucial element of the learning system since it affects the choice of the generative models used to model the concepts – apart from allowing learning from positive examples, these generative models have to support learning from negative examples as well. This latter process is what is called unlearning. In the following section we discuss a particular form of the generative models which are used in our implementations and allow flexible learning from positive, as well as from the negative examples.

7.2.4 Reconstructive Representations for Interactive/Online Learning

A popular approach to building a generative model from the observed data is to estimate the underlying probability density function (pdf), which *generated* the data. In particular, Gaussian Mixture Models (GMM) have been shown to approximate well probability density functions which may even be far from Gaussian [15]. Standard approaches to GMM estimation include Parzen estimators [15, 16, 17, 18], expectation-maximization-based (EM) algorithms [19, 20, 21], and variational-Bayes methods [22, 23], to name a few. However, extending the above techniques to online estimation is nontrivial, since they require all the data in advance. Most of the recent solutions for online estimation thus either impose temporal constraints on the incoming data [24, 25], assume a known prior over the model parameters [26], assume each observed sample is approximated by a single Gaussian with a known covariance [27], or assume the modes of the underlying pdf are well separated and Gaussian [28]. Recently, Kristan et al. [29, 14] have developed a framework for online estimation of the GMMs, which can be viewed as an *online* extension of the batch kernel density estimation [15], (OKDE). Among the existing approaches to online estimation of mixture models, OKDE makes the least assumptions about the incoming data and the shape of the approximated distributions. As a marked contrast to existing approaches, this is the only framework which also allows refining a GMM generative model from the negative examples through the process of unlearning. In this section we will briefly overview the learning/unlearning methodologies of the OKDE; for a detailed treatment, however, the reader is referred to [29, 14].

The online estimation in OKDE from positive observations proceeds as follows. A Gaussian kernel is assigned to each new observed sample and added to the existing mixture model. The variance of the kernel is determined using an online extension of the plug-in rule [15] for the bandwidth estimation. To maintain a low complexity, the mixture model is compressed from time to time into an equivalent mixture with a smaller number of components. This is achieved by iterating between a reduction step and an optimization step. The reduction step uses a reduced-set density estimation [30] to remove components from the mixture and the optimization step applies a Levenberg-Marquardt optimization to minimize the error between the original mixture model and its compressed equivalent. In this way, the OKDE produces a variable-bandwidth kernel density estimate of the underlying distribution with a low complexity.

Figure 7.2 shows results of an experiment in which a set of samples was sampled from a reference distribution which was a mixture of a uniform and a Gaussian distribution. These samples were then used one at a time in OKDE to approximate the reference pdf. For comparison, Figure 7.2 also shows the results for two standard batch kernel-density estimators: an optimal and a suboptimal. The optimal estimator used the solve-the-equation plug-in method [31] to estimate the kernels, while the suboptimal estimator used the Silverman's rule-of-thumb ([15], page 60). In terms of the integrated squared error (ISE) between the approximation an the underlying reference distribution, the OKDE outperformed the suboptimal batch KDE, and approached the accuracy of the optimal batch KDE (Figure 7.2b). Note from Figure 7.2c that the complexity of the mixture model produced by the OKDE remains low even after many samples have been observed, while the complexity of the mixtures produced by the optimal and the suboptimal KDE increases linearly with the number of samples.

As we have mentioned in the previous section, being able to learn from all-positive examples is not flexible enough for continuous learning in cognitive agents. Indeed, in a noisy environment, false examples will often get accidentally incorporated into the estimated mixture model and a mechanism for unlearning is required to *remove* them from the model. We will illustrate the unlearning framework of OKDE with an example of interactive learning of the concept of a *red color* (Figure 7.3); for the details the reader is referred to [14]. Assume that we present a robot with an object to which we refer as a *red fork* (see Figure 7.3a). The concept of a red color is constructed in form of a mixture model $p_{\text{red}}(x)$ (Figure 7.3b) from the sampled hue values (green dots in Figure 7.3a). By backprojecting $p_{\text{red}}(x)$ to the original image, we obtain the belief map in Figure 7.3c. Note that high beliefs are assigned to the color of the fork's handle and even higher to the color of its head. This is because the right-most mode in the $p_{\text{red}}(x)$ (Figure 7.3e) corresponds to the hue values from the forks yellow head. The agent would thus wrongly believe

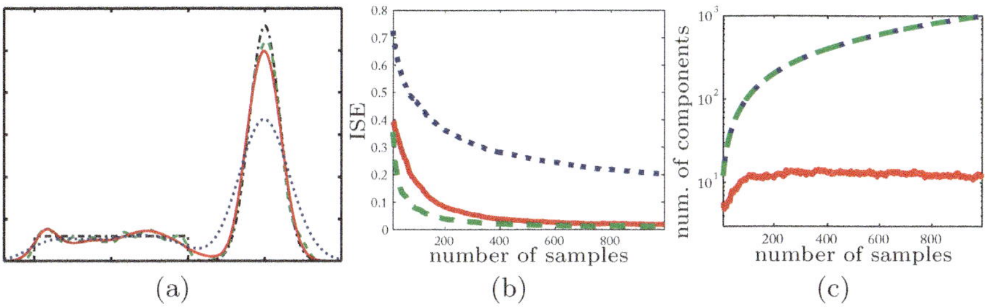

Fig. 7.2. Comparison between approximations of a pdf using the OKDE and two batch KDE methods. Approximations after observing 1000 samples (a). The ISE between the approximations and the reference model (b), and the number of components in the approximations (c), w.r.t. the number of samples. Results of the OKDE, the suboptimal batch KDE and the optimal batch KDE are shown in red full lines, blue dotted lines and green dashed lines, respectively. The reference pdf is depicted in (a) by black dash-dotted line.

that the yellow as well as the red hues make up the concept of the *red color*. To rectify this, we then present a yellow ball (Figure 7.3b) and say that its color is *yellow* and *NOT red*. As before, hue values are sampled from the ball and another mixture model, a negative example, $p_{\text{yell}}(x)$ is constructed (Figure 7.3f). The complement of this negative example $p_{\text{yell}}(x)$ (Figure 7.3g) can now be considered as an additional positive example, which assigns a low probability to those values which are likely to present the negative examples. Assuming that $p_{\text{yell}}(x)$ is independent from $p_{\text{red}}(x)$, the complement is fused with $p_{\text{red}}(x)$ by multiplication to yield a refined model $\hat{p}_{\text{red}}(x)$. The refined $\hat{p}_{\text{red}}(x)$, after an additional compression step, is shown in Figure 7.3h. Note that the mode corresponding to the yellow color has been attenuated, which is verified in the resulting belief image (Figure 7.3d). We see that now only the pixels on the fork's handle are believed to correspond the concept of the *red color*. This shows the flexibility of learning with OKDE: The agent was able to learn the concept of a red color from a highly ambiguous data under supervision in only two steps of interaction.

7.2.5 Experimental Results

The system presented in this section was primarily designed to work in interaction with a user and it forms a part of the PlayMate system. Fig. 7.1 shows a sample dialogue between the tutor and the system. The first part of the dialogue took place in the tutor-driven learning mode, when the tutor was teaching the system about the objects in the scene, while the second part of the dialogue took place using the tutor-supervised modes of learning, when the system took the initiative and asked the tutor for clarification when

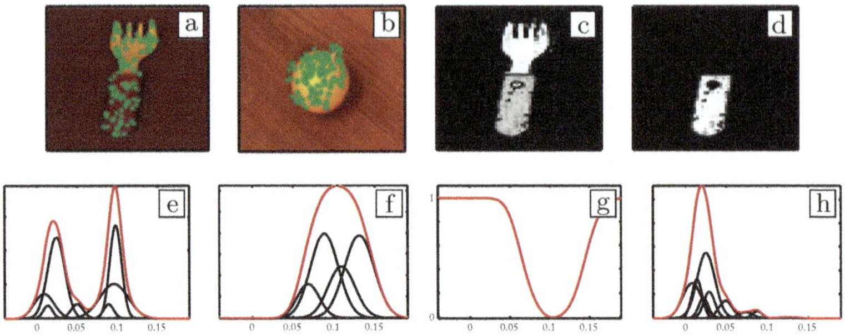

Fig. 7.3. Hue values are sampled from (a) to initialize the mixture model $p_{\text{red}}(x)$ (e). The mixture $p_{\text{yell}}(x)$ built from the hue values of a yellow ball (b) is shown in (f). The complement of $p_{\text{yell}}(x)$ is shown in (g) and the unlearnt compressed model $\hat{p}_{\text{red}}(x)$ is shown in (h). The belief images corresponding to $p_{\text{red}}(x)$ before and after unlearning, (e) and (h), are shown in (c) and (d), respectively. White colors correspond to high beliefs, while dark colors correspond to low beliefs. The mixtures in (b,d,f) are show in bright (red) lines, while their components are depicted in black lines.

needed. Additional examples and more detailed description of architectural issues are described in Chapter 9.

To comprehensively analyse the proposed learning modes, we have also built a simulation experimental setup. We performed experiments on images with known ground truth, and simulated the answers of the tutor. In this way extensive tests could be automatically performed and a reliable evaluation of the proposed methods were obtained. In the following we only briefly present the results; the detailed descriptions of the experiments can be found in [32, 33, 34, 29].

First, we present the results on learning visual attributes of objects. Basic shapes of various different colours and sizes were selected as test objects. Some of them are depicted in Fig. 7.4(a). We considered three visual attributes (colour, size and shape), and ten values of these visual attributes altogether (red, green, blue, yellow; small, large; square, circular, triangular, and rectangular). The main goal was to find associations between these attribute values and six extracted features (like the median of the hue over all pixels in the segmented region, or the area of this region, etc).

We incrementally updated the representations with the training images using different learning strategies. At each step, we evaluated the current knowledge by recognising the visual properties of test images. The evaluation measure we used is *recognition score*, which rewards successful recognition (true positives and true negatives) and penalises incorrectly recognised visual properties (false positives and false negatives).

The results (the curves of the evolution of the recognition score through time) are presented in Fig. 7.4(c), while Fig. 7.4(d) plots the frequency of communication between the system and the user, i.e. how often the user is required to provide information.

All the different learning strategies presented in subsection 7.2.2 were tested. It can be seen that Tutor driven learning (*TD*) performs best and yields very good recognition results. However, in this case the user has to completely describe the object at every learning step. Tutor-supervised learning (*TSc*, *TSl*) proved to be quite successful as well (in the beginning the system does not have a lot of knowledge, which would enable reliable autonomous recognition, so the first 15 training images were added in a tutor driven mode). Tutor supervised learning required significantly less interaction with the tutor; the average number of questions drops below 1 in the conservative and even to almost 0 in the liberal case. The exploratory approach (*EXc*, *EXl*), which does not involve interaction with the tutor, does not improve the model. So, as expected, there is a trade-off between the quality of the results and the autonomy of the system.

Exactly the same system was also used for learning simple spatial relations. We only changed the features that were to be extracted from the image (position of the object centers in the scene, distance between the objects, etc.). Using five such features, the learning framework was able to learn eleven spatial relationships (like 'to the left of', 'far from', or 'near'). The correctly assigned associations, along with the previously learned visual attributes, enabled the automatic detection of objects and the production of scene descriptions such as those presented in Fig. 7.4(b).

Next, we present the results of the experiment performed on a set of everyday objects (some of them are depicted in Fig.7.5(a)).

Fig. 7.5(b) shows the evolution of the overall accuracy of the concept recognition, which increases by adding new samples (solid line). The growth of the accuracy is very rapid at the beginning when new models of newly introduced concepts are being added, but still remains positive even after all models are formed due to refinement of the corresponding representations. Fig. 7.5(c) plots the average number of Gaussian components in the KDE models. One can observe that after a while this number does not grow any more, thus the model size remains limited. The models do not improve by increasing their complexity, but rather due to refinement of the underlying representations.

To demonstrate the unlearning algorithm, we incorrectly labeled every 10-th training sample in the first half of the incremental learning process. As a result, the underlaying KDE representations were corrupted and modelled also the feature values, which were not supposed to be encompassed. Therefore, the recognition results degraded (dotted line in Fig. 7.5(b)). However, by applying unlearning to the corrupted representations, they were successfully corrected, which resulted in a significant improvement of the recognition performance (dashed line). The recognition results after unlearning were very similar to those obtained in an error free learning process.

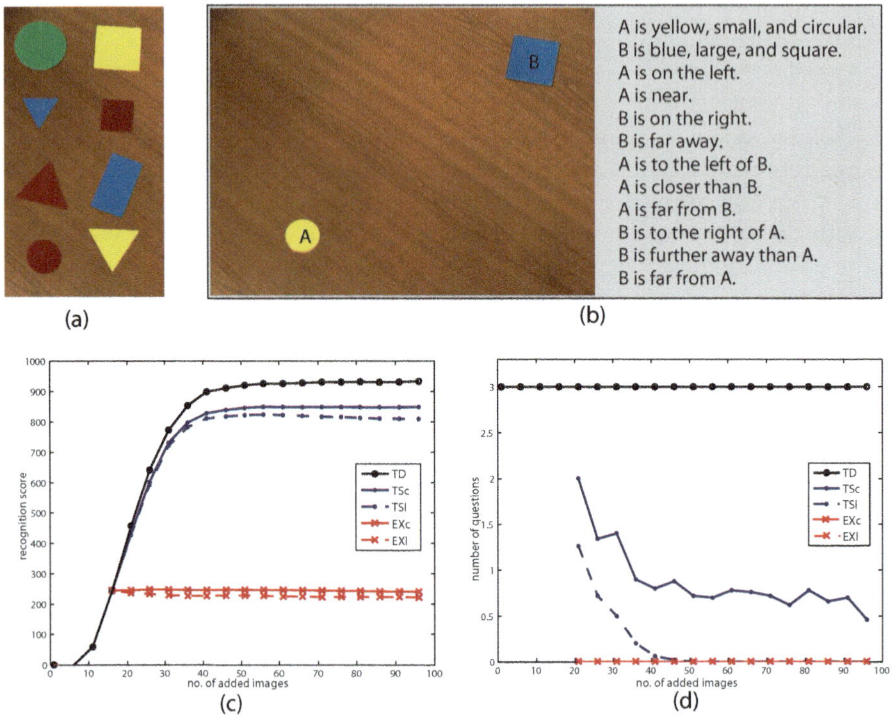

Fig. 7.4. (a) Input shapes. (b) Automatically generated scene description. (c) Recognition score. (d) Frequency of communication.

These results are further demonstrated on one particular example. Figs. 7.5(d-i) depict the evolution of the underlying KDE model of the feature 'Hue', which was associated with the concept 'Blue', over time. It shows how the initially simple model improves through time Figs. 7.5(d-f). It also shows the efficiency of the compression algorithm; the compressed model (Figs 7.5(g)) resembles the original one (Fig. 7.5(f)) at a very high degree of accuracy. Finally, Figs. 7.5(h,i) demonstrate the performance of the unlearning algorithm, which was able to undo the noise introduced into the model.

7.2.6 Discussion and Outlook

It this section we presented an interactive framework for continuous learning of visual concepts. The main goal is to learn associations between automatically extracted visual features and words describing the scene (visual attributes, spatial relations) in a user friendly, natural, open-ended, and continuous manner. The system facilitates interactive learning of basic visual concepts in a dialogue with a human tutor. The learning can be performed on different

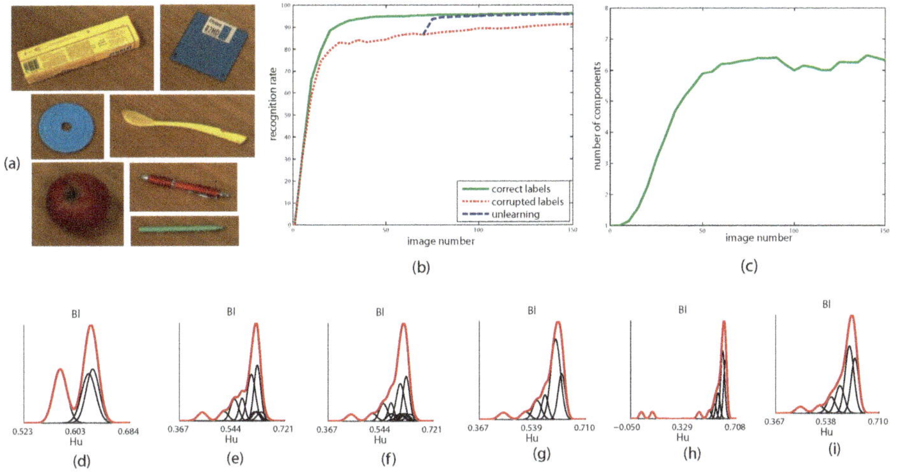

Fig. 7.5. Learning of basic object properties: (a) Seven everyday objects from the database. Results of incremental learning: (b) evolution of accuracy through time for incremental learning on correct data (solid line), on partially incorrect data (dotted line), and after unlearning (dashed line), (c) average number of components through time. Evolution of one KDE model representing the object property 'blue' through time: model after (d) 15, (e) 85, and (f) 120 added training samples. (g) Compressed final model. (h) Final model learned with incorrect labels. (i) Model after unlearning.

levels of supervision. The developed framework also supports unlearning, which enables recovery from the errors accumulated in the learned models.

In the previous subsection the methods proposed were evaluated in isolation by explicitly providing the ground truth information, which is supposed to be provided by other parts of a cognitive system setup. However, these methods are fully integrated in the overall system based on CAST, and play an integral role of the PlayMate system. Other parts of the system thus provide information needed by the learner (e.g., linguistic input) and use the information produced by the learner (e.g., recognition results). The entire system thus provides visual input and enables verbal and non-verbal communication with a tutor facilitating continuous, and interactive cross-modal learning.

The continuous learning framework we have developed is able to learn basic visual concepts (like colours, shapes and spatial relations). A natural extension of this system would be to make it fully scalable to enable it to learn more complex concepts and hierarchies of concepts. Furthermore, the system could also consider other modalities in a larger part as well. A very interesting research issue is introspection and detection of ignorance, and planning of new actions that the system should undertake to gather the knowledge which would be used to fill these gaps. This is a prerequisite for more efficient exploratory learning, which would require minimal supervision of a human tutor. The

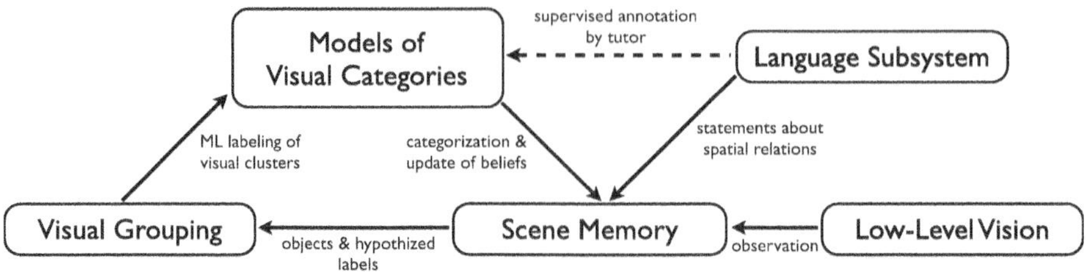

Fig. 7.6. System overview

current framework and the overall system we have developed is a solid base for such research and may serve as a good starting point and a tool in further investigations and development.

7.3 Cross-Modal Learning of Visual Categories

Section 4.1 already outlined a rough topology of existing work on visual categorization. One axis along which such methods can be aligned is how they are trained and which level of supervision they require. Interestingly, most today's object categorization methods use either supervised or unsupervised training methods. While supervised methods tend to produce more accurate results, unsupervised methods are highly attractive due to their potential to use far more and unlabeled training data. This section proposes a novel method that uses unsupervised training to obtain visual groupings of objects and a cross-modal learning scheme to overcome inherent limitations of purely unsupervised training. The method uses a unified and scale-invariant object representation by *Scale-Invariant Patterns (SIPs)* that allows us to handle labeled as well as unlabeled information in a coherent way. One of the potential settings is to learn object category models from many unlabeled observations and a few dialogue interactions that can be ambiguous or even erroneous. First experiments demonstrate the ability of the system to learn meaningful generalizations across objects from only a few dialogue interactions.

System Overview

Figure 7.6 shows an overview of the presented system, which is tighly related to the structure of this section. Section 7.3.1 describes how the vision system represents the visual input. This representation allows for an unsupervised visual grouping step as described in Section 7.3.3. To overcome the inherent limitations of unsupervised as well as fully supervised approaches, we describe how we combine different levels of supervision in Section 7.3.3. Section 7.3.4 describes the language system that parses an utterance to a logical form. Section 7.3.5 explains the spatial reasoning processes that associate the expressions with the visual observation. Finally we present two scenarios,

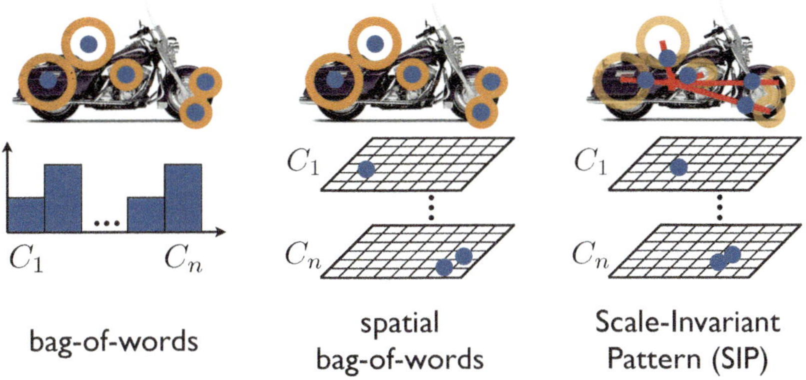

Fig. 7.7. The scale-invariant representation we use is derived from the common bag-of-words representation. Recently a spatial resolution was added to each bin. We improve the robustness of these type of representations by adding a scale-normalization step yet retaining the spatial information which yields what we call a Scale-Invariant Pattern (SIP). The figure visualizes how local featuers (blue dots) are entered in a spatially discretized histogram.

in which the system successfully resolved ambiguous information and even recovered from erroneous beliefs.

7.3.1 Object Representation by Scale-Invariant Patterns

In this section we describe how we encode visual patterns and objects as what we call *Scale-Invariant Pattern (SIP)*. SIPs are a key ingredient in our systems as they provide an embedding of visual patterns and objects into a vector space that facilitates clustering and recognition in a cheap yet effective manner. First, we describe the low-level feature description we base our representation on and then derive the actual representation.

When a new image is grabbed from the camera, SIFT descriptors [35] are extracted at Hessian-Laplace interest points [36]. While there exists a wide range of interest point and descriptor combinations, we opted for this particular combination based on evaluations on different categorization tasks [37]. Following a common philosophy in the field to visual categorization [38, 39, 40, 41, 42, 43], we first generate a visual codebook based on clustering of detected features. In our experiments we use a codebook with 1000 entries obtained by k-means clustering.

Based on this feature representation we derive one of the important ingredients of our system: the *Scale-Invariant Pattern (SIP)* representation. This representation is the basis not only for discovering objects in the scene, but also for visual grouping and object categorization. We adapt the scale-invariant representation from [43] which already has shown its versatility in weakly-supervised learning and ranking tasks. As visualized in Figure 7.7,

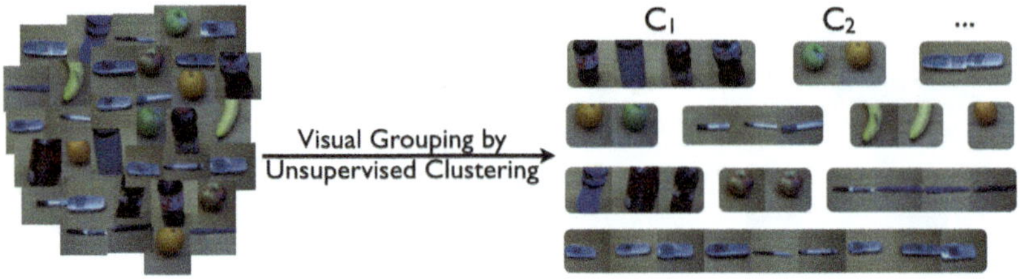

Fig. 7.8. Visual grouping of objects by clustering

this representation can be seen as a extension of the bag-of-words representation [38] by adding two spatial dimensions to each bin [41]. In order to obtain a scale-invariant representation, the feature positions (pos_x, pos_y) are normalized with respect to the center of the pattern (c_x, c_y):

$$(pos'_x, pos'_y) = \left(\frac{pos_x - c_x}{\sigma}, \frac{pos_y - c_y}{\sigma} \right) \quad (7.1)$$

where σ is the detected scale of the feature. For efficiency, we store these patterns as sparse vectors Ψ. The sparsity of the SIPs is one key to the efficiency we are aiming for while still maintaining a descriptive representation of the visual input.

7.3.2 Data-Driven Visual Grouping

Similarily to [44], we use an agglomerative clustering scheme (average linkage) to group object instances in an unsupervised manner. The objects are represented as SIPs (Sec. 7.3.1) which we will denote as sparse vectors Ψ in the following formulas. As described before we normalize these to unit length and we use the scalar-product to measure similarity between these objects. We prefer to use agglomerative clustering over k-means as we do not want to specify the number of visual clusters apriori. The threshold required for the agglomerative clustering scheme is set empirically to a constant value for all our experiments. Figure 7.8 visualizes the clusters C_1 to C_N obtained by our system given the observed objects displayed on the left. Although there are some confusions, we observe a good generalization across category instances. In order to obtain representatives $\boldsymbol{\Psi}_{C_l}$ for each cluster C_l, we compute a weighted sum of the observed patterns Ψ_k:

$$\boldsymbol{\Psi}_{C_l} = \sum_k p(C_l|\Psi_k)\Psi_k \quad (7.2)$$

In our implementation, we have chosen to use hard assignment of the SIPs to the clusters which renders the probability $p(C_l|\Psi_k)$ of assigning pattern Ψ_k to cluster C_l binary.

7.3.3 Combining Unsupervised and Supervised Learning

In this section, we present a model for visual category recognition that combines different levels of supervision in a joint model. The key ingredient is the SIP representation from Secion 7.3.1, which we use throughout. The last section formulated an unsupervised grouping process using this common representation.

Supervised Categorization

To provide basic functionality for our system, we describe how supervised categorization is implemented. Similarly to clustering in Section 7.3.2, we model each category A_i by a single representative $\boldsymbol{\Psi}^S_{A_i}$ (superscript S denotes the supervised model). This is done by summing over all training patterns available for that category

$$\boldsymbol{\Psi}^S_{A_i} = \sum_{j \in \mathbb{S}^{A_i}} \Psi_j, \tag{7.3}$$

where $\mathbb{S}^{A_i}$ denotes the indices of the SIPs that are labeled with category A_i in a supervised manner (e.g. "This is a bottle").

Incorporating Semi-supervision and Unsupervised Information in One Consistent Framework

We formulate the fusion of information obtained from supervised to unsupervised sources as an extension of the supervised case by assuming uncertainty about the correct labeling of the clusters C_l and their representatives $\boldsymbol{\Psi}_{C_l}$ from the unsupervised visual grouping step (Sec. 7.3.2):

$$\boldsymbol{\Psi}_{A_i} = \underbrace{\boldsymbol{\Psi}^S_{A_i}}_{\text{supervised}} + \underbrace{\sum_l p(A_i|C_l) \overbrace{\boldsymbol{\Psi}_{C_l}}^{unsupervised}}_{\text{semi-supervised}} \tag{7.4}$$

$p(A_i|C_j)$ encodes the belief that cluster C_l contains instances of category A_i. How this probability is computed from a few interactions and updated by associating spatial expression with visual observations is described in Section 7.3.5.

To perform classification in the supervised and semi-supervised case, we evaluate the proposed model $\boldsymbol{\Psi}_{A_i}$ as well as $\boldsymbol{\Psi}^S_{A_i}$ for an observed SIP $\bar{\Psi}$ by using histogram intersection. Intuitively, the intersection measures to which percentage the model explains the observation, which we interpret as the probability of belonging to the same class. In order to make models and observations comparable we normalize both to one. Bayes' rule is applied afterwards to obtain the model posterior:

$$p(\bar{\Psi}|\mathbf{\Psi}_{A_i}) = \sum \min(\bar{\Psi}, \mathbf{\Psi}_{A_i}) \tag{7.5}$$

$$p(\mathbf{\Psi}_{A_i}|\bar{\Psi}) = \frac{p(\bar{\Psi}|\mathbf{\Psi}_{A_i})p(\mathbf{\Psi}_{A_i})}{\sum_A p(\bar{\Psi}|\mathbf{\Psi}_{A_i})p(\mathbf{\Psi}_{A_i})} \tag{7.6}$$

$p(\mathbf{\Psi}_{A_i})$ is the category prior, which we assume to be uniform. We decide for the category label with the highest posterior:

$$\hat{A}_i = \underset{A_i}{\operatorname{argmax}}\, p(\mathbf{\Psi}_{A_i}|\bar{\Psi}) \tag{7.7}$$

7.3.4 Language System

In human-assisted visual learning, a human tutor provides the system with descriptions of the current visual scene. To relate these descriptions to the visual input, the system constructs a representation of the meaning of an utterance. While this section outlines the language system and how it interfaces with the other components in this cross-modal learning task, a more detailed description is postponed to Capter 8 in the context of the full dialogue system.

For this analysis of the utterances that accompany the visually perceived scenes, we use a Combinatory Categorial Grammar[45] parser[3]. The parser uses a CCG grammar to relate the syntactic structure of an utterance to the propositional meaning it expresses. Meaning is represented as an ontologically richly sorted, relational structure similar to a description logic formula [46], which makes it possible to use ontologies to mediate between linguistically expressed meaning, and the categories formed in the visual system. Using the hierarchical structure of ontologies, and the possibility to perform ontological inference over instances on these ontologies, provides a more general and better scalable approach to "visual grounding" of language than provided by the string-based approach proposed in e.g. [47], or previous ontology-based approaches such as [48, 49].

In our scenario, utterances are typically predicative copulative sentences in indicative mood (i.e. "X is Y"), which assert that a given predication ("Y") holds for the subject of the sentence ("X"). In our examples, the predication consists of a phrase that encodes a spatial relation (e.g. "left of the bottle" or "below the apple"). In the logical form, the subject is represented as the <Restr> of the state description that is denoted by the utterance, whereas the predication is represented as <Scope>.

We can thus easily derive the spatial configuration asserted in an utterance from its logical form representation. The following example shows such a logical form that is the result of the parsing process of the utterance "the mobile is left of the bottle":

[3] `http://openccg.sourceforge.net`

```
@b1:state(be ^
          <Mood>ind ^
          <Restr>(m1:thing ^ mobile ^
                  <Delimitation>unique ^
                  <Number>sg ^
                  <Quantification>specific_singular) ^
          <Scope>(l1:region ^ left ^
                  <Plane>horizontal ^
                  <Positioning>static ^
                  <Dir:Anchor>(b2:thing ^ bottle ^
                               <Delimitation>unique ^
                               <Number>sg ^
                               <Owner-of>+ ^
                               <Quantification>specific_singular)))
```

For entering utterances into the system, we connect the parser with a speech recognizer as well as a keyboard interface. In the experiments, we mostly use the keyboard interface as well as scripted input for larger evaluations.

7.3.5 Scene Reasoning

Modeling spatial relations as perceived by the human is a challenge in itself, as issues like reference frame and context have to be handled appropriately in situated dialogue systems [50]. Considering the scenarios and main focus of this section, we restricted ourselves to modeling four basic spatial relations $R \in \{\text{"leftof"}, \text{"rightof"}, \text{"above"}, \text{"below"}\}$. We employ triangular shaped distributions $p(\text{pos}(\Psi_i), \text{pos}(\Psi_j)|R)$ defined in 2d image coordinates, where objects are referenced by their patterns Ψ_i and $\text{pos}(\Psi_i)$ denotes their position in image coordinates. Although these distributions are represented as non-parametric kernel densities which lend themselves to online updating, we don't explore this option here and keep them fixed as pre-defined in the experiments.

Spatial Reasoning

We formulate the association of a spatial expression E extracted from an utterance (see Sec. 7.3.4) with two patterns Ψ_i and Ψ_j with positions $\text{pos}(\Psi_i)$ and $\text{pos}(\Psi_j)$ observed in scene S_k, as the problem of finding the most likely pair $\hat{P}_{i,j}$ of patterns: $\hat{P}_{i,j}^{(k)} = \text{argmax}_{P_{i,j}}\, p(P_{i,j}|E, S_k)$,where

$$\begin{aligned} p(P_{i,j}|E, S_k) &= p(\Psi_i, \Psi_j, \text{pos}(\Psi_i), \text{pos}(\Psi_j)|E, S_k) \\ &= p(\Psi_i|E, S_k)\; p(\Psi_j|E, S_k)\; p(\text{pos}(\Psi_i), \text{pos}(\Psi_j)|E, S_k), \end{aligned} \tag{7.8}$$

with

$$p(\Psi|E, S_k) = \sum_h p(\Psi|A_h)p(A_h|E, S_k). \tag{7.9}$$

As we don't model a complete category system yet, leave out contextual effects and assume certainty about the expression E referring to the categories A_{e_1} and A_{e_2} and the relation R, the equation simplifies to

$$p(P_{i,j}|E, S_k) = p(\Psi_i|A_{e_1})p(\Psi_j|A_{e_2})p(\text{pos}(\Psi_i), \text{pos}(\Psi_j)|R) \quad (7.10)$$

Finally, we insert the visual model from Eq. 7.4 to obtain a computational model:

$$p(P_{i,j}|E, S_k) = p(\Psi_i|\mathbf{\Psi}_{e_1})p(\Psi_j|\mathbf{\Psi}_{e_2})p(\text{pos}(\Psi_i), \text{pos}(\Psi_j)|R) \quad (7.11)$$

This formulation facilitates incorporating information and belief from previous interactions as well as learning from scratch. If no information about the visual categories is available $p(\Psi_i|\mathbf{\Psi}_{e_1})$ and $p(\Psi_j|\mathbf{\Psi}_{e_2})$ become uninformative and the system relies only on its notion of spatial relations $p(\text{pos}(\Psi_i), \text{pos}(\Psi_j)|R)$. This can lead to wrong associations. In the next section we present an example and show that the system can successfully deal with this issue.

7.3.6 Label Propagation and Conflict Resolution

In the following, we describe two scenarios, that show the capabilities of our system to propagate information, resolve ambiguities and recover from errors.

Scenario 1 - Label Forward Propagation

In the first scenario, one annotated example each for banana and mobile is presented to the system. Then the system observes the scene as shown in the screenshot in Figure 9(a) and the utterance "the can is above the mobile" is parsed. The red lines visualize the observed relations between objects in the scene. Very unlikely ones have already been pruned away by the system. By generalizing across category instances, system identifies "mobile" and "banana" correctly (with probabilities 0.69 and 0.66 respectively) while evaluating "mobile" model for the banana results in a low probability of 0.15. Consequently the most likely relation is inferred correctly and displayed in light green. A model for the category "can" is created and the observed mobile is added to the existing model for "mobile". In fact, the figure shows the state in which the acquired "can" model is already used for detection. The can is detected correctly, but also the bottle gets a high score for the "can" model, as it's the best explanation given the learned categories (banana, mobile, can).

Scenario 2 - Label Backward Propagation

In the second scenario, we show how the system can recover from erroneous beliefs and update its models accordingly. The system starts without any knowledge about visual categories. Figures 7.9 (b) show a screenshot displaying the scene as observed by the system, which is accompanied by the

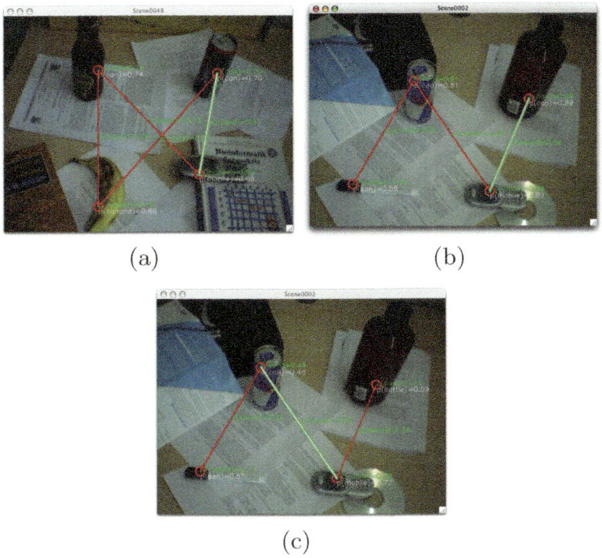

Fig. 7.9. (a) shows an example scenario for propagation of labels from known categories to unknown ones. (b) and (c) Show a scenario where the system updates associations (light green) to recover form an incorrect belief

utterance "the can is above the mobile". Using the same visualization as in the previous scenario, it can be seen in the left image that the most likely relation inferred by the system is wrong. Now we provide the system with supervised knowledge of the visual categories bottle and pen. Revisiting the scene in memory, the object probabilities get updated and the belief about the associated relation gets changed to the correct one as shown in the right image.

7.4 Learning Complex Actions

7.4.1 Introduction

In this section, we describe the development of an action analysis system for a cognitive robot. Our goal is to make the robot aware of the activities that occur in its environment so that it can react in an intelligent manner. This scenario has a different set of requirements from typical video analysis applications. Detecting and labeling motion patterns in videos is not sufficient and the robot needs to reason about the intention of the agent and the functional specification of the concepts of objects and situations. A more suitable representation is hence required in order to solve these cognitive tasks. We propose a novel action representation that addresses these issues based on the following principles.

Psychologically Realistic Modelling of Intentional Actions

A general approach to inferring the intention of an agent is to match the observed motion patterns against a set of typical behaviours known to the observer. The challenge here is to account for uncommon motion patterns. Human observers will first try to account for the variation by drawing additional evidence from the scene to check if something could have forced the agent to perform the action in a different manner. When the collected evidence fails to account for the uncommon movement, the inference is that the agent actually *intended* to perform the action in a different manner. In order to achieve this kind of reasoning in a robot, we propose to model intentional actions using a factorized hidden Markov model (FMM). By explicitly modelling the correlation between hand movements, object states and environmental context, FMM allows the robot to quickly generate an alternative explanation for uncommon hand movement, which is exactly the kind of reasoning we perform in almost any everyday event.

Integrating Probabilistic Action Rules with Markov Networks

In order to put object manipulation in the context of human activities, we propose to upgrade probabilistic action rules to model the correlation between the motion patterns, the properties of actions, its start situations and the action outcomes with probability distributions. These rules allows for functional specification of concepts that enable the robot to reason about the concept of objects (or situations) on which (or in which) the agent tends to perform certain kinds of actions, and the class of execution styles that lead to the action being executed successfully.

The remainder of this section proceeds as follows. We first describe our representation of actions and behaviours in Section 7.4.2. Inference and learning algorithms are discussed in Sections 7.4.3 and 7.4.4. Section 7.4.5 summarizes our experiment results, and we conclude in Section 7.4.6.

7.4.2 Action Representation

Let us consider a toy domain of color and shape games. In this domain, the agent sits near a table on top of which two colored flags are placed, one on the right and the other on the left. A colored geometrical shape (circular or squared) is then placed in front of the agent. In the color game, the agent moves the shape to the flag that has the same color as the shape. In the shape game, the agent moves the circular shape to the left flag and the squared shape to the right flag. In general, the object manipulation/movement varies from trial to trial, and may not be observed in its entirety. The agent may need to reach around an obstacle, or may fail to push the shape because the object surface is slippery. In order to infer the agent's intention, the observer therefore needs to take into account the surrounding context and make hypotheses about things that are ambiguous. Below, we present our action representation that supports these kinds of reasoning. We first describe our representation of object manipulation/movement using a hierarchical model, as shown in Fig. 7.10. We then describe how to put these movements in the context of game playing.

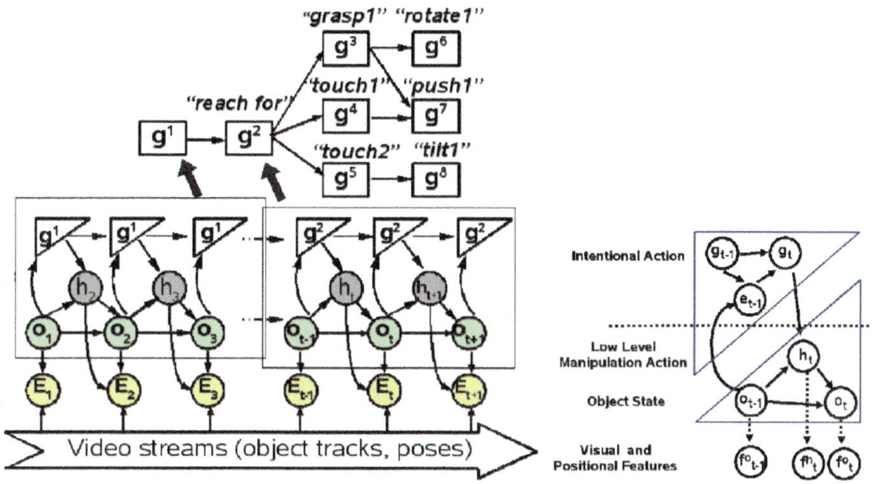

(a) (b)

Fig. 7.10. (a) System Overview, (b) Factorized HMM

Factorized, Hidden Markov Models

A complex manipulation consists of a sequence of simple actions, (e.g., "reach for the colored shape", "push the color shape to the left flag") as shown in Fig. 7.10 (a). Each simple action is associated with a goal (g), which is the desired pose of the target object. A common approach to modelling these motion sequences is based on hidden Markov models (HMM) [51, 52], where an action corresponds to a node in the HMM. Associated with a HMM node is a joint distribution of motion features that characterize the action. These motion features are computed at time t as $E_t = \{f^h, f^o\}$, where f^h denote hand motion features and f^o denote the features of object motion and pose.

In [53], we argue that reasoning about intentions is more realistically modelled by a factorized hidden Markov model (FMM), where we take a viewpoint of Markov Decision Process (MDP) [54]. In FMM, an action node is factored into hand maneuver (h) and object state (o), which are associated with f^h and f^o respectively. The links between h, o, and g inside the rectangular box on the left of Fig. 7.10 (a) represent the fact that the agent who aims to achieve $g^i = g^1$ chooses h_t at time t based on o_{t-1}, such that o_{t-1} transforms into o_t after the execution. This repeats until g^1 has been achieved (i.e., until o_{t-1} becomes similar to g^1). The agent then aims for another goal (e.g. g^2).

In Fig. 7.10 (b), an upper triangle node in Fig. 7.10 (a) is expanded to show that the value of g_t is conditioned on the value of the e_t node, which checks whether o_{t-1} has satisfied the current goal. We note that we use the triangle nodes g^1 and g^2 in Fig. 7.10 (a) instead of expanding them because the goals have not evolved during the execution.

Probabilistic Action Rules

In the game scenario, the evolution of the agent's behavior will depend on the situation that the agent is in. If the agent is playing a color game, the following might occur: "get the shape", "push the shape to the flag that has the same color as the shape", then "remove the shape". If we represent each of these steps by a rule that maps the initial state to the outcome state, action recognition can be reduced a constraint satisfaction problem, where the most likely sequence of actions is the one that satisfies the most constraints (as imposed by the rules). However, executing these actions may not lead to the expected outcomes. The agent may fail to push the shape because the shape is made of slippery material. A robot may execute "pick-and-place" more successfully than "pushing", where as it may be the other way around for a young child who has not mastered grasping. Reasoning about actions therefore depends on the situations. In order to handle the stochastic nature of an action, we represent an action by a probabilistic action rule that allows us to associate the transition dynamics of world states with probability distributions $P(s'|s, a)$, where $s, s' \in S$ (a world state described in terms of visual properties and relational features) and $a \in A$ (a set of actions). A typical probabilistic action rule $a(\bar{x})$ that involves a set of objects $\bar{x}$ has the form:

$$\forall \bar{x}.\Psi(\bar{x}) \wedge a(\bar{x}) \rightarrow \bullet \begin{array}{ll} p_1 & \Psi_1'(\bar{x}) \\ \cdots & \cdots \\ p_n & \Psi_n'(\bar{x}) \end{array} \tag{7.12}$$

where $\Psi(\bar{x})$ is the state before (antecedent) executing the action, and $\Psi_i'(\bar{x})$ is a possible outcome which is associated with a probability p_i, subject to a constraint that $\sum_{i=0,\ldots,n} p_i = 1$. By default each rule has an empty outcome (no change) that represents action failure and noise. We say that a rule covers a state $\Gamma(C)$ and action $a(C)$ if there exists a substitution σ mapping the variables in $\bar{x}$ to C.

In [55], P. Domingos et al. represent actions by probabilistic logic rules, which correlate (action) outcome states to the action properties specified in the antecedent. Our probabilistic rules upgrade their logic rules by integrating information about action processes into the rules, which enables the observer to make inferences about the kinds of execution styles the agent tends to perform, and the situations in which they are likely to succeed. We can restructure the terms in the antecedent into three groups: **deictic references**, **action context** (AC), and **outcome context** (OC). Each deictic reference consists of a variable V_i and restriction ρ_i which is a set of features that define V_i with respect to the variables $\bar{x}$ in the action, as well as to the other V_j such that $j < i$. In order to determine that an example is covered by a rule, we must ensure that these deictic variables can be resolved. Action context consists of features that condition the decision to act using a. These features impose restrictions on the properties of deictic variables; e.g. the types of objects, their colors and shapes. Outcome context consists of features that may cause variations in the outcome of the action rule. The following is an example of an action rule that describes an agent playing a color game. The rule contains three deictic variables: X (a geometrical shape), A (an agent) and Y (a flag). It also contains a restriction that X is located at the area $R0$ (front) of A. The AC part of the rule contains two features. The first represents the fact that there is no color variation between X and Y (i.e., they are of similar color). The second feature contains information that is provided by another modality (e.g., speech), which states that A is engaged in a color game. The OC and $Outcome$ parts indicate that the agent pushes X to the front ($R0$) of the flag Y with a success rate of 80%.

$$
\begin{aligned}
playColor(X) :\ & \{X : shape(X) \\
& \ A : agent(A), at(X, A.R0) \\
& \ Y : flag(Y)\} \\
:\ & \mathbf{AC} \rightarrow colorVariation(X, Y, 0), engageIn(A, colorgame) \\
:\ & \mathbf{OC} \rightarrow manipulatedWith(X, push) \\
:\ & \mathbf{Outcome} \rightarrow 0.8 : at(X, Y.R0) \\
& \qquad\qquad\qquad\ 0.2 : nochange
\end{aligned}
$$

7.4.3 Inference

Given a set of action rules $r_i, i \in \{0, ..., R\}$, the system can make inferences about unknown parameters by computing their likelihood from the observation. For example, in the game scenario, we can make inference about which game the agent is playing by comparing the likelihood of $r_i = playColor(X)$ and $r_j = playShape(X)$.

Computing the Likelihood of Action Rules

For each rule r_i, we obtain a set of possible substitutions, each of which maps the deictic references of the rule to the objects in the scene. For substitution σ^s, the rule r_i is instantiated by substituting all variables that appear in the rule according to σ^s. We then compute the joint probability of the terms in AC to obtain $P(AC)$ (i.e. the likelihood of the precondition(s) of the rule). If $P(AC)$ is larger than a threshold δ, we say that the rule can be instantiated with σ^s. For the instantiated rule, we verify that the agent executes the action according to the type of manipulation that would lead to the results specified in *Outcome*. This is done by matching the observed motion patterns with the appropriate FMM (factorized HMM). In our case, the manipulation type and outcome state specified in the *OC* and *Outcome* parts of the rule provide a template for creating this FMM. For *playColor*(X), an FMM of type "push X to Y.RO" is created. At time T, the system can check if the action rule has been completed, by computing $P(Outcome)$ in the same way that it computes $P(AC)$. If the agent has not completed the action, we can compute the likelihood of the rule as $P(AC)P(OC)$, where $P(OC)$ is the likelihood of the FMM. The outcome can then be predicted with the likelihood of $P(AC)P(OC) * 0.8$.

Computing the Likelihood of FMMs

As shown in Fig. 7.10 (a), object manipulation may consist of multiple actions. Since we only consider goal-based actions, let us represent an action i by $g^i \in G$, where G is the set of m actions known to the observer. To "push X to Y.RO", the agent may first reach for the object X (goal-based action g^i), then push it to $Y.R0$ (g^j), and retract the hand (g^k). This behaviour is encoded in a FMM by a $m \times m$ transition matrix Ω that models the probability distributions over the transitions between actions. Below, we first discuss the computation of $P(g^i_{0:t}|E_{0:t})$, or the likelihood of action i being observed from time frames 0 to t. We then discuss the computation of the likelihood of action sequences based on Ω. Let $E_{0:t}$ be $\{E_0, E_1, \ldots, E_t\}$, where $E_t = (f^h, f^o)_t$. As described in detail in [53, 56], we compute $P(g^i_{0:t}|E_{0:t})$ using Eq. 7.13.

$$P(g^i_{0:T}|E_{0:T}) = P(g^i_{0:T})P(E_0|o_0)\prod_{t=1}^{T}\sum_{h_t=h^u\in n(h)} P(E_t|h_t)P(E_t|o_t) \\ P(o_t|o_{t-1}h_t)P(h_t|o_{t-1},g^i_t) \tag{7.13}$$

We characterize action i by the hand maneuver types $h^u \in H^i$ that are conditioned on the states of object $o^v \in O^i$ as illustrated in Fig. 7.10 (b). In Eq. 7.13, the product of probabilities after $\sum$ is the likelihood that the agent is performing action i by executing a maneuver type h^u in response to the observed state of object at time t. Assuming that the state transition is deterministic, we only combine the likelihood of object state $P(E_t|o_t)$ with $P(h_t|o_{t-1}, g^i_t)$ and the likelihood of the hand maneuver $P(E_t|h_t)$. The combined probability is then propagated to the next time frame. The likelihood of action i therefore gradually increases when the observed hand maneuver matches the expectation, otherwise it decreases.

For a multi-step manipulation p (fmm^p), we characterize the transition patterns among m actions by a transition matrix Ω, where the element at row i and column j corresponds to $\omega_{ij} = P(g_t = i|g_{t-1} = j, e_{t-1} = 1)$. We can then derive the most likely sequence of actions from $E_{0:T}$ subject to Ω. This is done by searching for $t^i, t^j, ...t^k$ that maximizes $P(g^i_{0:t^i}g^j_{t^i+1:t^j}\cdots g^k_{t^{k'}+1:T}|E_{0:t^i:t^j:\ldots:T}, \Omega)$, where $g^i_{0:t^i}$ means action i (i.e. g^i) occurs from $t = 0$ to $t = t^i$, and k' is the action that occurs prior to k. From Fig. 7.10 (b), this can be computed as follow:

$$P(g^i_{0:t^i} g^j_{t^i+1:t^j} ... g^k_{t^{k'}+1:T} | E_{0:t^i:t^j:...:T}, \Omega) = P(g^i_{0:t^i} | E_{0:t^i}) P(e_{t^i} = 1 | E_{t^i}) \omega_{ji}$$
$$P(g^j_{t^i+1:t^j} | E_{t^i+1:t^j}) P(e_{t^j} = 1 | E_{t^j}) ... \omega_{kk'} P(g^k_{t^{k'}+1:T} | E_{t^{k'}+1:T}) \quad (7.14)$$

We compute $P(g^j_{t^i+1:t^j} | E_{t^i+1:t^j})$ using Eq. 7.13 and $P(e_{t^j} = 1 | E_{t^j})$ is the probability of the goal state of g^j being true at time t^j. The computation complexity of Eq. 7.14 (for short, we call this $P(fmm^p)$) can be exponential in time frame T, because $t^i, t^j, .., t^k$ can take any value from 0 to T. In [57], we present a variant of Viterbi's algorithm that efficiently computes Eq. 7.14 in linear time. Putting this into the context of action rule recognition, we let $P(OC) = P(fmm^p)$ for the rule that has p specified in the OC part. If the observed actions match the pattern of fmm^p, $P(fmm^p)$ will be high, which increases $P(AC)P(OC)$ accordingly.

7.4.4 Learning

In order to compute Eq. 7.13, we need the models for hand maneuvers $P(h^u|E)$, object states $P(o^v|E)$ and the policy for choosing hand maneuvers $P(h^u|o^v, g^i)$. In [53, 56] we present a supervised learning algorithm for these models. The user provides a set of segmented training video sequences, from which f^h and f^o are computed. Most discriminative features are then selected and clustered into groups based on a co-occurrence criteria. A classifier for h^u is then learned for each group of hand movement features (and similarly for o^v) based on the Bayes' assumption that all selected features f are independent given h^u. Finally, a noisy-OR generative causal network with links from o^v (cause nodes) to h^u (evidence nodes) is learned, and $P(h^u|o^v, g^i)$ is computed from the causal power associated with the links of this network, as shown in [58].

For a multi-step manipulation p, we need to learn Ω that is associated with the goal state of each step. In [57], we present an iterative algorithm similar to the learning algorithm in HMMs for learning Ω from a training data set Q, where each $q \in Q$ contains a single demonstration of behaviour. In HMMs the probability of a transition from one state to another is learned, whereas in our case, the probability of a transition from one goal state to another also depends on the kinds of movements that are associated with the goal states. In other words, the probability of "push X to Y then retract the hand" is different from "pick up X and move it to Y then retract the hand".

7.4.5 Experiments

In this section we present experimental results to test the performance of our FMM models.

Psychologically Realistic Modelling of Intentional Actions

Similar to our modelling of the correlation between (h, o, g), it has been shown by psychological experiments that a human infant appears to maintain a representation of actions that relate goal states, actions, and situation constraints

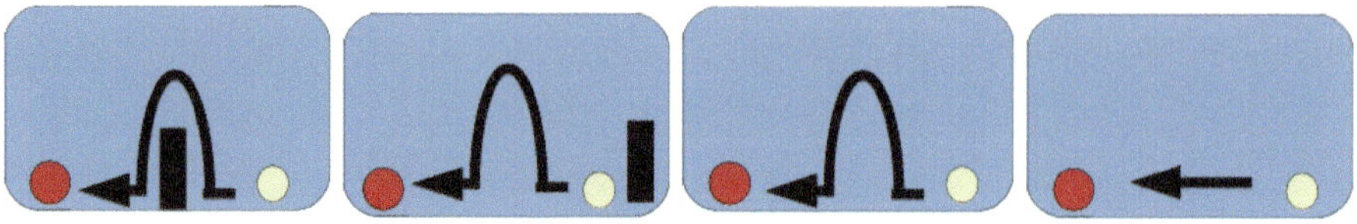

a) Habituation 1 b) Habituation 2 c) Test 1 d) Test 2

Fig. 7.11. The experimental events to test the understanding of goal-directed actions by one-year old infants

to one another via the rationality principle. In [59], infants were separated into two groups. Infants in the first group were habituated to events depicted in Fig. 7.11 (a) ("jumping over an obstacle"), while those in the second group were habituated to events depicted in Fig. 7.11 (b) ("jumping over nothing"). After habituation, they were shown events in Figures 7.11 (c) and (d), and the looking time was measured. The results have shown that on average the infants habituated to Fig. 7.11 (a) spent almost 5 seconds longer looking at Fig. 7.11 (c) than Fig. 7.11 (d). In contrast, the infants habituated to Fig. 7.11 (b) spent equal amount of time (around 1 sec difference) looking at Figures 7.11 (c) and (d). To the infants in the second group, "jumping over nothing" is quickly associated with the intention of the executor, whereas to the infants in the first group, the movement contradicts their expectation. In [53, 56], we have repeated this experiment by training two FMM-based action models to represent the two groups of infants, and use $-log(P(g^i|E_{1:T})$ as a measure of the degree of surprise on observing the test sequence $E_{1:T}$. Our results show that by taking into account the correlation between environmental context and hand movements, our action model has common properties with the action learning in human infants.

Recognition of Complex Actions

We have validated our complex action recognition approach on three types of action sequences. In a "putting" sequence, the agent puts a new object in front of itself. In a "getting" sequence, the agent moves an object that is already in the scene in front of it. In a "scrubbing" sequence, the agent scrubs (or pretends to scrub) a surface by pushing and pulling the object. We asked a user to perform each behaviour five times and use them to learn the transition matrices described in Section 7.4.4. In each trial, we changed the initial location of the object and the direction of scrubbing. In order to evaluate the learned behaviours, we collect two test videos of each behaviour. We segment actions in the videos, and compute the likelihood of each behaviour using Eq. 7.14. We compared the most likely behaviour with the ground-truth and obtained a 83.33% detection rate (i.e., five out of six behaviours are correctly detected). The failure occurs in one of the "putting" videos, which is caused by errors in object tracking.

Based on our probabilistic action rules, we have put these manipulation patterns in the context of color and shape games (Section 7.4.2). We tested our inference mechanism by asking a user to play color and shape games five times each. At the beginning of some trials, we put the left flag to the right and vice versa, and choose a different colored shape to play with. At the end of the trial, we compute the log ratio of $P(AC)P(Outcome)$ of the two rules to determine which game the agent has played. During the trials where $P(Outcome)$ is low, we compute, at each time frame, the log ratio of $P(AC)P(OC)$ of the two rules instead. We have obtained 100% accuracy at the end of all trials. During the trial, the performance has dropped to 78.57% due to tracking noise and the ambiguity in the detection of motion patterns (e.g., when the hand pauses). While being less reliable, reasoning about movement provides useful information for learning about failure.

7.4.6 Conclusion

In this section, we have proposed an integration of factorized hidden Markov models (FMMs) into probabilistic action rules as a representation of human behaviours. This integration allows us to reason about motion patterns as part of understanding about human intentions, as well as to relate functional concepts of an object to its properties. Our action learning and recognition system (ALRS) can be embedded into a reasoning apparatus of any intelligent system that requires information about the behaviour of agents. In the CoSy project, we embedded ALRS into a robotic system and used it to represent and reason about object manipulation during a game activity. Such information is then used by a planner to plan the robot's actions. We believe that our probabilistic action rules are also suitable for representing both the robot's and the agent's actions. Having a common representation would allow the robot to ground the action models to their own actions, and to bootstrap the models and acquire models that are specific to its configuration.

7.5 Functional Object Class Detection Based on Learned Affordance Cues

In recent years, computer vision has made tremendous progress in the field of object category detection. Diverse approaches based on local features, such as simple bag-of-words methods [38] have shown impressive results for the detection of a variety of different objects. More recently, adding spatial information has resulted in a boost in performance [41], and combining different cues has even further pushed the limits. One of the driving forces behind object category detection is a widely-adopted collection of publicly available data sets [60, 61], which is considered an important instrument for measuring and comparing the detection performance of different methods. The basis for comparison is given by a set of rather abstract, basic level categories [62].

These categories are grounded in cognitive psychology, and category instances typically share characteristic visual properties.

Fig. 7.12. Basic level (left) *vs.* functional (right) object categories

In the context of embodied cognitive agents, however, different criteria for the formation of categories seem more appropriate. Ideally, an embodied, cognitive agent (an autonomous robot, *e.g.*), would be capable of categorizing and detecting objects according to potential uses, and *w.r.t.* their utility in performing a certain task. This *functional* definition of object categories is related to the notion of *affordances* pioneered by [63].

Figure 7.12 exemplifies the differentiation between functional and basic level categories, and highlights the following two key properties: 1) functional categories may generalize across and beyond basic level categories (both a mug and a watering-can are *handle-graspable*, and so is a hammer), and 2) basic level categories can be recovered as *composite* functional categories (a mug is both *handle-graspable*, *sidewall-graspable*, and can be *poured* from).

7.5.1 Related Work

Attempts to detect objects according to functional categories date back to the early days of computer vision. [64] was among the first to suggest functional characterizations of objects as consequences of basic geometric properties. [65] pioneered a body of work on functional categorization of CAD-inspired face-vertex object descriptions by geometric reasoning, and was later extended by visual input for recognizing primitive shapes from range images of idealistic scenes [66]. [67] introduced an explicit mapping between geometric and corresponding functional primitives and relations, again restricted to a small class of parametric shapes. [68] added force feedback for distinguishing among different tools that afford piercing other objects. Only recently, [69] stepped into the direction of more realistic settings, recognizing previously unseen, real world objects, but specifically tailored towards grasp point prediction. The approach is based on training a logistic regression model on annotated synthetic images, combining 2D filter responses with 3D range data in a dense, multi-scale image representation.

In [70], we approach the challenge of functional object categorization from a completely different angle. First, we build our system on robust and well-established grounds in the field of object recognition, applicable to real-world images of cluttered scenes. We explore the capabilities of a widely adopted detection framework, based on a suitable geometric local feature representation.

Second, we choose to acquire functional category representations by observing few prototypical human-object interactions rather than explicitly modeling physical object properties. Naturally, the set of functional categories that our local feature-based vision system will be able to represent is restricted to those that are characterized by distinct visual features. As an example, consider the bent shape of a mug handle, which suggests to grasp the mug in a specific way. We call such distinct visual features *affordance cues*, and base our system for functional object category detection on the recognition of these cues. We report first results for the detection of two functional object categories learned by our system, and demonstrate their generalization capabilities across and beyond basic level categories. We show that our system supports the interpretation of these categories as composite functional ones.

7.5.2 Affordance Cue Acquisition

Given an observed human-object interaction featuring a single affordance cue (a video sequence plus tutor guidance), the purpose of the affordance cue acquisition sub-system is to obtain a visual feature-based representation of that cue. It proceeds by first estimating an accurate per-pixel segmentation of the *interaction region* (the region where tutor and object pixels overlap during interaction), and then extracting features in a local neighborhood around that region. Tutor guidance informs the system about the beginning and the end of an interaction. Figure 7.13 gives an overview of affordance cue acquisition, which is detailed in the following.

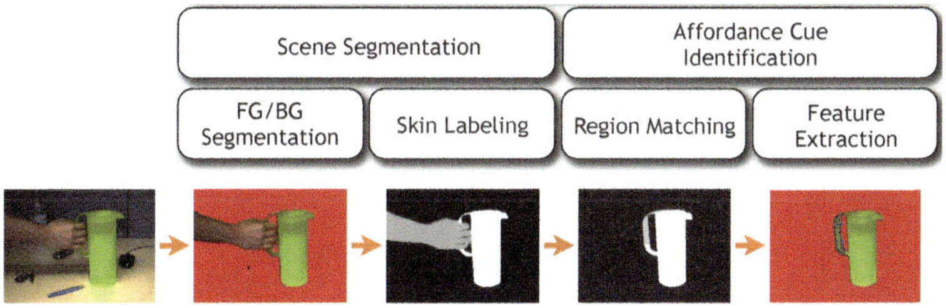

Fig. 7.13. Affordance cue acquisition overview

Foreground/Background Segmentation and Skin Labeling

We employ the *Background Cut* [71] algorithm for foreground/background segmentation, combining global and local Gaussian mixture color models with a data-dependent discontinuity penalty in a Conditional Random Field model [72]. In order to distinguish human tutor and manipulated object, we apply a likelihood ratio test on all pixels labeled as foreground by foreground/background segmentation. We build the ratio between the likelihood of a pixel originating from object color and the corresponding likelihood for skin color, using a pre-trained skin color model [73]. Figure 7.13 includes an example labeling (black denotes *background*, white *object*, and gray *skin*).

Region Matching

We determine the interaction region as the set of object pixels that has been occluded by the human tutor in the course of an interaction. We identify those pixels by choosing two frames from the interaction sequence, i) one during the interaction, and ii) one after (but with the object still visible). Then, the set of occluded object pixels is computed as the intersection of all skin-labeled pixels of frame i) with all object-labeled pixels of frame ii), transformed in order to equalize object pose differences between the two frames. The transformation is obtained by estimating the homography between frames i) and ii), using RANSAC. Initial point-to-point correspondences are established by Robust Nearest-Neighbor Matching of SIFT descriptors [74] on Harris-Laplace interest points [75].

Feature Extraction

Our representation of affordance cues is based on geometric local features called *k*-Adjacent Segments (*k*-AS) [76], and justified by the results of our evaluation of local features presented in Section 4.2.2. *k*-Adjacent Segments had initially been proposed in the context of shape-matching line drawings to real images [77]. *k*-AS detect distinct edge segments in an image, form groups of k such segments, and then encode their relative geometric layout in a low dimensional, scale-invariant shape descriptor. In our experiments, we consistently use $k = 2$, since *2*-AS features have shown a good discrimination/repeatability tradeoff [78]. We augment the groups returned by the *2*-AS detector by additional pairs of edge segments according to perceptual grouping criteria in the spirit of [79].

7.5.3 Functional Object Category Detection

A variant of the Implicit Shape Model (ISM) [80] serves as the basis for our functional object category detection system. We extend the original model in order to allow for independent training of several different affordance cues, and flexible combination for detecting composite functional categories. Figure 7.14 gives an overview of the ISM. Training an ISM for an affordance cue amounts to matching acquired affordance cue features to a previously built codebook, and storing the relative position (x, y) and size $(scale)$ of the object *w.r.t.* the feature occurrence along with matched codebook entries. For detecting an affordance cue in a previously unseen image, all features in a test image are again matched to the codebook. For every matched codebook entry, each stored feature occurrence probabilistically votes for a hypothesized object position in a generalized three-dimensional Hough voting space $(x, y, scale)$. The modes

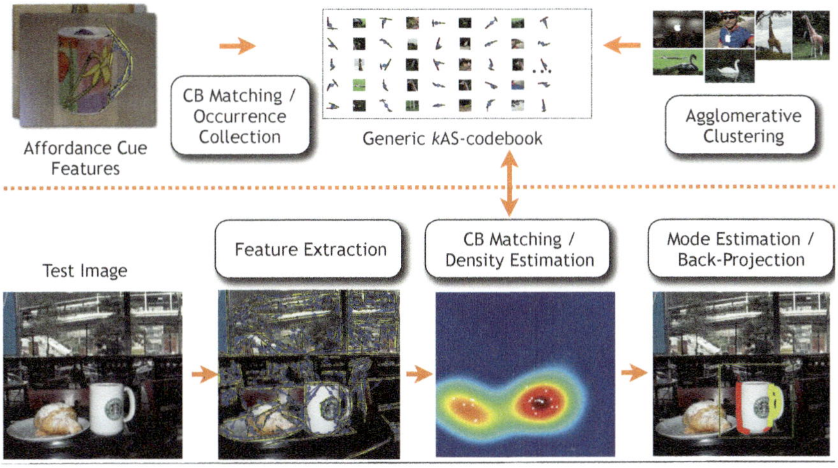

Fig. 7.14. Implicit Shape Model Overview

of the resulting vote distribution constitute hypotheses on possible affordance cue positions and scales, enabling edge-level localization by back-projecting the corresponding features into the image plane (see Figure 7.16).

7.5.4 Experiments

We report qualitative results for the detection performance of our system on a subset of the ETHZ Shape Classes data set [77], and a series of images from the PlayMate scenario.

The *handle-graspable* Category

We begin by giving results for the *handle-graspable* functional category (rows (a) to (c) of Figure 7.16), learned from affordance cue features of single images given in column (1). We observe that the models learned from either of the three mugs perform comparably well in detecting handle-like structures in the given test images, despite apparent appearance differences between the objects used for training and testing, and considerable background clutter.

Row (d) highlights the generalization of a *handle-graspable* model learned from mug (1)(c) over other object categories such as *coffee pot*, *vase*, and *electric water jug*. Image (5)(d) indicates the limitations of our approach. While the detector mistakenly fires on a circular sign in the background (false positive), it misses an obvious *handle-graspable* affordance cue on the white Thermos bottle (false negative). While the false positive can be explained by the limited information encoded by the *2*-AS features, the false negative may be attributed to predominant background structures.

The *sidewall-graspable* Category

Rows (e) and (f) of Figure 7.16 show the detection results for a second category, *sidewall-graspable*, again learned from single images. In row (e), a model has been learned from a bottle, and from a mug in row (f). The *sidewall-graspable* detector exhibits remarkable performance in the detection of sidewall-like structures in cluttered images, although it is slightly more susceptible to false positives than the *handle-graspable* detector, again due to the limitations of the employed features (see (e)(5)).

The *handle-graspable/sidewall-graspable* Category

We now combine both *handle-graspable* and *sidewall-graspable* affordance cues by training two independent ISM models, one for each cue, and joining their predictions for detection. In fact, the combination of both cues improves the detection performance of our system (example detections are given in row (g)). In particular, the *sidewall-graspable* affordance cue compensates for inaccuracies in the localization of *handle-graspable* features. By back-projecting features, the joint detector is able to to distinguish and accurately localize both of the two affordance cues, shown in yellow (*handle-graspable*) and red (*sidewall-graspable*).

Grasping

Figure 7.15 depicts an attempt to grasp a jug at its handle by a robot arm mounted onto our agent, after the system has acquired the corresponding affordance cue. Although actual grasping fails due to limited visual servoing capabilities, the agent manages to touch the jug at the correct position. We applied affordance cue acquisition independently for two cameras of a calibrated stereo rig, and obtained 3D coordinates by triangulation.

Fig. 7.15. Binocular affordance cue acquisition and resulting grasping attempt

7.6 Conclusion and Outlook

In this chapter we have explored several issues related to multi-modal learning for recognition and interaction of cognitive agent with its environment. In

Fig. 7.16. Example detections. Each row corresponds to a single experiment, unless otherwise stated. For each row (a) to (g), columns (2) to (5) give example detections for a system that has been trained solely on the highlighted affordance cue features in column (1). Line segments are plotted in yellow, and pairs selected as *2*-AS feature are connected by a blue line. Row (d) continues example detections of row (c), and row (g) depicts detections from a system trained on affordance cue features (1)(a) *and* (1)(f). Back-projected edges from the *handle-graspable* detector are plotted in yellow, those from the *sidewall-graspable* detector in red.

particular we have focused on (1) representations for different modalities, (2) levels of supervision in interactive learning, (3) exploitation of different modalities to improve learning and recognition, and (4) learning in embodied agent.

In Section 7.2 we have considered the problem of online, open-ended, learning of visual concepts. The goal was to learn associations between low-level visual features such as color, shape and spatial relations and the words describing the scene. Therefore, the developed system *associates* words, descriptions of the scene provided by the tutor, with the representations of the low-level visual features. Four general requirements for an open-ended learning system were defined and generative models called the *online* Kernel Density Estimates were explored in the context of online learning. The learning framework was embedded into the PlayMate system, within which we have studied different levels of supervision in tutor-agent learning: tutor driven, tutor supervised and explorative approach. Not surprisingly, experiments have shown that, with the tutor-driven learning, the robot achieved an almost perfect recognition score. However, this came at a cost of tutor's constant involvement – at each learning step, the tutor had to describe the visual input (e.g., image with objects) completely and without errors. In tutor-supervised learning, the robot would ask for the tutor's assistance only when it could not recognize the visual input with sufficient confidence. When unaware of its *ignorance*, the robot would often incorrectly recognize the visual data and then incorrectly learn (update its representations) from that. The incorrect updates of representations would eventually lead to the robot's reduced confidence about the recognition/classification of visual data. At that point the robot would start asking the tutor about the visual data more frequently and the tutor would provide the correct information through which the robot's representations would be corrected. Thus in a tutor-supervised framework, the tutor would act as a loose feedback control loop for the robot's learning. While the robot's recognition ability would decrease a bit in comparison to the tutor-driven learning, the tutor-supervised learning required significantly less interaction with the tutor and made the robot more autonomous in two ways: in the initiative to learn as well as in the initiative to engage verbal communication with the tutor when asking for clarification of the observed scene. The explorative approach, which did not involve any interaction with the tutor, produced the poorest results. This clearly indicates a trade-off between the quality of the results and the autonomy of the system – a learning autonomous system which knows too little and passively observes the environment, is likely to correctly learn only little or nothing. These results indicate a need for the system's ability to introspect, be more aware of its ignorance and plan ways to interact with the environment as a prerequisite for more efficient and exploratory learning.

A combination of supervised and unsupervised learning was also explored in Section 7.3 where a tutor-driven approach to cross-modal learning was treated probabilistically. Visual categories were encoded through a scale-invariant object representation called Scale-Invariant Patterns (SIPs). This representation allowed a coherent handling of labelled as well as unlabelled information. An important feature of SIPs is that they provide an embedding of visual patterns and objects into a vector space that facilitates clustering and recognition in an effective manner, which accounts for some of the general requirements for the open-ended learning system, which were defined in Section 7.2. The spatial relations ("left of", "right of", "below", "above") between objects were encoded a priori using triangular distributions which were modelled using Kernel-Density Estimates. We then considered a problem of learning an object category models from many unlabelled observations and a small number of dialogue interactions with the tutor. We allowed the information provided by the tutor to be at least ambiguous or even erroneous. Experiments have shown that the robot was able to generate meaningful generalizations across objects already from a few dialogue interactions. Since the process of learning was posed probabilistically, the uncertainty about categories was contained within distributions. During learning, the robot was thus able to propagate information, resolve ambiguities as well as recover from the errors. These results substantiate the findings about the tutor-driven learning discussed in Section 7.2.

Sections 7.2 and 7.3 indicate that for successful learning, constant tutor involvement is not required, however, the robot does require means of active interaction with the tutor or the environment to resolve ambiguities and to recover from errors. Since a natural way for the tutor to communicate with the robot is through actions (e.g., moving a hand and pointing to the objects, pushing objects), the robot does not only have to detect motion patterns, but has also be able to put them in the context of *understanding human intention*. This issue was addressed in Section 7.4, where learning of human actions and activities was studied in the context of tutor-driven learning. Intentional actions were presented by modelling the causal relations between the hand motion, object states, and the effects of actions, which were presented using factorized hidden Markov models. The particular representations, the probabilistic action rules, allowed the robot to reason about motion patterns as part of understanding about human intentions, as well as to relate functional concepts of an object to its properties. The developed action learning and recognition system was used within the PlayMate scenario to represent and reason about object manipulation during a human-computer interaction. The action recognition approach was validated on several action sequences. It has been shown, by reproducing a key result from the literature on action learning in children, that the proposed representation has properties that at least on some level of abstraction resemble action learning in humans. We believe that the developed probabilistic action rules are also suitable for representing both the robot's and the agent's actions. An attractive property of having a common representation would allow the robot to ground the action models to its own actions, and to bootstrap the models and acquire models that are specific to its embodiment.

A multi-modal learning in the context of embodied cognitive agent was addressed in Section 7.5. Ideally, an embodied, cognitive agent (e.g., an autonomous robot), would be capable of categorizing and detecting objects according to potential uses, and w.r.t. their utility in performing a certain task. In Section 7.5 we have therefore addressed learning of functional object categories from the perspective of state-of-the-art object detection. The functional category representations were acquired by observing prototypical interactions of a human with an object. In contrast to Section 7.4, where the aim was to *translate* the visual motion patterns into intentional actions, here the goal of observing an interaction was to characterize distinct visual features which determine a functional category of an object. For example, a human would grasp a mug by the handle, which would suggest that the bent shape of the handle characterizes the mug's functional category. Such distinct visual features are termed affordance cues, and the functional object category detection is thus based on the recognition of these cues. The detection system was built on robust and well established techniques of object recognition, applicable to real-world images of cluttered scenes. We have explored the capabilities of a widely adopted detection framework, based on a suitable geometric local feature representation. The functional object category detection was based on the Implicit Shape Model, which was extended to allow for independent training of several different affordance cues, and flexible combination for detecting composite functional categories. A result was an approach for tutor-driven acquisition, learning, and recognition of affordance cues in real-world, cluttered images. The approach was successfully applied to the task of functional-object detection from a cluttered scene. Clearly, the currently explored methodology is limited by the 2D nature of the used local features, however, it demonstrated another view of cross-modal learning thorough tutor interaction.

To summarize, in this chapter we have addressed several issues related to cross-modal learning, including the extent of tutor's involvement, relations between human actions and intentions, and the problem of online open-ended learning. In our quest to study these problems we have proposed different solutions, whose strength we have substantiated with extensive experiments. However, several open questions still remain. One is the scalability of the approaches—how to structure the complexity of the concepts? Is there a natural hierarchy of concepts that can lead to an increasing number of layered abstractions and sophistication? At which levels should different modalities interact and form associations? How to apply the proposed learning strategies to such representations? Can we improve the robustness of learning through combination of reconstructive and discriminative models of representation? An important research issue is also the question how can an artificial cognitive system perform introspection and detection of ignorance, which would subsequently lead to planning and execution of new actions that would help the system to gather information to fill these knowledge gaps. However, such capabilities, which represent a necessary prerequisite for more efficient exploratory learning, possibly with a minimal supervision of a human tutor, can only be researched, explored, and tested within an integrated artificial cognitive system for which the CoSy project provided an essential initial theoretical and experimental substrate.

References

1. Fidler, S., Skočaj, D., Leonardis, A.: Combining reconstructive and discriminative subspace methods for robust classification and regression by subsampling. IEEE Transactions on Pattern Analysis and Machine Intelligence 28(3), 337–350 (2006), `http://cognitivesystems.org/CoSyBook/chap7.asp#fidlerPAMI06`
2. Harnad, S.: The symbol grounding problem. Physica D: Nonlinear Phenomena 42, 335–346 (1990)
3. Ardizzone, E., Chella, A., Frixione, M., Gaglio, S.: Integrating subsymbolic and symbolic processing in artificial vision. Journal of Intelligent Systems 1(4), 273–308 (1992)
4. Chella, A., Frixione, M., Gaglio, S.: A cognitive architecture for artificial vision. Artificial Intelligence 89(1–2), 73–111 (1997)
5. Roy, D.K., Pentland, A.P.: Learning words from sights and sounds: a computational model. Cognitive Science 26(1), 113–146 (2002)
6. Roy, D.K.: Learning visually-grounded words and syntax for a scene description task. Computer Speech and Language 16(3), 353–385 (2002)
7. Steels, L., Vogt, P.: Grounding adaptive language games in robotic agents. In: Proceedings of the Fourth European Conference on Artificial Life, ECAL 1997, Complex Adaptive Systems, pp. 474–482 (1997)
8. Vogt, P.: The physical symbol grounding problem. Cognitive Systems Research 3(3), 429–457 (2002)
9. Bauckhage, C., Fink, G., Fritsch, J., Kummert, F., Lömker, F., Sagerer, G., Wachsmuth, S.: An integrated system for cooperative man-machine interaction. In: IEEE International Symposium on Computational Intelligence in Robotics and Automation, pp. 328–333 (2001)
10. Kirstein, S., Wersing, H., Körner, E.: Rapid online learning of objects in a biologically motivated recognition architecture. In: Kropatsch, W.G., Sablatnig, R., Hanbury, A. (eds.) DAGM 2005. LNCS, vol. 3663, pp. 301–308. Springer, Heidelberg (2005)
11. Steels, L., Kaplan, F.: AIBO's first words. the social learning of language and meaning. Evolution of Communication 4(1), 3–32 (2001)
12. Arsenio, A.: Developmental learning on a humanoid robot. In: IEEE International Joint Conference on Neural Networks, pp. 3167–3172 (2004)
13. Pollard, D.E.: A user's guide to measure theoretic probability. Cambridge University Press, Cambridge (2002)
14. Kristan, M., Skočaj, D., Leonardis, A.: Online kernel density estimation for interactive learning (submitted for publication), `http://cognitivesystems.org/CoSyBook/chap7.asp#KristanIMAVIS2008`
15. Wand, M.P., Jones, M.C.: Kernel Smoothing. Chapman & Hall/CRC (1995)
16. Scott, D.W., Szewczyk, W.F.: From kernels to mixtures. Technometrics 43(3), 323–335 (2001)
17. Goldberger, J., Roweis, S.: Hierarchical clustering of a mixture model. In: Neural Inf. Proc. Systems, pp. 505–512 (2005)

18. Zhang, K., Kwok, J.T.: Simplifying mixture models through function approximation. In: Neural Inf. Proc. Systems (2006)
19. Mc Lachlan, G.J., Krishan, T.: The EM algorithm and extensions. Wiley, Chichester (1997)
20. Figueiredo, M.A.F., Jain, A.K.: Unsupervised learning of finite mixture models. IEEE Trans. Patter. Anal. Mach. Intell. 24(3), 381–396 (2002)
21. Živkovič, Z., van der Heijden, F.: Recursive unsupervised learning of finite mixture models. IEEE Trans. Patter. Anal. Mach. Intell. 26(5), 651–656 (2004)
22. Corduneanu, A., Bishop, C.M.: Variational Bayesian model selection for mixture distributions. In: Artificial Intelligence and Statistics, pp. 27–34. Morgan Kaufmann, Los Altos (2001)
23. McGrory, C.A., Titterington, D.M.: Variational approximations in Bayesian model selection for finite mixture distributions. Comput. Stat. Data Analysis 51(11), 5352–5367 (2007)
24. Song, M., Wang, H.: Highly efficient incremental estimation of gaussian mixture models for online data stream clustering. In: SPIE: Intelligent Computing: Theory and Applications, pp. 174–183 (2005)
25. Arandjelović, O., Cipolla, R.: Incremental learning of temporally-coherent gaussian mixture models. In: British Machine Vision Conference, pp. 759–768 (2005)
26. Szewczyk, W.F.: Time-evolving adaptive mixtures, Tech. rep., National Security Agency (2005)
27. Declercq, A., Piater, J.H.: Online learning of gaussian mixture models - a two-level approach. In: Intl.l Conf. Comp. Vis., Imaging and Comp. Graph. Theory and Applications, pp. 605–611 (2008)
28. Han, B., Comaniciu, D., Zhu, Y., Davis, L.S.: Sequential kernel density approximation and its application to real-time visual tracking. IEEE Trans. Patter. Anal. Mach. Intell. 30(7), 1186–1197 (2008)
29. Kristan, M., Skočaj, D., Leonardis, A.: Incremental learning with Gaussian mixture models. In: Computer Vision Winter Workshop CVWW 2008, Moravske toplice, Slovenia, pp. 25–32 (2008), `http://cognitivesystems.org/CoSyBook/chap7.asp#kristanCVWW08`
30. Girolami, M., He, C.: Probability density estimation from optimally condensed data samples. IEEE Trans. Patter. Anal. Mach. Intell. 25(10), 1253–1264 (2003)
31. Jones, M.C., Marron, J.S., Sheather, S.J.: A brief survey of bandwidth selection for density estimation. J. Amer. Stat. Assoc. 91(433), 401–407 (1996)
32. Skočaj, D., Berginc, G., Ridge, B., Štimec, A., Jogan, M., Vanek, O., Leonardis, A., Hutter, M., Hewes, N.: A system for continuous learning of visual concepts. In: International Conference on Computer Vision Systems ICVS 2007, Bielefeld, Germany (2007), `http://cognitivesystems.org/CoSyBook/chap7.asp#skocajICVS07`
33. Skočaj, D., Ridge, B., Berginc, G., Leonardis, A.: A framework for continuous learning of simple visual concepts. In: Computer Vision Winter Workshop 2007, St. Lambrecht, Austria, pp. 99–105 (2007), `http://cognitivesystems.org/CoSyBook/chap7.asp#skocajCVWW07`
34. Skočaj, D., Kristan, M., Leonardis, A.: Continuous learning of simple visual concepts using incremental kernel density estimation. In: International Conference on Computer Vision Theory and Applications, Funchal, Madeira, Portugal, pp. 598–604 (2008), `http://cognitivesystems.org/CoSyBook/chap7.asp#skocajVISAPP08`

35. Lowe, D.: Object recognition from local scale invariant features. In: ICCV 1999 (1999)
36. Mikolajczyk, K., Schmid, C.: A performance evaluation of local descriptors. In: CVPR 2003 (2003)
37. Mikolajczyk, K., Leibe, B., Schiele, B.: Local features for object class recognition. In: ICCV 2005, Beijing, China (2005)
38. Csurka, G., Dance, C., Fan, L., Willarnowski, J., Bray, C.: Visual categorization with bags of keypoints. In: SLCV (2004)
39. Leibe, B., Seemann, E., Schiele, B.: Pedestrian detection in crowded scenes. In: CVPR 2005, San Diego, CA, USA (2005)
40. Sivic, J., Russell, B.C., Efros, A.A., Zisserman, A., Freeman, W.T.: Discovering objects and their locations in images. In: ICCV 2005, Beijing, China (2005)
41. Lazebnik, S., Schmid, C., Ponce, J.: Beyond bags of features: Spatial pyramid matching for recognizing natural scene categories. In: CVPR 2006, pp. 2169–2178 (2006)
42. Agarwal, A., Triggs, B.: Hyperfeatures - multilevel local coding for visual recognition. In: Leonardis, A., Bischof, H., Pinz, A. (eds.) ECCV 2006. LNCS, vol. 3951, pp. 30–43. Springer, Heidelberg (2006)
43. Fritz, M., Schiele, B.: Towards unsupervised discovery of visual categories. In: Franke, K., Müller, K.-R., Nickolay, B., Schäfer, R. (eds.) DAGM 2006. LNCS, vol. 4174, pp. 232–241. Springer, Heidelberg (2006)
44. Grauman, K., Darrell, T.: Unsupervised learning of categories from sets of partially matching image features. In: CVPR 2006, pp. 19–25. IEEE Computer Society, Washington (2006)
45. Baldridge, J., Kruijff, G.-J.M.: Multi-modal combinatory categorial grammar. In: EACL 2003, Morristown, NJ, USA (2003)
46. Baldridge, J., Kruijff, G.-J.M.: Coupling ccg and hybrid logic dependency semantics. In: ACL 2002, Morristown, NJ, USA (2001)
47. Roy, D.: Learning words and syntax for a scene description task. Computer Speech and Language 16(3)
48. Kruijff, G.-J.M., Kelleher, J.D., Berginc, G., Leonardis, A.: Structural descriptions in Human-Assisted robot visual learning. In: Proceedings of 1st Annual Conference on Human-Robot Interaction (2006)
49. Kruijff, G.-J.M., Kelleher, J.D., Hawes, N.: Information fusion for visual reference resolution in dynamic situated dialogue. In: André, E., Dybkjær, L., Minker, W., Neumann, H., Weber, M. (eds.) PIT 2006. LNCS (LNAI), vol. 4021, pp. 117–128. Springer, Heidelberg (2006)
50. Kelleher, J., Kruijff, G.-J., Costello, F.: Proximity in context: an empirically grounded computational model of proximity for processing topological spatial expression. In: Coling-ACL 2006, Sydney Australia (2006)
51. Brand, M., Oliver, N., Pentland, A.: Coupled hidden markov models for complex action recognition. In: IEEE Proceedings of Computer Vision and Pattern Recognition, Puerto Rico, USA (1997)
52. Wren, C., Pentland, A.: Dynamic modeling of human motion. In: Proceedings of the Third IEEE International Conference on Automatic Face and Gesture Recognition, Nara, Japan (1998)
53. Hongeng, S., Wyatt, J.: Learning causality and intention in human actions. In: Proceedings of IEEE-RAS International Conference on Humanoid Robots, Genoa, France (2006), `http://cognitivesystems.org/CoSyBook/chap7.asp#hong06`

54. Sutton, R.S., Barto, A.G.: Reinforcement learning : An introduction. MIT Press, Cambridge (1998)
55. Domingos, P., Richardson, M.: Markov logic: A unifying framework for statistical relational learning. In: Proceedings of the ICML 2004 Workshop on Statistical Relational Learning and its Connection to Other Fields, Banff, Canada (2004)
56. Hongeng, S., Wyatt, J.: Learning Causality and Intentional Actions. In: Rome, E., Hertzberg, J., Dorffner, G. (eds.) Towards Affordance-Based Robot Control. LNCS (LNAI), vol. 4760, pp. 27–46. Springer, Heidelberg (2008), `http://cognitivesystems.org/CoSyBook/chap7.asp#hong08a`
57. Hongeng, S., Wyatt, J.: Learning goal-based motion sequences of object manipulation, Tech. Rep. CSR-08-02, School of Computer Science, University of Birmingham (2008)
58. Glymour, C.: Learning causes : Psychological explanations of causal explanation. Minds and Machines 8, 39–60 (1998)
59. Gergely, G., Csibra, G.: Teleological reasoning in infancy: the naive theory of rational action. Trends in Cognitive Sciences 7(7), 287–292 (2003)
60. Everingham, M., Van Gool, L., Williams, C.K.I., Winn, J., Zisserman, A.: The PASCAL Visual Object Classes Challenge 2007 (VOC 2007) Results (2007)
61. Griffin, G., Holub, A., Perona, P.: Caltech-256 object category dataset, Tech. Rep. 7694, California Institute of Technology (2007)
62. Rosch, E., Mervis, C.B., Gray, W.D., Johnson, D.M., Braem, P.B.: Basic objects in natural categories. Cognitive Psychology
63. Gibson, J.J.: The theory of affordance, in: Percieving, Acting, and Knowing. Lawrence Erlbaum Associates, Hillsdale (1977)
64. Winston, P.H., Katz, B., Binford, T.O., Lowry, M.R.: Learning physical descriptions from functional definitions, examples, and precedents. In: AAAI 1983 (1983)
65. Stark, L., Bowyer, K.: Achieving generalized object recognition through reasoning about association of function to structure. PAMI 13(10), 1097–1104 (1991)
66. Stark, L., Hoover, A., Goldgof, D., Bowyer, K.: Function-based recognition from incomplete knowledge of shape. In: WQV 1993, pp. 11–22 (1993)
67. Rivlin, E., Dickinson, S.J., Rosenfeld, A.: Recognition by functional parts. Computer Vision and Image Understanding: CVIU 62(2), 164–176 (1995)
68. Bogoni, L., Bajcsy, R.: Interactive recognition and representation of functionality. CVIU 62(2), 194–214 (1995)
69. Saxena, A., Driemeyer, J., Ng, A.Y.: Robotic grasping of novel objects using vision. IJRR
70. Stark, M., Lies, P., Zillich, M., Wyatt, J., Schiele, B.: Functional object class detection based on learned affordance cues. In: 6th International Conference on Computer Vision Systems, ICVS (2008), `http://cognitivesystems.org/CoSyBook/chap7.asp#stark08icvs`
71. Sun, J., Zhang, W.W., Tang, X., Shum, H.Y.: Background cut. In: Leonardis, A., Bischof, H., Pinz, A. (eds.) ECCV 2006, Part II. LNCS, vol. 3952, pp. 628–641. Springer, Heidelberg (2006)
72. Lafferty, J., McCallum, A., Pereira, F.: Conditional random fields: Probabilistic models for segmenting and labeling sequence data. In: ICML 2001 (2001)
73. Jones, M.J., Rehg, J.M.: Statistical color models with application to skin detection. In: CVPR, pp. 1274–1280. IEEE Computer Society, Los Alamitos (1999)

74. Lowe, D.G.: Distinctive image features from scale-invariant keypoints. IJCV 60(2), 91–110 (2004)
75. Mikolajczyk, K., Tuytelaars, T., Schmid, C., Zisserman, A., Matas, J., Schaffalitzky, F., Kadir, T., Gool, L.J.V.: A comparison of affine region detectors. In: IJCV 2005 (2005)
76. Ferrari, V., Fevrier, L., Jurie, F., Schmid, C.: Groups of adjacent contour segments for object detection, Rapport De Recherche Inria
77. Ferrari, V., Tuytelaars, T., Gool, L.J.V.: Object detection by contour segment networks. In: Leonardis, A., Bischof, H., Pinz, A. (eds.) ECCV 2006. LNCS, vol. 3953, pp. 14–28. Springer, Heidelberg (2006)
78. Stark, M., Schiele, B.: How good are local features for classes of geometric objects. In: ICCV (2007), http://cognitivesystems.org/CoSyBook/chap7.asp#stark07iccv
79. Zillich, M.: Incremental Indexing for Parameter-Free Perceptual Grouping. In: 31st Workshop of the Austrian Association for Pattern Recognition (2007)
80. Leibe, B., Leonardis, A., Schiele, B.: An implicit shape model for combined object categorization and segmentation. In: Ponce, J., Hebert, M., Schmid, C., Zisserman, A. (eds.) Toward Category-Level Object Recognition. LNCS, vol. 4170, pp. 508–524. Springer, Heidelberg (2006), http://cognitivesystems.org/CoSyBook/chap7.asp#Leibe06b

8

Situated Dialogue Processing for Human-Robot Interaction

Geert-Jan M. Kruijff, Pierre Lison, Trevor Benjamin, Henrik Jacobsson, Hendrik Zender, Ivana Kruijff-Korbayová[1], and Nick Hawes[2]

[1] DFKI GmbH, Saarbrücken Germany
gj@dfki.de
[2] Intelligent Robotics Laboratory, School of Computer Science, University of Birmingham, Birmingham, UK
nah@cs.bham.ac.uk

8.1 Introduction

In CoSy, our robots were to be able to interact with human. These interactions served to help the robot learn more about its environment, or to plan and carry out actions. For a robot to make sense of such dialogues, it needs to understand how a dialogue can relate to, and refer to, "the world" – local visuo-spatial scenes, as in the Playmate scenario (9), or the spatial organization of an indoor environment in the Explorer scenario (10).

This is not a trivial problem. Language presents a powerful system to express meaning. Also, perception provides a cognitive system with rich experiences of the world. The fundamental problem for (situated) dialogue processing is how to relate linguistically expressed meanings to these experiences. If we look at this problem from the viewpoint of dialogue as a means of communication, then we can pose first of all the following two requirements. Any solution needs to be *efficient*, to allow a cognitive system to respond in a timely fashion, and *effective*, so that the system arrives at those meanings which are indeed likely to be correct in the given "context."

But, as a communication channel, or rather an information channel, spoken dialogue poses further requirements than just efficiency and effectiveness. Spoken input is typically noisy, with utterances in a spoken dialogue often being incomplete, or grammatically incorrect. We may only be able to construct partial representations of a meaning for an utterance. Furthermore, if we consider meaning to be about something, i.e. about a "referent" in a dialogue- or situated context, then this partiality goes beyond the boundary of a single utterance as meaning in a dialogue is typically contributed to such a referent over several utterances. A solution to modeling meaning in situated dialogue thus needs to allow for a *gradual resolution and construction* of

H.I. Christensen et al. (Eds.): Cognitive Systems, COSMOS 8, pp. 311–364.
springerlink.com

meaning representations, (with the possibility to "postpone" the resolution of ambiguities), and provide for meaning to *persist* over utterances, i.e. over the course of a dialogue. Finally, these requirements are based on a view of meaning being referential to a context, i.e. a dialogue context and a situated context. Meaning representations will need to indicate, and represent, this *referentiality*: how they purport to refer, and to what.

There exist several approaches to deal with each one, or several, of these issues. Roy & Reiter present a comprehensive overview of existing approaches to relating language and the world (up to 2005) in [1], and identify several important issues very similar to the requirements we raised above. One of the earliest systems which connected built utterance analyses to a visual world was Winograd's SHRDLU [2]. SHRDLU built linguistic analyses in an incremental, "left-to-right" fashion, and connected those to visuo-spatial representations of local scenes to help prune analyses. Among more recent approaches, the most developed are those by Gorniak & Roy, and Steels *et al.* Gorniak & Roy [3, 4] present an approach in which utterance meaning is probabilistically mapped to visual and spatial aspects of objects in the current scene. Recently, they have extended their approach to include action-affordances [5]. Their focus has primarily been on the aspect of how to relate language and information about the world, just like SHRDLU and the work by Steels *et al* [6, 7, 8]. The latter have developed an approach where the connection between word meaning and percepts is modeled as a semiotic network, in which abstract categories mediate between language and the visual world. Although many of these approaches use some form of incremental processing to gradually construct utterance meaning, connecting meanings to the social and physical context as they are construed, the (im)possibility to connect alternative meanings does not feed back into the incremental process to prune inviable analyses. This is where Scheutz et al [9, 10] take matters a step further. They present an approach for incremental utterance processing in which the analyses are referentially interpreted, and pruned if it is impossible to connect them in the situated contexts referred to.

The approach we describe in this chapter primarily advances on the above by addressing the requirements we raised earlier in a systematically integrated fashion. We postulate a *bi-directionality hypothesis*, a close coupling between how a cognitive system processes situated dialogue, and how it processes and represents experiences associated with a situated context to which dialogue refers. The purpose of this coupling is for these different modalities to interchange and interconnect information, to help guide processing. (Such a close coupling is inspired by how humans appear to process visually situated dialogue, see e.g. [11, 12, 13, 14].) More specifically, based on this hypothesis we address the requirements as follows.

Gradual construction. Utterance meanings are constructed incrementally, using Combinatory Categorial Grammar [15, 16]. Meaning is represented as an ontologically richly sorted, relational structure (using a decidable fragment of modal logic, [17, 18, 19, 20]). The ontological sorting is used as a mediating basis for binding linguistic content to content in other modalities, including visuo-spatial scenes, spatial maps, and actions ([21, 22, 23] and Chapters 2 and 7). In comparison to approaches as mentioned above we thus extend category-based mediation (the "what" dimension) with a spatio-temporal dimension of interpretation (the "where/when" and "how") covering several levels of spatial organization. In our approach we are not restricted only to visuo-spatial information about the currently perceivable scene.

Referentiality. The propositional meaning of an utterance is complemented with information on how that meaning is presented as referring to a larger context. Such referentiality is established directly, based on forms of anaphoric or deictic references, or more indirectly using a basic form of information structure (see Section 8.4). Structured discourse representation models [24] are used to model dialogue context, and represent referentiality. These models are complemented with a temporal representation of how referred-to events can be related ([25, 23], Chapter 6). Dedicated algorithms are used to resolve linguistic references to situated contexts such as visuo-spatial scenes (collection of visuo-spatial information on Binding working memory, [26]) and large-scale spatial organization [27] (subset of relevant locations in a map, see Section 8.5).

Persistence. The dialogue context model provides a persistent basis for maintaining how linguistic content has been resolved to content in a situated context. Structures relating different linguistic references to a specific referent can be combined to determine how information about that referent has been communicated over the course of a dialogue (see e.g. [28], and Chapter 7 for use in visual learning).

Efficiency & Effectiveness. Based on information about the current situated context (whether local, or at a larger scale, cf. also Section 8.5), and the current dialogue context, incremental comprehension of spoken utterances can prune unlikely word- and meaning-hypotheses. (In comparison, Scheutz et al only guide parsing.) The result is a combined performance of speech recognition and parsing of close to 90% correct interpretations of free speech on our domains (against a base-line performance of 68%).

Bi-directionality has a fundamental impact on how we conceive of linguistic meaning and its processing. Linguistic meaning arises as a reflection of how a cognitive system experiences, and structures its knowledge of, the world [29, 30]. Processing meaning is based on a continuous interchange of information between functions involved in computing aspects of meaning, be those functions related to experience or dialogue. The resulting view on modularity implies dialogue processing to be more like a "permeable glass box" than the traditional "black box." Interconnectivity is based on a model akin to pass-by-reference, to allow different functions to keep track of how content changes over time. Adopting this hypothesis, we show in this chapter how integration with information about situated contexts helps us to achieve better robust processing of spoken, situated dialogue processing on the Playmate and Explorer domains than if no situated information would have been used.

An overview of the chapter is as follows. In Section 8.2 we briefly discuss observations that have been made on how humans appear to process situated dialogue. These observations have in part inspired the functionality we consider in our approach. Then, Section 8.3 we first discuss the basics for contextualized language processing, showing how bi-directionality influences first of all the design of the processes and representations we adopt. We continue in Section 8.4 with looking into dialogues about what the robot can see, connecting dialogue meaning with an understanding of local visuo-spatial scenes. We discuss how we can talk about those scenes, for example when a human tries to teach the robot more about the objects it sees, and how bi-directionality helps focusing speech recognition, utterance analysis, reference resolution, and producing references to objects. The Explorer scenario provides us with a setting in which we go beyond the current scene, often discussing places we can visit, or where we can find objects in the world. Important is that we do not need to be in those places to discuss them. In Section 8.5 we present how we can go beyond the current situated context, and use information about the larger world around the robot to talk about other places. One interesting challenge bi-directionality helps addressing is in resolving and producing references to such places. Finally, in Section 8.6 we take meaning beyond having mostly a referential, *indexical* nature. We look into how particularly speech acts like questions and commands express intentionality, and how we can relate that to processes for motivation and planning. Bi-directionality enters the dialogue processing picture again by indicating which potential utterance interpretations correspond to possible plans, and which ones do not.

8.2 Background

Language provides us with virtually unlimited ways in which we can communicate meaning. This, of course, raises the question of how precisely we can then understand an utterance as we hear it. Empirical studies in various branches of psycholinguistics and cognitive neuroscience have investigated what information listeners use when trying to understand spoken utterances which are about visual scenes. An important observation across these studies is that interpretation *in context* plays a crucial role in the comprehension of utterance as it unfolds [11]. Following [13] we can identify two important dimensions of the interaction between the purely linguistic, dialogue context, and the situated context. One is the *temporal dimension*. The ways our visual attention are guided appear to be timed closely with how we proceed with understanding an utterance. In empirical studies we can witness this by for example eye movements. The second is the *information dimension*. This indicates that listeners not only use linguistic information during utterance comprehension, but also scene understanding and "world knowledge." Below we discuss aspects of these dimensions in more detail.

8.2.1 Multi-level Integration in Language Processing

Until the early 1990s, an often-adopted model of language comprehension was that of a modular, stage-like process. On this model, a language user would sequentially construct each level of linguistic comprehension – from auditory recognition all the way to pragmatic, discourse-level interpretation. As [31] observe, two hypotheses followed from this view. One hypothesis is that people first construct a local, context-*independent* representation of the communicated meaning. Only once this meaning has been completely constructed, it is interpreted against the preceding dialogue context (if any). Secondly, and related, is the hypothesis that dialogue context-related processing only enters the process of language comprehension at a relatively late stage.

Opposing these hypotheses is the view that language comprehension is an incremental process. In such a process, each level of linguistic analysis is performed in parallel. Every new word is related to representations of the preceding input, across several levels – with the possibility for using the interpretation of a word at one level to co-constrain its interpretation at other levels. (This may result in ambiguities to be resolved only at a later point.) A natural prediction that follows from this view is that interpretation against dialogue context can in principle affect utterance comprehension *as the utterance is incrementally analyzed*, assisting in restricting the potential for grammatical forms of ambiguity. [32, 11] phrased this as a *principle of parsimony*: those grammatical analyses are selected that for their reference resolution impose the least presuppositional requirements on a dialogue context.

Since then, various studies have investigated further possible effects of dialogue context during utterance comprehension. Methodologically, psycholinguistic studies have primarily investigated the effects of dialogue context by measuring *saccadic eye movements* in a visual scene, based on the hypothesis that eye movements can be used as indications of underlying cognitive processes [33, 34]. Alternatively, cognitive neuroscience-based studies use event-related brain potentials (ERPs) to measure the nature and time course of the effects of dialogue context on human sentence comprehension [35].

Both lines of study have found that lexical, semantic and discourse-level integrative effects occur in a closely time-locked fashion, starting already at the phoneme or sub-word level; see [36], and [37, 31, 38]. Particularly, a range of dialogue-level integrative effects have observed. Referential binding has been shown to play a role in the constraining various types of local syntactic ambiguities, like garden path-constructions [32, 11, 39], and relative clauses [40, 41]; [42, 37, 31]. These effects primarily concern a *disambiguation* of already built structures. Integrating semantic and dialogue-level information during utterance comprehension also has important *anticipatory* effects. [43, 44]; [45] observe how contextual information influences what lexical meanings can be anticipated, priming phonological understanding and lexical access. Contextual information can even override disprefered lexical meaning [46].

Anticipatory effects indicate that utterance comprehension is thus not only an incremental process of constructing and then disambiguating. Anticipation enables context-dependent phonological recognition, lexical retrieval, and syntactic construction - without there being a need to generate and test all combinatory possible constructions. Incrementality and anticipation based on multi-level integration appears to give rise to a process in which comprehension arises through a convergence based on constraining and co-activation. Dialogue context and the interpretative contexts which are delineated during utterance comprehension converge to become functionally identical [31]. As a result, ambiguity need not even arise, or is at least being much more limited a priori through context.

An important issue in all of the above remains of course the degree to which integrative effects indeed should commit to a certain understanding. Garden path sentences are a good example. They show that overcommitment risks the need for re-interpretation – an issue for *cognitive control* [47, 48, 49].

8.2.2 Language Processing and Situational Experience

We already noted before that humans integrate *linguistic* and *non-linguistic* information when processing an utterance. Below we discuss studies which investigate how categorical and contextual information from situation awareness can effect utterance comprehension. These studies use eye-trackers to monitor where people look at in a scene, and when.

[50] present a study revealing that listeners focus their attention on objects before these objects are referred to in the utterance. For example, consider a scene with a cat, a mouse, and a piece of cheese. When someone hears "The cat chases the mouse", her gaze already moves to the mouse in the scene before she has actually heard that word; similarly for "The mouse eats the cheese." Knowing that cats typically chase mice (not cheese), and that the argument structure of *chase* reflects this, the listener *expects* that the next object to be mentioned will be the mouse, and directs gaze to that object. We thus see an anticipatory effect arising from the online integration of lexico-semantic information (verbal argument structure), situational context (the present objects, and the reported action), and categorical knowledge (prototypical object-action relations).

Not only world knowledge can influence online utterance comprehension, also scene understanding can. For example, consider the situation in Figure 8.1. [33] show that, once the listener has heard "Put the apple on the towel ..." she faces the ambiguity of whether to put the (lone) apple onto the (empty) towel, or to take the apple that is on the towel and put it somewhere else. The ambiguity is revealed as visual search in the scene. Only once she has heard the continuation "... into the box" this ambiguity can be resolved. Interestingly, in [33] the listener cannot directly manipulate the objects. If this is possible (cf. Figure 8.1), [51] show that also reachability plays a role in comprehending the utterance. Because only one apple is reachable, this is taken as the preferred referent, and as such receives the attention. This underlines the effect *physical embodiment* may have on language comprehension.

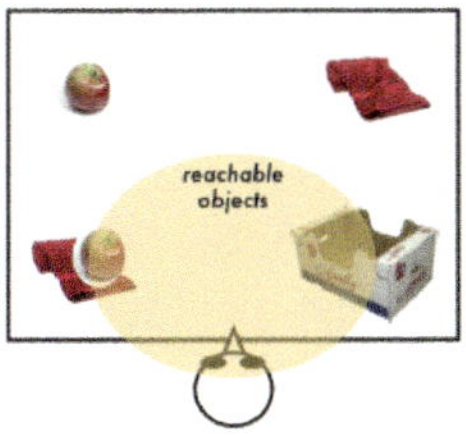

Fig. 8.1. Put, apple, towel, box

Scene understanding also concerns the *temporal projection* towards possible future events [52]. [12, 53] show how such projection can also affect utterance comprehension. These studies used a scene with a table, and beside it a glass and a bottle of wine. Investigated was where listeners look when they hear "The woman will put the glass on the table. Then, she will pick up the wine, and pour it carefully into the glass." It turns out that after hearing the "pouring" phrase, listeners look at the table, not the glass. Listeners thus explicitly project the result of the picking action into the scene, imagining the scene in which the glass is on the table.

These studies reveal that the interaction between vision and language is not *direct*, but *mediated* [12]. Categorical understanding plays an important role in the sensorimotoric grounding of language. This is further underlined by studies like [54, 55], following up on the idea of category systems as mediating between perceptual modalities and language [56, 57]. These studies show how categorical understanding gives rise to expectations based on affordances, influencing comprehension of spatial or temporal aspects of action verbs.

In conversational dialogue [14, 58] gaze has been shown to be automatically aligned in simple collaborative interaction. The time intervals between eye-fixations during production and comprehension of a referring expression are shorter than in monologue. This is further evidence for the relevance of visual common ground of interlocutors and how that accelerates the activation of jointly relevant concepts.

8.3 Talking

What does it take to make a robot talk? Specifically, what does it take to make a robot process situated dialogue?

In the context of CoSy, for a robot to talk it first of all needs to be able to process a speech signal, turning it into (possible) sequences of words. Then, turn those words into utterances, and assign a meaning representation to them. Depending on the referential status of content in these meaning representations, links are made to a model of the preceding dialogue, so that it can be established how meaning refers to things that were already said before (if at all), and how it moves the dialogue along.

That's one part – listening, comprehending utterances against a model of the dialogue context, and updating that context model as the dialogue continues. The other part is the talking part. Based on how the dialogue has developed so far, the robot should decide how to continue. Then, following up on this decision, it should see how to formulate the utterances to achieve that "goal", and formulate them such that they refer to the situations in a contextually appropriate way. It should be clear to the listener what the robot is referring to, talking about. And once the robot has said what it decided to say, it should of course again update the model of the dialogue context.

In this section we would like to focus on the comprehension side, and sketch how the production side is structured. We would like to start simple here, explaining the basics behind the approach we take to making robots talk. Explain the design decisions as they result from the bi-directionality hypothesis, why we do the things the way we propose to do them. Come the next sections, we will delve into more detail, or where necessary provide references to more technical discussions.

Adopting the bi-directionality hypothesis poses requirements both on how we design our processes, and our representations.

We consider the processes involved in comprehending situated dialogue to be "permeable glass boxes." While processing, it should be possible to take partial results, connect them with information from other processes, and then use the results to guide how to continue processing. One way to design such processes so is to make them *incremental*. In incremental processing, a process proceeds from the "beginning" towards the "end" of a representation it is to process, in a step-wise fashion. After each step, bi-directionality can be used to guide how to take the next step. Because linguistic representations are typically sequential, we can process them incrementally. Much of the benefits of bi-directionality consist therein that they can help processes focus on sensible analyses, discarding those which are not supported by the context.

Each process typically maintains several concurrent hypotheses. Particularly if we look at things from an efficiency point of view, there are several requirements bi-directionality raises for the nature of representations. First of all, we are looking at dialogue, a context in which interpretations develop over time. Utterances refer to the preceding context, adding to or correcting previous information. Connecting information across processes can be more efficient if these relations are clear. It identifies a history of what previously was already interconnected. The way we will address this requirement is by using discourse referents as permanent hooks to relate information to, both at utterance- and at dialogue-level.

Secondly, even though a process may maintain multiple hypotheses, this does not imply that they need not share certain similarities in how they interpret something. Representations should identify how alternative analyses are different, and where there are similarities, so that we can avoid having to check each hypothesis individually. We will address this requirement by packing multiple hypotheses into a single, possibly underspecified graph structure, and determining preference orders over alternatives.

Below we will explain these processes and representations in more detail. We start with representations, to make it clear what we are working with, and towards.

Representing an Utterance

Loosely speaking, a dialogue is an exchange of utterances between two or more "interlocutors." Usually, this exchange serves a particular purpose. In the context of human-robot interaction, that purpose usually relates to performing tasks in the real-world.

The thing is, whereas sentences in a text are usually complete, and grammatically well-formed, this need not be the case with utterances in spoken dialogue. Utterances are often incomplete or grammatically incorrect, and may include self-corrections. "Take the red uh ... no put that green one next to the ... you know, yes, the pyramid." This of course raises the question, what we should consider an utterance to be.

Most dialogue systems (still) consider an utterance to be like a sentence, and have a definable beginning and end. We adopt a more flexible notion than that. What we ultimately consider to be an utterance, depends on the context in which linguistically conveyed content is being used. As far as processing within our system is concerned, an utterance is a stream. There are marked points at which it can be further interpreted, either within the dialogue system or beyond it. At such "points," the representation of the utterance provides enough meaning to start off further forms of processing. Each further interpretation modality is thus free in considering when it works with meaning, and thus –ultimately– what it considers an "utterance" to be.

Which brings us to how we represent meaning. We represent meaning as an ontologically richly sorted, relational structure – a logical form [18, 20] in a decidable fragment of modal logic [17, 19]. The following is an example of a logical form:

$$
\begin{aligned}
&@w_1 : \text{cognition}(\mathbf{want} \wedge \langle \textsc{Mood} \rangle\, ind \wedge \langle \textsc{Tense} \rangle\, pres \wedge \\
&\quad \langle \textsc{Actor} \rangle\, (i_1 : \text{person} \wedge \mathbf{I} \wedge \langle \textsc{Num} \rangle\, sg) \wedge \\
&\quad \langle \textsc{Event} \rangle\, (p_1 : \text{action} - \text{motion} \wedge \mathbf{put} \wedge \\
&\qquad \langle \textsc{Actor} \rangle\, y_1 : \text{person} \wedge \\
&\qquad \langle \textsc{Patient} \rangle\, (m_1 : \text{thing} \wedge \mathbf{mug} \wedge \\
&\qquad\quad \langle \textsc{Delimitation} \rangle\, unique \wedge \langle \textsc{Num} \rangle\, sg \wedge \langle \textsc{Quantification} \rangle\, specific \wedge \\
&\qquad\quad \langle \textsc{Modifier} \rangle\, (r_1 : \text{q} - \text{color} \wedge \mathbf{red})) \wedge \\
&\qquad \langle \textsc{Result} \rangle\, (t_1 : \text{m} - \text{whereto} \wedge \mathbf{to} \wedge \\
&\qquad\quad \langle \textsc{Anchor} \rangle\, (r_2 : \text{e} - \text{region} \wedge \mathbf{right} \wedge \\
&\qquad\quad \langle \textsc{Delimitation} \rangle\, unique \wedge \\
&\qquad\quad \langle \textsc{Num} \rangle\, sg \wedge \\
&\qquad\quad \langle \textsc{Quantification} \rangle\, specific \wedge \\
&\qquad\quad \langle \textsc{Owner} \rangle\, (b_1 : \text{thing} \wedge \mathbf{ball} \wedge \\
&\qquad\quad \langle \textsc{Delimitation} \rangle\, unique \wedge \langle \textsc{Num} \rangle\, sg \wedge \langle \textsc{Quantification} \rangle\, specific)))) \wedge \\
&\quad \langle \textsc{Patient} \rangle\, (y_1 : \text{person} \wedge \mathbf{you} \wedge \langle \textsc{Num} \rangle\, sg) \wedge \\
&\quad \langle \textsc{Subject} \rangle\, i_1 : \text{person})
\end{aligned}
$$

Each node has a unique identifier with an associated ontological sort (e.g. $p1$: *action* – *motion* means that the handle $p1$ is of sort *action* – *motion*), and a proposition (e.g. **put** for $p1$). Nodes are connected through named relations. These indicate how the content of a single node contributes to the meaning of the whole expression. For example, "you" ($y1$) both indicates the one whom something is wanted of ($Patient$-relation from $w1$), and the one who is to perform the put action (*Actor*-relation from $p1$). Nodes carry additional features, e.g. $i1$ identifies a singular person.

Propositions and relations in such a representation are instances of concepts. This makes it possible for us to interpret logical forms further using ontological reasoning. We use this possibility in reference resolution, and in relating meaning representations to interpretations formed outside the dialogue system.

The relational nature of our representations provides us with several consequences. We build up our representations from elementary propositions as we illustrated above – sorted identifiers and propositions, features, and relations. An interpretation is thus simply a conjunction of such elementary propositions, and the more we can connect those elementary propositions, the more complete our interpretation becomes. This makes it relatively straightforward to represent partial interpretations. For example, for "take the red ..." receives the following interpretation:

$$
\begin{aligned}
&@t_1 : \text{action} - \text{motion}(\mathbf{take} \wedge \langle \textsc{Mood} \rangle\, imp \wedge \langle \textsc{Tense} \rangle\, pres \wedge \\
&\quad \langle \textsc{Actor} \rangle (a_1 : \text{entity} \wedge \mathbf{addressee}) \wedge \\
&\quad \langle \textsc{Patient} \rangle\, (m_1 : \text{thing} \wedge \\
&\qquad \langle \textsc{Delimitation} \rangle\, unique \wedge \langle \textsc{Num} \rangle\, sg \wedge \langle \textsc{Quantification} \rangle\, specific \wedge \\
&\qquad \langle \textsc{Modifier} \rangle\, (r_1 : \text{q} - \text{color} \wedge \mathbf{red})) \\
&\quad \langle \textsc{Subject} \rangle\, a_1 : \text{entity})
\end{aligned}
$$

The interpretation shows more than just the content for the three words. It also shows that "red" is expected to be the color of the "thing" which is supposed to be taken.

Characteristic for language is that it presents many ways in which we can say things – and interpret them. This inevitably means that we will usually get not just one, but multiple alternative interpretations for an utterance. To keep ambiguity to a minimum, we should look at to what extent these interpretations are indeed different. Where they show overlaps, we should ideally have to deal with those identical parts only once.

Using relational structure and elementary propositions enables us to do so. We represent alternative interpretations as alternative ways in which we can connect content, whereas identical content across interpretations is represented once. The procedure to create such "condensed" representations is called *packing*, after [59, 60]. Figure 8.2 illustrates the development of the packed packed representation for "here is the ball". At the first step ("take"), 9 logical forms are packed together, with two alternative roots, and several possible ontological sorts for the word "here". The second step reduces the number of alternative interpretations to one single logical form, rooted on the verb "be" with a "presentational" ontological sort. The possible meanings for the determiner is expressed at the dependent node of the "Presented" relation. At this point we have an *overspecified* meaning. Although the delimination is unique, we cannot tell at this point whether we are dealing with a singular object, or a non-singular (i.e. plural) object – all we know it has to be one or the other. This becomes determined in the fourth step ("here is the ball").

In the appendix to this chapter we present a detailed technical discussion of packing.

Representing the Interpretation of an Utterance in Context

The meaning of an utterance goes well beyond what is expressed just by the individual words that make it up. Meaning is about how the utterance relates to the context – the situation (indexically) and to the actions (to be) performed therein (intentionally). How it can be taken to refer to things we already talked about, to beliefs we have, to expectations which may be raised on the basis of what we are saying. How the utterance helps us to further the dialogue, helping to reach a goal – or not.

Referent resolution is the first step we take to relate content from the current utterance, to that of previous utterances in the dialogue context. The purpose here is to establish *co-reference relations*: relations between mentions referring to the same object(s) or event(s). Examples of references to previous objects are pronouns (e.g. "it"), or anaphoric expressions (e.g. "the red mug"). We are using a (simple) algorithm based on referent resolution in the segmented dialogue representation theory of [24].

For each index in a logical form, the algorithm determines potential antecedents in the preceding dialogue, using the model of the dialogue context the system maintains. There are two simple cases. One, we may be talking

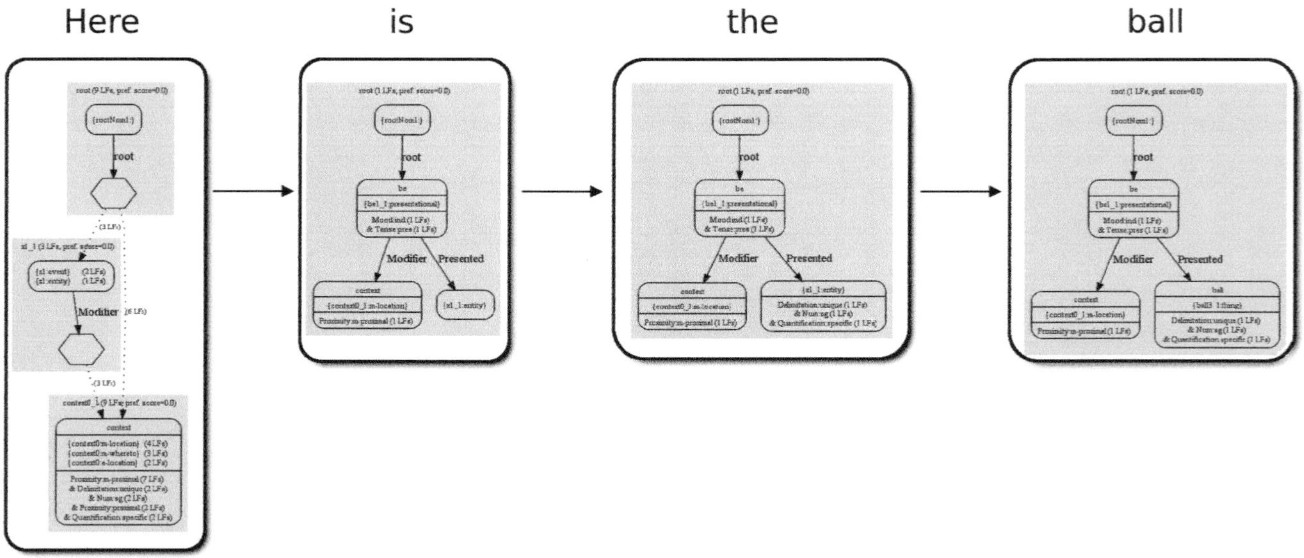

Fig. 8.2. Example of incremental parsing and packing of logical forms, for the utterance "Here is the ball". Four steps are shown.

about a something new. We then create a new (unique) referent identifier, say ant_n, and represent this as a *reference structure* [**NEW** : $\{ant_n\}$]. Two, there is a unique antecedent referent ant_i. We represent this as [**OLD** : $\{ant_i\}$], meaning there is a "discourse old" antecedent ant_i. In both cases we relate the index in the logical form (which only has naming uniqueness within the scope of the logical form) to the built structure.

Complications arise if a reference cannot be ambiguously resolved. A good example of such a situation arises when resolving deictic pronouns like "this". How a deictic pronoun needs to be resolved, depends on the dialogue- and the situated context. If the utterance is not accompanied by a gesture, the preference is to resolve the reference to a preceding antecedent in the dialogue. However, if the utterance is accompanied by a gesture, then this preference may be overridden. It may be that the gesture refers to an object which was mentioned before, just not most recently; or it may refer to an object which has not been talked about at all. To capture these possibilities, we allow for reference structures to specify preference orders over sets of old and new referents. For example, if a deictic pronoun can be resolved to several old antecedents, with ant_i the most preferred, or to a new referent ant_n, then we get

$$[\mathbf{OLD} : ant_i < \{ant_j, ..., ant_k\} < \mathbf{NEW} : \{ant_n\}].$$

Subsequently, information about grounding the utterance in the situated context then can help resolving this ambiguity (e.g. by providing support for a new referent). The example of deictic pronoun nicely illustrates the principle *bi-directional* nature of situated dialogue processing as implemented here. There is no strict pipeline of interpretation processes, invoked at incremental steps. Instead, interpretation processes interact to mutually constrain and complement the interpretations they form.

Another aspect of dialogue-level interpretation regards "speech acts", or dialogue moves. A dialogue move specifies how an utterance "functions in", i.e. contributes to furthering the dialogue. We determine an utterance's possible dialogue move(s) on the basis of the shape of the logical form, and expectations about possible moves to extend the current dialogue. Figure 8.3 illustrates a decision tree used to map logical form features to dialogue moves.

Once the dialogue move for an utterance has been determined, the utterance content, and its referent- and event structures are added to the dialogue context model maintained by the system. Figure 8.4 shows a snapshot of such a model. (We will elaborate on event structures in Section 8.6.)

Comprehending an Utterance in Context

When we try to comprehend an utterance, we analyze it at several linguistic levels. As we are dealing with spoken dialogue, the first step is the *automatic speech recognition* [ASR], which takes an audio signal stream as input and produces a word recognition lattice as output. This step is known to be particularly error-prone [61], for several reasons. The first one is the inherent

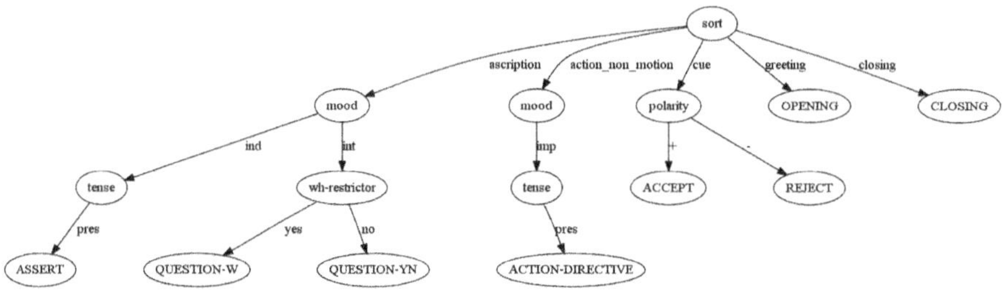

Fig. 8.3. Example of a decision tree for determining dialogue moves from LF form

noise present in the real-world environments in which our robots are deployed. Since we require the speech recognition system to be *speaker-independent*, we also have to deal with the wide variety of voices, accents and styles of speech of human speakers. And finally, natural spoken dialogue is also characterised by a high proportion of *disfluencies* (filled pauses, speech repairs, corrections, repetitions), and the production of many *partial* or *ill-formed* utterances, all of which negatively affect the performance of the speech recognition.

Our strategy for addressing this issue is to exploit *contextual knowledge* about the situated environment and the dialogue history to prime the utterance recognition. This knowledge is represented in the cognitive architecture as a cross-modal *salience model* of the situated context. It integrates both visual salience (objects perceived in the physical scene) and linguistic salience (previously referred-to objects within the current dialogue). The model is dynamically updated as the environment evolves, and is used to establish expectations about uttered words which are most likely to be heard given the context. The update is realised by continuously adapting the word probabilities specified in the statistical language model of the speech recognizer. We have shown that this approach yields a statistically significant improvement of the ASR performance compared to a baseline, non context-sensitive model [62].

As soon as the speech recognizer is able to suggest a (partial) recognition hypothesis for the utterance, a *word recognition lattice* is created and inserted into the working memory for subsequent analysis. A word recognition lattice is a packed representation for the set of potential recognition hypothesis, combined with their respective confidence scores. The set of recognition hypotheses can be easily retrieved by traversing the lattice. Figure 8.5 illustrates a typical example of word lattice.

This word recognition lattice is then further processed incrementally – the lowest, incremental level being that of grammatical analysis. For modeling natural language grammar we use the Combinatory Categorial Grammar (CCG) framework [15, 16]. CCG is a lexicalized framework: For each word, there are one or more lexical entries specifying a syntactic category, and a corresponding lexical meaning. A syntactic category defines how the word can be used

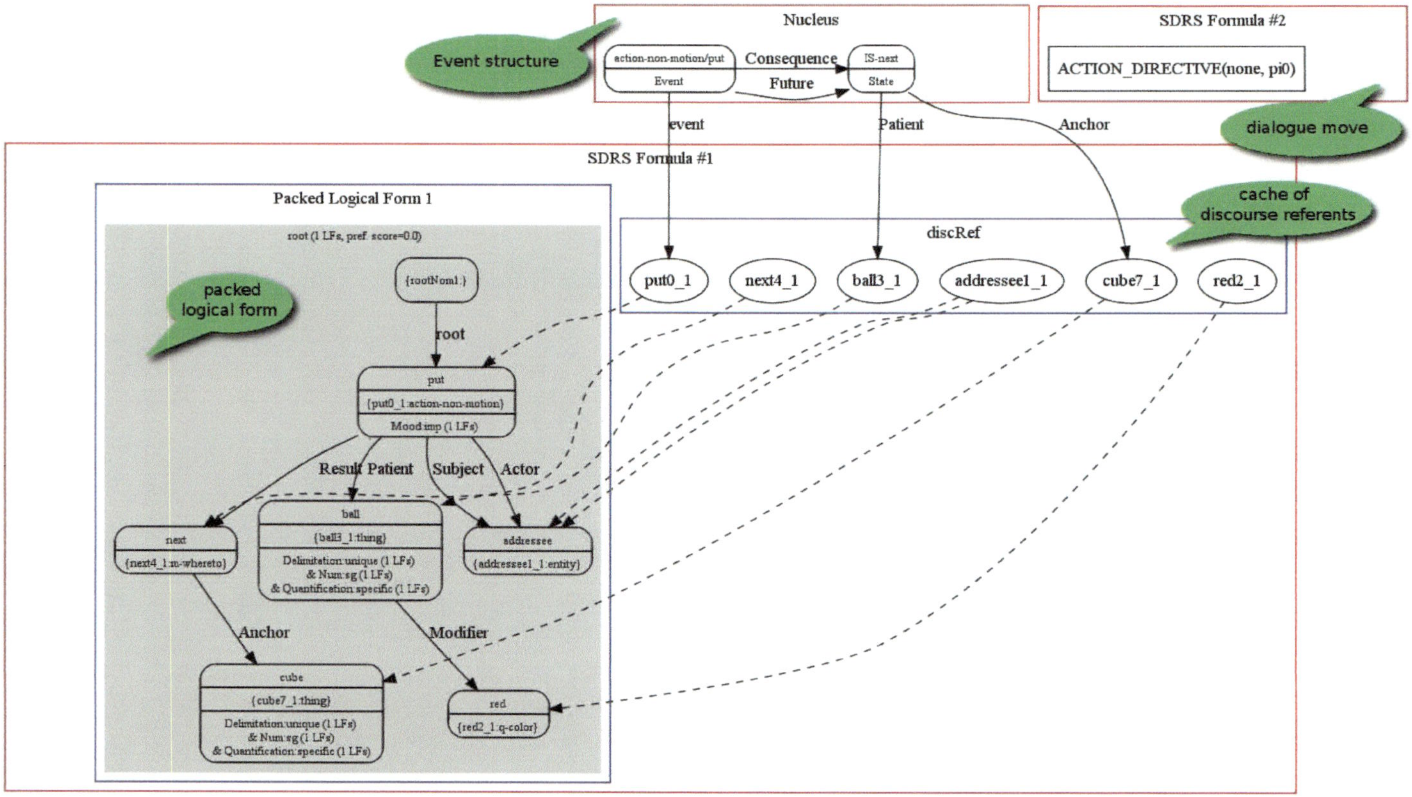

Fig. 8.4. System visualization of dialogue context model: Utterance content, coreference, event structures, and dialogue moves

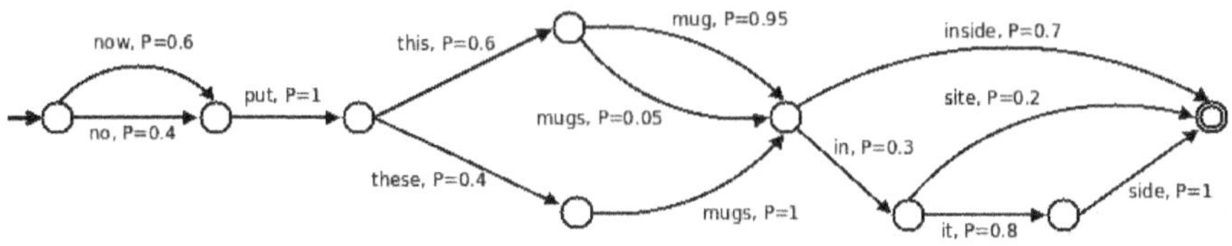

Fig. 8.5. A typical word recognition lattice

in forming a larger, grammatical expression. The lexical meaning specifies the word meaning. The meaning of an expression is built up compositionally, in parallel to its syntactical derivation.

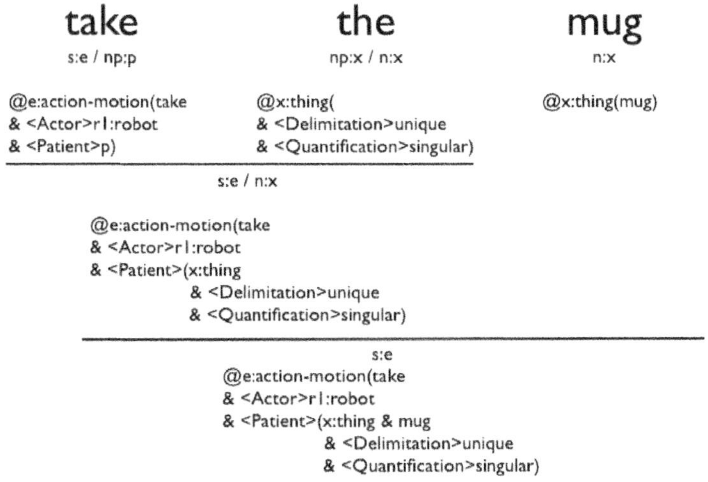

Fig. 8.6. Incremental analysis of "take the mug"

Figure 8.6 illustrates how meaning is built up in parallel to a grammatical derivation. The verb "take" has a syntactic category $s : e/np : p$. This means that it will yield a sentence s if it is combined to the right / with a noun phrase np. The indices e and p relate the syntactic material to the meaning being built: e provides a handle to the index for the verbal meaning, whereas p indicates that the noun phrase will provide the meaning for the Patient [20, 18].

The words "take" and "the" can be combined incrementally into an expression "take the", using function composition (the **B** rule in CCG, cf. [15]). The resulting syntactic category specifies that this expression requires a noun n to its right to yield a complete sentence. The meaning of the determiner "the" circumscribes that of the noun. Finally, "take the" and "mug" are combined into a complete expression, "take the mug".

We have factorized (incremental) grammatical analysis into several, interconnected functions: the incremental parsing process itself, packing/unpacking and pruning of incrementally construed analyses of utterance meaning, and context-sensitive lexical retrieval. Figure 8.7 illustrates the interactions between these different functions.

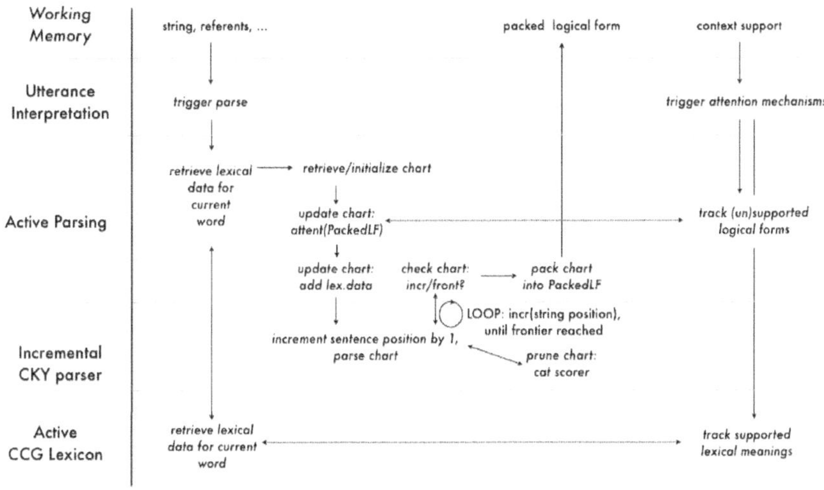

Fig. 8.7. Context-sensitive utterance interpretation at grammatical level: interactive processes for parsing and lexical retrieval, which can be primed by contextual information

Parsing begins by retrieving the lexical entries for the first word, and initializing the chart. A chart is a data structure in which all active and completed analysis are stored, marking for each analysis what part of the utterance (from beginning to some position x) it covers. Maintaining partial analyses makes it possible to re-use them at a later point, when constructing analyses that span more of the utterance. (This principle of re-using partial analyses sets chart-based parsing apart from e.g. backtracking, in which analyses are construed every time anew.) The chart is subsequently updated with the lexical entries for the first word, and a parsing process starts. Parsing is based on a bottom-up Early chart parser built for incrementally parsing Combinatory Categorial Grammar. Its implementation relies on basic functionality provided by OpenCCG[1].

Incremental chart parsing creates partial, and integrated analyses for a string in a left-to-right fashion. After each increase in position, the parser checks whether it has reached a *frontier*. A frontier is specified as a type of complete grammatical structure at the right branch of a grammatical

[1] `http://openccg.sf.net`

derivation. This enables us to specify whether the parser should return after every word, or e.g. after every phrase. At each frontier check, the chart is pruned using a *category scorer*. This scorer ranks the categories for the partial analyses construed so far, possibly pruning them if they are guaranteed not to lead to a complete analysis. (For example, in an incremental analysis, any category requiring an argument to the left \ preceding the beginning of the utterance will never be completed.)

Once incremental parsing stops, a packed logical form is construed, and provided to working memory. This packed representation of possible grammatical interpretations of an utterance provides the basis for further interpretation steps – for example, referent resolution. Depending on the exact fashion in which these processes are synchronized, the next phase of incremental parsing is triggered by the becoming available of further information on working memory (e.g. referents). In this case, the chart is retrieved, and updated with the lexical entries for the current word, and incremental parsing continues as described above.

The advantage of factorizing grammatical analysis into separate inference- and lexical retrieval processes is that the system can use information about the situated- and task-context to prime attention in both processes, possibly asynchronously (i.e. "opportunistically"). Activated categories for objects and events can help to restrict what lexical meanings are retrieved ("activated") for a word. Furthermore, based on what (partial) interpretations can be grounded in the context, unsupported interpretations (analyses) can be removed from the chart.

Picking Up the Right Interpretation

Even with the help of these contextual priming/pruning techniques, the outcome of the utterance comprehension process will nevertheless remain in many cases severely ambiguous and underspecified. This is not surprising: ambiguity is known to be extremely pervasive in natural language, at all processing levels (lexical, syntactic, semantic, pragmatic), and contextual priming/pruning alone cannot be expected to resolve all ambiguities. This means that most utterance will still yield tens, if not hundreds, of possible analyses. Without mechanisms for interpretations selection/filtering at our disposal, these ambiguities are inevitably going to hinder any further interpretation.

We therefore implemented a robust *parse selection* system able to determine the most probable analysis among a set of alternative interpretations. The parse selection is based on a statistical linear model which explores a set of relevant acoustic, syntactic, semantic and contextual features of the parses, and is applied to compute a *likelihood score* for each of them.

Our approach can therefore be seen as a *discriminative* approach to utterance interpretation: we first generate the possible analyses, and then discriminate amongs them according to various features.

The parameters of this linear model are estimated against an automatically generated corpus of ⟨utterance, logical form⟩ pairs. The learning algorithm is an *averaged perceptron*, a simple and efficient technique for parameter estimation which is known to give very good results for this task [63].

The parse selection can be formalised as a function $F : \mathcal{X} \rightarrow \mathcal{Y}$ where the domain $\mathcal{X}$ is the set of possible input utterances[2], and the range $\mathcal{Y}$ is the set of parses. We assume:

1. A function $\mathbf{GEN}(x)$ which enumerates all possible parses for an input x. In our case, this function simply represents the set of parses of x which are admissible according to the CCG grammar.
2. A d-dimensional feature vector $\mathbf{f}(x, y) \in \Re^d$, representing specific features of the pair (x, y). It incorporates various acoustic, syntactic, semantic or contextual features relevant for discriminating the parses.
3. A parameter vector $\mathbf{w} \in \Re^d$.

The function F, mapping an utterance to its most likely parse, is then defined as:

$$F(x) = \underset{y \in \mathbf{GEN}(x)}{\operatorname{argmax}} \ \mathbf{w}^T \cdot \mathbf{f}(x, y) \tag{8.1}$$

where $\mathbf{w}^T \cdot \mathbf{f}(x, y)$ is the inner product $\sum_{s=1}^{d} w_s \ f_s(x, y)$, and can be seen as a measure of the "quality" of the parse.

Given the parameters $\mathbf{w}$, the optimal parse of a given utterance x can be therefore easily determined by enumerating all the parses generated by the grammar, extracting their features, computing the inner product $\mathbf{w}^T \cdot \mathbf{f}(x, y)$, and selecting the parse with the highest score.

We present evaluation results of parse selection later on in the chapter, after we have discussed how language and visuo-spatial information can be combined.

Producing an Utterance in Context

Just like in comprehension we put context into the production of dialogue. Planning what to say and how to say it, is all influenced by context – dialogue-, situation-, and action contexts.

Producing one or more utterances is triggered by a communicative goal. This goal can arise within the dialogue system, for example to follow up in a purely communicative way (e.g. matching a greeting with a greeting), or from outside. The differentiation how communicative goals may arise enables us to put dialogue in the service of other modalities, e.g. to help clarify something the robot does not understand [64], and to achieve a continuum between planning action and interaction (as we will explain further in Section 8.6).

A communicative goal specifies a dialogue move, and content which is to be communicated. We formulate this goal as a logical form. As we describe in detail in [65], we then use a planner to expand (and possibly rewrite) this goal logical form into a logical form specifying the content for one or

[2] Or, in the more general case, a set of possible word recognition lattices.

more utterances. The planner uses a collection of systemic-functional grammar networks [66] to decide, on the basis of the provided content, the way this content can be related within the logical form and to the broader context, how to extend a logical form.

Relating content to context is particularly relevant in the case of generating referring expressions. This task of can be paraphrased as finding a description for an entity in the world (the *intended referent*) that refers to the intended referent and only the intended referent. This implies that the description must be chosen in a way that prevents it from referring to another entity in the current *context set*. All entities in the context set except the intended referent form the *contrast set*. The referring expression must thus distinguish the intended referent from the members of the contrast set. A referring expression is a noun phrase (NP) of any degree of complexity. In order to provide enough information to uniquely identify the intended referent, further attributes of the referent need to be expressed, for instance with adjectives or prepositional phrases, which in turn might contain a referring expression NP.

One of the best understood and widely accepted approaches for generating referring expressions, is the *incremental algorithm* of Dale and Reiter[67]. This algorithm needs a knowledge base that describes the *properties* of the domain entities through *attributes* and *values*. A special attribute is an entity's *type*. The algorithm is initialized with the *intended referent*, a *contrast set* (defined as the *context set* without the intended referent) and a list of *preferred attributes*. The algorithm tries to incrementally rule out members of the contrast set for which a given property of the intended referent does not hold.

In the course of this chapter we describe various instantiations of this algorithm, for producing references to aspects of local contexts (Section 8.4), and the larger spatial organization of the environment (Section 8.5).

Once content planning has yielded a complete logical form for one or more utterances, we provide content utterance by utterance to a realizer. This realizer uses the same grammar as the parser, to produce a set of possible surface strings expressing that content [68]. We use statistical models, trained over a corpus of "usual" utterances for our domain, to select the best realization [69]. This realization is then provided to the MARY speech synthesis engine, to produce audio output [70].

8.4 Talking about What You Can See

In the previous section we discussed how we model meanings for utterances, in a linguistic fashion. We put relations between concept instances, to see how they contribute to the overall meaning – and, by establishing how they relate to preceding utterances, how they contribute to the overall dialogue. Next, let's have a look at how to *situate* that meaning.

We begin by looking at how we could process situated dialogue about things you can see. Both the human and the robot are in the same place, and are talking about objects that are (roughly) in their field of view, often even within reach. Simply put, they are in the same room and that's all they talk about. We call this a *small-scale space* or *closed context.*

Looking at the literature, you will often find this problem of relating language to "the world" (which, really, mostly means small-scale space) referred to as symbol grounding. There is a linguistic symbol, representing some meaning, and it needs to be "grounded in" how the world is perceived. The degree to which this "grounding" determines the meaning of the linguistic symbol is one issue for discussion – the way we look at interconnecting content across modalities is described in more detail in Chapter 2.

But there is more to linguistic meaning than just that. Expressing something is an *act.* We convey meaning in a way that makes it clear not just how a listener should understand what we are saying, but also what she is to do with it. There is a purpose, an intention to saying something. This goes beyond trying to understand the dialogue move or speech act of an utterance. Such a move is –often– a reflection of the action to be undertaken in the real world. So when it comes to relating language to the world, we need to do more than connect symbols to perceptions. We also need to connect meanings to how these perceptions are to be acted upon, or dealt with.

Which brings us to another point we want to raise here. Not every bit of meaning is created and presented equal, in a contextual sense. As a dialogue progresses, we build up a collection of references to aspects of the real world. They form the common ground for the dialogue, a set of mutually held and agreed upon beliefs about the dialogue and the situated context in which that dialogue is set. When we use meanings which refer back to beliefs which are part of the common ground, such meanings provide "old information" on which we can build further, connecting "new information" to it. The point is, there is a difference in which new and old information should be grounded. Whereas the indexicality of old information is generally assumed to hold, for the meanings providing new information it needs to be established. And how that is to be done, couples back to the intentional aspect of an utterance.

Let us illustrate these ideas on an example, before we discuss how they are dealt with in our approach to situated dialogue processing. One popular setting for human-robot interaction is socially guided learning, in which a robot interacts with a human while trying to learn more about the world. As we describe in Chapters 7 and 10, we have explored several aspects of visual learning in such a setting. For example, when the human would like to explain more about a particular object to the robot, she could say something like "The red mug is big."

Now what would that mean?

The way she refers to the object in question, "the red mug," makes clear that she assumes that the robot knows what object she is talking about. It is presented as old information, something already talked about and identifiable not just in the dialogue context but also in the situated context. Next comes a bit of information about the size of the object. The property "big" is being attributed to the object, providing new information which the robot presumably did not yet know. Finally, the intention behind attributing such a new property to an already known object is to teach the robot. It should (still) ground "red mug" in its visual models of the scene, and try to update its models so as to be able to classify that object as "big." Separating old from new information thus first of all indicates, what we should be able to ground. The intention clarifies what to do with the new information. Whether or not the robot then succeeds in learning how to classify the "red mug" as "big" determines how it should react to the utterance. Instead, would we have have followed "standard" approaches to grounding, and not make any such distinctions (old, new; indexicality, intentionality), we would just have the robot try to connect all the meanings immediately to what it sees. The almost inevitable outcome of that would have been – "no." Not knowing that the property "big" could be applied to the object, the robot would not be able to ground the meaning "big red mug" to the visual object (using the proposition of applying the predicate "big" to the argument "red mug").

What You See and What You Mean

Building up a representation of the possible meanings for an utterance, we relate these meanings to the various models the robot maintains. These models can be of the environment, or of actions and plans set therein. We do so in a *mediated* way. We explained already before that we model meaning as ontologically sorted, relational structures – graphs which related concept instances through named relations. We can resolve mentions of these instances against the dialogue context model, so that we know how instances are being talked about and referred to over the course of a dialogue. Grounding such structures using mediation means we use ontologies to mediate the translation of the representations specific to language as a modality, into representations from which we can ultimately construct a-modal representations. (See also Chapter 2, on binding of modality-specific structures and the formation of unions as a-modal representations.)

Since our graph structures are directed and acyclic, we can use recursion over our structures. At each step, a node in the graph is translated into another graph structure on the basis of its ontological sort (as defined in the ontology used in the grammar). A translation can just cover this single node, or a subgraph governed by that node. It can exclude nominals in that subgraph from further processing, making it possible to collapse or omit parts of the graph. An example of that is the translation of spatial expressions, like "the box to the left of the ball:"

$@b_1$:thing(**box** $\wedge$
⟨DELIMITATION⟩ *unique* $\wedge$ ⟨NUM⟩ *sg* $\wedge$ ⟨QUANTIFICATION⟩ *specific* $\wedge$
⟨MODIFIER⟩ (t_1 : m − location $\wedge$ **to** $\wedge$
⟨ANCHOR⟩ (l_1 : e − region $\wedge$ **left** $\wedge$
⟨DELIMITATION⟩ *unique* $\wedge$ ⟨NUM⟩ *sg* $\wedge$ ⟨QUANTIFICATION⟩ *specific* $\wedge$
⟨OWNER⟩ (b_2 : thing $\wedge$ **ball** $\wedge$
⟨DELIMITATION⟩ *unique* $\wedge$ ⟨NUM⟩ *sg* $\wedge$ ⟨QUANTIFICATION⟩ *specific*))))

Such a representation is translated –more or less– into a structure there is a "Location:left-of" between the box and the ball.

Our approach is set apart from other approaches in several ways. First, we ground meaning as graph structures, not as individual words. Second, we are dealing with instances to which we assign discourse referents. Whenever we ground content, we maintain the connection between the discourse referent, and the content it is grounded to. Next time we have new content pertaining to that discourse referent, we update the translated content rather than that we would provide a whole new graph structure to be anchored in the situated context. Naturally, this connection is also used to inform dialogue about changes of information about content to which referents are attached – or can no longer be attached. Thirdly, any ambiguities present in a packed representation are propagated to content to be grounded. This is one point where it we make use of the mechanism for informing dialogue about the grounding possibilities of discourse referents, and any relations between them. As we will see below, ambiguities which cannot be grounded, can be pruned from the packed representation, not having any contextual support in the current context.

When mapping structures are based on ontological sort, content can be flagged as indexical, intentional, or both. As we describe in Chapters 2, 9 and 10, we assume architecture designs in which we can at least differentiate between a working memory for visuo-spatial information about the current context (the "binding" working memory), and a working memory for structures which will prompt the robot to take actions (the "motivation" working memory). Indexical content is put on binding working memory. (Currently, the architecture designs do not include a form of situated episodic memory. Where that to be the case, the resolved temporal reference of an utterance could be used to establish whether to indeed provide the content to the model of the current scene, or to store it with a future or past episode.)

We store intentional content on motivation working memory, with pointers to the indexical content it is related to. Intuitively, intentional content usually represents processes and ascriptions ("verbs"), with representations of objects they operate on including pointers to their indexical counterparts. For example, consider again the command "take the red mug," which results in the following logical form:

$$
\begin{aligned}
&@t1 : \text{action} - \text{motion}(\textbf{take} \ \wedge \ \langle \textsc{Mood} \rangle \, imp \ \wedge \ \langle \textsc{Tense} \rangle \, pres \ \wedge \\
&\quad \langle \textsc{Actor} \rangle (a_1 : \text{entity} \ \wedge \ \textbf{addressee}) \ \wedge \\
&\quad \langle \textsc{Patient} \rangle \ (m_1 : \text{thing} \ \wedge \ \textbf{mug} \\
&\qquad \langle \textsc{Delimitation} \rangle \, unique \ \wedge \ \langle \textsc{Num} \rangle \, sg \ \wedge \ \langle \textsc{Quantification} \rangle \, specific \ \wedge \\
&\qquad \langle \textsc{Modifier} \rangle \ (r_1 : \text{q} - \text{color} \ \wedge \ \textbf{red})) \\
&\quad \langle \textsc{Subject} \rangle \ a_1 : \text{entity})
\end{aligned}
$$

Translating this structure into content to be grounded, binding working memory will end up containing structures for the robot, and the red mug. On the intentional side, we will have a structure for the *take* action, identifying the robot as the Actor of the action, and the red mug as the Patient. The representations for the robot and the mug on motivation working memory refer to their corresponding representations on binding working memory, so that we can situate the intended action.

These translations are based directly on the content of the utterance. We have already seen that they make use of information from the dialogue context, namely resolved discourse referents, to establish what extra-linguistic content to connect to. This primarily concerns indexical information. On the intentional side, information about dialogue move is combined with utterance mood to provide further information about the kind of intention we are dealing with – for example, a command, an assertion, or a question. (See also 2 and 6 for how this can affect general motivation and planning in a cognitive system.)

We will deal with commands later on, in Section 8.6, and focus here on questions and assertions. Particularly in the scenarios we consider in 9 and 10, questions and assertions have in common that they involve a predication over an argument. If we say "The ball is red" we are effectively stating that we can predicate having a "red color" over the "ball" object. A question can also be taken to involve such a predication relation, only now we are quantifying that predication. We rephrase a question like "What color is the ball?" to predicate a color over "ball," quantifying that the value of the color attribute. Slightly more complicated, if we ask "Is the ball red?" we want to know whether we can indeed do so – turning quantifying the predication into a higher-order quantification over the truth of the predication (similar to situation theoretic semantics). For formal details, see [64].

Practically, we represent this information through additional relational structure on the working memory dealing with intentions. We connect the structure produced for the predicate to that for the argument, using two relations. One relation indicates whether we have a polar or factual question, i.e. whether we quantify over the truth of, or a value for, the predication. The other relation indicates what it is that the speaker is after – the intended belief state, resulting from answering the question. For example, for "What color is the ball?" we will connect the predicate "color" to the argument "ball" using a relation "Fact-Q", and a relation "SPEAKER-KNOWS:Colour." For polar questions like "Is the ball red?," the latter relation will also indicate the value of the color: "SPEAKER-KNOWS:Colour?red" i.e. the speaker knows whether the ball indeed has a red color.

The argument of SPEAKER-KNOWS can be a pointer to any type of elementary predication in a meaning representation. This makes it possible for to quantify over any type of information represented as predication – be that a sort ("Is this a ball?"), the value of an attribute, or a relation ("Is there a ball to the left of the box?"). Assertions only differ from questions in that we just introduce a relation stating that the hearer knows the asserted information (HEARER-KNOWS, with the same type of structure as SPEAKER-KNOWS).

How we subsequently evaluate whether we can achieve the desired belief state is dependent on the grounding of the various pieces of content. This is driven by the information status of content, as we explain next.

What You All Know, and What You Are Saying

We already differentiate content by its intentional or indexical nature. We make a further division into content which the speaker presents as "old information" belonging to the common ground, and that which is presented as new. Grammatically speaking, we are using a form of information structure [71]. We determine the information status of an object on the basis of its semantic features for delimitation and quantification, and how the content functions in the larger context of an intentional construction [18]. The basic idea is that if an object is established to be old news, we will immediately try to ground it. This way, it can provide the basis for forming the relevant situated context against which we need to consider the intention. Based on the intentional content, we will next evaluate whether the new information can indeed be grounded in relation to the already grounded information.

In Section 8.6 we will explain how this works out for commands like "put the ball to the left of the box." We will see there how information structure interacts with the intention of performing a "put" action on the mentioned objects, and the desired state of the ball being to the left of the box, to help establish how we should evaluate whether this action can indeed be performed. For our current purposes, we will focus on ascriptions – assertions such as "This box is red," or questions like "Is the red box to the left of the ball?"

As we saw above, we build additional structure for question- and assertion-interpretations. This structure reflects a desired update to an interlocutor's belief state. Evaluating a question or an assertion then boils down to establishing whether we can obtain that state. How we evaluate is guided by information structure. When translating meaning to indexical and intentional content, we determine the information status of objects on the basis of using semantic delimitation and quantification. We adopt the overriding exception that we will always interpret the argument of a SPEAKER- or HEARER-KNOWS relation to be "new." Any indexical information with an "old" information status will be written to binding working memory. We do this to establish the relation to common ground. In parallel, we represent the new information and any purely intentional information on the motivation working memory. Evaluation then consists therein to establish whether we can in principle update binding working memory with the new information, or check against the results of grounding whether the new information already holds or could be retrieved. A felicitous update may in this case provide the trigger for learning processes, as we discuss in Chapter 7.

The result is a model of information structure and its interaction with indexical and intentional content that is reminiscent of a dynamic semantics-based approach to information structure [71]. Where we differ is the use of multiple indexical and intentional contexts, and the evaluation of the update on one context relative to the intended use of information as stated in another context.

Using What You See to Rank Alternative Interpretations

As we already outlined in section 8.3, a *discriminative model* is used to assign a score to each possible semantic interpretation of a given spoken input. The discriminative model includes a wide range of *linguistic* as well as *contextual* features. The linguistic features are defined on the analyses construed at the different processing levels: the *acoustic* level (based on ASR scores), the *syntactic* level (based on the derivational history of the parse), and the *semantic* level (based on substructures of the logical form). As for the contextual features, they are defined using information from both the situated context (the objects in the visual scene) and the dialogue context (previously referred entities in the dialogue history).

Experimental Evaluation We performed a quantitative evaluation of our approach to parse selection. To set up the experiments for the evaluation, we have gathered a corpus of human-robot spoken dialogue for our task-domain, which we segmented and annotated manually with their expected semantic interpretation. The current data set contains 195 individual utterances along with their complete logical form.

Three types of quantitative results are extracted from the evaluation results: *exact-match*, *partial-match*, and *word error rate*. Tables 8.1, 8.2 and 8.3 illustrate the results, broken down by activated features, use of grammar relaxation, and number of recognition hypotheses considered.

Each line in the tables corresponds to a possible configuration. For each configuration, we analyse the accuracy results on different NBests, and give the precision, recall and F_1 value for each.

The first cell of the first line corresponds to the baseline: no grammar relaxation, no activated features, and use of the first NBest recognition hypothesis. The last line corresponds to the final results with all features, combined with the grammar relaxation mechanism.

Table 8.1. Exact-match accuracy results, broken down by activated features, use of grammar relaxation, and number of recognition hypotheses considered. For each configuration, we give the precision, recall, and F_1 value (all in percents).

Gram. Relax.	Activated Features				Nbest 1			Nbest 5		
	Sem.	Synt.	Ac.	Cont.	Pr.	R.	F_1	Pr.	R.	F_1
					40.9	45.2	**43.0**	14.4	13.9	14.2
				×	35.2	41.5	38.1	28.8	31.8	30.2
			×		42.8	46.3	44.5	38.1	47.1	42.2
			×	×	41.9	45.8	43.7	43.1	49.4	46.0
	×				59.0	54.3	56.6	30.3	51.3	38.1
	×			×	59.0	54.3	56.6	35.2	55.1	43.0
	×		×		59.0	54.3	56.6	58.3	65.4	61.6
	×		×	×	59.0	54.3	56.6	60.8	66.3	63.4
×					20.9	49.0	29.3	10.7	34.1	16.3
×				×	20.9	49.0	29.3	12.1	39.0	18.4
×			×		27.1	55.5	36.4	27.3	54.6	36.4
×			×	×	21.7	50.0	30.2	27.9	56.2	37.3
×		×			34.1	61.1	43.7	21.0	39.6	27.4
×		×		×	30.2	58.2	39.7	21.9	44.2	29.3
×		×	×		34.1	61.1	43.7	32.8	59.1	42.2
×		×	×	×	32.5	60.0	42.2	32.5	60.0	42.2
×	×				49.6	69.5	57.9	28.9	77.7	42.2
×	×			×	49.6	69.5	57.9	31.0	78.9	44.5
×	×		×		49.6	69.5	57.9	52.1	83.1	64.0
×	×		×	×	49.6	69.5	57.9	53.1	84.4	65.2
×	×	×			52.7	70.8	60.4	29.6	78.1	43.0
×	×	×		×	52.7	70.8	60.4	31.7	79.3	45.3
×	×	×	×		52.7	70.8	60.4	54.6	82.7	65.8
×	×	×	×	×	52.7	70.8	**60.4**	55.6	84.0	**66.9**

Table 8.2. Partial-match accuracy results, broken down by activated features, use of grammar relaxation, and number of recognition hypotheses considered. For each configuration, we give the precision, recall, and F_1 value (all in percents).

Gram. Relax.	Activated Features				Nbest 1			Nbest 5		
	Sem.	Synt.	Ac.	Cont.	Pr.	R.	F_1	Pr.	R.	F_1
					86.2	56.2	**68.0**	73.5	45.8	56.4
				×	85.5	56.0	67.7	81.3	54.2	65.1
			×		86.8	56.4	68.3	84.3	60.4	70.4
			×	×	86.2	56.2	68.1	85.4	60.4	70.7
	×				90.5	57.4	70.3	80.1	66.4	72.6
	×			×	90.5	57.4	70.3	83.3	67.2	74.4
	×		×		90.5	57.4	70.3	88.9	67.1	76.4
	×		×	×	90.5	57.4	70.3	89.5	67.2	76.8
×					75.7	73.3	74.5	71.4	81.9	76.3
×				×	73.7	72.8	73.2	71.7	78.7	75.1
×			×		75.3	73.2	74.2	74.6	73.1	73.8
×			×	×	72.7	72.5	72.6	74.6	74.1	74.4
×		×			80.9	74.6	77.6	76.2	72.1	74.1
×		×		×	80.2	74.4	77.2	78.7	76.2	77.4
×		×	×		80.8	74.6	77.6	80.3	74.5	77.3
×		×	×	×	80.4	74.5	77.3	80.3	75.5	77.8
×	×				86.5	75.8	80.8	80.7	88.4	84.4
×	×			×	86.5	75.8	80.8	80.0	88.3	84.0
×	×		×		86.5	75.8	80.8	86.2	86.7	86.4
×	×		×	×	86.5	75.8	80.8	86.3	87.2	86.8
×	×	×			88.1	76.2	81.7	79.3	88.2	83.5
×	×	×		×	88.1	76.2	81.7	81.7	88.5	85.0
×	×	×	×		88.1	76.2	81.7	87.5	85.4	86.4
×	×	×	×	×	88.1	76.2	**81.7**	87.6	86.0	**86.8**

Table 8.3. Word Error Rate results, broken down by activated features, use of grammar relaxation, and number of recognition hypotheses considered. For each configuration, we give the error rate (in percents).

Gram. Relax.	Activated Features				Nbest 1	Nbest 3	Nbest 5	Nbest 10
	Sem.	Synt.	Ac.	Cont.				
					20.5	26.9	29.7	25.9
				×	20.5	23.6	24.6	28.0
			×		20.5	19.7	19.6	19.7
			×	×	20.5	18.7	18.2	18.3
	×				20.5	24.6	25.6	31.2
	×			×	20.5	21.4	23.2	26.1
	×		×		20.5	18.3	18.4	18.1
	×		×	×	20.5	17.3	17.4	17.4
×					19.6	23.6	25.9	23.9
×				×	19.3	20.4	23.3	26.7
×			×		19.7	18.6	18.4	19.3
×			×	×	19.4	18.0	17.6	17.7
×		×			19.4	24.6	26.9	27.9
×		×		×	19.4	22.2	23.9	28.1
×		×	×		19.4	18.8	18.7	18.8
×		×	×	×	19.4	17.8	17.3	17.4
×	×				20.2	22.4	25.5	29.4
×	×			×	20.2	21.0	22.9	26.1
×	×		×		20.2	17.8	17.8	17.8
×	×		×	×	20.2	17.4	17.1	17.1
×	×	×			19.4	21.5	24.3	28.7
×	×	×		×	19.4	19.8	21.9	25.9
×	×	×	×		19.4	16.8	16.7	16.7
×	×	×	×	×	19.4	**16.5**	**15.7**	**15.7**

Two elements are worth noticing in the results:

1. In each of the three tables, we observe that no configuration is able to beat the results obtained with all activated features. In other words, it shows that all features types are playing a positive role on the task.
2. Likewise, we observe that taking into account more ASR recognition hypotheses has a positive effect on the results: the results obtained using five recognition hypotheses are substantially better than those obtained based only on the first hypothesis.

Comparison with Baseline Here are the comparative results we obtained:

- Regarding the exact-match accuracy results, the difference between the baseline results and the results with our approach (grammar relaxation and all features activated for NBest 10) is striking: the F_1-measure climbs from 43.0 % to 67.2 %, which means a relative difference of **56.3** %.
- For the partial-match, the F_1-measure goes from 68.0 % for the baseline to 87.3 % for our approach – a relative increase of **28.4** %.
- Finally, the decrease in Word Error Rate is also worth noting: we go from 20.5 % for the baseline to 15.7 % with our approach. The difference is statistically significant (p-value for t-tests is 0.036), and the relative decrease is of **23.4** %.

Using What You See to Figure Out What Is Meant

If we have a packed representation that includes alternative interpretations, any indexical ambiguity will end up as alternative relational structures on binding working memory. By monitoring which relational structures can be grounded in the current context, and which ones cannot, we can prune the set of interpretations we maintain for the dialogue. We thus handle examples such as those discussed in [10] through an interaction between binding, and dialogue processing. Below we provide a detailed example of resolving syntactic attachment ambiguities using the situated context. (Lexical ambiguities based in different semantic categories are resolved against visual categories.)

Fig. 8.8. Situated context for "put the ball near the mug to the left of the box."

Consider the visual scene in Figure 8.8, and the utterance "put the ball near the mug to the left of the box". Linguistically speaking, this utterance is ambiguous. There are several ways in which we can combine the modifiers "the ball", "near the mug", and "to the left of the box." Is the ball near the mug? Or is "near the mug" the place where the robot is to put the ball, with "the mug" supposedly being located left of the box?

On its own, the utterance is highly ambiguous. But, this ambiguity somehow vanishes when we consider the visual scene in Figure 8.8. Then it is clear that there is only one sensible way to understand the utterance. The ball is near the mug, and it should end up to the left of the box (as indicated by the arrow). The system achieves the same disambiguation effects through (incremental) pruning of linguistic interpretations, based on whether they can be grounded in the visuo-spatial situated context.

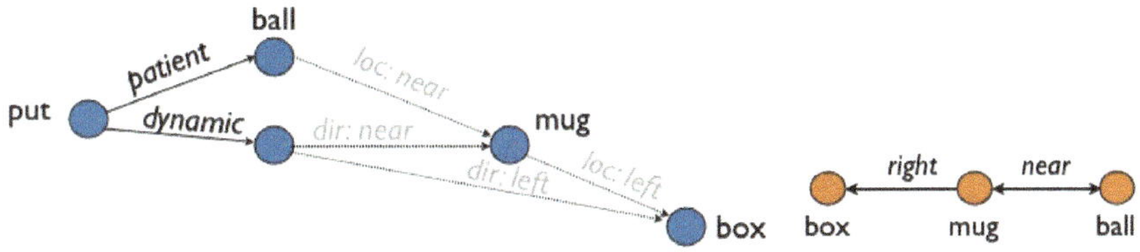

Fig. 8.9. Ambiguous (complete) packed logical form for "put the ball near the mug to the left of the box" (l.) and the spatial relations for the visual scene (r.)

Figure 8.9 (right) gives the spatial model for the visual scene in Figure 8.8 [26]. On the left is the (complete) packed logical form we obtain for "put the ball near the mug to the left of the box". Up to "put the ball near the mug" the modifier "near the mug" remains ambiguous between being the destination for where to put the mug, or specifying a location for "the ball." The visuo-spatial scene provides support for both interpretations, (although planning may prefer the locative reading, as the ball is already near the mug thus pre-empting execution of the action). As soon as "to the left of the box" is being processed, the spatial model invalidates the reading on which the mug is located left of the box. This resolves "to the left of the box" to be the destination of the put action, and (by grammatical inference over the resulting syntactic categories) "near the mug" to be the location modifier of "the ball."

Referring to What You See

A robot isn't just to understand what we are saying. It should also be able to produce dialogue which refers to the environment in meaningful and appropriate ways. In the context of small-scale space, what is particularly important is that the robot can refer to objects and the spatial relations between them.

This presents an interesting challenge. If the robot is to generate any form of spatial language, is needs to construct and maintain a model that explicitly marks the spatial relations between objects in the scene. However, the construction of such a model is prone to the issue of combinatorial explosion both in terms of the number objects in the context (the location of each object in the scene must be checked against all the other objects in the scene) and number of inter-object spatial relations (as a greater number of spatial relations will require a greater number of comparisons between each pair of objects. This becomes particularly problematic when we consider that a scene may be dynamic, requiring the robot to update its models.

We present in [72] a framework that addresses this issue. We provide a way to define the set of objects in the context that may function as a landmark, and then sequence the order in which spatial relations are considered using a cognitively motivated hierarchy of relations. Defining the set of objects in the scene that may function as a landmark reduces the number of object pairs that a spatial relation must be computed over. Sequencing the consideration of spatial relations means that in each context model only one relation needs to be checked and in some instances the agent need not compute some of the spatial relations, as it may have succeeded in generating a distinguishing locative using a relation earlier in the sequence.

A further advantage of our approach stems from the partitioning of the context into those objects that may function as a landmark and those that may not. As a result of this partitioning the algorithm avoids the issue of infinite recursion, as the partitioning of the context stops the algorithm from distinguishing a landmark using its target.

In Recapitulation

When it comes to talking about what you see, we discussed above several aspects in which the bi-directionality hypothesis turns up. The possible linguistic meanings we can provide for an utterance are connected to the way the situation is understood, which is coupled back to what meanings are established as contextually supported. We use this mechanism in post-filtering during incremental parsing, in parallel to predictive mechanisms such as parse selection and word lattice re-scoring, and during production in the generation of referring expressions.

We illustrated how we ground meanings, by looking at intentional and indexical aspects, and the information status of content. Instead of grounding all content wholesale word-by-word in visuo-spatial models, as is usually done, we first only ground meaning already part of the common ground, and then evaluate whether new information can be grounded in the sense as indicated by the intention of the utterance. This yields a situated form of dynamic, context-sensitive interpretation of linguistic meaning.

8.5 Talking about Places You Can Visit

Above we discussed how we process situated dialogue about small-scale space. The human and the robot are in the same location, and talk about things that are in view. Already there we faced the problem to determine what part of that space forms the current context – which objects, and what aspects of spatial organization, we can consider common ground.

This becomes an even bigger issue when we want to talk about *large-scale space* – that kind of "space which cannot be perceived at once" [73]. Discussing aspects of large-scale space, for example where a particular room is or where the robot could find a specific object, is typical for the Explorer scenario, see [74] and Chapter 9. Most of these referents will, however, not be in view for the interlocutors.

So, whereas in situated dialogue set in small-scale space we can call upon visuo-spatial content stored on a short-term 'binding' working memory, we need to go beyond that in the case of large-scale space. In this section we will discuss how we can integrate content from situated dialogue with ontological reasoning and conceptual spatial mapping [74, 75, 76].

8.5.1 Talking about Places

When it comes to talking about places, various Wizard-of-Oz studies have investigated how humans tend to inform robots about the spatial organization of an environment. For example, [77] discuss a study on how a human presents a familiar indoor environment to a robot, and [78] when a human talks with a robot wheelchair while being seated in it. These studies have yielded various important insights.

The experimental setup in [77] models a typical guided tour scenario. The human guides the robot around and names places and objects. One result of the experiment is the observation that people tend to employ many different strategies to introduce new locations. Besides naming whole rooms ("this is the kitchen" referring to the room itself) or specific locations in rooms ("this is the kitchen" referring to the cooking area), another frequently used strategy was to name specific locations by the objects found there ("this is the coffee machine"). Any combination of these individual strategies could be found during the experiments. Moreover, it has been found that subjects only name those objects and locations that they find interesting or relevant, thus personalizing the representation of the environment that the robot constructs.

In [78], the subjects are seated in a robot wheelchair and asked to guide it around using verbal commands. This setup has a major impact on the data collected. The tutors must use verbal commands containing deictic references in order to steer the robot. Since the perspective of the human tutor is identical to that of the robot, deictic references can be mapped one-to-one to the robot's frame of reference. One interesting finding is that people tend to name areas that are only passed by. This can either happen in a 'virtual tour' when giving route directions or in a 'real guided tour' ("here to the right of me is the door to the room with the mailboxes."). A robust conceptual mapping system must therefore be able to handle information about areas that have not yet been visited.

Next we discuss how we deal with the above findings, combining information from dialogue and commonsense knowledge about indoor environments.

8.5.2 Representing Places to Talk about

In Chapter 5, we present our approach to semantic modeling of space. In this approach, we use a multi-layered spatial map that represents space at different leveles of abstraction. The most abstract layer, the 'conceptual map', characterizes spatial units (e.g. rooms) by assigning them human concepts (e.g. "kitchen"), which can be used to resolve or generate linguistic expressions. The 'conceptual map' is represented as a Description Logics ontology, consisting of a concept taxonomy and a storage of instances, which form the T-Box and A-Box of a Description Logics reasoning framework.[3] The concept taxonomy is a hand-written common sense ontology representing various aspects of an indoor environment (different kinds of areas and other spatial structures, different kinds of objects, agents, and several different relations that can hold between any of these). During run-time the ontology is populated with instances of spatial units and objects through evaluation and interpretation of sensory data (e.g. laser range scans, and visual object detection). A conceptual map that is constructed only from sensory input, e.g. during an autonomous exploration of the robot's environment, will consist of instances of the abstract concept `Area` (corresponding to the units of the topological map layer), which are further specified by the appropriate sub-concepts `Room` and `Corridor` (based on the laser-based semantic place labeling method), and also instances of `Object`, further specified by their respective visual object

[3] We have used different 3rd party reasoners in our experiments, including RACER, Pellet, and Jena.

class, e.g. Couch or TV. On the basis of this object information, the reasoner can even further specify the area instances, for instance by inferring that a Room instance containing some KitchenObject instance (e.g. an instance of Coffeemachine) is an instance of the more special concept Kitchen.

Through this approach, the robot achieves a level of spatial understanding that is already compatible with the linguistic categories that humans use to refer to places in an indoor environment. The conceptual map, however, also holds information about the environment given by human users, for example in a 'guided home tour' interactive mapping set-up.

Our approach to interactive map acquisition accomodates the previously mentioned findings in studies on Human-Augmented Mapping [77] through the following properties:

References to whole rooms or specific locations are used to assert that the instance of the corresponding topological area is of the mentioned concept, even if the reasoner could not infer that knowledge on the basis of the robots own information.

References to specific objects, and thus omitting naming the whole room, will assert that an instance of the mentioned object type is present, which allows the reasoner to draw further inferences about the current topological area. In the above example, the user only points out that "there is the coffee machine". On the basis of its knowledge that the current area is a Room instance, which is asserted to contain a Coffeemachine instance, the reasoner now infers the new concept Kitchen.

Like this, our system can combine sensor-based information and information provided through dialogue with a human user. This allows the system to cope with otherwise incomplete information, and with highly personalized information. Our approach yields a conceptual representation of space that is suitable for understanding linguistic references to spatial entities, and for producing expressions that can be understood by human users.

8.5.3 Referring to Elsewhere

A conversational autonomous mobile robot will inevitably face situations in which it needs to refer to an entity (an object, a locality, or even an event) that is located somewhere outside the current scene. In technical terms, the robot must be able to produce a *referring expression* to an entity in large-scale space [27].

There are conceivably many ways in which a robot might to refer to things in the world, but many such expressions are unsuitable in most human-robot dialogues. Consider the following set of examples:

1. "the location at position ($X = 5.56, Y = -3.92, \theta = 0.45$)"
2. "the mug left of the plate right of the mug left of the plate"

3. "Peter's office no. 200 at the end of the corridor on the third floor of the Acme Corp. building 3 in the Acme Corp. building complex, 47 Evergreen Terrace, Calisota, Planet Earth, (...)"
4. "the area"

These referring expressions are valid descriptions of their respective referents. Still they fail to achieve their *communicative goal*, which is to specify the right amount of information that the hearer needs to uniquely identify the referent. First of all, robots are good at measuring exact distances, humans are not. So the robot should employ qualitative descriptions that make use of the same concepts as a human-produced utterance would. Second, specifying a referent with respect to another referent that is only identifiable relative to the first one leads to infinite recursion instead of the communicative goal. Finally, the robot might have a vast knowledge about facts and entities in the world, but it should not always try to uniquely separate the referent from all entities in the world. At the same time, it is necessary to provide enough information to distinguish the intended referent from those entities in the world that potentially distract the hearer. The following expressions *might* serve as more appropriate variants of the previous examples:

1. "the kitchen around the corner"
2. "the red mug left of the china plate"
3. "Peter's office"
4. "the large hall on the first floor"

The fact that these *might* (or *might not!*) be successful referring expressions points to the importance of knowing what the given context in a situation is. This is especially the case for a mobile robot that operates and interacts in large-scale space. It is thus an important basis to endow the robot with a spatial representation that resembles the way humans conceive of their environment. But it is not enough; the robot must also be able to determine which entities in the world might act as *potential distractors* with respect to the hearer's knowledge.

In the following paragraphs we will show how our multi-layered conceptual spatial map provides a suitable knowledge base for Dale and Reiter's *incremental GRE algorithm*[67]. Furthermore, we will propose a method for a proper construction of the *context set* for successfully referring to entities in large-scale space.

The instances in the ontology are the *entities* of the world model. The conceptual hierarchy provides the taxonomical *type* information of the instances that the GRE algorithm requires. Furthermore, a number of concepts such as `Office`, `Kitchen`, `Corridor`, `Table`, etc. are marked as basic level categories, cf. [79] and [80]. The relations between instances are the *attributes* that the algorithm can use to further specify a referent. In terms of the Dale and Reiter algorithm, we currently use the following list of attributes, ordered by their preference: ⟨ type, topological inclusion, ownership, name ⟩.

Type

We represent an entity's type as the (asserted and inferred) concepts of the corresponding instance. Through ontological reasoning, we can retrieve an instance's most specific concept, its basic level category, and all the instances of a concept.

Topological Inclusion

If the current context spans topological units at different hierarchical levels (cf. Figure 8.10) it is important to specify the intended referent with respect to the topological unit that contains the referent, e.g. when referring to "the kitchen on the 3rd floor", or "the table in the lab". In the ontology the transitive property `topoIncluded(X,Y)` and its inverse property `topoContains(Y,X)` represent topological positions of entities. By constructing a query to the reasoner that only returns those 'topological containers' of an entity that don't contain any other entities which in turn also contain the entity, we assure to only take into account direct topological inclusion despite the transitivity of the ontological properties.

Ownership

Areas in an environment are often referred to by identifying their owners, e.g. "Bob's office". In our ontology instances of `Area` can be related to a `Person` instance via the `owns(X,y)/isOwnedBy(Y,X)` relation pair. People are instances of the ontological concept `Person`. The name of a person is represented as a string datatype property.

Name

As names are usually (locally) unique, e.g. "the Occam meeting room", or "office 120", they are definitely a highly discriminating attribute for the GRE task. However, names do not seem to be a preferred category for referring to rooms as they seldom contain more useful information than a generic NP + PP referring expression, e.g. "the meeting room on the first floor next to the large hall". On the contrary, such a generic referring expression might even bear additional useful information. Moreover, remembering the inherently artificial name for an entity might involve a higher cognitive load than processing the information encoded in a more generic referential description. For other scenarios though, such as an information desk agent at a hospital, or any other institution in which there is a specific naming scheme, such as e.g. encoding floor number and department, and numbering them in sequential order, the name feature can conceivably be placed in a higher-ranking position in the preference list. In our ontology names for areas are represented as a string datatype property.

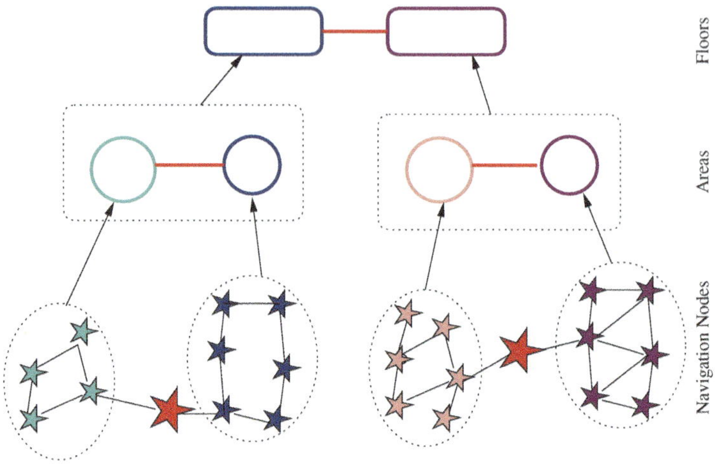

Fig. 8.10. A topology of places, rooms and floors. Stars depict navigation nodes that denote free and reachable space for our robotic system. The set of navigation nodes is partitioned into distinct spatial areas, such as e.g. rooms. Areas in turn can belong to a floors, which are on the next level of abstraction. Using *topology abstraction*, we construct an appropriate context set for the GRE task.

Determining the Appropriate Contrast Set

In order to successfully identify a referent it is important to determine a correct and appropriate contrast set. If the contrast set is chosen too small, the hearer might find it difficult to uniquely identify the intended referent with respect to his or her knowledge. If, on the other hand, a too large contrast set is assumed, the generated referring expression might violate *Grice's Maxims*, here the Maxim of Quality, in that it contains too much unnecessary information.

Since the contrast set is defined relative to a context set, the crucial task is hence to determine which part of the environment constitutes the current context. For one, the context needs to include the intended referent. The context must also include the current *referential anchor*, i.e. what is considered the current position of the two interlocutors. In the simple case, this referential anchor is the physical location where the dialogue takes place. But as a dialogue evolves, the referential anchor moves through space and time. Consider the following example dialogue:

> Person A: "Where is the exit?"
> Person B: "You first go down this corridor. Then you turn right. After a few steps you will see the big glass doors."
> Person A: "And the bus station? Is it to the left?"

As can be seen, any utterance in such a collaborative dialogue is grounded in previously introduced discourse referents, both temporally and spatially.

Assuming the last mentioned discourse referent (or the physical location of a conversation) as the referential anchor, the question remains which other entities constitute the current discourse context. In other words: when referring to things, places, and actions in large-scale space, what possible distractors must one rule out in order for a referential description to be successful?

It is a widely accepted theory that humans tend to represent large-scale space in terms of topologies, rather than using exact measures. Following this view, we claim that the context for a dialogue situated in large-scale space can be determined on the basis of a topological representation.

Figure 8.10 shows a topological segmentation of an indoor space like the one used in our robotic system. The smallest unit in this representation is a graph-like structure of 'place nodes' (distinct places of approx. 1m in diameter that can be reached by the robot) and 'object nodes' (places from which objects are visible). These nodes are linked by edges that denote accessibility or visibility of one node from another. Through a number of processes, cf. Chapter 5, this graph is segmented into distinct areas, corresponding to e.g. rooms, corridors, or special regions within larger spatial structures. This segmentation into areas yields a first topological abstraction of space, in which the information about containment and reachability of its units is preserved, but metric distances don't play a role.

Our process of topology abstraction for determining the context set is depicted in Figure 8.11. It can be paraphrased as "Start with the referential anchor and check whether the intended referent is a member of the set of the referential anchor and its child nodes. If so, this set is the referential context. If not, construct the set of the referential anchor's parent nodes and their children, and check again. Repeat this procedure of topological abstraction until the intended referent is a member of this growing context set." With respect to the ontological representation of the conceptual spatial map, the function *topologicalChildren(x)* corresponds to a query that matches all instances i for which `topoContains(x,i)` applies. *topologicalChildren(x)* is defined as the set of instances i for which the direct, intransitive variant `direct-topoContains(x,i)` is true.

This means that if an object is located in the same room as the user and the robot, only local landmarks should be considered potential distractors. Likewise, if the robot is to produce a referring expression to a room on a different floor, all known entities inside the building will form the context. Using topological inclusion as the most preferred attribute (after type) will then essentially function as an early pruning of the hierarchically ordered context set.

8.5.4 Understanding References to Elsewhere

A conversational robot should not only be able to produce meaningful speech, but also must be able to understand verbal descriptions given by its users. Similar to the challenge of generating referring expressions to entities in large-scale space, a dialogue system for a mobile robot will have to deal with its

```
Require: r = intended referent; a = referential anchor
  Initialize: CONTEXT = {}
  CONTEXT = CONTEXT ∪ topologicalChildren(a) ∪ {a}
  if r ∈ CONTEXT then
    return CONTEXT
  else
    Initialize topological containers to check: CONTAINERS = {a}
    while r ∉ CONTEXT do
      if CONTAINERS = {} then
        return failure
      end if
      for each c ∈ CONTAINERS do
        for each p ∈ topologicalParents(c) do
          CONTAINERS = CONTAINERS ∪ {p}
          CONTEXT = CONTEXT ∪ topologicalChildren(p)
        end for
      end for
    end while
    return CONTEXT
  end if
```

Fig. 8.11. Topological abstraction algorithm for context generation.

user's referring expressions. The robot essentially needs to match a complex nominal construction with its internal knowledge base. Analogous to the task of generating referring expressions, an appropriate context against which to compare the referential description is crucial.

The first step is to translate the semantic interpretation of an utterance into a query to the ontology reasoner. This is being done through the architectural binding subarchitecture. A Logical Form is translated into a proxy structure, i.e. a number of proxies with well-defined 'concept' features, and labelled relations between them. The subarchitecture that holds the conceptual spatial mapping and reasoning functionality then reads the full relational proxy structure and converts the provided features into attribute-value pairs in the representation of the ontology. The relations are also reconstructed in the ontology language. Iteratively, an ontological description of the referring expression is generated. This description will then serve as a query to the reasoner. Upon such a query the reasoner will return all instances in its knowledge base that fulfill the criteria specified by features and relations.

In order to not overgenerate possible referents, the resulting query needs to be evaluated against a subset of the full knowledge base. The relevant subset is the discourse context. Following the approach described above, the query is first evaluated against the child nodes of the current discourse anchor and the discourse anchor itself. If the reasoner does not find any instances that satisfy the description, the context set is increased using the method of topological abstraction until at least one possible referent can be identified within the context.

8.6 Talking about Things You Can Do

So far we have seen how bi-directionality figures in situated dialogue processing, when talking about things and space. We are using information about the situated context to predictively filter interpretations in speech recognition and incremental parsing, and later on use grounding of content to further zoom in on those interpretations which are contextually supported. Furthermore, going back and forth between what we know about the scene, the environment, and what we would like to say, we can generate referring expressions which are contextually appropriate.

But how about action? And where do we draw the line between action and interaction, in situated dialogue? We know from human communication that the very situatedness of such dialogues allows us to "overload" actions, giving them also a communicative function. If we want to provide similar possibilities for human-robot interaction, we need to consider how action planning, and dialogue planning, should interact. This takes the bi-directionality even further, from sharing content to sharing decision making processes.

Below we first discuss how action planning and situated dialogue processing interact at content-level, and then close with a discussion on the spectrum between planning for action and interaction.

Things You Can Do

When we talk about performing an action, it's about more than just the act itself. There are the objects involved, who is to do something. There is what we assume as outset, and what we expect to be the outcome of the action. And it is all these aspects that somehow need to match up in context – in the dialogue context as well as in the situated context.

What we do is to interpret the action further. We follow [25] and assign the action a so-called *event nucleus*. The event nucleus models the action as an event with temporal and causal dimensions. It models what needs to be done before this action could be performed, and what would result from performing this action – in as far as we can determine this linguistically, of course. In [23] we discussed detail how we formally model the event nucleus. Here we will focus on the basic intuitions behind these models, and discuss their interaction with the ideas about indexicality, intentionality, and information structure we discussed already earlier.

Figure 8.12 provides the basic model we use for an event nucleus. Like all our other representations, it is a relational structure, with variables which are ontologically sorted. The sorts model a temporal ontology of types of events [25], whereas the variables themselves gives us the possibility to explicitly refer to these aspects of an event. We resolve any explicit or implicit temporal references to these variables, so that we can establish the temporal relations between events. After all, the ways in which events may be related need to match the order in which we have been talking about them. For example, consider we have a sequence of instructions like "First take the mug, then put

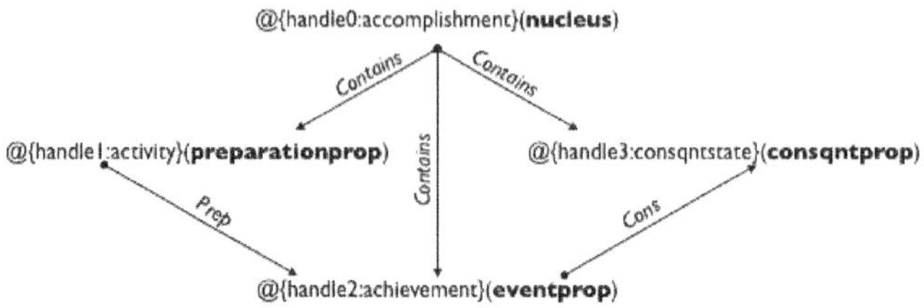

Fig. 8.12. Event nucleus, from [23]

it to the left the plate." This sequence already indicates that we first need to do the taking, then putting. Using the event nuclei associated with these actions, we can establish a much richer interpretation though. The consequence of the "take" action is that we have the mug – which is exactly what needs to be the case for the "put" action. Together with the fact that these actions are to apply to the same object ("the mug" / "it"), we can associate "take" as a preparatory action to "put." Finally, we can establish that being "near the plate" is an intended state for the mug – the goal we are trying to achieve with the put action.

When we look at the nature of the content we provide for this intended state or goal, we again see there is a need to differentiate between the information status of content. We consider content which specifies the kernel (or the root) of the content for the intended state to be "new." Thus, if we have "put [the mug] to the left of the plate" we start out with considering the mug and the plate to be "old" and part of the identifiable common ground. The intended location of the mug, being left of the plate, is new. Before grounding it in binding working memory, we need to establish whether there is a suitable location given the action(s) we want to perform.

We thus follow in principle the same approach as we do for questions and assertions. Considering the command as intentional content, we provide it to motivation working memory with pointers to the indexical content for "mug" and "plate." Action planning in turn tries to establish a suitable location and plans for executing the action, as described in more detail in [81]. Bi-directionality between planning, motivation, binding working memory and dialogue processing subsequently provides feedback on the basis of which we post-filter out any potential linguistic interpretation for which we cannot find a suitable, situated plan, and provide communicative feedback on the command.

Between Saying and Doing

In situated dialogue, there is no clear division between action and interaction. As a simple example, consider reacting to a put-command like the one above. The robot could listen. Then provide an elaborate response to make clear it has understood: "Okay, let me put the mug to the left of the plate." And then do it. This is possible, but it stands in some contrast to what humans tend to do. There, you often see that someone says "fine" and moves the mug around. Or she even just does the latter. The action *is* the feedback, performing it gets overloaded with a communicative function of providing feedback.

What we can observe here is the next step in following out the bi-directionality hypothesis. For most of this chapter, we have considered bi-directionality in situating the linguistic content construed in a dialogue, be that during comprehension or production of utterances. Here we go one small step further, and take bi-directionality to the level of decision processes. We go from how to understand communication, to how to direct it in a given situation.

This builds forth on the bi-directional links we have already established, at an intentional level, between action planning and situated dialogue processing. We pointed out how dialogue meaning has two interrelated dimensions, namely an indexical and an intentional one. These dimensions determine how meaning gets related to content in other modalities in a cognitive architecture. Notably, intentional content is used within the motivational subsystem of the architecture to establish plans for action.

It is on this basis we make a first attempt to establish a spectrum of intentions between planning for action and interaction. Dialogue planning can –naturally– generate dialogue moves from purely linguistic intentions, e.g. those arising at engagement level [82]. At the same time, it can handle intentions which originate outside the realm of dialogue. Typically, these intentions stem from an action plan created on the basis of a communicated intention. The intentions are always accompanied by indexical content, to which they apply. For example, an externally raised need to inquire with the user about a property of a particular object will be modeled as a "question" intention together with references to the object and the property. We can integrate such an external intention as a continuation in the dialogue context model, by using the links we maintain between dialogue referents, and the cross-modally connected indexical and intentional content they figure in. We provide several examples of how this works out in a cross-modal clarification context in [64].

8.7 Conclusions

We started out with the idea that, when it comes to processing, context is the first and last that situated dialogue is about. It drives what we want to say, it drives how we want to say or understand something, it drives how we want to communicate. The way context comes into play, we hypothesize, is through bi-directionality. Processes can mutually influence each other by sharing information – about the content they are forming, and the decisions they are proposing to make.

Throughout this chapter, we have discussed how that works out. We started out with just talking about ... talking. How we propose the linguistic aspects of processing situated dialogue can be set up, and how information about what is salient in the current context (be that the situated context or the dialogue one) can act as predictor for how to recognize speech or how to interpret an utterance, and how it naturally influences how we want to refer to aspects of the environment.

We then continued by gradually investigating more and more aspects of bi-directionality in situated dialogue processing. We looked at dialogue about the current visual scene. We saw how relating linguistic content to that scene requires distinguishing the indexical aspects of *what* you are talking about, from the intentional aspects of *how* you mean what you are saying – against the background of "old" and "new" information, relative to a common ground already formed. And the other way round, feeding back into the language system, we saw that alternative interpretations could be pruned in post-filtering on the basis of whether they could be grounded in the situated context.

Next, we moved perspective from the scene in front of us to the larger environment around us – where not everything we want to talk about can be seen all at once. This required to establish connections with further models for situation awareness, and ontological reasoning to establish how aspects of the environment could be referred to. Bi-directionality made it possible to complement linguistic content with further information necessary to resolve what it is someone is referring to.

Finally, we put bi-directionality into action. We looked at how communicated content about actions could be enriched with event information, so that we could establish when, how and what was to be done. We established bi-directional links between planning, motivation and dialogue processing. This provided the possibility to create plans on the basis of what was being talked about – and, again, to use information about possible and impossible plans to weed out irrelevant linguistic interpretations. But, bi-directionality between action and dialogue can mean even more than that. We saw that, if we consider action and interaction to be a spectrum, rather than two isolated forms of acting, we can also consider bi-directionality at the level of decision processes.

The chapter contributes an approach to situated dialogue processing in human-robot interaction. The approach has been fully implemented and integrated in the systems described in 9 and 10. We have been able to show that the system achieves better performance on understanding situated dialogue in our domains by making use of information from dialogue- and situated contexts, than if it were not. At a deeper level, the chapter contributes insights in what appears to be required if a system is to process situated dialogue; more particularly, what requirements on processes and representations need to be addressed if information is indeed to be exchanged and interconnected. Crucial is a basis for establishing a common categorical and referential ground in connecting content across modalities, to deal with always-present ambiguities in efficient ways, and to provide means for persistence in maintaining these connections over time to deal with changes in content and context.

Ultimately, when it comes to situated dialogue, what we really understand about dialogue – is how we understand how we can experience.

References

1. Roy, D., Reiter, E.: Connecting language to the world. Artificial Intelligence 167(1-2), 1–12 (2005)
2. Winograd, T.: A process model of language understanding. In: Schank, R., Colby, K. (eds.) Computer Models of Thought and Language, pp. 152–186. Freeman, New York (1973)
3. Gorniak, P., Roy, D.: Grounded semantic composition for visual scenes. Journal of Artificial Intelligence Research 21, 429–470 (2004)
4. Gorniak, P., Roy, D.: Probabilistic grounding of situated speech using plan recognition and reference resolution. In: Proceedings of the Seventh International Conference on Multimodal Interfaces, ICMI 2005 (2005)
5. Gorniak, P., Roy, D.: Situated language understanding as filtering perceived affordances. Cognitive Science 31(2), 197–231 (2007)
6. Steels, L., Baillie, J.-C.: Shared grounding of event descriptions by autonomous robots. Robotics and Autonomous Systems 43(2-3), 163–173 (2003)
7. Steels, L.: Semiotic dynamics for embodied agents. IEEE Intelligent Systems 21(3), 32–38
8. Steels, L.: The symbol grounding problem has been solved. so what's next? In: De Vega, M., Glennberg, G., Graesser, G. (eds.) Symbols, embodiment and meaning. Academic Press, New Haven (2008)
9. Scheutz, M., Eberhard, K., Andronache, V.: A real-time robotic model of human reference resolution using visual constraints. Connection Science Journal 16(3), 145–167 (2004)
10. Brick, T., Scheutz, M.: Incremental natural language processing for HRI. In: Proceeding of the ACM/IEEE international conference on Human-Robot Interaction (HRI 2007), pp. 263–270 (2007)
11. Altmann, G., Steedman, M.: Interaction with context during human sentence processing. Cognition 30(3), 191–238 (1988)
12. Altmann, G., Kamide, Y.: Now you see it, now you don't: Mediating the mapping between language and the visual world. In: Henderson, J., Ferreira, F. (eds.) The Interface of Language, Vision, and Action: Eye Movements and The Visual World, pp. 347–386. Psychology Press, New York (2004)
13. Knoeferle, P., Crocker, M.: The coordinated interplay of scene, utterance, and world knowledge: evidence from eye tracking. Cognitive Science
14. Hadelich, K., Crocker, M.: Gaze alignment of interlocutors in conversational dialogues. In: Proc. 19th CUNY Conference on Human Sentence Processing, New York, USA (2006)
15. Steedman, M.: The Syntactic Process. The MIT Press, Cambridge (2000)
16. Baldridge, J., Kruijff, G.: Multi-modal combinatory categorial grammmar. In: Proceedings of EACL 2003, Budapest, Hungary (2003)
17. Blackburn, P.: Representation, reasoning, and relational structures: a hybrid logic manifesto. Logic Journal of the IGPL 8(3), 339–625 (2000)
18. Kruijff, G.: A categorial-modal logical architecture of informativity: Dependency grammar logic & information structure, Ph.D. thesis, Faculty of Mathematics and Physics, Charles University, Prague, Czech Republic (April 2001)
19. Areces, C.: Logic engineering. the case of description and hybrid logics, Ph.D. thesis, Institute for Logic, Language and Computation, University of Amsterdam, Amsterdam, The Netherlands (October 2000)

20. Baldridge, J., Kruijff, G.: Coupling CCG and hybrid logic dependency semantics. In: Proc. ACL 2002, Philadelphia, PA, pp. 319–326 (2002)
21. Kruijff, G., Kelleher, J., Hawes, N.: Information fusion for visual reference resolution in dynamic situated dialogue. In: André, E., Dybkjær, L., Minker, W., Neumann, H., Weber, M. (eds.) PIT 2006. LNCS (LNAI), vol. 4021, pp. 117–128. Springer, Heidelberg (2006), http://cognitivesystems.org/cosybook/chap8.asp#Kruijff/etal:2006-PIT
22. Jacobsson, H., Hawes, N., Kruijff, G., Wyatt, J.: Crossmodal content binding in information-processing architectures. In: Proceedings of the 3rd ACM/IEEE International Conference on Human-Robot Interaction (HRI), Amsterdam, The Netherlands (2008), http://cognitivesystems.org/cosybook/chap8.asp#jacobsson+hawes+kruijff+wyatt_2008
23. Kruijff, G., Brenner, M.: Modelling spatio-temporal comprehension in situated human-robot dialogue as reasoning about intentions and plans. In: Proceedings of the Symposium on Intentions in Intelligent Systems, Stanford University, Palo Alto, CA. AAAI Spring Symposium Series (2007), http://cognitivesystems.org/cosybook/chap8.asp#Kruijff/Brenner:2007
24. Asher, N., Lascarides, A.: Logics of Conversation. Cambridge University Press, Cambridge (2003)
25. Moens, M., Steedman, M.: Temporal ontology and temporal reference. Journal of Computational Linguistics 14, 15–28 (1988)
26. Kelleher, J., Kruijff, G., Costello, F.: Proximity in context: an empirically grounded computational model of proximity for processing topological spatial expressions. In: Proceedings of ACL/COLING 2006 (2006), http://cognitivesystems.org/cosybook/chap8.asp#Kelleher/etal:2006
27. Zender, H., Kruijff, G.: Towards generating referring expressions in a mobile robot scenario. In: Language and Robots: Proceedings of the Symposium, Aveiro, Portugal, pp. 101–106 (2007), http://cognitivesystems.org/cosybook/chap8.asp#zender/kruijff:2007-gre
28. Kruijff, G., Kelleher, J., Berginc, G., Leonardis, A.: Structural descriptions in human-assisted robot visual learning. In: Proc. 1st Annual Conference on Human-Robot Interaction, HRI 2006 (2006), http://cognitivesystems.org/cosybook/chap8.asp#Kruijff/etal:2006-vision
29. Bloom, P.: How children learn the meanings of words. The MIT Press, Cambridge (2000)
30. Stone, M.: Intention, interpretation and the computational structure of language. Cognitive Science 28(5), 781–809 (2004)
31. Van Berkum, J., Zwitserlood, P., Brown, C., Hagoort, P.: When and how do listeners relate a sentence to the wider discourse? evidence from the n400 effect. Cognitive Brain Research 17, 701–718 (2003)
32. Crain, S., Steedman, M.: On not being led up the garden path: The use of context by the psychological syntax processor. In: Dowty, D.R., Karttunen, L., Zwicky, A.M. (eds.) Natural language parsing: Psychological, computational, and theoretical perspectives. Cambridge University Press, Cambridge (1985)
33. Tanenhaus, M., Spivey-Knowlton, M., Eberhard, K., Sedivy, J.: Integration of visual and linguistic information in spoken language comprehension. Science 268, 1632–1634 (1995)
34. Liversedge, S., Findlay, J.: Saccadic eye movements and cognition. Trends in Cognitive Science 4(1), 6–14 (2000)

35. Van Berkum, J.: Sentence comprehension in a wider discourse: Can we use erps to keep track of things? In: Carreiras, M., Clifton Jr., C. (eds.) The on-line study of sentence comprehension: Eyetracking, ERPs and beyond, pp. 229–270. Psychology Press, New York (2004)
36. Allopenna, P., Magnuson, J., Tanenhaus, M.: Tracking the time course of spoken word recognition using eye movements: Evidence for continuous mapping models. Journal of Memory and Language 38(4), 419–439 (1998)
37. van Berkum, J., Hagoort, P., Brown, C.: Semantic integration in sentences and discourse: Evidence from the n400. Journal of Cognitive Neuroscience 11(6), 657–671 (1999)
38. Van Petten, C., Coulson, S., Rubin, S., Plante, E., Parks, M.: Time course of word identification and semantic integration in spoken language. Journal of Experimental Psychology: Learning, Memory, and Cognition 25(2), 394–417 (1999)
39. Altmann, G.M.: Ambiguity in sentence processing. Trends in Cognitive Sciences 2(4)
40. Spivey, M., Trueswell, J., Tanenhaus, M.: Context effects in syntactic ambiguity resolution: discourse and semantic influences in parsing reduced relative clauses. Canadian Journal of Experimental Psychology 47(2), 276–309 (1993)
41. Spivey, M., Tanenhaus, M.: Syntactic ambiguity resolution in discourse: Modeling the effects of referential context and lexical frequency. Journal of Experimental Psychology: Learning, Memory, and Cognition 24, 1521–1543 (1998)
42. van Berkum, J., Brown, C., Hagoort, P.: Early referential context effects in sentence processing: Evidence from event-related brain potentials. Journal of Memory and Language 41, 147–182 (1999)
43. Tanenhaus, M., Magnuson, J., Dahan, D., Chambers, G.: Eye movements and lexical access in spoken-language comprehension: Evaluating a linking hypothesis between fixations and linguistic processing. Journal of Psycholinguistic Research 29(6), 557–580 (2000)
44. Dahan, D., Tanenhaus, M.: Continuous mapping from sound to meaning in spoke-language comprehension: Immediate effects of verb-based thematic constraints. Journal of Experimental Psychology: Learning, Memory, and Cognition 30(2), 498–513 (2004)
45. Van Berkum, J., Brown, C., Zwitserlood, P., Kooijman, V., Hagoort, P.: Anticipating upcoming words in discourse: Evidence from erps and reading times. Journal of Experimental Psychology: Learning, Memory, & Cognition 31(3), 443–467 (2005)
46. Nieuwland, M., Van Berkum, J.: When peanuts fall in love: N400 evidence for the power of discourse. Journal of Cognitive Neuroscience 18(7), 1098–1111 (2006)
47. Botvinick, M., Braver, T., Barch, D., Carter, C., Cohen, J.: Conflict monitoring and cognitive control. Psychological Review 108(3), 624–652 (2001)
48. Hommel, B., Ridderinkhof, K., Theeuwes, J.: Cognitive control of attention and action: Issues and trends. Psychological Research 66, 215–219 (2002)
49. Novick, J., Trueswell, J., Thompson-Schill, S.: Cognitive control and parsing: Reexamining the role of Broca's area in sentence comprehension. Cognitive, Affective, and Behavioral Neuroscience 5(3), 263–281 (2005)
50. Altmann, G., Kamide, Y.: Incremental interpretation at verbs: Restricting the domain of subsequent reference. Cognition 73(3), 247–264 (1999)
51. Chambers, C., Tanenhaus, M., Magnuson, J.: Actions and affordances in syntactic ambiguity resolution. Jnl. Experimental Psychology 30(3), 687–696 (2004)

52. Endsley, M.: Theoretical underpinnings of situation awareness: A critical review. In: Endsley, M.R., Garland, D.J. (eds.) Situation awareness analysis and measurement. Lawrence Erlbaum, Mahwah (2000)
53. Kamide, Y., Altmann, G., Haywood, S.: The time-course of prediction in incremental sentence processing: Evidence from anticipatory eye-movements. Jnl. Memory and Language 49(1), 133–156 (2003)
54. Glenberg, A., Kaschak, M.: Grounding language in action. Psychonomic Bulletin & Review 9(3), 558–565 (2002)
55. De Vega, M., Robertson, D., Glenberg, A., Kaschak, M., Rinck, M.: On doing two things at once: Temporal constraints on actions in language comprehension. Memory and Cognition 32(7), 1033–1043 (2004)
56. Glenberg, A.: What memory is for. Behavioral & Brain Sciences 20, 1–55 (1997)
57. Barsalou, L.: Perceptual symbol systems. Behavioral & Brain Sciences 22, 577–660 (1999)
58. Pickering, M., Garrod, S.: Toward a mechanistic psychology of dialogue. Behavioral and Brain Sciences 27, 169–225 (2004)
59. Oepen, S., Carroll, J.: Ambiguity packing in constraint-based parsing: Practical results. In: Proceedings of the 6th Applied Natural Language Processing Conference (ANLP 2000), pp. 162–169 (2000)
60. Carroll, J., Oepen, S.: High efficiency realization for a wide-coverage unification grammar. In: Proceedings of the International Joint Conference on Natural Language Processing (IJCNLP 2005), pp. 165–176 (2005)
61. Moore, R.K.: Spoken language processing: piecing together the puzzle. Speech Communication: Special Issue on Bridging the Gap Between Human and Automatic Speech Processing 49, 418–435 (2007)
62. Lison, P., Kruijff, G.: Salience-driven contextual priming of speech recognition for human-robot interaction. In: Proceedings of ECAI 2008, Athens, Greece (2008), `http://cognitivesystems.org/cosybook/chap8.asp#Lison/Kruijff:2008`
63. Collins, M.: Parameter estimation for statistical parsing models: theory and practice of distribution-free methods. In: New developments in parsing technology, pp. 19–55. Kluwer Academic Publishers, Dordrecht (2004)
64. Kruijff, G., Brenner, M., Hawes, N.: Continual planning for cross-modal situated clarification in human-robot interaction. In: Proceedings of the 17th International Symposium on Robot and Human Interactive Communication (RO-MAN 2008), Munich, Germany (2008), `http://cognitivesystems.org/cosybook/chap8.asp#Kruijff/etal:2008`
65. Kruijff, G.: Context-sensitive utterance planning for CCG. In: Proceedings of the European Workshop on Natural Language Generation, Aberdeen, Scotland (2005), `http://cognitivesystems.org/cosybook/chap8.asp#Kruijff:2005`
66. Bateman, J.: Enabling technology for multilingual natural language generation: the kpml development environment. Journal of Natural Language Engineering 3(1), 15–55 (1997)
67. Dale, R., Reiter, E.: Computational interpretations of the gricean maxims in the generation of referring expressions. Cognitive Science 19(2), 233–263 (1995)
68. White, M., Baldridge, J.: Adapting chart realization to CCG. In: Proceedings of the Ninth European Workshop on Natural Language Generation, Budapest, Hungary (2003)
69. White, M.: Efficient realization of coordinate structures in combinatory categorial grammar. Research on Language and Computation 4(1), 39–75 (2006)

70. Schröder, M., Trouvain, J.: The german text-to-speech synthesis system mary: A tool for research, development and teaching. International Journal of Speech Technology 6, 365–377 (2003)
71. Steedman, M., Kruijff-Korbayová, I.: Discourse and information structure. Journal of Logic, Language and Information 12, 249–259 (2003)
72. Kelleher, J., Kruijff, G.: Incremental generation of spatial referring expressions in situated dialog. In: Proceedings of the 21st International Conference on Computational Linguistics and 44th Annual Meeting of the Association for Computational Linguistics, pp. 1041–1048 (2006), http://cognitivesystems.org/cosybook/chap8.asp#Kelleher/Kruijff:2006
73. Kuipers, B.: Representing knowledge of large-scale space, Ph.D. thesis, Massachusetts Institute of Technology, Cambridge, MA, USA (1977)
74. Kruijff, G., Zender, H., Jensfelt, P., Christensen, H.: Situated dialogue and spatial organization: What, where.. and why? International Journal of Advanced Robotic Systems 4(2), http://cognitivesystems.org/cosybook/chap8.asp#Kruijff/etal:2007-JARS
75. Zender, H., Jensfelt, P., Mozos, O.M., Kruijff, G., Burgard, W.: An integrated robotic system for spatial understanding and situated interaction in indoor environments. In: Proc. of AAAI 2007, Vancouver, BC, Canada, pp. 1584–1589 (2007), http://cognitivesystems.org/cosybook/chap8.asp#Zender/etal:2007-AAAI
76. Zender, H., Jensfelt, P., Mozos, O.M., Kruijff, G., Burgard, W.: Conceptual spatial representations for indoor mobile robots. Robotics and Autonomous Systems, special Issue From Sensors to Human Spatial Concepts 56(6), http://cognitivesystems.org/cosybook/chap8.asp#Zender/etal:2008
77. Topp, E.A., Hüttenrauch, H., Christensen, H., Severinson Eklundh, K.: Bringing together human and robotic environment representations – a pilot study. In: Proc. of the IEEE/RSJ International Conference on Intelligent Robots and Systems (IROS), Beijing, China (2006)
78. Shi, H., Tenbrink, T.: Telling rolland where to go: Hri dialogues on route navigation. In: Proceedings of the Workshop on Spatial Language and Dialogue (5th Workshop on Language and Space), Delmenhorst, Germany (2005)
79. Brown, R.: How shall a thing be called? Psychological Review 65(1), 14–21 (1958)
80. Rosch, E.: Principles of categorization. In: Rosch, E., Lloyd, B. (eds.) Cognition and Categorization, pp. 27–48. Lawrence Erlbaum Associates, Hillsdale (1978)
81. Brenner, M., Hawes, N., Kelleher, J., Wyatt, J.: Mediating between qualitative and quantitative representations for task-orientated human-robot interaction. In: Proceedings of the Twentieth International Joint Conference on Artificial Intelligence, IJCAI 2007 (2007), http://cognitivesystems.org/cosybook/chap8.asp#Brenner/etal:2007
82. Sidner, C.L., Lee, C., Kidd, C., Lesh, N., Rich, C.: Explorations in engagement for humans and robots. Artificial Intelligence 166(1-2), 140–164 (2005)

A Packing Algorithm

A *packing* mechanism [59, 60] is used by the incremental parser to efficiently represent and manipulate logical forms accross the communication subarchitecture. A packed logical form [PLF] represents content similar across the different analyses of a given input as a single graph, using over- and under-specification of how different nodes can be connected to capture lexical and syntactic forms of ambiguity.

After each incremental step, the resulting set of logical forms is compacted into a single representation, which can then be directly manipulated by various processes, in order, for example, to prune unsupported interpretations. It can also be *unpacked*, ie. the original logical forms can be completely regenerated (this is done by traversing the packed structure).

The packed representations are made of two basic elements: *packing nodes* and *packing edges*. A packing node groups a set of nominals sharing identical properties and named relations under a particular subset of the logical forms. Packing edges are responsible for connecting the different packing nodes together, thus ensuring the correspondence between the packed structure and the set of logical forms it represents.

The packing of logical forms is performed in two main steps:

1. An initial PLF is first constructed on the basis of the set of logical forms (*Step 1* of algorithm 6). To this end, each logical form is traversed and its nominals are used to populate the packed structure.
2. The resulting structure is then compacted by merging particular substructures (*Step 2* of algorithm 6).

A.1 Example

The Figures 8.13-8.15 below exemplify a simple case of packing operation. The parsed utterance is "Take the ball to the left of the box". Two distinct readings can be derived, depending on the interpretation of the phrase "to the left of the box". In the first reading (LF_1 in the figure 8.13), the robot is asked to take the ball and put it to the left of the box - the phrase is thus seen as indicating the *direction* of the move. In the second reading (LF_2) however, "to the left of the box" indicates the *location* of the ball to take.

Figure 8.14 illustrates the application of the first step of the packing operation. A packing node - drawn in the figure as a square - is created for each nominal. A packing edge is constituted for each relation found in the logical forms. As shown in the figure, some packing edges are shared by both logical forms, whereas others are only evidenced in one of them. An example of the first case is the edge between "take" and "robot", which shared by the two logical forms LF_1 and LF_2. The edge between "take" and "left" illustrates the second case: it is only evidenced in LF_1.

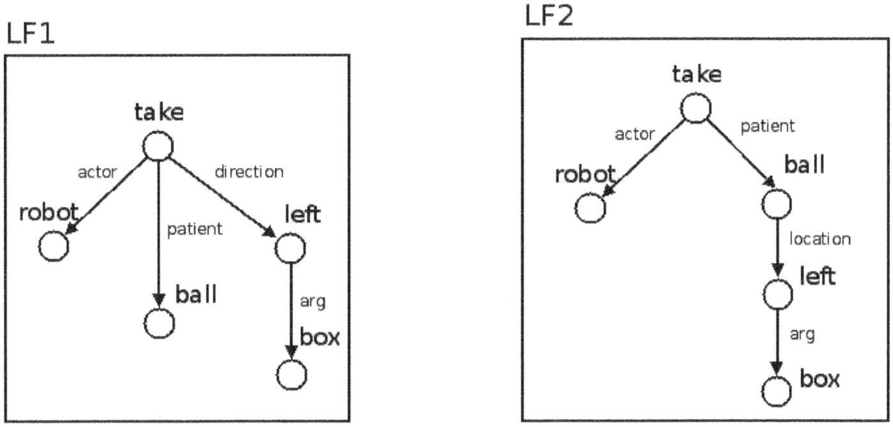

Fig. 8.13. The two initial logical forms LF_1 and LF_2 retrieved from parsing the utterance "Take the ball to the left of the box"

In the example we present here, all packing edges have only one packing node target. In the general case however, several distinct targets can be specified within the same edge.

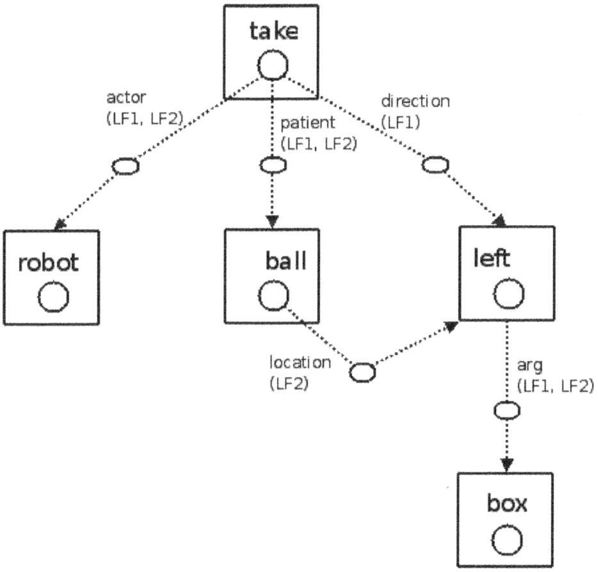

Fig. 8.14. The resulting packed logical form, before compacting

During the second step, the packed structure is compacted by merging packing nodes. The criteria to decide whether two packing nodes can be merged is the following: if (1) two packing nodes are connected by a packing edge, and if (2) the logical form identifiers for the head node, the edge and the target node are all identical, then the two packing nodes can be merged. For example, the packing node surrounding "take" and the one surrounding "robot" can be merged, since the two nodes and the edge between them are present both in LF_1 and LF_2.

The compacting operation is repeated until no more merges are possible. In our case, illustrated in the figure 8.15, we are left with two packing nodes, one rooted on the nominal "take", and one on "left".

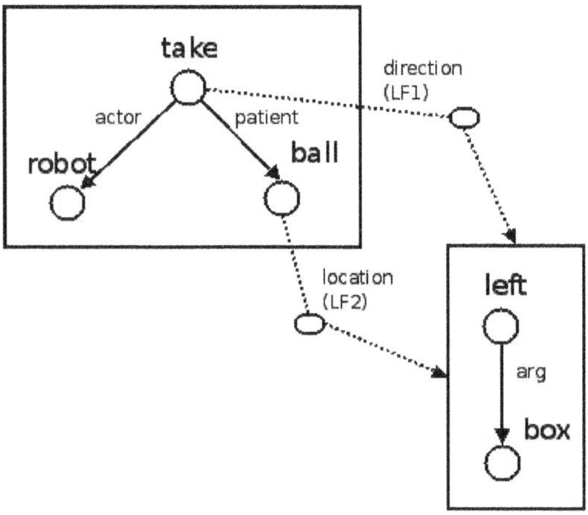

Fig. 8.15. The final packed logical form, after compacting

A.2 Data Structures

We present below the informal specifications of the various data structures used to construct PLFs. See figure 8.17 for a graphical representation.

PackedLogicalForm:

- id: packed logical form identifier
- packingNodes: set of packing nodes
- root: root packing node

PackingNode:

- id: packing node identifier
- packedNominals: set of packed nominals
- lfIds: set of LF identifiers, enumerating the logical forms in which the nominals included in the packing node are present
- root: root nominal

PackedNominal:

- id: packed nominal identifier
- sort: ontological sort
- prop: logical proposition
- features: set of packed features
- relations: set of internal relations
- packingEdges: set of outgoing packing edges

PackedFeature:

- feature: name of the feature
- value: value of the feature
- lfIds: set of the LF identifiers, enumerating the logical forms in which the feature holds

PackingEdge:

- id: packing edge identifier
- head: head nominal
- mode: edge label
- packingNodeTargets: set of packing node targets

PackingNodeTarget:

- lfIds: set of LF identifiers, enumerating the logical forms in which the edge exists
- target: packing node targeted by the edge

Fig. 8.16. Data structures used to construct PLFs

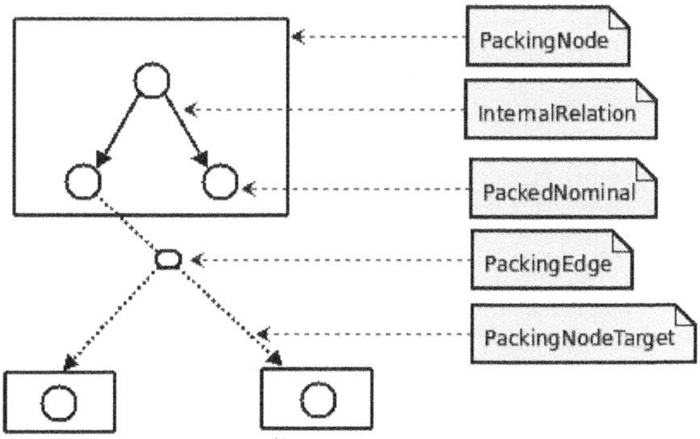

Fig. 8.17. Graphical representation of the data structures

A.3 Pseudo-code

We finally describe the details of the algorithms used in the packing mechanism we implemented.

Algorithm 6. : `Pack(LFs)` - Packing of a set of logical forms

Require: LFs is a set of logical forms (describing the same utterance)

```
% Step 0: Initialization
rootNominal ← ⟨ rootSort, 'root', ∅, ∅, ∅⟩
rootNode ← ⟨ {rootNominal}, ∅, rootNominal ⟩
packingNodes ← {rootNode}
PLF ← ⟨ packingNodes, rootNode ⟩

% Step 1: Construction of the packed logical form
for lf ∈ LFs do
   AddLFInformation(lf, PLF)
end for

% Step 2: Merge of the packed logical form
PLF = MergePackedLogicalForm(PLF)

return PLF
```

Algorithm 7. : `CreateNewNode(nom)` - using the information in `nom`, create (1) a new packing node, (2) a new packed nominal inside it and (3) new packing edges connected to the latter.

Require: A well-formed nominal nom

```
newEdges ← ∅
for every relation rel in rels(nom) do
   % A packing edge is defined with a head nominal, a mode ("edge label"), a set of packing
   node targets, and a set of logical form identifiers
   newEdge ← ⟨ head(rel), mode(rel), {target(rel)}, {lfId(nom)}⟩,
   newEdges ← newPackingEdges ∪ {newEdge}
end for

% A packing nominal comprises an ontological sort, a logical proposition, a set of features,
a set of internal relations, and a set of outgoing packing edges
newNom ← ⟨ sort(nom), prop(nom), feats(nom), ∅, newEdges ⟩

% A packing node is a triple comprising a set of packing nominals, a set of LF identifiers,
and a reference to the root nominal
newPackingNode ← ⟨{newNom},{lfId(nom)}, newNom⟩

return newPackingNode
```

Algorithm 8. : `AddLFInformation(lf, PLF)` - Add the information contained in `lf` to the packed logical form.

```
Require: lf is a well-formed logical form

  for every nominal nom in nominals(lf) do

    if there is no packing node in PLF which encapsulates a packed nominal with the ontological
    sort sort(nom) and the logical proposition prop(nom) then

      % We create a new packing node and its related substructures
      newPackingNode ← CreateNewPackingNode(nom)

      % We add the packing node to the PLF structure
      packingNodes(PLF) ← packingNodes(PLF) ∪ {newPackingNode}

    else
      % We update the existing nominal and its dependent edges
      let pNom = the packed nominal with sort(nom) and prop(nom)
      let pNode = the packing node encapsulating pNom

      pNode ← IntegrateNominalToPackingNode(nom, pNode)
    end if

    if nom is the root nominal in lf then
      % We establish a connection between the root node and the current one

      let packingNode = the packing node which encapsulates nom in PLF

      Add a packing edge between root(PLF) and packingNode
      lfIds(root(PLF)) = lfIds(root(PLF)) ∪ {id(lf)}
    end if

  end for

  return PLF
```

Algorithm 9. : `IntegrateNominalToPackingNode(nom, pNode)` - integrate the information contained in `nom` to the existing packing node `pNode`

```
Require: A well-formed nominal nom
Require: A well formed packing node pNode which already encapsulates a nominal with the same
   ontological sort and logical proposition as nom

  let pNom = the nominal encapsulated in pNode

  for every relation rel in rels(nom) do
     if ∃ edge ∈ edges(pNom) where mode(rel) = mode(edge) then
        % If there is already a packing edge with same mode, add one packing node target and
        the LF identifier
        targets(edge)←targets(edge) ∪ {target(rel)}
        lfIds(edge) ← lfIds(edge) ∪ {lfId(nom)}
     else
        % Else, we create a new packing edge
        newEdge ← ⟨ head(rel), mode(rel), {target(rel)}, {lfId(nom)}⟩
        edges(pNom) ← edges(pNom) ∪ {newEdge}
     end if
  end for

  % Update the features in the nominal, and the LF identifiers in the packing node
  feats(pNom) ← feats(pNom) ∪ {feats(nom)}
  lfIds(pNode) ← lfIds(pNode) ∪ {lfId(nom)}

  return pNode
```

Algorithm 10. : `MergePackedLogicalForm(PLF)` - compact the PLF representation by merging nominals

```
Require: PLF a well formed packed logical form

  while there are packing nodes in PLF which can be merged do
     for every packing node packingNode ∈ PLF do
        for every nominal nom ∈ nominals(packingNode) do
           for every edge edge ∈ edges(nom) do
              if edge has only one packing node target then

                 let LFS_head = set of logical forms identifiers in packingNode
                 let LFS_edge = set of logical forms identifiers in edge
                 let LFS_target = set of logical forms identifiers in target(edge)

                 if LFS_head = LFS_edge = LFS_target then
                    % If the set of logical forms shared by the two packing nodes (and the
                    packing edge between them) is identical, then they can be merged in one
                    packing node

                    let targetNom = the head nominal of target(edge)

                    Merge packingNode and targetNom into a single packing node

                    Transform edge into an internal relation (in the merged packing node) be-
                    tween nom and targetNom

                 end if
              end if
           end for
        end for
     end for
  end while
  return PLF
```

Part III

Integration and Systems

9

The PlayMate System

Nick Hawes[1], Jeremy L. Wyatt[1], Mohan Sridharan[1], Marek Kopicki[1], Somboon Hongeng[1], Ian Calvert[1], Aaron Sloman[1], Geert-Jan Kruijff[2], Henrik Jacobsson[2], Michael Brenner[3], Danijel Skočaj[4], Alen Vrečko[4], Nikodem Majer[5], and Michael Zillich[6]

[1] Intelligent Robotics Laboratory, School of Computer Science, University of Birmingham, Birmingham, UK
{nah,jlw,mzs,msk,sxh,ijc550,axs}@cs.bham.ac.uk
[2] DFKI GmbH, Saarbrücken, Germany
{henrikj,gj}@dfki.de
[3] Albert-Ludwigs-Universität Freiburg, Department of Computer Science, Freiburg, Germany
brenner@informatik.uni-freiburg.de
[4] VICOS Lab, University of Ljubljana, Slovenia
{danijel.skocaj,alen.vrecko}@fri.uni-lj.si
[5] Technische Universität Darmstadt, Darmstadt, Germany
nmajer@mis.informatik.tu-darmstadt.de
[6] Automation and Control Institute, Vienna University of Technology, Austria
zillich@acin.tuwien.ac.at

9.1 Introduction

Research in CoSy was scenario driven. Two scenarios were created, the PlayMate and the Explorer. One of the integration goals of the project was to build integrated systems that addressed the tasks in these two scenarios. This chapter concerns the integrated system for the PlayMate scenario.

The work described here on the PlayMate scenario is concerned with understanding, at a systems level, the problems that an intelligent system must face if it must interact with humans in an object rich environment. In particular the goal is to understand how a robot can interact with a human in a space in which they both manipulate objects, and in which they can talk about those objects. This requires many abilities. The robot must be able to understand and generate references to objects, actions with them, their properties and spatial relationships. It must understand how actions alter the relationships between objects, be able to recognise actions the human performs with objects, and it must be able to learn about the effects of actions on objects, both by discovery and by watching others. If there are several opportunities for action it must choose between a number of potential goals at any one time.

H.I. Christensen et al. (Eds.): Cognitive Systems, COSMOS 8, pp. 367–393.
springerlink.com

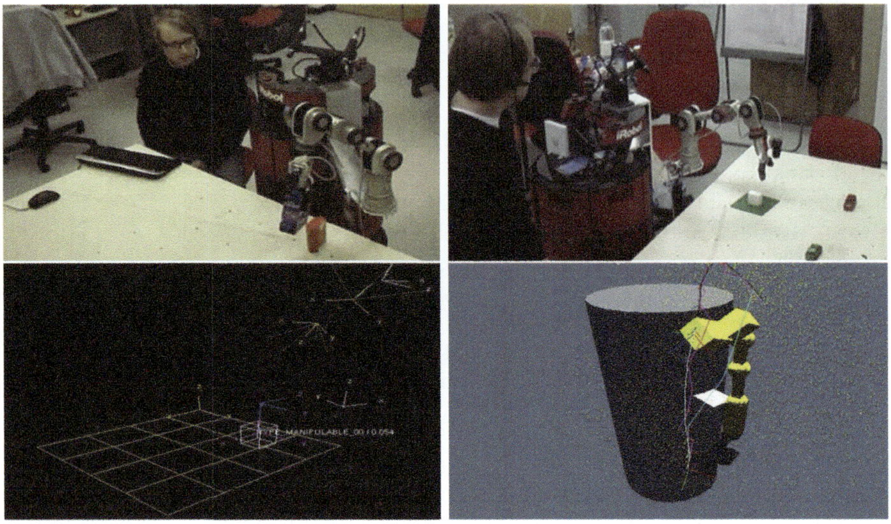

Fig. 9.1. The PlayMate robot with two of the object sets that it uses. The lower panels show the robot's visualisation of its environment, and the manipulation planner.

Finally, since it is working with a human, the human may change the world while the robot is acting. Thus the robot must meet many of the requirements that we outlined in Chapter 2, and utilise many of the technologies that were described in the chapters leading up to this one. This chapter describes how these technologies were integrated to solve some of the tasks that exist in the PlayMate scenario. It describes both the complete PlayMate system, and the major innovations that cut across the PlayMate and the Explorer. It is important to note that the system described here is the last of a series of systems [1, 2, 17, 18]. Each of these systems was been used as an experimental platform, and the lessons learned at each stage were used to revise the cognitive architecture (see Chapter 2), and the systems level engineering approach.

To help us in our exposition we will use a single example that runs through the chapter and shows the various problems that we encounter in building a robot with multiple modes of sensing and action. The script for this example is given in Figure 9.2. The PlayMate robot is stationary, but positioned next to a table which holds a number of objects. It has two head mounted cameras, and an arm with a parallel-jaw gripper and a camera on the wrist. The objects are everyday cardboard packages such as small cereal boxes, or brightly coloured shapes (triangles, squares, circles) with differently coloured handles as seen in Figure 9.1.

The ability to handle this script requires that the robot be able to solve a number of problems. First, *it must be able to understand not just the content, but the role of an utterance in the interaction*, when the human says `"This`

```
Human (H) puts a blue square down on the table
H: "This is a blue square."
Robot (R): "OK."
H puts down a red square to the right of the blue square.
H: "This is a red square."
R: "OK."
H picks up the red square and puts down a red triangle to the right
of the blue square.
R: "What is the thing to the right of the blue square?"
H: "It is a red triangle."
R: "Ok."
H picks up the red triangle and the blue square and puts down a
blue triangle.
H: "What colour is this triangle?"
R: "It is blue."
H picks up the blue triangle and puts down a green triangle on the
right and a red circle on the left.
H: "What shape is this thing?"
R: "Do you mean the green thing or the red thing?"
H: "The red thing."
R: "It is a circle."
H: "Put the circle to the right of the triangle."
R picks up the circle and puts it to the right of the triangle.
H puts a red triangle on the table.
H pushes the red triangle across the table to the red circle.
H: "What is the game?"
R: "We are playing the colour game."
```

Fig. 9.2. An example script of an interaction between the PlayMate and a human

is a blue square." The robot must have a way of knowing that it should learn something. Second, *the robot must be able to bind information from one modality to that from another*, so that when the red square appears it understands which visual object is being referred to when the human refers to "a red square". When it is learning it also needs to identify which object it is learning about. *It must also have ways of finding information that is required by one modality (e.g. language) from another (e.g. vision).* So that if the human asks a question — "What is the thing to the right of the blue square?" — it can produce the answer. Alternatively, since the robot is required to learn about the relationships between two modalities (vision and language) it must also have the ability to learn associations between aspects of the two, i.e. to perform multi-modal or cross-modal learning. If the information available to the robot is insufficient for it to be able to perform its task it must acquire more. In other words *the robot must plan clarification processing to resolve ambiguities.* An example of this is when the human asks a question but the referent of the question is unclear — "What shape is this `thing?"` — so that the robot has to ask which thing is being referred to. Also, *if the robot is unsure about something it should be able to raise its own questions, so that it can learn.* An example of this is when the robot realises that it doesn't know the red triangle, it is motivated to ask its own question. In understanding the references to objects the robot has to have and use a mapping between quantitative and qualitative notions of spatial arrangement. When the human tells the robot to put the circle to the right of the triangle the robot has to map from a qualitative notion to a metric one (it has to put the circle in a particular place). Whereas in order to be able to make the utterance `"What is the thing to the right of the blue square?"` it must perform the mapping in the other direction, from metric to qualitative.

However, looking at this script, and comparing it to the script for a system such as SHRDLU [3] or Shakey [4] from the 1970s it would seem to be little different. It is however, important to note that SHRDLU was not embodied. Systems such as Shakey at the time were embodied, but used a similar interaction model to SHRDLU which abstracted away most of the difficult parts of the interaction model — although they did this for very good reasons. Figure 9.3 ilustrates the difference between the interaction model that underpinned Shakey and SHRDLU and the one that must necessarily be used to address some problems raised by the script above.

The upper panel of Figure 9.3 represents events and actions, including speech acts, as having duration, and the world as being subject to periods of continuous change punctuated by periods of stasis. This model is referred to here as the Multiple Strand Continuous Change (MSCC) model of interaction. It also models sensing and sensory processing of the robot as on-going during actions of the human or the robot. In the lower panel is a characterisation of the interaction model that underpinned both Shakey and SHRDLU. In this second interaction model — which will be referred to as the Single Strand Discrete Change (SSDC) model — actions have no temporal duration, as sensing is performed only on a static world, and utterances are received as complete text strings. The second interaction model makes a small number of key assumptions which do not hold in general, but which must be addressed in both the script for the PlayMate, and in intelligent robotics more generally. Before we commence the dicussion, note the difference between what is meant here by an *interaction model* and what we call a *representational model.* A interaction model of a system is here a model of how the system interacts with its environment. A representational model is a model that the system may have internally. When we refer here to the internal state, or the internal or world model of the robot, we mean a representational model.

The assumptions of the SSDC interaction model are as follows:

- Events in the world have no duration.
- Events are modelled by analysing the static world states that precede and follow them.
- There is a single processing strand.

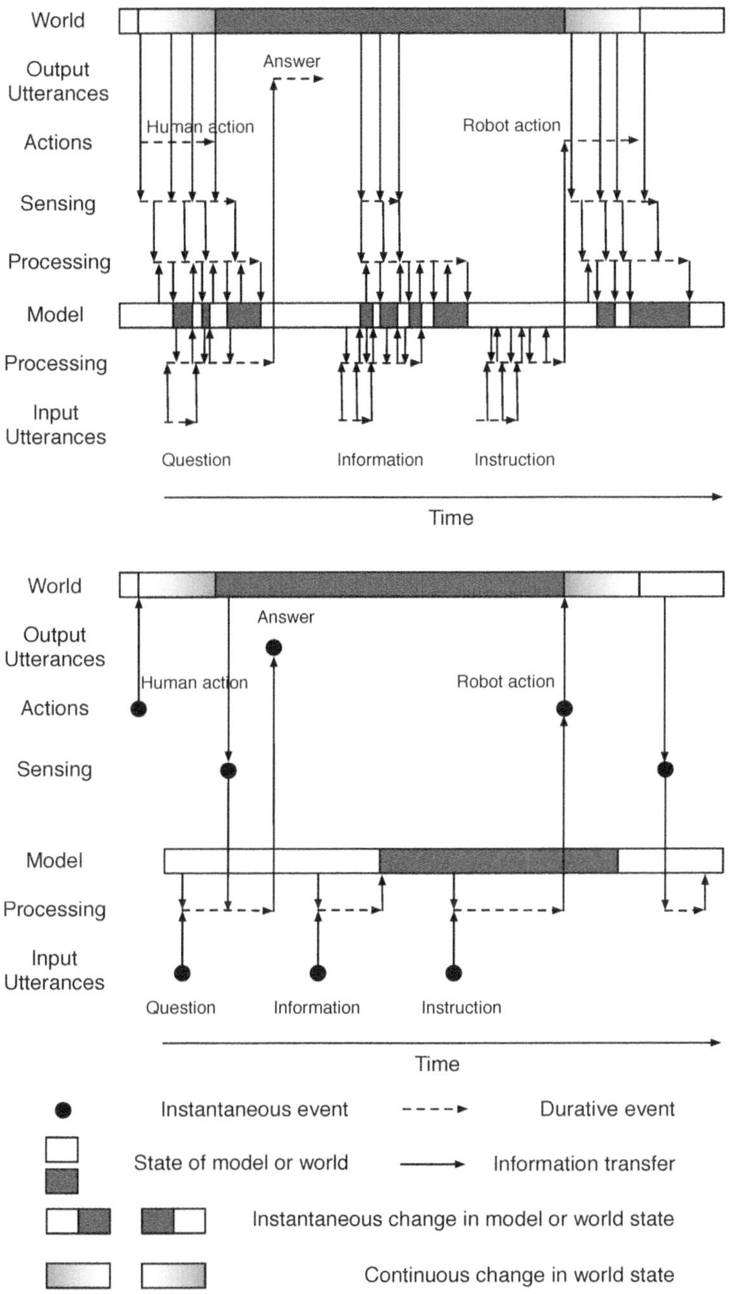

Fig. 9.3. Two interaction models of change and processing. In the upper panel is the interaction model that our PlayMate task requires, and below is the interaction model that underpins the approaches of Shakey and SHRDLU.

- There are no parallel events.
- Perception can be performed reliably enough without cross-modal inference.

The first of the assumptions in the SSDC interaction model is that *the temporal duration of an event is essentially irrelevant.* Events cause transitions between states. This is familiar in AI from the viewpoint of a representational model: the robot's internal model of the external world can take one of a number of discrete states, and when the world is sensed the internal model will transition from one state to another. In addition, the internal model represents the robot's and human's actions as a function modelling these state transitions, which it can use for planning or reasoning. The SSDC interaction model assumes these discrete states and durationless transitions when designing the system's processing. This assumption breaks down if the robot has to process the durative and continuous change in the world induced by an action of the human or robot. This breakdown is caused by the fact that the SSDC interaction model doesn't attempt to process the sensory changes induced by actions until after they have ended. This fits with the second assumption, that *events are modelled by analysing the static world states that precede and follow them.* This means that the robot can't handle the processing of continuous change in our script as when the human is pushing an object through the scene. An additional drawback of this approach in Shakey was that the world had to be allowed to "settle" before sensing was performed. This model does not work where continuous feedback is required, or where dynamic control becomes important. This is the case when the robot is moving and tracking objects.

The classic way that the SSDC model is extended to environments with change is to ensure that the rate of sensing and processing in the robot is much higher than that in the world. If this holds it permits the third assumption of the SSDC model: *there is a single strand of processing.* This model of processing is realistic only where the robot has a single pipeline of incoming information, i.e. *there are no parallel events.* This is not necessarily the case in our script. It could be that the person starts to speak while they are moving an object. In addition, for the single strand processing model to work, the processing to be performed must relatively simple. Analysing a changing visual scene, understanding the speech sginal, and analysing the meaning of the utterance at the same time is not simple. Processing information in a single strand means that the interaction can only proceed in steps: there must be total ordering of the events in the SSDC model, which will not happen typically when executing our script in a natural manner. The consequence of this is that to handle multiple concurrent, durative events, and respond in reasonable time, processing must be done in parallel. In other words there must be Multiple Strands.

A further assumption of Shakey was that perception was a process of translating information from sensory information into a single representation. In the single strand interaction model there is no opportunity for information from one kind of sense to assist the interpretation of another sense. Thus it is implicitly assumed that *perception can be performed reliably enough without cross modal inference.* In fact, because correct interpretation of sensory information is so hard to do reliably, it is wise to integrate information from many sources as a way of increasing reliability. As soon as this is accepted we need a way of ensuring that these strands interact correctly. Under a continuous change model it is not possible to guarantee the time that any particular piece of sensory processing will take. This means in turn that if processing strands need to swap information then the time it takes to process information in one strand relative to another can matter.

In summary, while the script appears at first sight to be quite similar to those handled by SHRDLU and Shakey, when executed in a natural way by a human, the assumptions of the interaction model that underpinned those systems, and many subsequent ones (including models inspired by human cognition such as [5]), break down. To tackle the PlayMate script in a natural way, is in essence to tackle a different script that is more appropriately modelled by the MSCC model that underpins CAS and the PlayMate and Explorer designs. The script detailed above is one of a large number that the PlayMate can handle, but it illustrates many of the key issues. In the next section (Section 9.2) we will give a sketch of the components and shared representations that the PlayMate employs. In that section we focus on two of the generic sub-systems that are employed in all activities: the binding system, and the system for motivation and planning. This system sketch will provide a framework for understanding the content of Section 9.3 where we describe the cross-cutting issues and the innovative solutions employed in the PlayMate. These issues include those raised in the discussion of the script above, but we highlight our approach to cross-modal learning, incremental processing, disambiguation and clarification, and mediation between different levels of abstraction.

9.2 System Overview

In the previous section we outlined the kinds of tasks the PlayMate robot can be engaged in, and some of the requirements arising from those. In Chapter 2 we also outlined the requirements that arise generically on the architecture for both the PlayMate and Explorer scenarios. In this section we describe a specific system, built using CAST [19], which makes much stronger architectural commitments than CAS itself. We also give an overview of the system structure as it stands at the end of the project. The PlayMate and Explorer share much of the same structure as one another, and we refer to the architectural instantiation that is common across the PlayMate and the Explorer as PECAS (the PlayMate/Explorer CoSy Architecture Sub-Schema). As part of

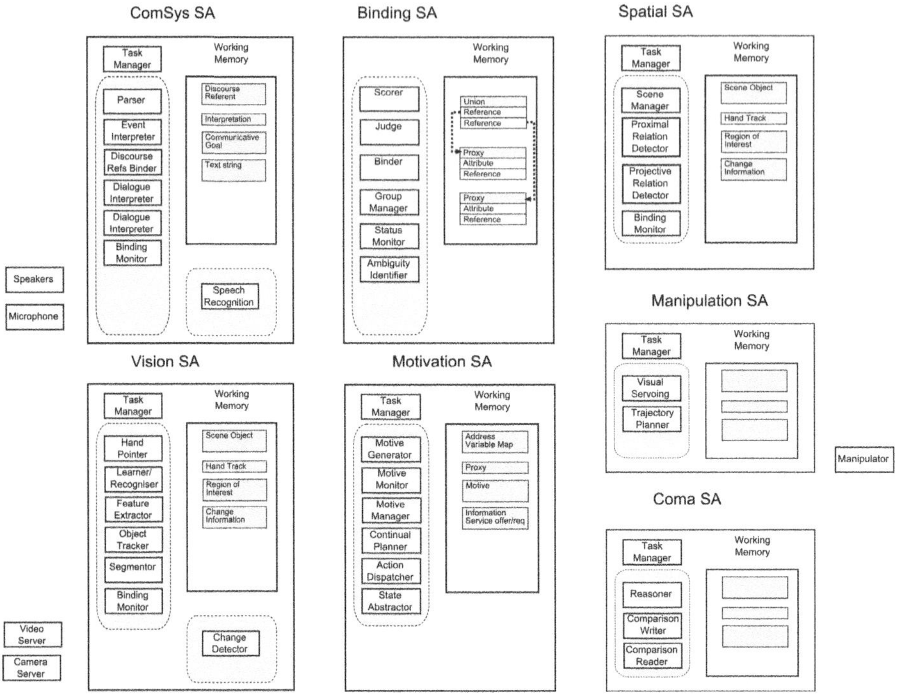

Fig. 9.4. An overview of the PlayMate System. There are seven subarchitectures, each of which corresponds to a functional and development unit.

the description of PECAS we will describe how the system structure supports principled solutions for a number of possible tasks.

The PlayMate system is an architectural instantiation of the CAS schema. It follows a broadly *functional decomposition* contra much recent work in robotics, but utilises *parallelism* and *incrementality* to build up its representation of the environment efficiently.

Following this broadly functional decomposition, there are seven subarchitectures (SAs) in the PlayMate. These are concerned with *visual processing* (Vision SA), *Communication* (ComSys SA), *spatial representation and reasoning* (Spatial SA), *manipulation* (Manipulation SA which includes relevant visual processing), *ontological representations and reasoning* (CoMa SA), *binding* of information between modalities (Binding SA), and control of *motivation and planning* (Motivation and Planning SA). In the PlayMate, central roles are played by the last two of these subarchtictures. In the latter part of this chapter we will show how we use these to control most of the reasoning processes and flow of control, thus provided general control solutions for a variety of potential tasks (question answering, question asking, cross-modal learning, manipulation, handling ambiguities). We now briefly sketch the components employed in each subarchitecture and the representations that they share on their working memories. The reader should refer to Figure 9.4 to aid understanding. Since the CoMa subarchitecture is used only peripherally in the examples presented here, and is explained in Chapter 10 (Section 10.3.2), we omit further description of it here.

9.2.1 Vision SA

The visual system operates in two ways: it has the ability to analyse static scenes (e.g. producing descriptions of objects and their poses), and to analyse dynamic scenes (e.g. recognising actions with objects). When new objects enter the field of view and then cease moving as they are placed on the table the resulting static scene is analysed. A simple *change detector* is used to determine whether the scene is, or has stopped changing. If the scene ceases changing, regions of interest (ROIs) in the scene are segmented from the background, and information about them is written to the Vision SA's working memory. It is important to note that all the information is exchanged between components on the working memory, and that most of this is captured in two types of structure. The first of these structures captures information about ROIs in the image, and the second captures hypotheses about objects that may exist in the world. In the final system we rely on there being a one-to-one correspondence between ROIs and objects. Clearly this is not always the case, for instance due to occlusion and partial occlusion. In our work on planning of visual routines we have developed methods to split ROIs into smaller regions that are more likely to correspond to single objects [6, 7]. However, to overcome this in general we would need to treat segmentation in a rather more sophisticated way, an example of which is the kind of iterated recognition and segmentation employed in the Implicit Shape Model [8]. Given our simplifying assumption of a one-to-one mapping between ROIs and objects, the *segmentor* creates, for each ROI, not just a structure representing the ROI, but also one representing the object. These are linked, using the references that CAST allows between items on working memories, and each of them is a slot and filler structure that allows components to work in parallel. These slots for the object include descriptions of the object's category (by the *categoriser* component), global shape classification and colour classification. The ROI slots include the segmentation of the object within the ROI, and features calculated by the *feature extractor*. This calculates, for example, local shape and colour features for each region of interest. This split in the level of abstraction is quite useful: ROI structures typically store quantitative local visual features, and object structures capture the more abstract classification of the whole object into a smaller number of qualitatively different classes. This correspondence between *local-quantitative* and *global-qualitative* is part of the process by which we gradually produce abstract representations suitable for planning and high-level reasoning. The other key notion from CAS that is displayed here is that the components in the visual system *incrementally refine*, in *parallel*, these *shared representations* of what is in the scene.

When a static scene starts changing again, either because the robot or the human intervenes, dynamic scene analysis starts. In the case of objects, all the objects that have been segmented from the previous static scene analysis are tracked by the *object tracker*, giving their full pose. In addition we separately track hands that appear with a *hand-tracker*, and analyse not only their position, but also any pointing gestures that they may make. These pointing gestures can be used to manipulate the salience of different objects or areas within the scene. Hand information is stored in a separate *hand* structure on the visual working memory. When objects and hands are being tracked information about their pose is constantly updated on the working memory. Many quantitative spatial features are also extracted from the tracking information that describe the relative pose of hands and objects, and these are used by the *event analyser* to classify simple and complex actions as described in [9] and Chapter 7. This allows the system to recognise some of the actions that are performed by the human with an object.

In addition to the ability to describe regions of the scene, objects in the scene, and actions performed with those objects, the visual system also permits learning of object properties. The *learner/recogniser* learns correlations between the low level features associated with ROIs (e.g. colour and shape features) and abstract classifications associated with objects (e.g. the ball being red). To learn this requires a feedback signal. There are many possible feedback signals, but in our case we utilise the information from language. This cross-modal learning is described in more detail in [10, 11], in Chapter 7 (Section 7.2), and also in Section 9.3.1 below.

9.2.2 Communication SA

The Communication subarchitecture (or ComSys) has a number of components concerned with the understanding and generation of natural language utterances. The main discussion of the techniques employed occurs in Chapter 8, but it is worth highlighting some points here. First a *speech recogniser* is used to convert speech to a text string. But even at this early stage the entities that exist as recorded on the binding subarchitecture can be used to bias the likelihoods of recognising certain words[1]. This is an important example of how tentative binding can be used to assist early processing of information by cutting down the space of hypotheses. This process of using binding to limit interpretations is also carried out by the *parser*. This uses a combinatorial categorial grammar with semantics expressed in a modal logic. As described in Chapter 8, this produces logical forms that correspond to semantic interpretations of an utterance. In the ComSys we have the ability to perform interpretation incrementally, i.e. as the elements of the text string arrive from recognition. Whether parsing is performed incrementally or not, the possible

[1] Our current implementation this is slow, and so, even though it improves reliability, it was therefore not used in our experiments.

bindings of the discourse referents in the possible semantic interpretations cut down the interpretations that can be considered. Each utterance is interpreted as it arrives, and both the content of the utterance and its intent are modelled. The resulting interpretations of all the utterances in the dialogue to date are grouped together in what we call a *packed logical form.* We think about this as a forest of trees, where each tree corresponds to the logical form from a single utterance. The dialogue model includes a packed logical form, information on all the discourse referents, a representation of events as well static scenes, and contextual information. All of this information is communicated to the binding and motivation subarchitectures by the ComSys binding monitor. This creates proxies representing both the indexical and intentional content of the utterances in the dialogue. In rough terms the indexical content (information about entities in the world) is used by the binder to link with information from other modalities, particularly vision. Meanwhile the intentional content (information about the purpose of the utterance) is used by the motivation and planning subarchitecture to raise goals for activity elsewhere in the system. The ComSys is able to signal to the overall system about whether information is intended by the speaker to be new information, e.g. in an utterance that provides a description of an object to be learned. This is useful when it comes to cross-modal learning between language and vision.

9.2.3 Manipulation SA

Because our primary integration goal in the PlayMate scenario was to understand the integration and architectural issues some of our components implement standard techniques, rather than each pushing forward the boundaries of each sub-field. This is true for manipulation where we utilise standard approaches to grasping and path planning for the manipulator. The manipulation is given the identity of objects it must manipulate, and their desired target locations. It plans trajectories through space to a suitable location to commence grasping by using a probabilistic road map planner with optimisation of the planned trajectories to minimise energy costs and path length. Having reached the location to commence grasping a local reactive controller utilises information from a visual analysis of the area in front of the manipulator. The manipulator has a camera on the hand, and this provides a high-resolution image of the objects to be grasped. Because of the mechanical restrictions imposed by two fingered manipulators (we use a parallel-jaw gripper on an arm with only 5 degress of freedom) we have restricted our objects to always have graspable surfaces in the form of handles, or to objects that naturally afford grasping (e.g. upright oblong boxes of a suitable width). Given these objects the visual analysis must still correctly identify the graspable surface from within the image. This analysis is performed using edge detection and anytime perceptual grouping methods [12]. These analyse the image and find candidate graspable surfaces. The surface that best matches the hypotheses from the visual sub-system about the location and size of the object is chosen, and a power grasp is executed on the object.

9.2.4 Spatial SA

The robot may receive information about the spatial configuration of objects from either language or vision. From language we have a representation that is essentially qualitative. Utterances that the PlayMate might typically be expected to handle include several different types of spatial references, such as *projective* references, e.g. `"The box to the left of the red ball"`, `"The square in front of the triangle"` or *proximal* ones, e.g. `"The triangle near the red thing"`. Both of these types of reference are essentially qualitative. The visual system can also produce a representation of spatial relations, but this is essentially metric. The spatial subarchitecture captures both metric and qualitative representations of space, which it derives from visual input. This information will be combined with information from language about space in the binder. How the mapping from these metric to qualitative representations is performed is described in Section 9.3.3. But note here that this process of abstraction is central to the representational problems the PlayMate faces. In addition, any solution intended to support communication must also take account of the context sensitive way in which humans use spatial references. The spatial SA has a component called the *scene manager* which monitors changes in the information on the visual working memory. When a new object appears there it creates a record called a *location structure* of the corresponding metric location on its own working memory. These location structures point back to the visual information from which they were derived, creating the kind of *processing trail* that was discussed in Chapter 2. Because, during a dynamic scene, the position of objects can change quickly the actual location information recorded in the spatial working memory is only updated using the record on the visual working memory when the visual scene ceases changing. The set of the locations that hold during a period of stasis are grouped by reference in a *scene structure*. This grouping is also performed by the scene manager, which creates an ordered stack of scenes. This means that the spatial system has memory, which is necessary to enable the system to reason about the changes induced by previous trains of action. The spatial system also builds a qualitative representation of the spatial relations between scene locations. Two components search for *proximal* relations and *projective* relations between these locations, and create *scene relations* which are also referenced by the corresponding scene structure. These relations are detected in a context sensitive manner using potential field models that are described in detail in Section 9.3.3. The qualitative spatial description thus derived is related to descriptions from language by the binding subarchitecture. Our approach to binding is described in the next sub-section.

9.2.5 Binding

We have already described the binding problem in Chapter 2. We described how CAS allows two levels of solution to the binding problem. Within subarchitectures the results of processing by components are bound by design time decisions about the structure of the representations that reside on the working memory. We referred to this as *implicit binding.* However, as explained in Chapter 2, this is not enough in many cases. If in the PlayMate scenario there are two objects on the table, and we refer to `"the blue circle"` we need to relate structures on the communication working memory to those on the visual working memory. We cannot use implicit binding to relate these two structures. This is because those structures will have been created at different times. Also, because there are many ways of referring to an object, and many ways of interpreting a reference — especially ambiguous references, which may require accommodation or reinterpretation of the scene — the binding of structures from across modalities must be done at run time, and done in context sensitive way that produces an interpretation of the scene that is consistent across both modalities [13].

Because we have chosen a largely functional decomposition for the PlayMate systems we have built, explicit binding is a cross-modal affair: we only have to perform explicit binding on entities across modalities. This is not to say, however, that binding can occur only after the results of modality specific processing have been completed. In Chapter 8 we described how we use *possible bindings* between entities in the visual scene and discourse referents in the utterances to incrementally interpret the utterances. This use of *incremental binding* means that binding becomes a way of achieving cross-modal inference [14]. This is a powerful feature of the architectural approach taken in CoSy.

However, it would be reasonable to hypothesise that there are several different ways that binding could be performed within a system like the PlayMate. Our approach is carried out using a *central mediator* and *amodal representations* that are *abstractions* from the modality specific representations. This is a very old approach, with a pedigree that in some sense goes all the way back to the Shakey project. An alternative approach would be to perform explicit binding pair-wise across subarchitectures, i.e. to perform *distributed explicit binding* as opposed to our approach of *centralised explicit binding.* But, as we argued in Chapter 2 (Section 2.5.1), centralised explicit binding provides the functionality we require at a lower computational cost whilst also being a more modular and extensible solution.

We now describe how our binding process works for the case of some simple steps in our example. The discussion should be followed with reference to Figure 9.5. It is important to note that to simplify the discussion we treat the steps of the process as if they were synchronous. In fact all the processing runs asynchronously. First recall the script:

```
Human (H) puts a blue square down on the table
H: "This is a blue square."
```

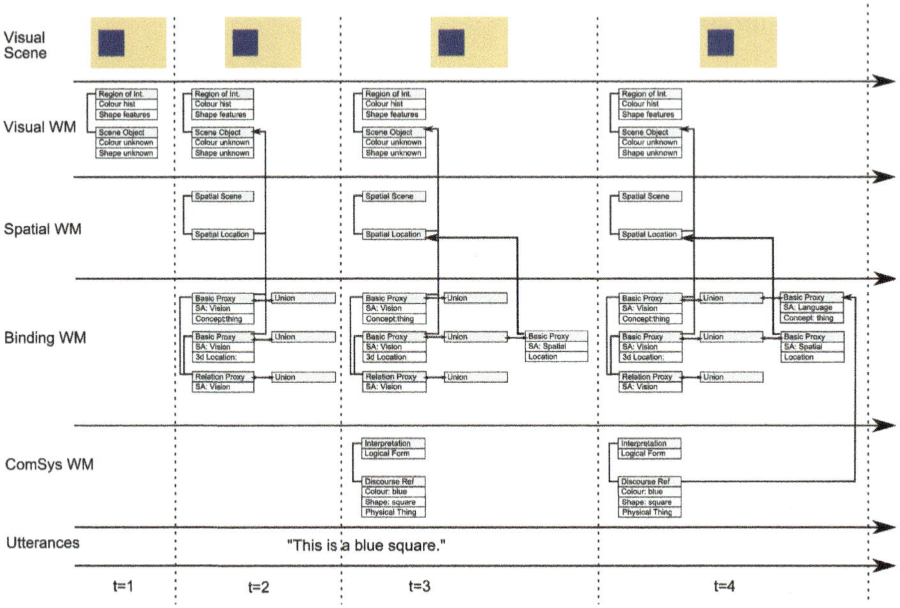

Fig. 9.5. A timeline for a simple binding between a visual object and an utterance. In a real system run these events will occur asynchronously and thus in a relative order that depends on the timing of the utterance and the placement of the object. For simplicity we have presented the system state here at chosen fixed points. Arrowed links indicate one way references, while non-arrowed connections indicate two way references between representations. The rows represent the working memory contents for each subarchitecture listed on the left.

This simple example requires that the robot understands that the sole object in view is the object being referred to. In one sense to bind these two information items (the scene object in the visual working memory and the discourse referent in the communication system working memory) is trivial, the robot must simply associate the only visual object there is with the only discourse referent there is: `"blue square"`. But to do the binding in a way that is general requires that the robot understand that binding them is permissible because of what is known about them. In this first step, since the robot does not yet understand from visual input alone that the object is blue and square, the binding must occur purely on the basis that since they are both structures that represent physical things they could be the same. Since no other binding is possible, they are bound.

Figure 9.5 shows this process. As described in Chapter 2, within each subarchitecture a binding monitor component looks for new entities in that subarchitecture's working memory. When a new information structure is created, the monitor writes an abstract amodal representation of that structure's contents to the working memory of the binder. Not all information creates structures on the binder, and a crucial design decision in any specific system is which pieces of information we regard as relevant changes. Once created, this abstract representation is called a proxy. A binding monitor may write several proxies to the binder to capture different aspects of the information. In our example, after time $t = 1$ when the *scene object* is created in the visual working memory, the visual binding monitor creates three proxies. The first proxy represents the physical object, the second proxy the location of this object, and the third proxy the relation between the first two proxies — in this case the simple fact that the object is at the location. Each proxy is essentially a bag of features, which are attribute-value pairs. In our example, the first proxy has the feature that it's a *type* of physical thing, and the second that it is a location. Note that the abstraction process into these amodal descriptions of the objects is entirely decided by the binding monitor.

Also by time $t = 2$ a separate entity has been created for the spatial location in the spatial subarchitecture. By time $t = 3$ a proxy for this has also been created on the binding working memory. Recall that the purpose of the binder is to decide which proxies are related to which other proxies. If proxies are related it groups them into a set called a *union*. The precise algorithm for matching is described in Chapter 2, but in the basic idea is that if enough common features of proxies have values that match, and if there are no features in common that mismatch, the proxies are bound into a union. To restrict binding the binder also assumes that different proxies from the same subarchitecture can't be bound. In other words the binding process trusts that the modality specific sub-systems generate the right proxies.

In our example, since the spatial proxy with a location and the visual proxy representing that location match (the locations are the same) they are bound into the same union. Since the spatial information here was derived from vision this seems redundant. In fact it is not since we could equally derive spatial information from language, and the spatial sub-system does not tell the binder the origin of it's representations. Thus the binder is blind to the processing trails that exist from the proxies back to the source data.

During this processing the human has made the utterance: `"This is a blue square."`. Parsing leads to a logical representation of the semantic content of the utterance, together with a record of the discourse referent. The binding monitor for the communication subarchitecture creates a proxy from the discourse referent, as seen at time $t = 4$. This proxy contains information that the referent is blue, square, and that it is a physical thing. Since the only feature this proxy has in common with the visual proxy for the object is *type*, and the types match (they are both physical things), the proxies are bound into the same union. So if one modality provides a richer description of an entity than another modality this does not prevent binding. In our example the binding process ends here.

One key design decision is the abstract set of features and values employed for binding. These form our core amodal language. In our case, not only are these features chosen carefully, but they are related in a type network or ontology. This type network allows matching of entities lower in the type hierarchy with entities higher up. Thus if the visual system sees a toy cow on the table, but the human refers to it as an animal the type network allows matching. This kind of reasoning about types and their relations is performed by the Conceptual Mapping subarchitecture. This is mostly used in the Explorer scenario, so we do not discuss it here, but it allows what we call *ontology mediated binding*. This is another powerful aspect of our approach.

We have now covered the basic notions of binding as they are realised in the PlayMate system. We have introduced the ideas of *implicit*, *explicit* [13], *incremental* [14] and *ontology mediated* binding [1]. We have described in detail how *explicit binding* occurs. We have emphasised that it relies on the ability to abstract from modality specific representations into a common amodal representation. Our implementation of this binding approach is also asynchronous, and does not depend on information arriving from different modalities in any particular order. This is important when trying to integrate representations that change at different rates. As mentioned in Chapter 2, an important aspect of the abstract features we have chosen for the PlayMate is that they are temporally stable. We now turn to the other central system: the motivation and planning subarchitecture.

9.2.6 Motivation and Flow of Control

The motivation and planning sub-system plays the role of receiving potential system goals from other sub-systems, and then deciding which ones the whole system should pursue (motive management) and how (planning). We wanted to achieve this in as flexible and general purpose a way as possible. For overall system behaviour the planning occurs at an abstract level, using representations that are essentially those employed by the binder. Within other subarchitectures (manipulation, communication and vision) domain specific planners fill out the details of individual steps of the high-level plan.

The motivation sub-system has a *monitor* component. This is different to monitors elsewhere in that it essentially keeps an up to date local record of entities on the binding working memory that can be used for planning. It watches the proxies and unions that appear on the binding working memory and keeps a list of them - grouped by union - with pointers to the proxies themselves. This list is referred to as the *Address-Literal Map* (ALM), because it maps from working memory addresses (i.e. system content) to literals used in the planning process. The ALM is used by the planner to access information about the world via the binder.

New system goals are adopted in a three-stage process. First a sub-system such as the communication system or the visual system posts a proxy structure to the motivation subarchitecture working memory. These proxy structures contain information relevant to deciding what kind of system goal might be raised. This allows sub-systems to raise potential system goals in parallel and asynchronously. In the second stage the *motive generator* component then reads each proxy structure as it arrives, and decides whether or not to generate a *motive structure* on the working memory. In the third stage the motive structure is picked up by a *motive manager* and a decision is made about whether to turn that motive into a system goal. If so then the motive will be turned into a goal posted to the motivation working memory, which will make reference to variables listed in the ALM. This three-stage process allows local sub-systems to raise potential system goals, for the system to filter these and to maintain a resulting set of motives, and then to choose which of these motives will be acted upon now. This allows the system to switch goals if urgent new motives arrive, and not to forget old motives that were of low priority, but which can be acted on later. Motives can therefore be thought of as very similar to desires in a BDI framework.

This system goal, posted by the motive manager, will in turn be picked up by the continual planner (as described in Chapter 6). Before planning can begin the planner asks for the information from the proxies (referred to by the ALM) in a form suitable for planning. This translation process is carried out by a *state generator* component. Following the planning process an *action dispatcher* sends action requests to different subarchitectures, and checks that the subarchitectures claim they have been executed. The planner, being continual, also includes an execution monitor which checks the achievement of the plan steps via the *state generator*, the ALM and the proxies referenced by the entries in it.

To illustrate the process of raising and acting on motives we return to our example. In the following sections we will show how the binding and motivation approaches described above provide a general way to tackle tutor driven learning, self-driven learning, clarification, following instructions and answering questions posed by humans.

9.3 System Level Control of Information Flow

In the following sections we will go through the steps of our example, using this as a way to describe the commonalities in how the PlayMate handles different types of activity at a system level. The example script requires that the PlayMate be able to handle cross-modal learning, question answering, question asking for clarification, and mediation between qualitative and metric representations of space. Since the component technologies have been described in detail in various chapters we will not focus on the details of those here, but on how the system level flow of information allows the right sub-systems to be active at the right time.

9.3.1 Cross Modal Learning

In Chapter 7 we described methods for cross modal learning at an algorithmic level. In particular we described a continuous learning algorithm that is used for learning associations between vision and language. Specifically it learns the correlations between the abstract qualities of objects (colour and shape names) or relations between objects (names for spatial relations), and features calculated directly from low-level image properties (such as colour features or local shape features). The framework described in Chapter 7 allows for the learning to be either tutor driven or tutor supervised. To recap, in the first approach the tutor drives the learning process by choosing the descriptions of the objects that provide the qualitative labels. In tutor supervised learning, the learner raises queries about the visual objects presented in order to obtain qualitative descriptions. In our example the first exchange is tutor driven:

```
Human (H) puts a blue square down on the table
H: "This is a blue square."
```

Whereas the second exchange is an example of tutor supervised learning:

```
H picks up the red square and puts down a red triangle to the
right of the blue square.
R: "What is the thing to the right of the blue square?"
H: "It is a red triangle."
R: "Ok."
```

At a systems level the challenge is to decide when to engage in one type of learning or another. In addition the system must distinguish between when learning activity is required, versus another kind of activity (question answering, physical action). To do this we use the motivation system described above. Sub-systems raise potential system wide activities, and the motivation sub-system chooses between them. Learning also raises a second problem for our broad approach. When a description given to the robot by the human is intended for learning, how does the learner know which object is being referred to? In the section on binding above (Section 9.2.5) we described how we handle this, binding only on common features, and where necessary using the ontology to handle binding across the type hierarchy. To understand how the motivation and binding solutions work together to enable cross-modal learning at a systems level we return to our simple example of `"This is a blue square"`. To aid the discussion Figure 9.6 shows a timeline that begins where the time-line from Figure 9.5 ends. As described in Section 9.2.6, when the utterance is processed a proxy telling the system about the intent of the utterance is posted on to the working memory of the motivation subarchitecture (time $t = 1$). This proxy contains the information that the ComSys has identified as new. The *motive generator* has a number of templates for creating motives from proxies posted by other sub-systems. In this case it raises a motive and posts it onto the motivation working memory ($t = 2$). Motives are essentially possible system goals, expressed in the language of the planner. The goal arising from the statement `"This is a blue square"` is:

```
(for all ?sa (want-colour-info ?sa) (K ?sa (colour v1:blue))
and
(for all ?sa (want-shape-info ?sa) (K ?sa (shape v1:square)))
```

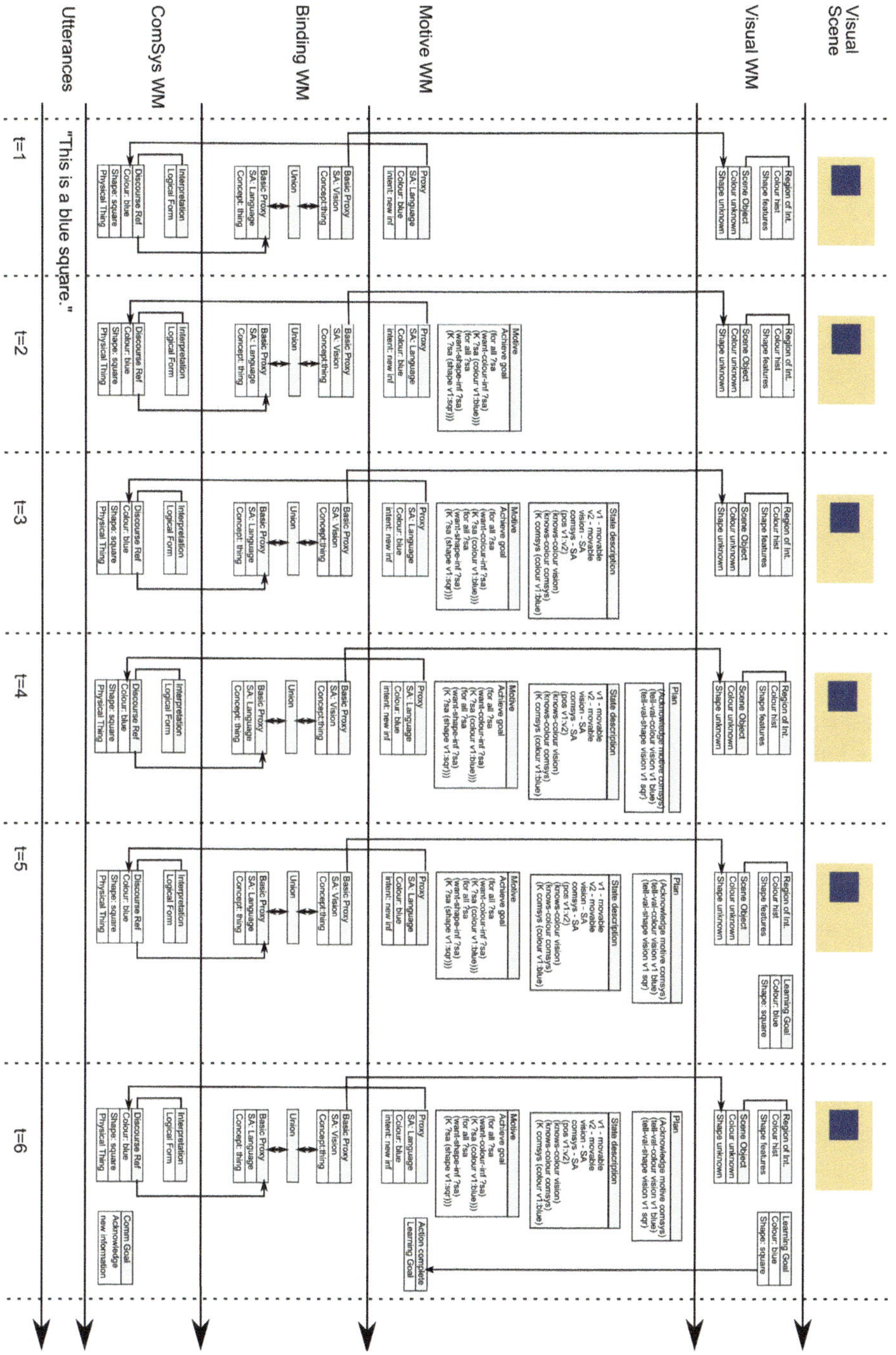

Fig. 9.6. The extended timeline for learning activity for the "This is a blue square." example

where `?sa` refers to any subarchitecture, and `K` is a modal operator for believes, such that `(K agent fact)` means that some agent believes some fact. As described in Chapter 6, our planning language allows us to make epistemic states into system goals. In addition we model the architecture as a multi-agent system in which the subarchitectures are agents. In order to take advantage of this, all the other subarchitectures register with the motivation and planning subarchitecture which kinds of beliefs they are concerned with. These *information service offers/wants* are stored during the entire run of the system on the motivation subarchitecture working memory. To support cross-modal learning both the ComSys and Vision subarchitectures register that they can offer information about colour and shape. They also register that they want to be told information about colour and shape that may be provided by other sub-systems. In the case of cross modal learning the goal created essentially says that all agents that want information of the type that has just arrived should believe this new information. If the motive is adopted by the *motive manager* the goal is adopted by the *planner*. It requests an abstract description of the state suitable for planning, which is in turn supplied by the *state generator* and posted onto the working memory ($t = 3$). This state includes information about the general service offers and wants of different subarchitectures. It also represents the specific beliefs that subarchitectures have. As mentioned previously, the objective of binding is partly to provide stable entities to support planning. The state generator essentially uses the abstract state descriptions used by binding, and re-represents them in the language of the planning domain. A plan is produced ($t = 4$) and its steps are executed one by one ($t = 5, t = 6$). After each plan step has been executed the continual planner we employ performs execution checking, again using the abstracted state description provided via *binding* and the *state generator*. The plan in this case is very simple:

```
(acknowledge motive comsys)
(tell-val colour vision v1 blue)
(tell-val shape vision v1 square)
```

It simply asks the motivation system to tell the ComSys that its request is being dealt with, and then tells the vision system about the colour and shape of the object. This plan has two important features. First it does not decide whether or not the human needs an acknowledgement. That decision is left to a separate dialogue management and planning process. This means that our continual planner is isolated from some aspects of the system's activity. The second point is that the discourse referent `blue square` and the visual object were bound explicitly, but the new information to support learning was sent to vision, avoiding the binder. Thus in this case learning is mediated centrally, but information exchange occurs between modalities, not via a central world model.

In our second learning exchange recall that the learning is driven by the robot:

```
H picks up the red square and puts down a red triangle to the
right of the blue square.
R: "What is the thing to the right of the blue square?"
H: "It is a red triangle."
R: "Ok."
```

This is handled at a system level in a very similar way to the tutor driven learning. When the visual system realises that it is unsure of the qualitative colour and shape labels for the red triangle it posts a proxy to the motive sub-system. Rather than containing new information this proxy is marked as representing a question. This time the motive generator creates a goal that the vision subarchitecture wants information on the colour and shape of a visual object (a simplication of what it "is"):

```
(K vision (colour v1 ?c))
(K vision (shape v1 ?s))
```

This goal is then planned for as previously, and a plan will be produced of the following form:

```
(acknowledge-goal-accepted vision)
(ask-val colour motive comsys v1 ?c)
(ask-val shape motive comsys v1 ?s)
(tell-val colour motive vision v1 ?c))
(tell-val shape motive vision v1 ?s))
```

In this case the planner realises that the ComSys can be asked about the colour and shape of the object and that it can then send this information directly to the vision sub-system to support learning. It is of course perfectly possible that a motive for learning would be raised on behalf of vision while the speaker was making an utterance intended to support learning. In this case the system should be able to catch this by realising that a simpler plan will suffice.

9.3.2 Clarification and Question Answering

We now briefly show how both clarification and question answering requests can be handled using essentially the same mechanisms described above for tutor driven and tutor supervised learning. Recall that in the dialogue there is the following exchange:

```
H picks up the red triangle and the blue square and puts down
a blue triangle.
H: "What colour is this triangle?"
R: "It is blue."
```

In this case the motive proxy structure will be created by the ComSys, and converted into a goal for the ComSys to know what the colour of the triangle is. In this case the plan will be:

```
(acknowledge-goal-accepted comsys)
(tell-val colour motive comsys v1 blue)
```

Because the visual system has analysed the scene, and confidently classified the object's colour this information has been stored in binding. Thus the planner has access to this. The discourse referent for `"this triangle"` has also been bound to the visual object. This means that the planner doesn't need to ask vision. Essentially the reason for this boils down to the fact that in our current system visual processing is data driven, so information flows directly from vision to the binder. In Chapter 2 we outlined how the visual system could have task driven processing. This would make it much more like language, in that vision would need to be asked, so that the plan would need an extra `ask-val` step as did the one for tutor-supervised learning. Note that both cases are handled by the planning process, rather than by the designer.

So far we have seen that our system handles two kinds of learning, and question answering in essentially the same manner. Finally we see that clarification requests can also be handled in a similar way. In the next step of the dialogue the human places two objects on the table:

```
H picks up the blue triangle and puts down a green triangle
on the right and a red circle on the left.
H: "What shape is this thing?"
R: "Do you mean the green thing or the red thing?"
H: "The red thing."
R: "It is a circle."
```

In this dialogue the utterance from the human is ambiguous. Perhaps the robot should assume that the most salient, or recently deposited object is the topic of the query. However, lacking such information it must clarify the ambiguous reference. In our case this occurs in the process of the system's attempt to raise a motive to answer the question: `"What shape is this thing?"`. There will be three relevant proxies on the binder: one from vision for the green triangle (`v1`), one from vision for the red circle (`v2`), and one from language for the discourse referent for `"thing"` (`v3`). Since there is not enough information from the binder to bind the linguistic proxy to either of the others the question cannot be directly answered. An important design decision is how this is handled. On the one hand the ambiguity could be written into the planning state, together with actions that have the effect of removing ambiguities. This presents problems in that expressing the possible bindings in terms of planning operators is extremely challenging. An alternative is for the motive generator to perform simple reasoning to establish that the goal will be unachievable without further information. This is the approach we take. Thus when presented with an ambiguity that leads to an unachievable goal the motive system instead posts a goal to resolve the ambiguity. When this is resolved it posts the goal to answer the original question:

Disambiguation goal:
`(K binder (colour v3 ?c))`
Task goal (posted if the disambiguation goal is achieved):
`(K comSys (shape v3 ?s))`

The first goal is to enable the binder to get the information it needs to distinguish which object the referent should be bound to. The reasoning to generate the distinguishing features is performed by the binder itself. When `v3` has been bound to one of `v1` or `v2` the next goal of the ComSys knowing the shape of the object can be adopted. The plan for the initial goal in this case will be:

```
(acknowledge-goal-accepted comsys)
(ask-val colour motive comsys v3 ?c)
```

which will cause the ComSys to generate the clarification question: `"Do you mean the green thing or the red thing?"` as a way of obtaining the required colour information to complete the binding of `v3`. When the answer is: `"The red thing."` `v3` and `v2` are bound, and the second goal is planned for:

```
(tell-val shape motive comsys v2 ?s)
```

A comparison of the relationship between these different types of control flows is given in Figure 9.7.

Proxy from SA	Motive	Plan
Proxy SA: ComSys New Inf Colour v1:blue	Goal: for all SAs who want this type of information to know it: (for all ?sa (want-colour-inf ?sa) (K ?sa (Colour v1:blue)))	Plan: (acknowledge-goal-accepted comsys) (tell-val colour motive vision v1 blue)
Proxy SA: ComSys Question Colour v1:?c	Goal: for the SA that asked for this information to know it: (K ComSys (Colour v1 ?c))	Plan: (acknowledge-goal-accepted comsys) (tell-val colour motive comsys v1 blue)
Proxy SA: Vision Question Colour v1:?c	Goal: for the SA that asked for this information to know it: (K Vision (Colour v1 ?c))	Plan: (acknowledge-goal-accepted vision) (ask-val colour motive comsys v1 ?c) (tell-val colour motive vision v1 ?c)
Proxy SA: ComSys Question Shape v3:?s	Goals: for the binder to know the colour of object v3 to identify it, so that the ComSys can be told its shape: (K Binder (Colour v3 ?c)) (K ComSys (Shape v3 ?s))	Plan: (acknowledge-goal-accepted comsys) (ask-val colour motive comsys v3 ?c) (tell-val shape motive comsys v3 ?s)

Fig. 9.7. Tutor driven learning, question answering, tutor supervised learning and clarification of ambiguity are all handled in a similar framework. Essentially each begins with a proxy posted on the motivation and planning SA working memory. Then a motive/goal is created, which takes into account the desired knowledge state of the various agents. Finally the planner creates a plan which includes knowledge generating actions. These actions may cause specialised planning processes to be run in other sub-systems.

9.3.3 Mediating between Qualitative and Quantitative Representations

In the final step of our example the human gives the robot an instruction to move one of the objects:

```
H: "Put the circle to the right of the triangle."
R picks up the circle and puts it to the right of the
triangle.
```

To act on the kinds of action commands we are interested in, the robot must be able to translate from the qualitative linguistic spatial description of the location to place the object, to both a geometric description of the location that can be used by the manipulation system (i.e. a geometric point positioned in the robot's world), and a logical description for the planning domain (i.e. a symbolic representation of this point and its relationships with other points). This translation involves constructing geometric models of the semantics of spatial terms. In our approach we use potential field models to create abstract spatial relationships from metric information in the spatial subarchitecture [81]. These potential field models are simple, but are able to take into account the context when interpreting spatial relationships. For instance, the notion of the nearness of two objects is mediated by the proximity and salience or size of other objects that may act as landmarks. The potential field models thus let us extract qualitative spatial relations that correspond to the terms as used by humans. To move from qualitative to metric we simply take the mode of a field with respect to some landmark. This is how we generate actual goal locations for objects when given instructions like the one in the final step of our example.

9.4 Conclusions and Discussion

In this chapter we have looked at some of the issues involved in building an integrated robot architecture for situations where the robot and a human can talk about and manipulate a set of objects in a shared space. Working from requirements we explained how the PlayMate architectural solution provides a specialisation of CAS. Essentially the PlayMate is a functional instantiation of the CAS architecture schema. This is only one possible instantiation among many. The functional approach, while still deeply unfashionable in parts of robotics has some advantages. In particular it is one way that we can produce abstract enough representations to support planning of system wide activity. While we do embrace an abstract world model we wish to emphasise that the approach here is not that of Shakey or similar. There are a number of design choices that we have embraced and from which we believe others could benefit. These are:

- **Shared Representations**: while systems like Shakey employed a central representation and a serial processing model we have used parallel processing of distributed representations and explicit sharing of representations. Brooks criticised approaches that pass representations as being subject to a "strong as the weakest link" problem [16]. Sharing representations can overcome this weakness because by refining shared representations we can leverage many sources of information to refine our hypotheses and render our conclusions more reliable. This is already a feature of statistical inference in AI. We have taken the same principle and employed it at an architectural level. We have also explored one set of approaches for parallel and incremental refinement. The sharing, rather than the point-to-point transmission, of representations, and the parallel, rather than serial refinement, make the model very different to that of early representation-heavy systems like Shakey.
- **Abstraction**: In our view the ability to coordinate system wide behaviour requires that we have quite abstract, stable representations that are suitable to support symbolic planning. This was another criticism of the Shakey approach: that such abstract representations are hard to produce and brittle in the face of change. In the PlayMate we employ ideas like *processing trails* to allow the stable and abstract representations to point back to rapidly changing information without being corrupted by this change. In addition we employ a variety of levels of abstraction: within vision for instance we maintain both low level visual features, and qualitative descriptions of objects; and in the spatial system we store both metric and qualitative descriptions of spatial relations. Finally we have looked for ways that we can move in both directions: from metric to qualitative and back again.
- **Binding**: Once the benefits of representations that are distributed across many sub-systems are accepted there is a price to be paid. This is the binding problem: it will not always be possible to tell which representations from one sub-system are related to those from another without performing some inference. We can think of our approach to explicit binding as being a way of tying together elements across a field of evolving representations rather than as a single monolithic central representation. We have shown in different parts of this book how a cognitive system with such a distributed set of representations can benefit from strategies such as *incremental binding*, *implicit binding* and *ontology mediated binding*. We have employed all these in the PlayMate system to good effect.
- **Processing control:** Once we think about a distributed system composed of many processing agents, we need to address the issue of how that processing is controlled, and how information flows through the system. How can the system produce coherent behaviour from these many parts? Rather than take an approach in which coordination emerges we have pursued one in which it is the result of informed decision making. This idea — that we can model the internal informational effects of processing — is a

very powerful one. The idea need not be restricted to planning approaches as here: it is perfectly reasonable to suppose that learning strategies could be employed to acquire some aspects of processing control. In this chapter we have focussed on the way that a variety of different activities can be modelled using planning operators.

Overall in this chapter we have presented, at a systems level, a synthesis of old and new approaches from AI and robotics to building complete cognitive systems. We draw on, and synthesise, several traditions: representational approaches to AI, aspects of parallel processing from robotics, and ideas from statistical machine learning. The problems their synthesis posed resulted in new architectural ideas. Their synthesis in the PlayMate in particular has been an important driver in preventing us from shying away from hard issues we would otherwise have ignored.

References

1. Kruijff, G.-J.M., Kelleher, J.D., Hawes, N.: Information fusion for visual reference resolution in dynamic situated dialogue. In: André, E., Dybkjær, L., Minker, W., Neumann, H., Weber, M. (eds.) PIT 2006. LNCS (LNAI), vol. 4021, pp. 117–128. Springer, Heidelberg (2006), `http://cognitivesystems.org/cosybook/chap9.asp#Kruijff/etal:2006`
2. Hawes, N., Sloman, A., Wyatt, J., Zillich, M., Jacobsson, H., Kruijff, G.-J., Brenner, M., Berginc, G., Skočaj, D.: Towards an integrated robot with multiple cognitive functions. In: Holte, R.C., Howe, A. (eds.) Proceedings of the Twenty-Second AAAI Conference on Artificial Intelligence (AAAI 2008), pp. 1548–1553. AAAI Press, Vancouver (2008), `http://cognitivesystems.org/cosybook/chap9.asp#Hawes/etal:2007a`
3. Winograd, T.: Procedures as a representation for data in a computer program for understanding natural language, AI Tech. Report 235, MIT (February 1971)
4. Nilsson, N.J.: Shakey the robot, Technical Note 323, Stanford Research International (April 1984)
5. Langley, P., Choi, D.: A unified cognitive architecture for physical agents. In: Proceedings of the Twenty-First National Conference on Artificial Intelligence (2006)
6. Sridharan, M., Wyatt, J., Dearden, R.: HiPPo: Hierarchical POMDPs for Planning Information Processing and Sensing Actions on a Robot. In: International Conference on Automated Planning and Scheduling, ICAPS (2008), `http://cognitivesystems.org/cosybook/chap9.asp#Sridharan/etal:2008a`
7. Sridharan, M., Dearden, R., Wyatt, J.: E -HiPPo: Extensions to Hierarchical POMDP-based Visual Planning on a Robot. In: The 27th PlanSIG Workshop (2008), `http://cognitivesystems.org/cosybook/chap9.asp#Sridharan/etal:2008`
8. Leibe, B., Leonardis, A., Schiele, B.: Robust object detection with interleaved categorization and segmentation. Int. J. Comput. Vision 77(1-3), 259–289 (2008), `http://dx.doi.org/10.1007/s11263-007-0095-3`, `http://cognitivesystems.org/publications/cosyBib2008.asp#Leibe05c`

9. Hongeng, S., Wyatt, J.: Learning causality and intention in human actions. In: Proceedings of the 6th IEEE-RAS International Conference of Humanoid Robots (Humanoids 2006). IEEE, Los Alamitos (2006), http://cognitivesystems.org/cosybook/chap9.asp#hong06
10. Skočaj, D., Kristan, M., Leonardis, A.: Continuous learning of simple visual concepts using incremental kernel density estimation. In: International Conference on Computer Vision Theory and Applications, Funchal, Madeira, Portugal, pp. 598–604 (2008), http://cognitivesystems.org/cosybook/chap9.asp#skocajVISAPP08
11. Fritz, M., Kruijff, G., Schiele, B.: Cross-modal learning of visual categories using different levels of supervision. In: The 5th International Conference on Computer Vision Systems (2007), http://cognitivesystems.org/cosybook/chap9.asp#fritz07
12. Zillich, M.: Incremental Indexing for Parameter-Free Perceptual Grouping. In: 31st Workshop of the Austrian Association for Pattern Recognition (2007), http://cognitivesystems.org/cosybook/chap9.asp#zillich2007incremental
13. Jacobsson, H., Hawes, N., Kruijff, G.-J., Wyatt, J.: Crossmodal content binding in information-processing architectures. In: HRI 2008: Proceedings of the 3rd ACM/IEEE International Conference on Human Robot Interaction, pp. 81–88. ACM, New York (2008), http://doi.acm.org/10.1145/1349822.1349834, http://cognitivesystems.org/cosybook/chap9.asp#Jacobsson/etal:2008a
14. Lison, P., Kruijff, G.: Salience-driven contextual priming of speech recognition for human-robot interaction. In: Proceedings of ECAI 2008, Athens, Greece (2008), http://cognitivesystems.org/cosybook/chap9.asp#Lison/Kruijff:2008
15. Brenner, M., Nebel, B.: Continual planning and acting in dynamic multiagent environments. Journal of Autonomous Agents and Multiagent Systems (Accepted for publication), http://cognitivesystems.org/cosybook/chap9.asp#Brenner/etal:2008
16. Brooks, R.A.: Intelligence without representation. Artificial Intelligence (47), 139–159 (1991)
17. Hawes, N.: CoSy PlayMate Year 3 (2007), http://cognitivesystems.org/cosybook/videos.asp#PlayMateY3
18. Hawes, N.: CoSy PlayMate Year 4 (2008), http://cognitivesystems.org/cosybook/videos.asp#PlayMateY4
19. Hawes, N., Zillich, M., Wyatt, J.: BALT & CAST: Middleware for Cognitive Robotics. In: Proceedings of IEEE RO-MAN 2007, Jeju, Korea, August 2007, pp. 998–1003 (2007), http://www.cognitivesystems.org/publications/hawesetal07cast.pdf

10

The Explorer System

Kristoffer Sjöö[1], Hendrik Zender[2], Patric Jensfelt[1], Geert-Jan M. Kruijff[2], Andrzej Pronobis[1], Nick Hawes[3], and Michael Brenner[4]

[1] Royal Institute of Technology (KTH), Centre for Autonomous Systems, Stockholm, Sweden
{krsj,patric,pronobis}@csc.kth.se
[2] DFKI GmbH, Saarbrücken, Germany
{zender,gj}@dfki.de
[3] Intelligent Robotics Lab, School of Computer Science, University of Birmingham, Birmingham, UK
nah@cs.bham.ac.uk
[4] Albert-Ludwigs-Universität Freiburg, Department of Computer Science, Freiburg, Germany
brenner@informatik.uni-freiburg.de

10.1 Introduction

In the Explorer scenario we deal with the problems of modeling space, acting in this space and reasoning about it. Comparing with the motivating example in Section 1.3 the Explorer scenario focuses around issues related to the second bullet in the example. The setting is that of Fido moving around in an initially unknown (Fido was just unpacked from the box), large scale (it is a whole house so the sensors do not perceive all there is from one spot), environment inhabited by humans (the owners of Fido and possible visitors). These humans can be both users and bystanders. The version of Fido that we work with in the Explorer scenario can move around but interaction with the environment is limited to non-physical interaction such as "talking". The main sensors of the system are a laser scanner and a camera mounted on a pan-tilt enabling Fido to look around by turning its "neck". Figure 10.1 shows a typical situation from the Explorer scenario.

The construction of spatial models from sensor data in the context of mobile robotics has been studied extensively in the literature. Simultaneous localization and mapping (SLAM) is by now a mature technology and it is not primarily in this field that the Explorer makes contributions. The kind of maps typically created by SLAM are, as discussed in Chapter 5, focused on providing the robot with means to localize and determine how to move from one place to another. While this is still required by the robot in the Explorer scenario, it is not the primary focus. Besides the obvious challenges in

H.I. Christensen et al. (Eds.): Cognitive Systems, COSMOS 8, pp. 395–421.
springerlink.com

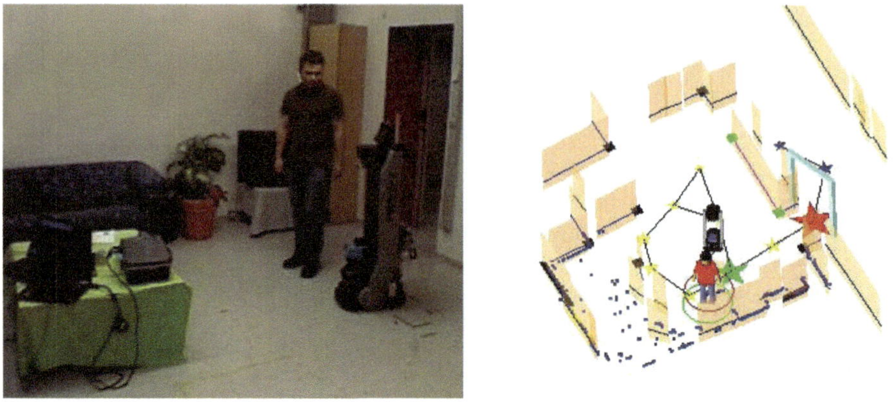

Fig. 10.1. The user shows the Explorer robot where the living room is. The robot's visualization of a similar situation can be seen on the right-hand side.

creating an integrated system, one of the motivations for the Explorer, as with the PlayMate, is to study the problems that occur when an intelligent robot must interact with humans in a rich and complex environment. In this case we focus on the design of models of space that are able to facilitate such interactions. In order to do this, the representation used by the robot must support the anchoring of spatial concepts shared between the robot and humans. Such spatial concepts may not be needed by and may not, for that matter, be available to a robot acting on its own. However, when communicating with a human, they play a key role in generating a shared understanding of space. For example, instead of the robot talking about an area delimited by a certain polygon it is more natural to talk about a certain room.

In the Explorer scenario spatial models are built using input from sensors such as laser scanners and cameras but equally importantly also based on human input. It is this combination that enables the creation of a spatial model that can support low level tasks such as navigation, as well as interaction. Even combined, the inputs only provide a partial description of the world. By combining this knowledge with a reasoning system and a common sense ontology, further information can be inferred to make the description of the world more complete. Unlike the PlayMate system, all the information that is needed to build the spatial models are not available to it sensors at all times. The Explorer need to move around, i.e. explorer space, to gather information and integrate this into the spatial models. Two main modes for this exploration of space have been investigated within the Explorer scenario. In the first mode the robot explores space together with a user in a home tour fashion. That is, the user shows the robot around their shared environment (Fido needs to know where things are and what stuff is called in his new home). This is what we call the *Human Augmented Mapping* paradigm. The second mode is fully autonomous exploration where the robot moves with the purpose of covering space. In practice the two modes would both be used interchangeably to get the best trade-off between autonomy, shared representation and speed.

Another important aspect of the Explorer scenario is the ability to perform tasks autonomously. If robots like Fido is ever going to take the steps from toys to utilities they need to do something for us besides providing entertainment. Since the Explorer system does not have manipulation skills tasks are somewhat limited. The focus in the Explorer is not on performing a particular task to perfection, but rather acting within a flexible framework that alleviates the need for scripting and hardwiring. We want to investigate two problems within this context: what information must be exchanged by different parts of the system to make this possible, and how the current state of the world should be represented during such exchanges.

One particular interaction which encompasses a lot of the aforementioned issues is giving the robot the ability to talk about space. This interaction raises questions such as: how can we design models that allow the robot and human to talk about where things are, and how do we link the dialogue and the mapping systems?

10.1.1 Related Work

There are a number of systems that permit a robot to interact with humans in their environment. Rhino [1] and Robox [2] are robots that work as tour guides in museums. Both robots rely on accurate metric representations of the environment, and both have quite limited communicative capabilities. In the Explorer the communication with humans and reasoning about space are central elements. Examples of robots with more elaborate dialogue capabilities are RoboVie [3], BIRON [4], GODOT [5], WITAS [6] and MEL [7]. BIRON is endowed with a system that integrates spoken dialogue and visual localization capabilities on a robotic platform. This system differs from ours in the degree to which conceptual spatial knowledge and linguistic meaning are grounded in, and contribute to, situation awareness. In contrast, in our system, information from dialogue and situated contexts can be combined during processing of utterances [8]. Furthermore, whereas RoboVie and BIRON use finite state machines to model dialogue behavior, we combine information states [9], like GODOT; together with a task-oriented perspective, as WITAS or MEL. One more things that the sets the Explorer system aside from the above system is that the integration mechanisms themselves are as important or maybe more important than the performance of the end product. That is, we want to study how to integrate a large set of components in a cognitve system in a flexible and scalable way rather then creating a system that can perform certain tasks well.

10.1.2 Outline

The outline of the rest of this chapter is as follows. In Section 10.2 we give an overview of the Explorer system, focusing mainly on outlining the differences to the PlayMate instantiation presented in the previous chapter. In

Section 10.3 we describe how the spatial model is acquired, how it can be used for conceptual reasoning and finally how cross-modal knowledge is represented and shared across the system. Section 10.4 details the specifics of planning in the Explorer domain. Finally, in Section 10.5, we describe in detail an example task and how different parts of the system contribute to the completion of that task. We present some conclusions in Section 10.6.

10.2 System Overview

This section gives an overview of the system structure used in the Explorer scenario. Figure 10.2 shows the subarchitectures (SAs) used and some of the more important data structures published in the working memories of the different SAs. Comparing with Figure 9.4, which describes the instantiation of the PlayMate scenario, we see that the scenarios share four SAs: ComSys SA for *communication* with the user, Binding SA for *binding* of information between modalities, Motivation and Planning SA for *motivation and planning* and the Conceptual map SA, for *ontological representations and reasoning.*

The SAs that have been removed from the PlayMate scenario are Manipulation SA, Spatial SA and Vision SA. There is no manipulation in the explorer scenario which eliminates the need for a dedicated SA. The Spatial SA in its current form deals with spatial relations in a tabletop scene; in the Explorer scenario the environment is large-scale and the Spatial SA is replaced by the Navigation SA, which handles *motion control* and the three lowest levels of the *spatial model* (see Chapter 5), and the Place SA which provides capabilities for *place recognition and categorization.*The Conceptual Mapping SA takes a more prominent role in the Explorer scenario than it does in the PlayMate scenario. Here it is used to represent large-scale space at an abstract level, allowing for natural language expressions to be related to places in the world – and their respective representations in the robot's lower level maps. The requirements on the visual processing system also differ considerably between the PlayMate and the Explorer. The camera is mobile in the Explorer which violates some of the assumptions in the Vision SA. The Object SA, dealing with *object detection* reuses some of the components from the Vision SA.

In the remainder of this section we will briefly outline the composition of those SAs in the Explorer that do not exist in the PlayMate.

10.2.1 Navigation SA

The Navigation SA hosts the three lowest levels of the spatial model, i.e., the metric map, produced by the SLAM Process; the navigation graph, and the topological map; both of which are produced by the NavGraphProcess. The metric map, represented as a line map, and the navigation graph are published in the working memory for other components to use. The SLAM Process also updates one structure for the current metric position and one for the current topological position of the robot. Each node is assigned an `AreaID` representing

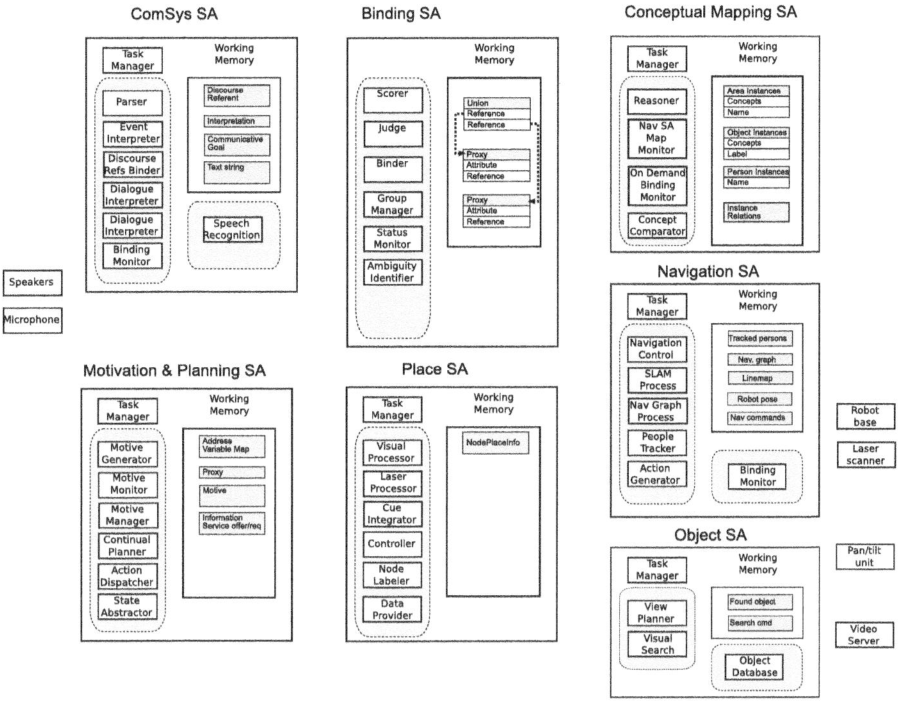

Fig. 10.2. An overview of the Explorer System. There are 7 sub-architectures, each of which corresponds to a functional and development unit.

the topological area it belongs to. Some nodes are gateway nodes (doorways), and are connected to nodes with differing `AreaID`. This implicitly defines the topological map: All navigation nodes with the same `AreaID`, taken together, correspond to one node (area) in the topological map, and each gateway node corresponds to an edge connecting two different areas.

Detecting and Tracking People

Detecting humans and keeping track of them is one of the key capabilities of a robot that aspires to interact with humans. In the Explorer system we use relatively simple means to realize this and make the assumption that there are not too many people close to the robot. For people detection we use a method similar to that used in [10, 11] which detects motion using laser scanner data. A new person is hypothesized when motion occurs far enough from any existing person. Each new person hypothesis is given a unique `PersonID`. Association between detected motion and a hypothesis is based on nearest neighbor matching. Tracking is accomplished by associating detected motion to hypothesis and using these as measurements in a Kalman filter that estimates the position and velocity of each person. Once a person disappears from the field of view of the sensor the hypothesis is removed. If the same person reappears they will be given a new `PersonID` since the system is not able to identify individual people. Using the camera in combination with the pan-tilt for actively acquiring and using the appearance of the person could possibly deal with the problem.

Motion Control

Mobility is one of the most significant differences between the Explorer and the PlayMate. The Navigation Control module is based on the Nearness Diagram method [12] and executes the low level "go-to" commands. The target location is defined based on the current task which might be to follow a person, move to a specific point in space, etc. This module uses the navigation graph for the purpose of path planning by finding a path from the current robot position via the closest node in the graph, through the graph to the node closest to the goal and finally to the goal location.

10.2.2 Object SA

The Object SA collects the components involved in the finding of objects. It consists of a module for view planning and one for visual search. The view planning component creates a plan for which nodes to visit, in what order and in what direction to look given the assumption that objects can be found in places where the metric map registers obstacles. The visual search can be performed using a pan-tilt-zoom camera where an attention mechanism gradually guides the robot to zoom in closer and closer on object hypotheses where finally a SIFT based method is used for recognition. In the absence of a camera with zoom, the SIFT based matching algorithm can be used directly.

The Object SA can be used in two modes: one to perform active object search in the current region, which engages the above-mentioned modules, and another where the images are continuously processed to detect objects. In both cases the objects that are found are published on the working memory. This is detected by the Navigation SA, whereupon it in turn extends the spatial model with the new objects. This then propagates the information onwards to the Conceptual Mapping SA.

10.2.3 Place SA

The Place SA is responsible for assigning nodes and areas to one of predefined semantic place categories (e.g. an office, a corridor etc.) as described in Chapter 5. It gathers sensory data from a laser scanner and a camera and processes them in parallel using dedicated visual and laser processing components. The results for both sensors are integrated by a cue integration component, which provides beliefs about a place category for the current viewpoint. This information is, in turn, integrated temporally and spatially in a component responsible for assigning place labels to the nodes and areas.

Each area is initially categorized as unknown until the Place SA delivers a reliable classification result. The place categorization information is pulled from Place SA by the Navigation SA whenever the current node is changed in the navigation graph. The interaction with the Navigation SA is carried out via the structure `NodePlaceInfo` which contains the established place category for an area (and a node).

The area categorization provides important information when reasoning about space. As an example, object search is expensive and the place categorization can help to speed it up by allowing for a more selective search for objects. If the task is to populate the map with objects and it is known that the current area is a kitchen the search can be focused on typical kitchen objects.

10.2.4 Conceptual Mapping SA

The subarchitecture for Conceptual Mapping maintains a symbolic representation of space suitable for situated action and interaction. It represents spatial areas, objects in the environment, and abstract properties of persons in a combined A-Box and T-Box reasoning framework. Section 5.6 gives details on the underlying knowledge representation and knowledge processing principles.

For our implementation, we use the Jena reasoning framework[1] with its built-in OWL reasoning and rule inference facilities. Internally, Jena stores the facts of the A-Box and the T-Box of the ontology reasoner as RDF triples.[2] The knowledge base can be queried through SPARQL queries.[3] This reasoning framework is wrapped inside a CAST component, the *Reasoner*, which handles all necessary communication with the reasoning framework through SPARQL queries. There are a number of other components that mainly manage the interaction with other subarchitectures, namely the *Nav SA Map Monitor*, the *On Demand Binding Monitor* and the *Concept Comparator*. In Section 10.3.2 we will describe the run-time behaviour of the individual components of this subarchitecture.

10.2.5 The Robot Platforms

The Explorer system was developed and tested on two similar mobile robot platforms, Minnie from KTH and Robone from DFKI, seen in Figure 10.3. Both platforms are equipped with a SICK laser scanner, Minnie with an LMS 200 and Robone with an LMS 291. The laser scanner is the main sensor for the Navigation SA. Both robots also have a pan-tilt unit with a camera. Though Robone has a Videre stereo camera setup, only one of the cameras was used in the scenario.

[1] `http://jena.sourceforge.net`
[2] `http://www.w3.org/RDF`
[3] `http://www.w3.org/TR/rdf-sparql-query`

Fig. 10.3. The mobile platforms on which the Explorer scenario was developed and tested. Left: Minnie from KTH, Right: Robone from DFKI.

10.3 Spatial Modeling and Reasoning

In this section we will look closer at spatial modelling, spatial reasoning and maintaining relations between entities in the spatial model.

10.3.1 Map Acquisition

When building spatial models for human-robot interaction it is natural to adopt an interactive scheme for the acquisition process as well. We use the paradigm of *Human-Augmented Mapping (HAM)* [13, 14].

Human Augmented Mapping

The acquisition process using HAM can be described as a guided tour scenario. The user takes the robot on a tour of the environment and provides labels for areas and objects of importance. Wizard-of-Oz studies have investigated the interaction between human and robot in HAM [15, 16]. The findings concern, for example, the type of dialogue typically used and strategies to introduce new locations.

In a typical HAM scenario, the user walks up to the robot and initiates the mapping process with a command like "follow me!". The robot continuously tracks the position of the user and follows them through the environment. As the robot moves, the spatial model is built from the sensor data.

One important aspect of our implementation of HAM is that the interaction is not master/slave-like, with the user initiating all interactions. The robot is also able to engage in, for example, clarification dialogues when it encounters contradictory or ambiguous information. In [17] we describe in detail a HAM case study. In the example illustrated there, the robot detects a false door when passing by a table and a trash bin placed close together (see Figure 10.4). This makes the robot segment space into a new area. However, as it moves on it finds itself reentering a part of space classified as belonging to the previous area. This should not be possible without passing a door which leads to a contradiction; the robot can then ask if it really passed through a door recently.

Fig. 10.4. Left: The user activates the robot at its recharging station. Right: The robot passes through an opening between a table and a trash bin interpreting the narrow opening as a door.

Throughout the HAM session, the user can query the spatial knowledge of the robot. The robot will be able to provide more and more precise descriptions of space as more information comes in. Although the HAM paradigm is very useful for acquiring a shared representation for spatial knowledge, it is rather time-consuming for the user. It is therefore natural that the user walks through the environment relatively quickly and introduces the main features. The robot could then revisit the environment later when not assigned a specific task and extend the model with more detected objects, better covered space, etc.

Autonomous Exploration

In addition to the HAM scheme for map acquisition, the Explorer also possesses the ability to do autonomous exploration. The Explorer uses a frontier-based strategy [18] for autonomous exploration. Briefly put, the algorithm maintains a representation of the world where space is classified as `FREE`, `OCCUPIED` and `UNKNOWN`. The frontiers are defined as the borders between `FREE` and `UNKNOWN` space.

Exploration is considered complete when there are no more reachable frontiers. Exploration can be configured to be confined to one area. In this case, the exploration frontiers are considered unreachable if they require passing a door, i.e., changing area. When detecting a door, the robot will back up into the area from which it came and select a new frontier to explore.

Because the way the navigation graph is built requires the robot to move, exploration is not only about having the sensor see all parts of the room, but the robot needs to move there as well. The rationale behind requiring the robot to travel a path to make it part of the navigable space, i.e. a part of the navigation graph, is that some obstacles may not be detected by a sensor such as a laser scanner. Upward-facing IR-sensors for detecting tables are examples of sensors that could detect obstacles which the laser cannot see. Hence, in order to force the robot to actually visit all the space physically, we limit the range of the sensor for the purposes of updating the occupancy grid, which is used to define the frontiers, to 2m. This way, the robot will move across a large part of a room, even if it is able to see all parts of it from the door.

10.3.2 Acquiring the Conceptual Map

Upon start-up, the system comes with a rich conceptual ontology consisting of taxonomies of indoor area types, of commonly found objects, and of different spatio-topological relations that can hold between area instances and object instances. Moreover, the ontology contains a concept for persons and a relation that denotes ownership. This conceptual knowledge is held in the reasoner's T-Box. In case the system is started with a blank map, the A-Box of the reasoner is empty. Otherwise, it will contain area, person and object instances, and their relations, as contained in the loaded map. It is worth mentioning that the exact positions of persons are not represented in the conceptual map, just their ownership relations, which hold irrespectively of their current whereabouts.

As the robot learns about the world (either through interaction with the user, or through its autonomous map acquisition skills), this knowledge is added to the A-Box of the reasoner. To this end, the subarchitecture for Conceptual Mapping contains a component that constantly monitors the working memory of the Navigation subarchitecture, the *Nav SA Map Monitor* (see Figure 10.2). Whenever the Navigation subarchitecture identifies a new topological area, it creates a working memory entry for it. The working memory entries for areas on the *Navigation Working Memory* include a field that contains the most specific area category that can be autonomously extracted from sensory data. Initially, the working memory entries for areas will contain the neutral category "area". As soon as the Place subarchitecture for place categorization has reliably determined a more specific category (e.g. "corridor", or "office" etc.), it will overwrite the corresponding working memory entry for that area. The *Nav SA Map Monitor* is notified whenever an area working memory entry is created or an existing one is modified. In these cases, a

```
(area0 rdf:type Corridor),
(area1 rdf:type Library),
(area2 rdf:type Office),
(area3 rdf:type Office),
(...)
(nick rdf:type Person),
(nick name Nick), (nick owns area2),
(...)
(obj1 rdf:type Book),
(obj2 rdf:type Mug),
(...)
```

Fig. 10.5. RDF triples in the A-Box of the conceptual map (namespace URIs omitted)

new instance of the area's category is created in the reasoner, or a given one is modified respectively. A similar information flow is implemented for visually detected objects. Figure 10.5 shows a part of the A-Box in the Explorer example scenario.

So far we have only described how the Conceptual Mapping Subarchitecture reflects knowledge present elsewhere in the system, e.g. in the Navigation Subarchitecture. However, one of the main roles of this subarchitecture is to infer new or more specific knowledge based on partial information. The Description Logic definitions of the concepts in the T-Box express properties that form necessary and sufficient conditions for being instances of that concept. By combining and reasoning over instances and their relations, the reasoner can infer more specific concepts for those instances. In our current system, the reasoner can infer subconcepts for room instances based on the objects they contain, according to the principles described in Section 5.6.

The other components of the Conceptual Mapping SA, namely the *On Demand Binding Monitor* and the *Concept Comparator*, are used to make the information inside the Conceptual Mapping SA available to other subarchitectures. In the current system, the *On Demand Binding Monitor* registers its competences with the Planning & Motivation Subarchitecture, and upon request, contributes relevant data to the Binder working memory. The competences offered are *spatial reference resolution* and the possibility to provide *typical locations* for objects. We will detail the properties of these competences below. The *Concept Comparator* compares two **Concept** Binding Features according to their taxonomical relation in the T-Box of the *Reasoner*. The comparator will return `true` if the two concepts are ontologically equivalent, or if the concepts are in a taxonomical subsumption relation. It will return `indeterminate` if at least one of the concepts is unknown. Otherwise, i.e. if and only if both concepts exists but are not hierarchically related, the comparison result will be `false`.

10.3.3 Cross-Modal Spatial Knowledge Sharing

To enable different modalities to support each other with high-level knowledge, a number of protocols were implemented for the publishing of data on the binder.

Current Spatial Context

The spatial context denotes high-level data on the spatial state of the robot and of other entities in its current vicinity. This defines a class of binding proxies that are used by the rest of the system to reason and plan. The following proxies are always present on the binder:

- A *robot* proxy representing the physical robot itself
- An *area* proxy for the area the robot is currently in
- A *position* relation connecting the robot proxy with its area

In addition, the following are represented as appropriate:

- A *person* proxy for each person currently being tracked by the people tracking module
- *area* proxies for each person
- *position* relation proxies between the above
- An *object* proxy for each object belonging to an Area that is being represented
- *position* proxies connecting each object and its corresponding Area
- *Close* relation proxies between the robot and persons that are near to it

The Robot Proxy

There is always exactly one robot proxy on the binder. The only feature of this proxy is its **Concept**: `robot`. The proxy is designed to bind to proxies generated by ComSys, representing the listener – the "You" in a dialogue – and does so on the basis of its **Concept** feature (using concept comparators provided by the Conceptual Mapping SA).

Area Proxies

As described in Section 5.5.2, the navigation subsystem divides space into areas, based on door nodes. At the level of planning and reasoning, these areas constitute the basic units of spatial location. On the binder, the location of objects, persons and the robot itself is represented by position relation proxies connecting the entity with an area proxy.

Area proxies have the sole binding feature **AreaID**, a number that uniquely identifies the area to the navigation subarchitecture. This feature identifies the proxy as an area proxy, and also provides the information necessary for another subarchitecture to create a navigation command to move to the area in question.

Object Proxies

An object proxy has the feature **Concept**, describing the particular class of object that it belongs to – for example, `book` or `mug`. The concept string corresponds to the argument that is used to issue a search command to the Object SA; thus, the feature can be used both for binding and for executing a plan.

Person Proxies

Person proxies store the following binding features:

- **Concept**: always set to `person`
- **Location**: last observed metric position
- **PersonID**: unique identifier

The **Concept** feature provides basic binding control, making sure (through the Conceptual Mapping SA:s comparators) that only other types of person bind to the proxy. The **Location** provides the planner with an exact target for approaching a previously seen person in order to initiate a dialogue. It also helps in binding: as the system is incapable of distinguishing between individuals, it uses position to adjudicate binding. If a newly detected and a previously detected person proxy have locations that are nearer than a certain threshold, they are bound. Obviously, this makes the strong implicit assumption that people are immobile.

Position Relation Proxies

The position proxy is a relational proxy, denoting the spatial relationship "X is in Y", where X is the proxy at the From side and Y the proxy at the To side of the relation. It has the following features:

- **Label**: always set to `position`
- **OtherSourceID**: set to the ID of the navigation subarchitecture; negated (see below)
- **TemporalFrame**: `PERCEIVED` for directly perceived entities; `ASSERTED` for others

The **OtherSourceID** feature is compared to the **SourceID** feature of other proxies on the binder, as one criterion on whether the proxies should bind or not. Since it is negated, this prevents the relation proxy from binding to other position proxies generated by the navigation subarchitecture.

TemporalFrame indicates the currency of the location information. While a person is being tracked, its position is regarded as perceived; if it goes out of perceptual context, yet remains on the binder (due to being bound to another subsystem's proxy), its status is changed to asserted.

This is so that the system can clean up old person proxies: if the asserted proxy binds to a perceived one (such as when the robot returns to a person it has previously had a dialogue with), the newer proxy "takes over" from the older one and the latter can be removed.

Objects' positions are considered perceived indefinitely after they are detected, since object detection is a discrete event, unlike the continuous tracking of persons.

The robot's position is always considered perceived.

Closeness Relation Proxies

Whenever a person is near to the robot, a relation proxy is created between the two, with the features:

- **Label**: always set to `close`
- **OtherSourceID**: set to the ID of the navigation subarchitecture; negated
- **TemporalFrame**: always `PERCEIVED`

The closeness relation between robot and person is considered a prerequisite for initiating dialogue with the person. That is, if the planner is to plan a dialogue action, it must first ensure that closeness holds, by moving the robot up to the person if necessary.

As with position proxies, the "close" proxy has an **OtherSourceID** feature to prevent binding between the "close" proxies themselves. The relation is also always considered perceived; if a person ceases to be tracked by the robot, it is no longer considered "close" even though it may still be on the binder (with an asserted position).

Supplementing Non-spatial Context

The preceding paragraphs define the current spatial context of the robot; that is, the entities that are perceptually relevant to it. In addition, there may be spatial proxies that are not part of this context, yet still remain on the binder. This is done in order to supplement other contexts, such as dialogue. Accordingly, any spatial proxies that are bound to proxies from other modalities will not be removed as they go out of spatial context. Any person or object that is thus sustained will also sustain its containing area and the relation between the two. Similarly, sustained area proxies will also sustain their contents in terms of objects.

For example, a person proxy will usually go out of context and will be removed when the person is no longer being tracked (having passed from the sensor scope of the robot). However, if this person proxy was bound to e.g. a ComSys proxy representing the speaker in a conversation, the spatial person proxy will not be removed. Consequently its position relation proxy and the proxy of the area where the person was last seen also have their removal

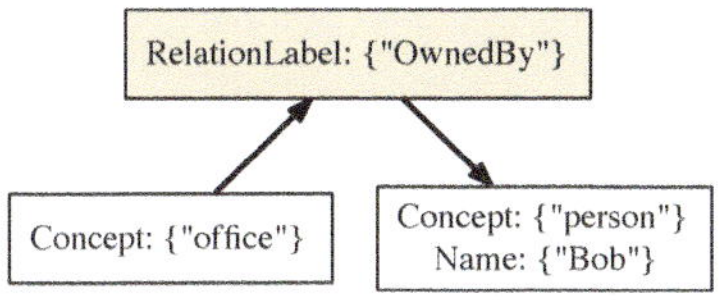

Fig. 10.6. Proxy structure generated by ComSys for "Bob's office"

from the binder suspended. This allows e.g. the planner to access cross-modal information relevant to its plan as it executes it or performs re-planning.

In addition to retaining spatial proxies that are bound to other modalities' proxies, it is sometimes necessary to conversely supplement other modalities' proxies by creating spatial proxies that will bind with them. For example, when ComSys registers the mention of an area that is not currently part of the spatial context, the Conceptual Mapping SA may resolve this mention to a known area instance (see below). It will then provide a proxy of its own, containing an **AreaID** feature of the known area; detecting this, the Navigation SA will in turn create a proxy for that area (provided it hasn't already got one on the binder).

As in the case of the direct spatial context, an area that is put on the binder in this manner will be accompanied by proxies for all objects known to be in that area, as well as position relations linking the objects to the area.

Situated Resolution of Referring Expressions

When the user gives the robot an order that involves a reference to a location, ComSys will generate proxies that reflect the given verbal description. Figure 10.6 shows an example of such a proxy structure. The planner, however, can only send a navigation command to the Navigation SA that contains a goal location that is specified in terms of an `Area-ID`.

The *On Demand Binding Monitor* of the Conceptual Mapping SA offers the competence to resolve such a structural description of a location to an **AreaID**. Whenever the planner needs to resolve an **AreaID** feature for a proxy structure, it will send a request for resolution to the *On Demand Binding Monitor*. This component will then try to find an instance in the conceptual map that matches the structural description represented by the proxies. Internally, the proxy structure is translated into a SPARQL query to the A-Box of the *Reasoner*. Figure 10.7 shows an example of a SPARQL representation for "Bob's office".

The results of that query are then transformed into proxies and put on the binder. The proxies generated by the *On Demand Binding Monitor* of the Conceptual Mapping SA will contain the most specific concepts as **Concept** features, any other additional information stored in the A-Box, such as name information, and most importantly the `Area-ID`.

```
SELECT ?x0 ?x1  WHERE {
      ?x0 rdf:type Office. ?x1 owns ?x0.
      ?x1 rdf:type Person. ?x1 name 'Bob'.
      ?x1 owns ?x0.
}
```

Fig. 10.7. SPARQL query for "Bob's office"

On the binder the original ComSys proxies and their counterparts from the Conceptual Mapping SA are bound to a common union. The *On Demand Binding Monitor* then reports `task done` back to the planner, which then can continue to try to find a plan that satisfies the user's command.

Providing Default Assumptions to the Planner

Sometimes the user gives the robot a task that involves an object whose current location is unknown to the robot. In such a case, the system can make use of the conceptual knowledge represented in the Conceptual Mapping SA. As mentioned earlier, the concepts in the ontology are defined through necessary and sufficient conditions, mostly involving the existence of certain objects in certain places. Here we make use of this encoded implicit knowledge to form assumptions about where certain objects can typically be found. For example, the concept of a library is defined as follows:

```
Class(Library complete Room
  restriction(hasObject valuesFrom(LibraryObject) minCardinality(5))
)

SubClassOf(Library restriction(hasObject someValuesFrom(Book)))
```

This means that `Library` is equivalent to the anonymous concept that fulfills the necessary and sufficient properties of being a subconcept of `Room` and containing at least 5 `LibraryObject` instances. Moreover, a `Library` has the necessary condition of containing `Book` instances. The latter one, however, is not a defining (i.e. necessary and sufficient) property because books can also be found elsewhere.

In any case, such a definition allows the system to form a hypothesis that `Books` and other `LibraryObjects` can be expected in a `Library`. Consequently the *On Demand Binding Monitor* registers its competence to provide typical locations to the Planning & Motivation SA.

When queried by the planner to provide a typical location for a given object, a SPARQL-query to the T-Box of the *Reasoner* is constructed and executed. Figure 10.8 shows a query for the typical location of books.

If a typical location is found, the *On Demand Binding Monitor* creates a proxy for the location that contains its most specific concepts. In our example this would be `Library`. It also creates a proxy for the concept in question (`Book`

```
SELECT DISTINCT ?defaultLoc WHERE {
      ?defaultLoc rdfs:subClassOf ?blankNode.
      ?defaultLoc rdfs:subClassOf oe:Area.
      FILTER (!isBlank(?defaultLoc)).
      FILTER (?defaultLoc != owl:Nothing).
      FILTER (isBlank(?blankNode)).
      ?blankNode owl:someValuesFrom ?defaultObjClass.
      Book rdfs:subClassOf ?defaultObjClass.
}
```

Fig. 10.8. SPARQL query for the typical locations of books

in our example), and a Location relation between them with the additional restriction that it has a **TemporalFrame** feature with value `TYPICAL`. This denotes that the presence of such an object is not guaranteed to hold at a specific point in time, but can be assumed to be typically the case. It is then up to the planner to use this information to execute an action, e.g., a visual search in such a place where the object in question is likely to be found, in order to instantiate this "typical" knowledge with perceived information.

10.4 Planning

The Motivation and Planning Subarchitecture is general-purpose, and has therefore been reused (with different configuration information) in both the Explorer and PlayMate systems (see Chapter 9 for a more detailed description). The data the subarchitecture reasons about, however, is highly domain-specific. In the Explorer scenario, this includes *goals* referring to spatial concepts, *beliefs* about object positions and spatial relations, and *actions* to perceive and manipulate the environment of a mobile robot. The goals, beliefs and actions are represented in the symbolic planning language MAPL (see Chapter 6). The symbols used are maintained by the Address-Variable Map (AVM, see Chapter 2) such that they can be mapped back to binding proxies and, consequently, to component-specific representations later during plan execution.

In the Explorer scenarios described in this chapter, the main motivation for the robot to act is usually provided by an extrinsic source, i.e. a human user gives the robot a task. The ComSys generates appropriate proxies on the Motive working memory which are then translated into a planning goal using the AVM.

Planning starts with the creation of an initial planning state from the contents of the current state of the binder. The task of the planner is then to determine a sequence of actions whose execution from this state will achieve the goal. Interestingly, the planning problems arising in the Explorer scenario are characterised by a large degree of uncertainty and incompleteness in knowledge: the map may still be incomplete, object locations might be unknown and the observability of the environment is severely limited. As explained in

Chapter 6, the planner actively tries to reduce such gaps in the robot's knowledge by using a *continual* planning approach: the planner repeatedly switches from planning to execution in order to gather additional knowledge. Based on the information gained, the planner first revises its current state and, subsequently, its plan. It can then execute this plan further (possibly switching back to planning later again) until a goal has been achieved.

A plan consists of both external (physical) actions and internal (i.e. processing) actions. Physical actions are often scenario-specific. For the Explorer they include:

- *Follow person*: Track and follow a human
- *Approach person*: Move to the proximity of a person
- *Gain attention*: Attract the attention of a human (e.g. say "Excuse me")
- *Move*: Move to a given area in the map
- *Object search*: Perform object search for a given object
- *Inform*: Verbalize and transfer information on an object's position to a human that is close-by

Internal actions are mostly concerned with the task-driven extension of the planning state, i.e. the querying of subarchitectures for additional information that may be relevant for the problem at hand. In the current Explorer scenario, it is mainly the Conceptual Mapping Subarchitecture that is queried for default information, e.g. about where books are usually found. While this kind of information could be provided to the planner in the initial state, this would lead to an information overload: there is just too much information that the different parts of the system could provide to the planner – often at high processing costs, yet relevant only to few tasks (e.g. visual information that has to be extracted from camera data). Thus, instead of generating all this potentially irrelevant data in advance, the continual planner will determine possibly relevant sources of information on its own as part of the initial planning phases in the continual planning process. When these information gathering actions have been executed, a new plan is generated in the next planning phase that exploits the new information.

If a subarchitecture provides some behaviour (sensing, acting, reasoning, etc.) that is to be used by the planner, it needs to define two interfaces to this action for the planner:

- A MAPL action in the planning domain including preconditions, parameters and effects
- An action dispatcher, which translates a request from the planner to execute a specific MAPL action, i.e. with all parameters instantiated, into whatever format is required to set the subarchitecture in action.

During plan execution, the planner calls the appropriate action dispatchers to map the MAPL actions into the local representation used by the executing subarchitecture. For example, an action that in a MAPL plan is described as

```
PhysicalAction motmon0: approach-person motmon4 area_id_0
```

```
Human (H) approaches robot (R) who is idling in the corridor.
H: 'Find me the Borland book.'
R: 'OK.'
R turns and moves off, going to the door into the library. It
enters the door.
R moves about the room, turning to face different directions at
each location, searching for the Borland book.
R detects the book.
R moves back out into the corridor and up to H.
R: 'The Borland book is in the library.'
```

Fig. 10.9. An example of the Explorer performing an object localization task

contains the planner symbols `motmon0`, `motmon4` and `area_id_0`, which correspond to the robot proxy, the user person proxy, and the area proxy on the binder, respectively. The action dispatcher uses the AVM to find the proxy corresponding to `motmon4`, extract the **Location** feature for the person and issue a low-level movement command to the Navigation SA to move to that location.

10.5 Scenario: Find Object

Here, we describe in detail an example of a task performed by the robot, and how the different parts of the system contribute to fulfilling this task.

The robot is ordered by a user to "Find me the Borland book". It moves off to look for the book, visually locates it, and returns to report its findings. The externally apparent features of the task are described in Figure 10.9.

Initial Binding State

Before the order "Find me the Borland book" is spoken, the following entities are represented on the binder:

- The robot
- The area corresponding to the corridor (with **AreaID** #1)
- A person
- A Position relation connecting robot and area (TemporalFrame `PERCEIVED`)
- A Position relation connecting person and area (TemporalFrame `PERCEIVED`)
- A Close relation connecting robot and person (TemporalFrame `PERCEIVED`)

Note that no objects are present, nor are other areas except the current area. A snapshot from a system run illustrates the initial state (Figure 10.10).

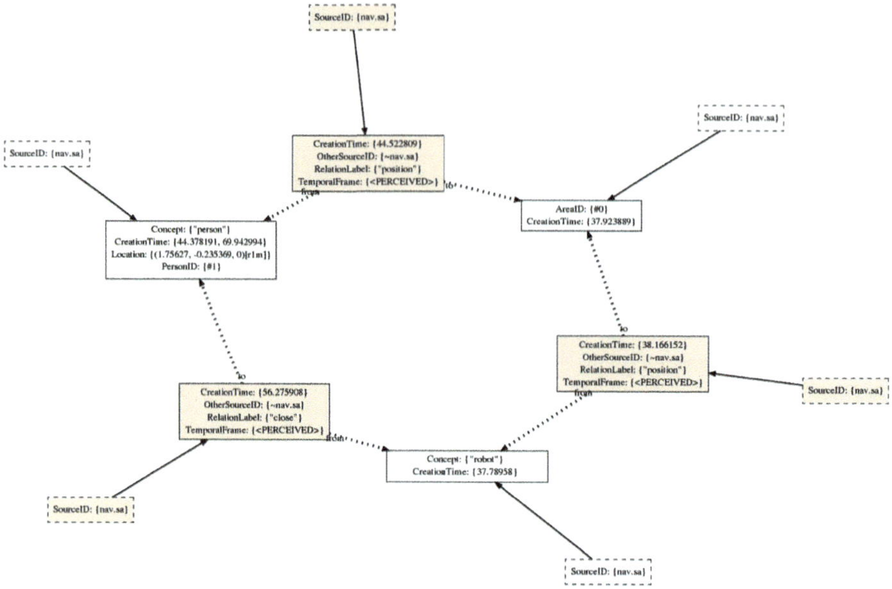

Fig. 10.10. Initial binding state

Processing Utterance

As ComSys receives and interprets the user's phrase "Find me the Borland book", it adds the following proxies to the binder, corresponding to the different parts of the utterance.

- The robot itself, being the recipient of an order, is represented by a proxy with **Concept** `addressee`, which binds to the robot proxy already present.
- The word "me", referring to the speaker, generates a "person" proxy identified by the **Name** feature `I`.
- The expression referring to the book is represented by a "Borland_book" proxy, which is not bound to any other proxies at this point.

Motive Generation

The phrase "Find me the Borland book" is used to generate the motive that will provide the planner with a goal state, as described in Chapter 2: ComSys writes proxies to Motivation SA working memory corresponding to the semantic interpretation of the command. The motive generator then creates a motive structure, which the motive manager turns into a planning goal.

The planning goal resulting from the above command is an epistemic one, saying, in words, that the user needs to know the position of the Borland book.[4] In the planning language MAPL (see Chapter 6) this goal is represented as follows:

```
(:goal (K motmon4 (perceived-position motmon6)))
```

where the symbols `motmon4` and `motmon6` represent the user and the book, respectively. These symbols are maintained by the Address-Variable Map (AVM, see Chapter 2) which allows the planner to refer back to their respective source modalities during execution.

Continual Plan Creation

Planning starts with the creation of an initial planning state from the contents of the current state of the binder. The planner will determine a sequence of actions whose execution from this state will achieve the goal.

Planning in both the Explorer and the PlayMate system is a continual process, i.e. the plan is revised repeatedly during its execution. Plan revision occurs due to external reasons (exogenous events, unexpected execution results) or because internal state changes enable the planner to fill in details in its plan that have been deliberately postponed before.

The early phases of continual planning in this Explorer scenario can be described as means-end reasoning to determine necessary information for more detailed planning. For example, the planner first forms a very abstract plan which, in words, can be expressed as "determine the possible position of the book by querying some person or subarchitecture who supposedly knows, then verify this information by actually going there and identifying the book. Finally, go back and relate that information to the user." The continual planning process thus first queries the internal Conceptual Map Subarchitecture which provides information about default locations of books. This new information will trigger a plan revision (for details of this process, see Chapter 6) that leads to a new, more detailed, plan that involves concrete movement to the library and a search for the object there. If execution of this plan fails, e.g. because the book is not in the library, this is detected during plan monitoring and leads to another plan revision. The planner will then rely on sources of information other than Conceptual Mapping, e.g. humans (except the user). For the planner, querying a human and querying another subarchitecture about some information are identical. Both behaviours are planned as `ask-val` actions, the only difference to the planner being the addressee and the state variable the query is about. Likewise, providing information to another agent is realised by the same planning operator `tell-val`, regardless of whether this agent is a human or another subarchitecture.

While humans and other external agents are mostly treated the same as internal components, the planner makes one important distinction: external agents will usually be given *acknowledgements* when requests have been accepted and when achieved. To that end, the planner implements the Continual

[4] This is a convenient interpretation of "Find me the Borland book" because the robot cannot pick anything up.

Collaborative Planning algorithm presented in Section 6.7. The realisation of acknowledgements is not determined by the planner and is usually realised by the ComSys.

The plan created initially upon receiving the order "Find me the Borland book" consists of the following steps:

1. Acknowledge command acceptance to user
2. Have the Conceptual Mapping SA provide a default position for the book
3. Search the position provided for the book
4. Tell the user the perceived position of the book

Plan Execution and Revision

The first actions in the plan (acknowledging the new command and querying for default information about the position of the book) are dispatched to the appropriate execution modules (ComSys and Conceptual Mapping SA, respectively). As a result, ComSys produces the utterance "OK" and the Conceptual Mapping SA augments the binder state with the information that the Borland Book, being a book, will typically be found in a library. This is represented by a proxy of **Concept** **`borland_book`** and a proxy of **Concept** **`library`**, connected by a relation **`position`**, with `TYPICAL` status.

The Conceptual Mapping SA also volunteers the specific information it has on libraries; namely, the fact that it knows about an area that is a library, with a specific **AreaID** (#1). It publishes this information in the form of another proxy with **Concept** `library` and **AreaID** #1. On the binder, this information is bound together into a structure whose meaning is "The Borland book will typically be in area #1" (see Figure 10.11).

The addition of this new knowledge to the planning state will lead to the expansion of the assertion `object-search-abstract` which abstracted from the actual position of the book as long as there was no hypothesis for its position. When such a hypothesis exists, this assertion is no longer allowed to be used (see Chapter 6), i.e. the planner is forced to reason in a more detailed fashion with respect to the book position. After replanning, a more concrete plan is produced:

1. Move to area #1
2. Search the current room for the book
3. Get back to user (area #0)
4. Tell the user the perceived position of the book

Note that by now, the planner also is certain that for finding the book the robot must move away from the user. Thus it includes an action for getting back to him later in the plan.

Using the correct **AreaID**, the planner issues a navigation command "Go to area #1" to the Navigation SA, and the robot moves into the library. The planner then proceeds to issue an object search command to the Object SA.

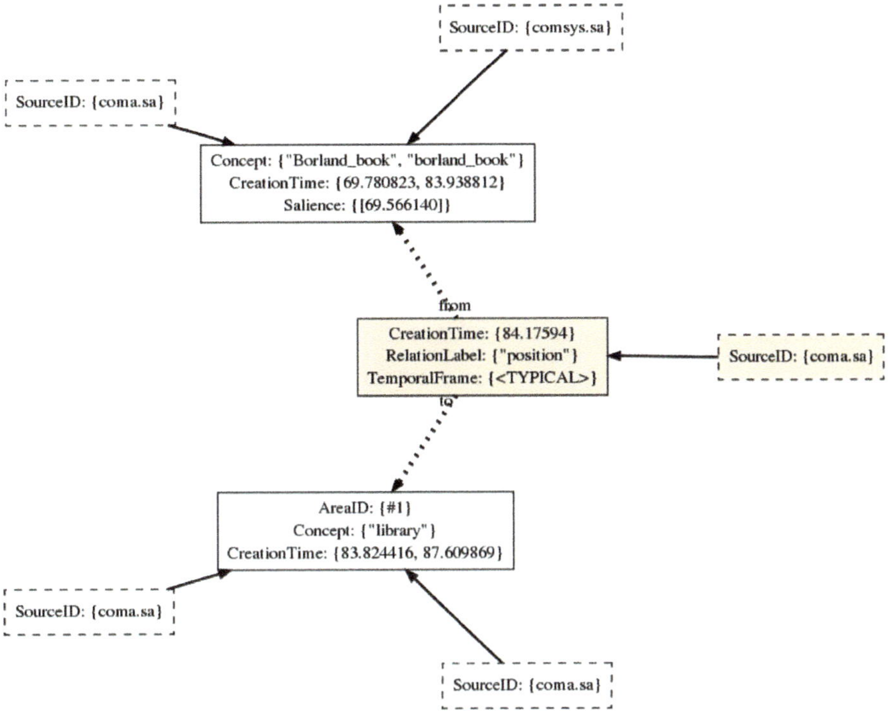

Fig. 10.11. Binding representation of hypothetical position of Borland book

The robot searches the room as described in Section 5.7.1. Once the object is found, the Navigation SA adds it to the navigation graph. Since it is part of the current spatial context, it is also exported to the binder in the form of an object proxy, connected to the room's proxy by a new position proxy. This position proxy has `PERCEIVED` temporal frame.

The new proxies bind to the old complex, resulting in the structure in Figure 10.12. Plan monitoring verifies that the search action was successful. Because the perceived book proxy (with source ID `nav.sa`) has been bound to the proxy from ComSys, the Planner can surmise that the position of the former applies also to the latter. Thus, the robot now knows the position of the Borland book that the user mentioned, which means the precondition now holds for "Tell the user the perceived position of the book". Consequently, plan execution goes on to the "Move to the user" action. If object search had failed, this would have invalidated the plan and triggered replanning.

The person proxy on the binder put there by the Navigation SA remains, even though it is no longer being tracked, because it is bound to the user's ComSys proxy (see Figure 10.13). Thus, the planner can read the **Location** feature of the person from the binder, and issue a navigation command to

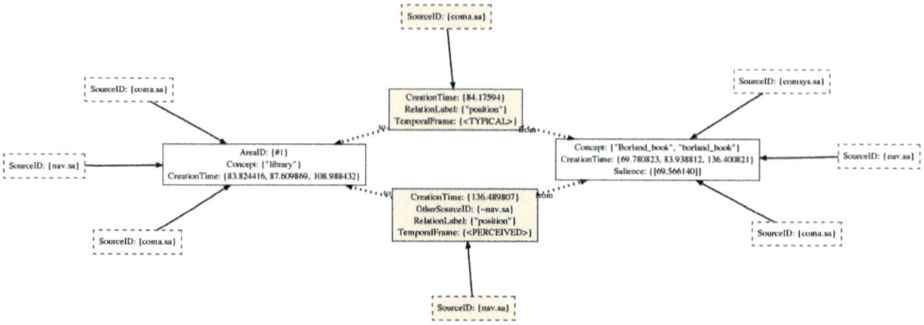

Fig. 10.12. Visually confirmed information about the book is added to the binder

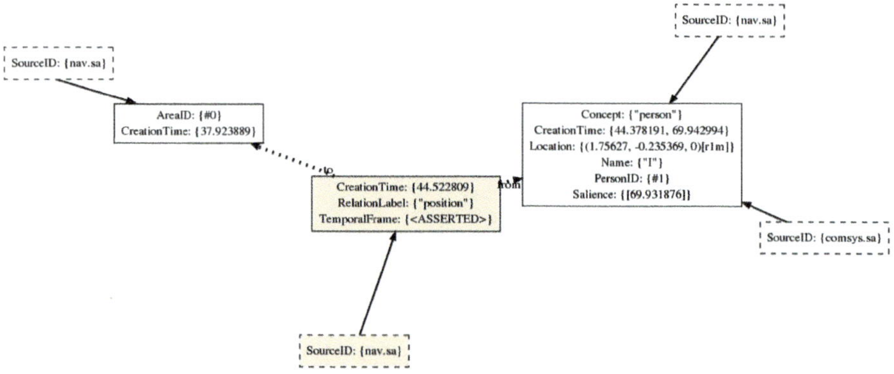

Fig. 10.13. Retained binding data about the user

the Navigation SA to go to this location. Once there, it calls upon ComSys to formulate a response to the user's request, using the binder contents (Figure 10.12) to do so.

10.6 Conclusions

In this chapter we have presented one more instantiation of the CAS architecture schema in the form of the Explorer system. The same software framework and many of the components are common with the PlayMate instatiation. While the PlayMate and the Explorer share the underlying framework and many components, the scenarios/tasks are in fact quite different. This provides evidence that CAS/CAST are general tools and can be used for a wide variety of problems.

In this chapter we have also shown how we can share representations across different modules by employing abstraction and various strategies for binding knowledge, that is associating knowledge from one sub-system with that from another. We showed how the system can bind not only with explicitly perceived knowledge but also with ontological knowledge. This provides a mechanism to, for example, fall back to knowledge about the typical location of objects and thus provide a working hypothesis for a task even when the location of an object is not known.

The abstraction of knowledge also provided the means to move away from the hard-coded coupling between user input and action and use a much more flexible solution using a planner.

With the Explorer system, we also wanted to investigate how to create spatial representations which could bridge the gap between low-level robot control, and the qualitative ways in which humans tend to understand space. We approached this problem from a system point of view, looking at how the combination of different information sources could help to bridge that gap – building up a more comprehensive sense of space.

Following out that approach, we have found that integrating different modalities leads to significant synergies in building up a more complete understanding of the spatial organization of an environment, particularly towards a semantic understanding. Synergetic effects could be observed in information sources complementing each other, and in disambiguating interpretations. These synergies happen over time, and have highlighted an important requirement for spatial knowledge representation and reasoning. Namely, knowledge must not, and cannot, be irrevocable. Spatial reasoning appears to be inherently non-monotonic. The robot needs to be able to retract earlier inferences, to prevent that erroneously acquired or asserted knowledge leads to irrecoverable errors in inferred knowledge.

Synergies only arise when we integrate many components. And that integration brings not only more complete knowledge and more capabilities, it also increases complexity and presents problems due to the fact that the real world is unpredictable to some extent. For example, in a scenario where the robot continuously interacts with a user and is facing her/him most of the time, the information content of the sensor input suffers as the user occupies a large part of the field of view. In our case, the camera was mounted on a pan-tilt unit and could have been used to actively look for objects and build a metric map using visual information while following the user. However, this conflicts with the use of the camera to indicate the focus of attention on the user. As a result, most of the time the camera only sees the user and not the environment. The user's presence not only disturbs the visual object recognition but also influences the performance of the multi-modal place classification. From this practical issue, we can derive again more fundamental issues too. Integration should not only lead to synergy, but also to robustness and plasticity. In forming more complete interpretations, dependencies between modalities should not be static. In situ, a robot should be able to resort to alternative means for perception. And over time, a robot should be able to complete its incomplete observations, e.g. through autonomous exploration.

In addition to such practical issues, the experiments we ran in real environments highlighted new requirements for the system. For example, spatial referencing needs to be improved in both directions of the communication and using several modalities. This would allow the user to indicate a specific object through, e.g., gesture or gaze direction when saying “This is X”.

References

1. Burgard, W., Cremers, A., Fox, D., Hähnel, D., Lakemeyer, G., Schulz, D., Steiner, W., Thrun, S.: Experiences with an interactive museum tour-guide robot. Artificial Intelligence 114(1-2)
2. Siegwart, R., et al.: Robox at expo.02: A large scale installation of personal robots. Robotics and Autonomous Systems 42, 203–222 (2003)
3. Ishiguro, H., Ono, T., Imai, M., Maeda, T., Kanda, T., Nakatsu, R.: Robovie: an interactive humanoid robot. Int. J. Industrial Robotics 28(6), 498–503 (2001)
4. Haasch, A., Hohenner, S., Huewel, S., Kleinehagenbrock, M., Lang, S., Toptsis, I., Fink, G.A., Fritsch, J., Wrede, B., Sagerer, G.: Biron - the Bielefeld robot companion. In: Prassler, E., Lawitzky, G., Fiorini, P., Haegele, M. (eds.) Proc. Int. Workshop on Advances in Service Robotics, pp. 27–32. Fraunhofer IRB Verlag, Stuttgart (2004)
5. Bos, J., Klein, E., Oka, T.: Meaningful conversation with a mobile robot. In: Proceedings of the Research Note Sessions of the 10th Conference of the European Chapter of the Association for Computational Linguistics (EACL 2003), Budapest, Hungary (2003)
6. Lemon, O., Bracy, A., Gruenstein, A., Peters, S.: A multi-modal dialogue system for human-robot conversation. In: Proceedings of the Second Meeting of the North American Chapter of the Association of Computational Linguistics (NAACL 2001), Pittsburg, PA (2001)
7. Sidner, C., Kidd, C., Lee, C., Lesh, N.: Where to look: A study of human-robot engagement. In: Proceedings of the ACM International Conference on Intelligent User Interfaces (IUI), pp. 78–84 (2004)
8. Kruijff, G., Lison, P., Benjamin, T., Jacobsson, H., Hawes, N.: Incremental, multi-level processing for comprehending visually situated dialogue in human-robot interaction. In: Proceedings of the Symposium on Language and Robotics (LANGRO 2007), Aveiro, Portugal (2007), `http://cognitivesystems.org/cosybook/chap10.asp#Kruijff/etal:LANGRO2007`
9. Traum, D., Larsson, S.: The information state approach to dialogue management. In: van Kuppevelt, J., Smith, R. (eds.) Current and New Directions in Discourse and Dialogue. Kluwer Academic Publishers, Dordrecht (2003)
10. Lindström, M., Eklundh, J.-O.: Detecting and tracking moving objects from a mobile platform using a laser range scanner. In: Proc. of the IEEE/RSJ International Conference on Intelligent Robots and Systems (IROS 2001), Wailea Maui HI, USA, vol. 3, pp. 1364–1369 (2001)
11. Wang, C.-C., Thorpe, C.: Simultaneous localization and mapping with detection and tracking of moving objects. In: Proceedings of the IEEE International Conference on Robotics and Automation (ICRA 2002), vol. 3, pp. 2918–2924 (2002)
12. Minguez, J., Montano, L.: Nearness diagram navigation (ND): Collision avoidance in troublesome scenarios. IEEE Transactions on Robotics and Automation 20(1), 45–59 (2004)

13. Topp, E.A., Christensen, H.I.: Topological modelling for human augmented mapping. In: Proc. of the IEEE/RSJ International Conference on Intelligent Robots and Systems (IROS 2006), Beijing, China (2006)
14. Kruijff, G.-J.M., Zender, H., Jensfelt, P., Christensen, H.I.: Situated dialogue and understanding spatial organization: Knowing what is where and what you can do there. In: Proc. of the 15th IEEE International Symposium on Robot and Human Interactive Communication (RO-MAN), Hatfield, UK, pp. 328–333 (2006), `http://cognitivesystems.org/cosybook/chap10.asp#kruijff/etal:2006-roman`
15. Topp, E.A., Hüttenrauch, H., Christensen, H., Severinson Eklundh, K.: Bringing together human and robotic environment representations – a pilot study. In: Proc. of the IEEE/RSJ International Conference on Intelligent Robots and Systems (IROS), Beijing, China (2006)
16. Shi, H., Tenbrink, T.: Telling rolland where to go: Hri dialogues on route navigation. In: Workshop on Spatial Language and Dialogue (5th Workshop on Language and Space), Delmenhorst, Germany (2005)
17. Kruijff, G.-J.M., Zender, H., Jensfelt, P., Christensen, H.I.: Situated dialogue and spatial organization: What, where... and why? International Journal of Advanced Robotic Systems, special section on Human and Robot Interactive Communication 4(2), `http://cognitivesystems.org/cosybook/chap10.asp#kruijff/etal:jars`
18. Yamauchi, B.: A frontier-based approach for autonomous exploration. In: Proc. of the 1997 IEEE International Symposium on Computational Intelligence in Robotics and Automation, Monterey, CA, pp. 146–151 (1997)

11

Lessons Learnt from Scenario-Based Integration

Nick Hawes[1], Michael Zillich[2], and Patric Jensfelt[3]

[1] Intelligent Robotics Lab, School of Computer Science,
University of Birmingham, Birmingham, UK
n.a.hawes@cs.bham.ac.uk
[2] Automation and Control Institute,
Vienna University of Technology, Vienna, Austria
zillich@acin.tuwien.ac.at
[3] Royal Institute of Technology (KTH), Centre for Autonomous Systems,
Stockholm, Sweden
patric@csc.kth.se

11.1 Introduction

From the very start the CoSy project set out to demonstrate and evaluate its progress in implemented, integrated systems. Chapters 9 & 10 set out both the two scenarios we chose to integrate around, and the contributions we made by studying problems following an integrative, rather than isolationist, methodology. However, these contributions did not come without a cost. Following an integrated systems methodology (and therefore delivering a genuinely integrated project) demands a large input in terms of person hours, a demand which is regularly underestimated in the planning phase (both of whole projects and of development cycles). In CoSy we put in an extremely large amount of time and effort into the "integration process." At some point or other almost everyone associated with the project wrote code that was used in a demonstrator system. From undergraduates and masters students, to postgrads and postdocs, up to PIs and other faculty members, we all bought into the collective ingenuity or insanity required to produce a state-of-the-art intelligent robot. It is rare that so many people from so many different disciplines work together to integrate at this scale. Whilst many of us have built integrated systems before (some of which could do more within a single domain) none of us have worked to put so much from many different fields into a single system.

So, what can we learn from our experiences? The short (and almost universally appreciated) answer is that integration is *hard*. Very hard. There is not just one reason why this is the case, instead there are many interrelated issues that make the kind of problems we are often faced with hard. These issues

H.I. Christensen et al. (Eds.): Cognitive Systems, COSMOS 8, pp. 423–438.
springerlink.com

are explored in Section 11.2. Having been through the good, the bad and the ugly of the integration process over four years, we have a lot of informative (if not unique) first-hand experience. The role of this chapter is to attempt to share this experience with others, with the hope that they can learn from both our successes and our failures. We have distilled our experiences into a collection of lessons. These are presented in Section 11.3.

Although the rest of this book is a scholarly work, this chapter is equal parts memoir, training tool and self-help guide. We are a collection of scientists who must tackle large engineering challenges in order to demonstrate theoretical advances in our fields. Within our fields we apply various evaluation methods to supply evidence to support our conclusions. Faced with the task of doing the same for our conclusions on the integration process, we are left only with a collection of subjective experiences. So, whilst we suggest that our recommendations are taken with a pinch of salt, please consider them for what they are: the findings of scientists trying to understand their own behaviour[1].

It is worth asking whether our experiences in the CoSy project qualify us for commenting on the difficulties of systems engineering for science. Perhaps we were all just painfully bad engineers and adopted flawed practices from the start. Whilst all projects have their weaknesses, on average we'd like to think our engineering abilities were at least above average[2]. To bolster our defence we would like to present the evidence of the complexity of the systems we were working on. Chapters 9 & 10 present overviews of systems that include components of massively varying levels of sophistication (from background subtraction, to multi-level intentional action recognition), maturity (from brand new components, to off-the-shelf techniques wrapped in CAST) and experimental purpose (from demonstrating the power of small changes to an algorithm, to justifying the existence of a collection of components). Making all of these components, and their developers, work together in perfect harmony requires an effort that goes beyond basic software engineering.

11.2 But Is It Implemented?

At the heart of the difficulties of doing integrated systems science is a tradeoff: innovation vs. realisation (or science vs. engineering). It is crucial that any new advance is demonstrated in an implemented system; it is often not enough to demonstrate the theoretical potential of a new idea, it must be implemented and evaluated *in silico* before it will pass peer review. Conversely, it is not just enough to demonstrate that some piece of software or hardware works in practice; there must be a theoretical underpinning to justify its contribution to knowledge. The best work in intelligent systems science, and in AI and robotics in general, has a good balance between the theoretical and practical: not only does the work represent a theoretical advance, its

[1] Some people would argue that CoSy is similarly inspired.

[2] But we would say that, wouldn't we!

implementation is solid enough for its contributions to be repeatedly demonstrated in a practical setting. Examples of recent component science that has had this balance include the Fast Forward planner [1] and the SIFT vision algorithm [2]. In terms of system science it is very rare to find whole systems (rather than single components) that are available to the community. This is in part due to the difficulties caused by differences in software and hardware between groups, and the additional effort required to make complex systems of interdependent components available in a usable state. Additionally, the lack of systems, and system science, that is available for testing by the community is potentially due to the relative youth of the field.

At the moment we are still in the early stages of developing a science of building intelligent systems. This means that our theories that can be tested by integration (as opposed to the component theories that can be tested in isolation) are still taking shape, and we are still learning what questions to ask about how we put the parts together. In opposition to this stands "scenario-based integration" (as described in Section 1.4.8) where we know the kinds of behaviours we'd like our system to generate, but we're free to specify how they're achieved. Whilst this is a compelling approach to setting integrated system challenges, it has the potential to focus minds on the wrong element of what is being delivered at the end of the process (a particular implementation vs. new knowledge).

Related to this is the issue of reliability. For a system to really satisfy the requirements of a scenario (and to be a compelling demonstrator at a project review) it must be engineered to a standard where it can perform a task repeatedly without crashing (physically or virtually). This arguably requires a great deal more effort than building a system to be run purely in a lab to generate results for a research paper. Herein lies one of the difficulties of integration: if we are doing science (and we are employed as scientists after all!) then why should we spend our valuable time engineering a system to be a convincing demonstrator when a lesser effort is required to test our hypotheses? One answer to this is that we *are* interested in intelligent systems as systems that can operate successfully in the long term. Our scientific aims should not just address snapshots of behaviour (e.g. can it do this), but how collections of behavioural competences work together in an intelligent system over the lifetime of that system. A second answer is that any science of artifacts is dependent on engineering to support any demonstration of progress. Obviously it is possible to have solid scientific ideas demonstrated with unreliable engineering and vice versa, but it is easier to evaluate *and* demonstrate the science when the engineering is transparent.

11.3 Lessons

From our experiences of the integration process we have created a series of lessons that we have learnt. We present them here in a rough order of importance. Although some readers may place more emphasis on one or more of the lessons, or prefer them in a different order, it is probably worth considering all of them when considering proposing, starting or approving an integrated systems project.

11.3.1 Integrate Ideas First

The most fundamental issue in integration is that the ideas underlying the work must be integrated before a line of code is written. This means there must be a clear reason for building a particular system in a particular way before the engineering begins. Generally this means the scientific aims of the integration (whether the system is an answer to a hypothesis itself, or whether it is a tool to support the exploration of other hypotheses) must be set out from the beginning. Although this may sound painfully obvious, it is not always the case in large-scale projects. Many such projects promise integrated systems as a deliverable or demonstrator, without truly understanding the reasons for building such a thing (or the cost they will have to pay in terms of person-hours). We refer to this as integration for integration's sake. In this way some researchers end up being assigned integration tasks they are not happy with, or equipped or employed for, and the project suffers as a whole. This lesson generalises to "think before you act", a statement which sounds simple in theory, but is surprisingly hard in practice. This is made doubly hard in integrated systems projects where most contributors have dual loyalties: to their component science (which often exists independent of the project) and to the system as a whole.

A second element of this first lesson is that integrating ideas first means that a more concerted effort can be made to only integrate ideas that are a good match for each other, or clearly all fit into a single system. One of the side-effects of integration for integration's sake, where the end product of the process is a working system rather than new knowledge, is that separate ideas which work well in isolation (often as clearly defined units of work from separate workpackages) are roughly bolted together to show that they all contribute to the same end. However, this type of integration often shows nothing about the individual pieces of work that wasn't demonstrated in isolation, and, because the integration process was shallow and aimed only at demonstrating integration in any form, nothing is learnt about the science of building integrated systems.

In the CoSy project we were lucky to be part of a project where ideas were (mostly) integrated early on, and where these ideas often came from the synergistic effects of working on a multi-disciplinary project with scientific goals. However, we did have our "integration for integration's sake" moments and we have been part of, and talked to researchers from, other projects which were less dedicated to doing science in integrated systems. Our experiences support this lesson in two additional ways. First, our integrators were much more motivated when the engineering work was being performed based on ideas that were clear from the start and had some relationship to scientific integration. Second, and perhaps most importantly, when there is a good match between the ideas that are the basis for the integration, the interplay between theory and practice can be very productive (e.g. as in the interplay between the CAS architecture schema and our ideas of cross-modal binding). In such cases problems in engineering can sometimes be related to problems in the original ideas, and progress can be made across both domains when a solution is found. It is also the case that concepts originating from the engineering can work their way back into the theory if they prove powerful enough. We experienced this when we discovered that some kind of change notification system is an essential part of the theory of CAS as well as its implementation. Such progress cannot be ruled out through misguided integration efforts too, but it seems that real advances in integrated systems require a certain relationship between the theory and practice, and shallow integration efforts are less likely to foster this relationship.

One important effect of idea integration is that it reduces the instances of a common misconception that occurs when integrating ones own piece of work (e.g. in form of a software component) with work of others: the X is easy fallacy. This describes the assumption that the other parts of an integrated system are essentially solved, rather than being work in progress (as they should be for most ambitious research projects). Underlying this misconception is often the feeling that the other problems within an integrated system are somehow easier to solve than somebody's ones own work. This is a view which stems from a lack of understanding of the difficulties and open research issues in other fields. Whilst it is impossible to overcome this lack of understanding for all people and all fields (we'd all have to become experts in everything), by integrating ideas at the outset of a project at least a superficial understanding of the key problems related to the project can be gained by all involved parties.

To illustrate this, consider the following, overly optimistic, statement that you might hear during a system design meeting: "... and then the vision component puts all the objects found on the table into working memory ...". Needless to say (for a vision researcher) this is a very ill-posed problem without an existing solution. What happens if an object cannot be detected? What about false positives? What about partly occluded objects? What happens if an object is not actually on the table but is currently being grasped and carried by the robot? What if the scene is currently changing? And what is an object anyway? Are markings, scratches and sticky tape on the table objects too? Do you expect three objects or three-thousand? Could the rest of the system handle three-thousand objects? Although this example features vision, similar assumptions can easily be made (and have been made) about many other disciplines.

If all parts of a system are designed under the assumption that other parts will simply work flawlessly in the required manner, the result will be a very brittle system that works in small number of restricted cases. It is important to keep in mind that nothing is easy. Part of integrating ideas (and people) from diverse fields is building an awareness of the possibilities and limitations of methods within different fields. This requires that individual researchers occasionally take the difficult decision to say "No - we can't and won't do that. It's impossible (within the time frame of the project, with the given personnel resources)". Even if the echo would be "But I saw this paper by group X (or video on YouTube) - they solved it!", it is always worth asking which sub-sub-problem of the big problem did this group solve, and how was the solution integrated with the rest of its enclosing system (as we unfortunately often see an inverse relationship between the power of a deployed solution and the strength of its integration with other systems).

11.3.2 Integrate People Second

Given that you've learnt from the previous lesson and have wisely chosen and integrated your ideas for an intelligent system, then you have to get it built. The next lesson to learn is that it is not only the ideas that must be integrated for your efforts to succeed, your people must be integrated too. The integration of people primarily means the establishment of communication channels between the people designing and building the system[3]. For building an integrated system based on new scientific ideas, this means more than just having an email address or Skype ID for everyone else in the team. For such engineering tasks it is essential that tight feedback loops are created and maintained between the people that have the design ideas (including the science that is to be done), and the people doing the implementation work. Often these are the same people, but it is important that, as integration across ideas is crucial to the project's success, this theoretic integration informs the engineering as much as possible. One reason that this is important is that the problems with the theoretical integration often emerges during early engineering work. Problems with the designs should be fed back quickly to all relevant parties, rather than worked around in software.

The ultimate aim of people integration is for each member of the integration team to have a shared understanding of the problems they are solving and the methods they are employing. Such an understanding is necessary because decision made during the implementation of one part of the system can have inadvertent effects on other parts of the system. Although careful planning may help to avoid this, it is really only an appreciation of the reasoning underlying the plans that can keep people pulling in the same direction. If people integration is successful, then they will not only have a shared understanding, but also shared goals in terms of the whole system. Such shared goals should ultimately mean that researchers and engineers move away from only caring about one part of the system (vision, manipulation, language etc.), and start

[3] Although this will have had to have happened to some degree during the ideas phase, communication channels may only exist between team leaders or people with a history of previous interactions.

to care about the interplay of these systems and the effects they have on overall system performance. Naturally there is a balance to be found in this. Whilst too little group investment in shared goals will hamper the integration process, too much investment in them may distract from the component scientific goals that drive much project work (especially PhD work).

From experience it seems apparent that integrating people around shared goals prevents them from getting disenfranchised during the integration process. From a management standpoint this means that people stay motivated, work for each other, and take responsibility for developments across the whole project. Such behavioural characteristics are very difficult to instill in a top-down manner (particularly in scientists who do not take direction easily), but develop naturally through interpersonal relationships around shared challenges (via social, and peer, pressure). To draw a parallel with the previous lesson: whilst the science of integrated systems will develop from the synergies between ideas that are traditionally studied in isolation, the integrated systems themselves will best emerge from the synergies between people.

11.3.3 Choose Your Tools Wisely

The next lesson concerns one of key practical elements of the creation of an integrated system: the speed at which ideas can be turned into functioning code. Although this speed is influenced by many things (including the integration of ideas and people, plus their underlying abilities as scientists, engineers and communicators), the tools that are used during the integration process can have a massive impact (both positively and negatively). A tool may be anything that is used in the workflow from thought to software; anything from a mailing list to a programming language, from a software library to robotic hardware. Although it is to be expected that the tools available and adopted differ from site to site in a large project, we will focus on the tools which are chosen for project-wide use. As our experience is mostly with software tools, we will focus on these at the expense of hardware choices (which can have a massive impact across projects).

For us, the ultimate tool is something which allows us to achieve some goal (e.g. implementing an algorithm or behaviour, communicating an idea, finding a bug, etc.) which would not have been possible, or would have been a lot harder to achieve, without the tool. Ideally, any tool will not make any other (unrelated) goal harder to achieve, or otherwise hinder progress. Such an ideal is rarely realised, with most tools either reducing the space of development options available at a later stage, requiring new concepts or practices to be learnt, or (perhaps most insidiously) old practices to be changed. There is often a relationship between the amount of these practical 'hoops' an engineer is willing to jump through, and the (perceived) benefit of using the tool in the first place.

In CoSy we enforced some tool choices across the consortium. These were Subversion[4] for software versioning, the use of the CMake[5] and Apache Ant[6] build systems, the use of the C++ and Java programming languages, and the use of the CAST architecture toolkit [3] (which brought with it the implicit use of the OmniORB CORBA libraries[7] and Boost[8] shared pointers). Some tool conventions emerged through consensus and collaboration across the consortium. These were a mailing list purely for integration developments, Doxygen[9] for documenting code, Bugzilla[10] for bug-tracking, common directory structures and software release practices. Other tools were adopted selectively by single partners and their use was accommodated (but not necessarily widely taken up) by the consortium. These were OpenCV[11] for image-processing and as a simple windowing toolkit; QT[12] and SWT[13] for more complex interfaces; various forms of run scripts, configuration files, and configuration options to components; and Matlab[14] and Python[15] as additional programming languages (which required specialised integration mechanisms). In hindsight almost all of these things (whether software libraries or development processes) should be considered during the planning of an integration process, and standardised and enforced as required[16]. However there is a tension between enforcing standard tools and practices which allow the majority to make progress with the least amount of overhead, and giving talented individuals the freedom to tackle their problems in the ways that come naturally to them.

To develop this theme further we can consider two examples from the above list: OpenCV and CAST. At some point in time almost every developer of a component from the visual subarchitecture needed to pop up a window to show an image (showing the current frame, a region of interest, or some result of processing). When doing this in relative isolation (i.e. developing their component with the bare minimum of other components) using OpenCV for this was simple and efficient. However, OpenCV was not designed for use in multi-threaded systems, so when two or more components with OpenCV visualisations were placed together in the system we often saw crashes when one

[4] http://subversion.tigris.org
[5] http://www.cmake.org
[6] http://ant.apache.org
[7] http://omniorb.sourceforge.net
[8] http://www.boost.org
[9] http://www.stack.nl/ dimitri/doxygen
[10] http://www.bugzilla.org
[11] http://opencv.willowgarage.com/wiki
[12] http://trolltech.com/products
[13] http://www.eclipse.org/swt
[14] http://www.mathworks.com/products/matlab
[15] http://www.python.org
[16] The types of tools and the issues they address should be considered, not necessarily these particular instances.

of them tried to open a window. This simple example demonstrates a number of things that can be generalised to more complex examples. First, the importance of forward planning in terms of tool choice. If the individual users of OpenCV had considered the future uses of their components (including debugging them as part of the whole system, i.e. testing them in an integrated context) they may have made a different choice. Second, it demonstrates how a single choice made in a couple of components can effect software across the whole project: if a third party ran two OpenCV-using components without knowledge of their contents, they would experience behaviour that the component authors would not have experienced on their own. Third, and perhaps incidentally, it shows how tools for debugging and visiualisation (in this case the ability to display results in a window) are an important element of any software project. Because such tools should be simple and accessible developers tend to use tools they are already familiar with (hence the use of OpenCV here). If a project is intending to standardise software tools across its members, such common debugging elements (including logging and image display) should be one of the focus points of this effort. In complex integrated systems debugging happens continually, and is often required with every new combination of components. As such it should be made as easy, and as uniform, as possible.

The second example of tool use in CoSy is CAST, our architecture schema toolkit. CAST was used from year 2 onwards in the project as the sole software middleware for components in our integrated systems. From its inception CAST always had two purposes. It's practical role was to make writing multi-language, distributed, component-based architectures as simple as possible (after some initial bad experiences with an arbitrary collection of other tools). Under the cover of this practical tool (i.e. something that should make people's lives easier) we also used CAST as an embodiment of our architectural theory (see Chapter 2), thus enabling us to integrate these important ideas from an early stage. However, in the early stages (year 2 of the project) CAST was a less than useful software tool. It was initially hard to learn, configure and debug, and contained numerous, occasionally obscure, bugs. This allowed us to witness first-hand the adverse effects of enforcing a bad software choice on a whole project: that first year we used CAST we almost certainly made less progress than we would've done without it. However, over the remaining years of the project (and afterwards) we tracked down bugs and refactored CAST to be much more useful as a development tool. Whilst this was pure engineering, making our tools better ultimately allowed us to do better science (in this case build more complex integrated systems in less time from heterogeneous parts). Watching CAST progress from a hindrance to a help demonstrated the power of choosing the right tools.

When tools pervade throughout a system they not only aid its development, they also start to shape its design. Although a special case (as it was intended to have this effect from the start), CAST demonstrates this. Before we used CAS or CAST in our integrated systems, our planning meetings for the integrated systems were often ultimately unproductive, as we mixed up various interrelated issues (communication patterns vs. the information being communicated, system decomposition vs. desired functionality, even requirements vs, designs). Even after we proposed the schema as a theory this did not improve greatly. However, once people started manipulating CAS concepts (working memory entries, change events etc.) in software, the theory, and its implications and constraints, started to seep back into their heads. This led to integration meetings where people started with a common system design vocabulary, and were free to discuss representations, algorithms, component interactions and behaviour at various levels of detail, within a fixed global framework. Whilst CAST will not be the correct choice for every project, we found having a common tool (and therefore common vocabulary) that supports both system design and implementation (something which most component-based middleware would not) invaluable.

11.3.4 Find the Scenario Sweet-Spot

As stated previously, all our integration work was based around scenarios, i.e. examples of system behaviour. Although we had guiding scenarios from the start (first the Fido scenario from Section 1.3, then later the PlayMate and Explorer scenarios from the preceding chapters), these were too abstract to serve as meaningful targets for integration. Therefore, for each round of integration work (roughly each year of the project) we had to select more detailed sub-scenarios from these abstract scenarios to use as targets[17]. This selection process was the main process by which we chose which bits of component science would be integrated into the two main demonstrator systems. This choice is crucial because the demonstrators and their target behaviours represent the public face of a project's ambitions to the wider community (i.e. they are a project doing X and Y). As such, making a particular set of behaviours prominent in an integration scenario can provide high visibility for the component science, with commensurate pressure on the relevant partners to deliver both good components and good integration. In turn this drives both the integrated systems, and the progress of the whole project, in particular ways. This is because a particular set of behaviours will require supporting efforts from the rest of the consortium. In the best case these supporting efforts will be in line with the scientific aims of the relevant partners in the consortium (e.g. the marriage of planning and communication to tackle the problem of clarification [4]). In the worst case the supporting efforts will either require pure engineering work from other partners (e.g. integrating standard solutions to solved problems), or work on problems that are not within a partner's field of expertise or have a difficulty that is beyond the scope of the overall system behaviour (cf. the X is easy fallacy). These issues all arise without even discussing the scientific aims of the partners choosing the scenarios or the

[17] See Chapter 12 for further discussions on scenarios and surrounding issues.

project they're working within (we'll take it as read that these will play a major influence on the decisions). All this demonstrates what a critical decision (in terms of science, progress and project politics) the integration scenarios are for an integrated systems project. In CoSy we spent a significant amount of time discussing and choosing scenarios, working on systems as a result of our choices, and having both positive and negative experiences during this process. This process has provided us with the following insights.

Perhaps the most crucial element of a scenario is how difficult it is to actually achieve given the time and resources available to the project. There are two interrelated issues here: how difficult it is to solve all the necessary sub-problems of the scenario then integrate them in a system, and how difficult it is to decompose the scenario into sub-problems in the first place. The latter is not necessarily an indicator of the former, but it does need to be performed if the scenario is to be tackled. In terms of scenario difficulty there appears to be a sweet-spot where the scenario is difficult enough to be challenging, but not so difficult that it is impossible given the resources available. Ideally the difficulty should be uniform across the subfields tackled by the scenario. In other words, each partner should need to extend the state of the art in their fields, but shouldn't be expected to have to tackle extremely difficult unsolved problems within a limited time-frame. Setting this kind of difficulty level is easier said then done. As we are engaged in research it is often difficult to predict the real difficulty of a task, including the time it will take to complete, and whether a solution exists at all. It is also very difficult to estimate the time it will take to engineer a particular solution into an integrated system. It is often this part that takes a lot longer than anticipated (e.g. after a system has been prototyped in Matlab, turning it into a component in an integrated system can take at least as long again).

Aside from scientific progress, the difficulty of a scenario has a real impact on the morale of the researchers involved in the project. If the work involved in the scenario looks trivial and too easy then it is hard to get people interested in working on it. It is also difficult to motivate the same people into contributing to integration work early on in the integration cycle; if things look easy then they are more likely to be left until the last minute. Conversely if the scenario looks too hard then it is harder to motivate researchers as they may feel their efforts are best placed elsewhere (where they might have a better chance of success) as either their own parts look unachievable, or they don't believe the components they are supposed to integrate with will be completed successfully.

Choosing a scenario purely by picking something which is just slightly beyond the state-of-the-art is a dangerous long-term process. This is because without taking into account longer-term aims it is impossible to determine whether or not your overall research program is heading in the right direction or not. This dictates that any scenario should be embedded within a long-term roadmap which links it into a partially-ordered series of other scenarios leading to a desired future system (see Section 12.1 for further discussion).

However a scenario is chosen, the process of choosing it must actively involve all of the people who will ultimately be contributing to the integrated system. This is because, as stated previously, the choice will dictate their work for the next period. The process of defining of a scenario that spans several disciplines is inevitably one rooted in compromise. What makes a scenario challenging and interesting for one partner can often make it either trivial or impossible for another. As such, the process of choosing a scenario requires a constant balancing of interests across all parties, with each contributor required to compromise on some aspects of the scenario. It is worth noting that in CoSy, as in other projects, the scenarios only dictate the nature and sophistication of the component science that will be integrated together. It is generally expected that partners will take the component science to a level of greater sophistication than is shown in the demonstrators (for evaluation and publication etc.). This means that although the scenario sets the general tone of the work, it doesn't set any limit on how far the work can be taken.

We often encountered difficulties in defining scenarios when working on the PlayMate at the intersection of work on vision, communication, planning and manipulation. If we take a stereotyped view of each disciple (where taking this view is part of the problem) we consider that vision provides us with some type of scene description, communication works out what the human wants the robot to do with the objects in the scene, planning decides how this should be done, and manipulation does it. If you take any part of this process in isolation the researchers will want to solve a problem of a particular complexity. For example, communication researchers will probably want to use natural descriptions of the objects in the scene (e.g. categorical ones) and perhaps complex spatial prepositions to provide unambiguous descriptions of the objects' start and end positions (e.g. "put the mug in the big box"), and planning researchers may want the robot to be given an instruction with multiple different solutions with different costs, or other constraints in construction. However, in robotics both these disciplines must take into account the limitations of the sensor and effector modalities being used. In the above examples the assumptions being implicitly made of vision (at least view-independent categorical perception and the ability to see things that are within other things) and manipulation (the ability to manipulate a mug and place it inside another object, where both could have an arbitrary relationship to the robot and each other) actually place the scenario (in this unconstrained state) way beyond the state-of-the-art (in integrated systems at least) of the two disciplines. Assumptions like this most often occur due to the X is easy fallacy (as described in Section 11.3.1). This happens when researchers have a stereotyped view of what other disciplines are about, and what their proponents should provide to an integrated system (e.g. vision is about whole objects, planning is about controlling the entire system, manipulation can pick and place anything etc.). When researchers are forced to reconsider these stereotypes not only does a better theoretical integration emerge, but interesting, hitherto unconsidered, uses for particular approaches emerge (e.g. see the uses of planning in Section 2.5.4 and Section 6.9).

To make scenarios feasible in situations like this, we have found that we must go through a process of adding detail to descriptions and then altering (often simplifying) the scenario based on the abilities the descriptions require. This process can be used on either the behaviour of the scenario participants (the robot and humans) or the environment the scenario is taking place within. For example, we often had to be very explicit about the nature of the objects involved in a scenario (so that they could work with both the limited manipulation abilities of the PlayMate and categorical perception routines described in Chapter 4), the scenes in which the robot would encounter these objects in (e.g. how much occlusion, how many objects, any restrictions on positions to support grasping etc.), and how the robot was expected to interpret these scenes (e.g. what kind of information should vision make available to the rest of the system). This in turn influenced the descriptive language used by the robot's communication system. Given limited manipulation and 3D vision capabilities we also had to be explicit about what physical actions the robot was expected to perform. This often meant that interesting planning problems (such as object stacking or construction) had to be ruled out. This was also the case with the Explorer: no on-board arm restricted the scenario tasks in general, an inability to reidentify people made certain long-term interactions difficult, and the limitations of appearance-based vision restricted the objects that the Explorer could locate to ones with prominent texture features (resulting in the "Borland book" example).

The aforementioned process of adding detail to scenarios is crucial. It is easy to assume that both the proposer of a particular scenario and their collaborators understand what is meant by a particular statement (e.g. "then the robot see and picks up the shape described by the human"), but there are often innumerable caveats, conditions, and additions needed to implement such a behaviour in practice. By drilling down into the detail at design-time (e.g. how does it see the shape, what information is returned in what format, who is responsible for each part of the process, what are the limitations on what can be seen, etc.) the whole team can try to anticipate problems before they occur during implementation (e.g. "oh, but my technique is not view-independent" or "but how are the actions going to be triggered by the planner?") and spot gaps in the expertise and engineering (e.g. "but who's going to implement that bit, it's not my field really"). This process can be quite tedious, but we have found it invaluable in really fleshing out and specifying our integration work[18].

[18] As with integrating ideas first this is really just a "think before you" act lesson. As with all the previous lessons it sounds like common sense, and something that every good scientist and engineering should do. But as with previous lessons it is one of those things that is very hard to do in a multi-site, multi-discipline project. However, in this type of project careful planning becomes even more critical as the opportunities for misunderstanding are even greater than usual.

Finally, it is worth adding that once you have spent a great deal of time choosing your scenario and specifying it in great detail, you should feel free to deviate from the detail where appropriate when actually implementing the system! As we discussed in Section 11.2 AI and robotics both have a theoretical and a practical element, with the former being refined by the latter. This means that as you integrate, you often discover that your initial plans were overspecified or otherwise incorrect, or that there is just a better (more natural, efficient, elegant etc.) way of solving a particular problem that you didn't foresee until faced with the code. This is a perfectly acceptable way of proceeding with an integrated system (as long as the proposed changes are communicated back to the team, cf. the second lesson). In fact, it could almost be said that if you don't deviate from your initial scenario in some way you weren't being ambitious enough in the first place.

11.3.5 Beware the Modularity Mantra

Our final lesson takes us away from the planning process and addresses one of the fundamental problems in system design and implementation: modularity. Often an integrated system will consist (as in our case) of a collection of modules of some sort. Note that it is not at all clear what the granularity of these modules should be. Quite certainly it is not one module per site or research partner (that would be a very unlikely coincidence!). It also need not be one module per algorithm. Keep in mind that cross-module communication is typically immensely expensive compared to normal code, not so much in runtime but in *coding complexity* (concurrency, asynchronous messages, locks, data consistency checks etc.). This will almost certainly be the case irrespective of the particular communication layer you are using.

It will often make sense to put related algorithms into one module, maybe even to the point where everything is just one monolithic piece of code. Just because modularity is desirable in many applications and processes does not mean that a ruthless modular decomposition of an intelligent system actually makes sense (although in our case it did). Collaboration need not happen solely across module boundaries. It is a very simplistic notion (and proven inefficient in many integration efforts) to assume that partner A writes module X, partner B writes module Y, then they define clear interfaces and everything works. Unfortunately, it is never this simple in practice. If the collaboration is to be worth anything, it can be expected to be deep and complex. Such depth can never rarely be simply captured by "clear interface definitions" and standard software engineering practice. However, there is one exception to this: if partner B is just required to package an off-the-shelf algorithm into a module. This might be a necessary endeavor at the beginning of the project (e.g. to plug a gap in capabilities identified by detailed analysis of the integration scenario). Note though that this is not actually *collaboration*, as partner B does not get anything in return. This gives us another motivation for examining scenarios in detail: to limit the occurrences of this type of interaction (which can be frustrating for partner B).

So, it is important to carefully consider the granularity of modules, what it is that defines module boundaries, and what dependencies arise across these boundaries[19]. Often in the initial planning stages of a project these modules will be defined on an ad hoc basis, using the aforementioned stereotyped view of a system and its component functionality (e.g. "there is a planning module, a language module, a vision module ..." etc.) or on the availability of existing code (e.g. "there is a SIFT module, an edge detector module, and a colour-based segmenter module ..." etc.). Such choices may come from practical concerns (providing a starting point for a design and characterising the types of behaviours the system will be involved in) but lack any theoretical justification. Therefore such coarse initial choices should be reconsidered as the project gains a greater understanding of the problems it is trying to solve and the methods it has at its disposal.

11.4 Conclusion

It is hard to draw concrete conclusions from the lessons in this chapter, and our integration experiences on the CoSy project as a whole. Our subjective experiences (within the project and discussing these issues with the intelligent robotics community) lead us to believe that we are not the only group facing problems in integrating our component science into intelligent systems based on coherent principles (i.e. ideas that are integrated first). However, the precise problems, their causes and effects, have not been considered, let alone formalised, enough for a consensus to emerge. So, although the lessons and thoughts presented here won't offer immediate solutions to those with similar problems, documenting our experiences and making them public are important first steps towards developing this necessary part of the science of intelligent systems. As such, it is advisable not to study our experiences with too critical an eye, or treat our lessons as set in stone. Rather, treat them as a work in progress, and try to take their general direction as something you can follow in your own integration work. This latter point is important. We must try to learn from our integration experiences as a community. The time spent engineering for scientific ends in integrated systems projects can occupy a large proportion of the available project person-months. This time must not be continually misspent in the same ways. Remember, integration is hard. Integration across multiple international partners is even harder. As such it should be afforded the respect that other hard problems are.

If we had to provide one succinct statement to take away from this chapter then it would be "cogitate before you integrate". Applying this maxim to

[19] One way of measuring the information-processing costs of modularity in an architecture for an intelligent system is presented in 2.5.3.

theoretical and practical integration tasks at the personal, workpackage and project level may take up time when you'd rather dive head-first into an interesting problem, but it might save you at least that much time again when the first unforeseen problems start appearing.

References

1. Hoffmann, J., Nebel, B.: The FF planning system: Fast plan generation through heuristic search, JAIR
2. Lowe, D.G.: Distinctive image features from scale-invariant keypoints. International Journal of Computer Vision 60(2), 91–110 (2004)
3. Hawes, N., Zillich, M., Wyatt, J.: BALT & CAST: Middleware for cognitive robotics. In: Proceedings of IEEE RO-MAN 2007, pp. 998–1003 (2007), http://cognitivesystems.org/cosybook/chap11.asp#Hawes/etal:2007
4. Kruijff, G., Brenner, M., Hawes, N.: Continual planning for cross-modal situated clarification in human-robot interaction. In: Proceedings of the 17th International Symposium on Robot and Human Interactive Communication (RO-MAN 2008), Munich, Germany (2008), http://cognitivesystems.org/cosybook/chap11.asp#Kruijff/etal:2008

Part IV

Summary and Outlook

12

Cross-Disciplinary Reflections: Philosophical Robotics

Aaron Sloman

University of Birmingham
axs@cs.bham.ac.uk

12.1 Introduction

This chapter reports work done mostly by one member of the team – a philosopher with substantial AI programming experience, whose primary interests were in the very long term goals of the project, summarised in Chapter 1, including the goal of shedding light on problems solved by biological evolution, and who was not directly involved in the coding but who interacted closely with people who were, and with people outside the project, in several related disciplines. The majority of the work reported here is concerned with requirements, and gaps between those requirements and the current state-of-the-art in AI/Robotics, and related disciplines. A key feature of this work is its emphasis on study of aspects of the 3-D environment we and other animals inhabit, with which a Fido-like intelligent domestic robot (described in Chapter 1) would need to interact. This is an essential part of a strategy for developing a roadmap to bridge the gaps in the long term.

From the start, the CoSy project emphasised the need to study *requirements*, as a prerequisite for producing designs. This analysis supported the suspicions in the original proposal, namely that current state-of-the-art designs and implementations were nowhere near meeting the long term requirements. So our goal became to develop a methodology for identifying the gap more precisely, and a strategy for bridging the gap (or gaps). As an aid to these tasks, we organised several interdisciplinary events, including a tutorial at IJCAI'05, a Symposium at AISB'06 and the "Meeting of minds" workshop in Paris, in 2007. Most of this work did not fit into the formal work-plans and deliverables, but results were published in workshop and conference papers,

H.I. Christensen et al. (Eds.): Cognitive Systems, COSMOS 8, pp. 441–484.
springerlink.com

contributions to collections, the euCognition wiki, and online presentations and discussion papers,[1] along with web sites for the events we organised.[2]

Some of the results of that study are presented below, suggesting directions for future work and implications for other disciplines, including study of humans and other animals. The project's goal of producing and demonstrating a sequence of implemented, integrated systems starting in year 1 (see Section 11.1) required many detailed decisions to be taken before long term requirements could make much progress. As a result, the long-term requirements analysis proceeded in parallel with most of the design and implementation work, and only loose connections were possible. Had the project been funded for 10 or 15 years, without the need to produce publications and demonstrations every year from the start, things might have been different.

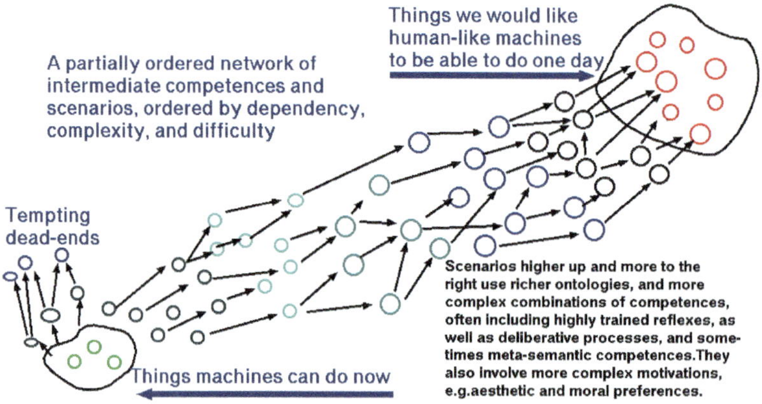

Fig. 12.1. How to develop a long term research roadmap based on a partially ordered network of scenarios developed by backward-chaining (presented at euCognition roadmap meeting Jan 2007)

As work on the two streams (requirements analysis and implementation) progressed, it became clear that the aforementioned gap was even greater than we had anticipated. This created a tension between the "safe" approach of taking existing techniques and attempting to combine them, where possible with additions, and the "risky" approach of trying to find ways to reduce some of the huge gaps between current techniques and animal/human competences.

[1] See [1] [2] [3] [4] [5] [6] [7] [8] [9] [10] [11] [12] [13] [14] [15]) [16] [17] [18] [19] [20] [21] [22] [23] [24] [25] [26] [27] [28] [29] [30] [31] [32]
Length restrictions allow only a subset of the ideas to be presented here.

[2] http://www.cs.bham.ac.uk/research/projects/cosy/conferences
http://www.cs.bham.ac.uk/research/projects/cogaff/gc/aisb06
http://www.cs.bham.ac.uk/research/projects/cosy/conferences/mofm-paris-07

Attacking the gaps head-on would have required most of the four years to be spent on the study of long term requirements, and attempting to decompose those requirements in a backward chaining process of the sort described in [4] and depicted in Figure 12.1. Instead, a fairly small subset of the work in the whole project focused on long term requirements: most of the design and implementation work was an attempt to extend the current state of the art, as reported in preceding chapters, especially the state of the art in integrating different kinds of functionality.

The latter work did include some requirements analysis, e.g. for architectures and tools for integration of subsystems (Chapters 9-11). The specific short-term demonstrator goals were grown, modified and constrained through experience of trying to put pieces together by extending techniques originally developed for modules running in isolation. As a result of our growing appreciation of the gaps between the state of the art in AI and robotics and the functionality required for the futuristic 'Fido' scenario, most of the effort in the project went into feasible extensions of component capabilities, along with integration into a system that combined the components. Some of what that left out is presented in the rest of this chapter. (Choosing scenarios to work on raises many practical problems, some of which are discussed in Chapter 11.) Many of the detailed requirements became visible as a result of reflecting on what our implementations could not do, illustrating the importance of implementation-based requirements analysis! For this purpose rapid-prototyping tools without a prior commitment to any particular architecture are essential, as discussed in [33, 34] and Chapter 11.

12.2 Must an Intelligent Robot Use Language?

Since preverbal children and many animals that do not use language can interact with complex environments, including environments in which processes are occurring, e.g. during nest-building, fighting, manipulating sources of food or shelter, eating things that do not come ready carved up into bite-size chunks, etc., that shows that the ability to use a human language is not a prerequisite for such competences. However, it can be argued (as in [15]) that both pre-verbal humans and other intelligent animals must use forms of representation *internally* that support structural variability, context-sensitive compositional semantics, as well as the ability to introduce substantive extensions to the ontology. They must also be suitable for use in perception, planning, various kinds of learning, the expression of motives and preferences, and the generation and control of action.

Such features, especially structural variability and compositional semantics, are normally thought of as key features of languages used for *communication*, but they must have existed earlier in *internal* "languages". What forms those representation used, and still use, is an open question. There are no obviously correct candidates, though they are unlikely to have the grammatical or logical structures of languages that evolved for communication rather than for representing percepts and controlling actions. Neither do we claim that the internal languages are fixed at birth: they may be extended by bootstrapping and debugging processes, including substantive (non-definitional) ontology extension [17]. So our arguments are not endorsements of Fodor's theory in [35].

Since some researchers object to the use of the word "language" to label something that is not used for communication, we call this notion of language that covers both forms of representation used for communication and forms used internally for perceiving, thinking, etc. 'Generalised Language' (GL), discussed in more detail in [36] [15]) [27].

The argument that GLs used internally precede the use of verbal languages for communication both in evolution and in individual human development has many implications, including implications concerning requirements for future intelligent robots.

Looking at videos of pre-verbal infants and toddlers helped to draw attention both to gaps in their understanding of various aspects of the environment and to the depth and variety of their visual and manipulative competences, despite those gaps. Before a human child starts learning to talk there is already a deep understanding of, and interest in, many structures and processes in the environment [38], and those competences and interests are required for the language learning process. Learning a language is part of learning how to achieve collaborative goals in a shared, partially understood, 3-D environment, using an exosomatic ontology (defined below in Section 12.5.1), not just learning mappings between acoustic signals (spoken words) and other sensory signals. Because our robots were nowhere near human toddler competences in vision and manipulation, and did not have the rich internal information-processing formalisms ("internal languages") postulated above (and in [36, 15], their language learning processes described in previous chapters had to be totally different from human language learning, and much more artificial and restricted. It is to be hoped that this deficit can be remedied in future research.

12.3 The Role of the Environment

Thirty years ago, [39] suggested that work in AI could clarify or solve many philosophical problems, but our analysis of requirements in the CoSy project revealed further implications of the fact that important aspects of human and animal intelligence were evolutionary responses to the challenge of interacting with and manipulating movable, reconfigurable, 3-D objects of varying complexity in an extended environment, only part of which is perceivable at any time. A core feature of the challenge is perception of concurrently changing 3-D spatial (geometric and topological), causal and functional relationships between both whole objects and parts of objects, where some of the changes involve independently movable limbs and hands. Some of the requirements are discussed in [30]. Most of the requirements deriving from that challenge appear not to have been noticed by roboticists, vision researchers or psychologists, and philosophers have not realised their significance for philosophy of science and philosophy of mathematics (see [21, 5]).

A very early robot with manipulative capabilities was the Edinburgh robot Freddy_II [40] developed around 1973, which could assemble two different objects (a toy car and a toy boat) from parts initially piled or scattered randomly on a table.[3] Its speed and versatility were severely limited[4] yet recent robots have not reached some of Freddy's competences, even though many hardware and software components required have separately developed enormously. Why not?

Processes of learning and development in humans and some other animals depend on rich interactions with the environment in early months and years that lay a foundation on which many other aspects of human intelligence depend in later life. Many biologists study animal behaviours, and many developmental psychologists study spatial competences in infants and young children (e.g. the book by Gibson and Pick [41]). However, the methods of experiment accepted as producing significant results are so restrictive that most of the cognitive richness of processes required for interaction with manipulable objects, described in [20], and below, goes unnoticed. Likewise, limitations of current tools and techniques cause AI researchers to ignore most of the complexity in the environment, as do many researchers working on digital companions for the elderly or disabled [14]. Examples of what they ignore (e.g. differences between tracking moving 2-D image features and perceiving a 3-D process) are given later in Section 12.12.2.

Without a rich and deep, mostly culture-neutral, biologically rooted, understanding of space, time, and the 3-D physical environment in which animal activities are embedded, a robot is unlikely to be able to learn to talk, think, and perform tasks like a human adult – though it may cope with a very restricted subset in a very brittle way, as in many robot demos.

12.4 Analysing Requirements Is Very Hard

Analysis of requirements is a task whose complexity is largely unnoticed. Many researchers think it is sufficient to define some goal in terms a very high level description e.g. "recognition of everyday objects in everyday situations", "coping flexibly with domestic tasks", "engaging in natural conversation about some topic". Sometimes, words like "reliability", "flexibility", "robustness", "intelligence", "autonomy", "versatility", "extendability", and "maintainability" are used to indicate design features, even though such words refer to very

[3] While this chapter was in preparation, a remarkable video of Freddy_II was made available here `http://en.wikipedia.org/wiki/Freddy_II`

[4] In 1973, 384KBytes of memory was a luxury, computer speeds were measured in kilocycles, and it could take several minutes just to find the bounding contours in an image, ruling out concurrent perception and action, or visual servoing.

different characteristics in different contexts. For example, the requirements for *robustness* are very different in an operating system, a word processor, a theorem prover, a medical expert system, and a lawnmower. (Discussed in more detail in [42].)

Sometimes benchmark tasks (e.g. fixed sets of images for training and testing) are used to replace imprecise requirements. But the benchmarks often lack "ecological validity", directing research down narrow paths and diverting attention from the problems the benchmarks were intended to characterise.

Close examination of problems with which various environments confront humans and other animals reveals richness and diversity of problems and solutions that usually escape notice. The diversity of possible solutions indicates a need to study trade-offs between alternative competences and alternative designs, rather than performance metrics and bench-marks (see Section 12.11).

By collecting partially ordered sets of scenarios (ordered by both difficulty and dependency), we can identify short, medium and long term challenges to be met in specifying designs, as indicated in Figure 12.1. Some of the scenarios should include multiple interacting competences with the interactions described in film script detail. Systematically varying features of the scenarios, can demonstrate the inadequacy of designs tailored to *limited* sets of examples.

This task of generating scenarios is very close to the process of conceptual analysis in philosophy, where theories (e.g. about the nature of desire, intention, attention, perception, belief, understanding, etc.) need to be tested by production of examples. In principle, this is an area where philosophers and AI researchers should be able to interact, partly because good philosophers have already developed the ability to think up examples to challenge theories.

We developed templates for scenarios,[5] and a scenario-generation methodology based on a 2-D grid of types of competence against types of object, with complexity as a third dimension, summarised in [2]. Unfortunately the task of generating and analysing demanding scenarios proved difficult for researchers who had never previously done anything like it, especially while facing great challenges in their own sub-fields. Eventually, scenarios limited by short-term feasibility were developed, described in Chapters 9, 10 and 11.

12.5 Robotics and Philosophy of Science

12.5.1 Ontologies and Laws

A project like CoSy illustrates limitations of conventional philosophy of science. Chapter 2 of [39][6] explained how science is an attempt to understand both the form and the content of the world, where the form has two aspects:

[5] E.g. in `http://www.cs.bham.ac.uk/research/projects/cosy/scenarios`

[6] Online at `http://www.cs.bham.ac.uk/research/projects/cogaff/crp/chap2.html`

(a) what sorts of things are possible (an ontology), and (b) how those possibilities are limited, e.g. in *laws* such as "All As are Bs", which rules out the possibility of something being an A and not a B. But that presupposes an ontology that includes the possibility of A things and B things.

Conventional philosophy of science emphasises (b), not (a), whereas the deepest scientific advances are of type (a), substantially extending our ontologies, and thereby allowing new questions and theories to be formulated, e.g. adding atomic theory, or evolution by natural selection to our ontology. Such advances require new concepts, extending the pre-existing ontology substantively, i.e. adding new concepts that cannot be defined in terms of old ones, though they may be connected via what Carnap called "meaning postulates" in [43]. Substantive kinds of ontology extension must occur in children [15].

Deep research in cognition and robotics requires researchers to extend their scientific ontologies if they are to produce more intelligent machines. Following [30] these can be labelled "designer ontologies", in contrast with the ontologies required by robots (or animal being modelled!) "application ontologies". Intelligent robots will also need mechanisms capable of substantive (non-definitional) ontology extension. Mechanisms for extension of sensorimotor ontologies by dimensionality reduction, presented in Chapter 3, may be useful, but cannot add new dimensions, e.g. required for interpreting 2-D motions as projections of 3-D rotations. Similarly, mechanisms using sensorimotor statistics to induce new concepts useful for prediction, will not be enough: contrary to "Symbol-grounding" theory, which is a serious impediment to progress, as explained in [17]. In particular, what a child develops and what a robot will need includes "exosomatic" concepts referring not to sensorimotor patterns, but to objects and processes in the environment that could exist independently of the observer.

12.5.2 No "Right" or "Best" Designs

Philosophers often try to specify *necessary* conditions for something to be a mind. This leads to shallow and unsatisfactory theories. Attempting to identify *one* best design for intelligent systems would be like physicists attempting to study only the substances that exist in some particular spatio-temporal region, e.g. Rome in 1630. Instead, there is a broad class of *possible* active information-processing systems, a class that includes myriad varieties of organism produced on this planet by biological evolution, and probably even more produced in other parts of the universe, or in the future on earth, and some that are possible, but never will be produced.

Different designs can be evaluated in relation to different sets of requirements. Sets of requirements for organisms are referred to as "niches". So the task is to understand the space of sets of designs, the space of sets of requirements and the relationships between them (Figure 12.2). Since individuals can develop and since species can evolve, there are also trajectories within those spaces and we need to understand how those trajectories work.

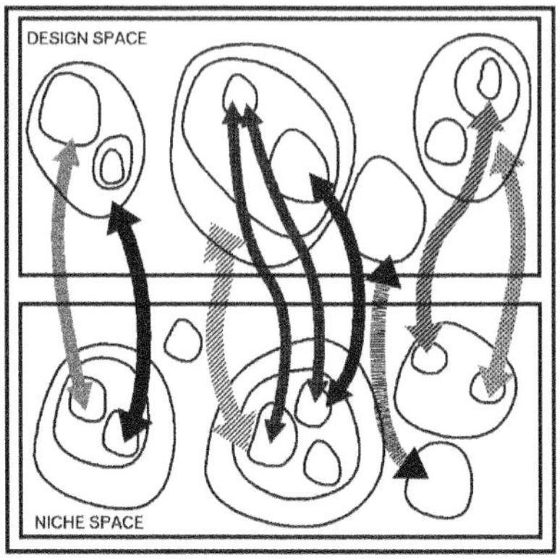

Fig. 12.2. The space of possible designs, the space of possible niches (sets of requirements) and the varied relationships and trade-offs between them. Both spaces have many discontinuities. There are trajectories of various sorts through both spaces, including development and learning done by individuals, evolution of species, social/cultural evolution, and in the case of artificial systems design changes. In an ecosystem there will be complex feedback loops involving trajectories in both design space and niche space.

This broad-minded approach to both philosophy and AI, presented in [44, 45, 46], is hard for most researchers, especially as progress is inevitably slow. Studying the full spaces in depth is impossible, but we can explore limited regions ("neighbourhoods") in design space and niche space ([47]).

Detailed specification of such regions cannot be done in a research proposal: it is the *result* of research. As the work on integration in CoSy progressed, decisions had to be taken about what was and what was not being addressed, such as whether the robot should be able to perceive *processes* (like its hand moving), whether to include recognition of objects or perception of 3-D structure, since they required very different mechanisms, whether it should interact sensibly with more than one human at a time, which kinds of failure in performing tasks it should be able to detect and remedy, which aspects of verbal interaction should be capable of influencing visual processing, or vice versa. Some choices between project sub-goals, and the design problems they led to, were not visible to participants before the project started. Such "invisibility" is reduced as more researchers gain experience in integrated projects.

12.5.3 A Science of Explosive Diversity

The problem of description arises partly because individual designs for "complete" working systems can vary enormously: in their architectures, e.g. in the variety of components they contain; the forms of representation they use; the kinds of information they acquire, manipulate and use; the variety of connections between sub-systems; and whether the architecture is static or grows itself, as certainly happens in humans, but not yet in CoSy. This diversity makes it very hard to compare designs, especially when represented in complex diagrams using arbitrary diagrammatic notations. We need a better way to talk about such complexity.

Much of the variation between designs is closely related to the different challenges posed by the kinds of environment that different systems need to interact with, and a detailed study of interesting designs must be linked to a detailed study of the relevant environments, a point emphasised also by Neisser [48] and Gibson [49] (discussed further below). [12] illustrates some aspects of the complex feedback between evolution of designs and evolution of niches, from microbes in chemical soups to articulated animals surrounded by diverse rigid and non-rigid 3-D structures, and also other intelligent systems.

12.5.4 Individual Variability

Not only the diversity of designs, but also the diversity of states and processes possible for *instances* of a design needs to be studied. Instances of more complex designs are capable of more diverse states, processes, and forms of development over time, illustrated by the vast diversity of human minds. We mention some high level concepts that may help, in Section 12.7.

Some organisms, often labelled "precocial", remain largely unchanged throughout their life (apart from parameter adjustments within a fixed framework), or follow patterns of change common to all members of their species (e.g. microbes, insects, and probably most other invertebrates) whereas others, the so-called "altricial" species, start highly incompetent, and develop under the influence of complex feedback from the environment.

A paper (with J.Chappell) was presented at IJCAI 2005 [6] arguing that the precocial-altricial spectrum is just as relevant to robots as to animals. A sequel was an invited journal paper [1], presenting ideas about multi-layered bootstrapping processes based on a combination of features of the genome and specific features of the environment revealed by exploratory play, as in Figure 12.3. These ideas are still being developed, but have already had some influence. We propose to apply these ideas in research on primate competences.

12.5.5 The "Designer Stance" in Biology

Robotic research exposes questions not normally asked by researchers in animal behaviour. Animal behaviour researchers should adopt what McCarthy [50] calls "the designer stance" and ask "What mechanisms, forms of representations, and architectural features, would I need to put into a robot to enable it to do *that*". Robotics-inspired questions can draw attention to previously unnoticed fine details of competences generating behaviours, and may also transform some nature-nurture debates, e.g. by showing that the demands of interacting with a complex, structured, manipulable 3-D environment may be

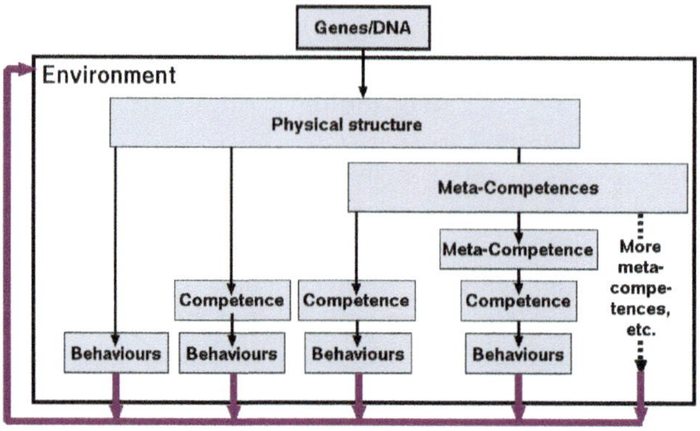

Fig. 12.3. Varieties of control routes from the genome to competences and behaviours: on the left most details are determined by the genome, whereas towards the right there are more complex staggered or layered processes of learning to learn, producing sequences of more sophisticated competences and meta-competences. (Based on [1].)

at least as important in driving evolution of cognitive powers as the need for social interaction, which has gained far more attention.

E.g. the work on altruism in young humans and chimpanzees in [38] investigates motivation, but takes for granted the cognitive competences mentioned above in Section 12.2, required for all of: perceiving what is going on, inferring somebody's intention (e.g. "trying to get books into a cupboard"), planning actions to achieve that intention, deciding to execute actions to unblock a precondition (walking forward and opening the door), and performing the actions. If all that, including the use of meta-semantic capabilities, can occur in pre-verbal children, and in simpler forms in chimpanzees and other animals, that raises deep questions about the pre-verbal forms of representation available to animals. There are also deep questions about where those representations come from – are they innate, or, as seems more likely, since evolution is unlikely to have provided an innate concept of a cupboard door, are they a result of developmental processes including processes driven by meta-competences that cause exploratory behaviours that produce learning as indicated in Figure 12.3, explained in [1].

12.5.6 Should Requirements Refer to Laws of Behaviour?

For animals with very large collections of capabilities, changing motives, plans and beliefs, and frequently changing environments, there may not be any *laws* that characterise their behaviour – only *possibilities* that need to be explained. So although some members of the cosy team studied psychological research literature, as reported in Chapter 8, it is important not to assume that laboratory results provide evidence of laws to which human-like robots had to conform – they merely show what sometimes happens, especially when the subjects are from a single culture, restricted age ranges, and numbers of subjects are often well below 100.

Even for a particular design there may not be well defined *laws* predicting characteristics or behaviour, only a rich space of possibilities. There may, however, be laws concerned with "low level" features of transducers and implementation mechanisms, and there will be some limitations on what is possible for any particular implementation of a design, for instance limitations of processing speeds, and capacity limits. Nevertheless, how any particular individual instance of an "altricial design", behaves within the limits permitted by the architecture and mechanisms used, far from conforming to exceptionless laws will typically depend in complex ways on its goals, preferences, interests, what the individual has learnt, what it was doing previously, etc. – conditions that can vary enormously across individuals and from time to time for each individual. In addition for deterministic dynamical systems with nonlinear feedback prediction can be impossible if measurements of initial states have bounded precision.[7] So the kind of research that is appropriate to a design-based science of cognitive systems, natural and artificial, will not fit the popular conception of science as mainly a search for exceptionless laws (or even high probability generalisations) but will require us primarily to attempt to understand varieties of possibilities that particular designs support in combination with particular contexts and personal histories [39, Ch 2].

12.6 Environment-Neutral Requirements and Limitations

Some requirements and limitations that arise when they are not satisfied are not concerned with specific types of environment: they are "topic-neutral". Other requirements will be discussed later. The main topic-neutral requirements and limitations are concerned with architectures and with forms of representation available or missing.

12.6.1 Forms of Representation

Limitations due to forms of representation available include: if full predicate logic is not available, it will be difficult to express or reason about non-singular propositions such as "There is no tiger in the room" or "Every person in the room is taller than at least one other person in the room" or the corresponding questions, or goals (e.g. "Find chairs for everyone"). However "compiled" versions of some of these can be expressed as procedures (e.g. a procedure for fetching chairs, with appropriate stopping condition). This will enable a robot to do things, but not to describe or think about what it is doing.

[7] See `http://www.ecmwf.int/research/predictability/background`

Systems lacking modal operators (e.g. "possible", "necessary", "impossible", "contingent") may be unable to represent the difference between an *empirical* generalisation such as that pools of water sometimes merge while being counted and a *necessary* truth such as that counting a fixed set of objects in different orders must give the same result. (See Section 12.13.) Without modal operators it will also be impossible to represent what is *possible* or *impossible*, and therefore affordances will be inexpressible. (See Section 12.12.5. Words of natural language related to affordances, such as "graspable" refer to what is possible, and therefore involve an implicit modal operator.)

The forms of representation available and the architecture can also constrain what *questions* can be formulated (internally or externnally), a topic discussed in more detail in an early CoSy requirements deliverable [30]. Being able to formulate questions that can generate information seeking goals is an important aspect of being an autonomous learner.

Being able to think about or communicate with other intelligent individuals requires the use of forms of representation that support meta-semantic competences: the ability to refer to things that refer. This is also required for certain kinds of introspection and self-understanding. Meta-semantic competences require the ability to make a distinction between representations whose function is to refer to the world and those whose function is to represent what someone or something else is referring to, possibly erroneously. This is generally referred to by philosophers as "referential opacity". It is not clear when human children have that competence or whether other animals have it, though the ability to understand or tell stories, or engage in "make believe" play requires it. We did not attempt to provide this level of sophistication in our robots so in a sense they could not think about other individuals as having information processing capabilities, including beliefs, intentions, desires, etc. This necessarily limited the forms of interaction that were possible. For example, without such meta-semantic competence a robot cannot consider whether a human lacks information, or whether been misperceived, or a communication misunderstood.

There is no general agreement on how referentially opaque forms of representation should be dealt with. Many researchers hope that an extension to logic, e.g. using new logical operators, will suffice. I suspect that the best solution is to extend architectures, to allow the same form of representation to have different roles in different parts of the system, so that a form of representation may be taken as believed to be true if it occurs in one part of the system, and as a specification of someone else's belief if it occurs in another part of the system. Similar remarks can be made about differences between propositions, questions, goals, conjectures, memories, fantasies, etc.

Another important feature of a form of representation may be the way it can be used to control search in certain classes of problem. Being restricted to Fregean (logical, applicative) forms of representation, without any subsystems able to reason with analogical or other forms of representation may be a serious handicap for some classes of problem, as exlained in [51] and [39, Chapter 7].

Moreover, in some cases probabilistic representations are more useful than categorical representations. In other cases hybrids are useful. Often it is useful to combine many detailed information items with histograms that can reveal otherwise unnoticed global patterns – e.g. most edge-features in one part of an image are horizontal, and vertical in another. We did not attempt to produce a catalogue of forms of representation that could be used to select optimal candidates for various parts of the system. Instead, in most parts of cosy, forms of representation were used that were traditional for work of that sort in AI. It may be useful at some future date to investigate whether this has restricted progress (either in CoSy or in the whole field of robotics).

12.6.2 Architectures

Limitations directly related to architectural features include: whether a system can do certain tasks in parallel or not; whether certain behaviours are interruptable; whether there are some low level "cognitively impenetrable" components (Pylyshyn in [52]) whose functioning is not affected by other sub-systems; whether some competences can be modified (e.g. debugged or speeded up) as a result of self-monitoring during performance; whether information about the environment is represented on different scales (e.g. local and global); whether information is stored about what has recently happened (episodic memories); whether there are mechanisms that check for consistency within and between percepts, previously acquired generalisations, episodic memories; whether information changes can trigger side effects via constraint propagation (as suggested in connection with vision, below in Section 12.12.6 and Figure 12.8); whether there is an "alarm" system (or several) that trigger rapid reorganisation in response to a detected threat or opportunity (Section 12.10). This is not a complete list.

Since the environment endures over time with some changing and some static features, while sensory contents change on much faster time-scales, it can be useful to have an architecture in which the main source of information about the environment for most of the system is *not* the contents of sensory signals but an enduring, incrementally updated representation of the environment. In some cases this should be a-modal so that different sensory sources can be used in parallel adding or modifying different features, and acting as cross-checks.

Both motion of perceived objects and the perceiver's own motion (including saccades) entail such requirements for a visual architecture, discussed by Trehub in [53]. In particular the mapping between an enduring scene representation and image pixels will be constantly changing, though not the mapping between the scene representation and the *optic array*, if the viewpoint is fixed. However if information at different levels of abstraction is represented, the very notion of registration becomes blurred, though a sort of registration is indicated in figure 6 of [39, Chapter 9]. Further complexities are required if manipulation or motion of either objects viewed, or the viewer, causes different parts of an object to be visible. The requirement to represent hidden parts (e.g. the far side of a cube after rotation, or the contents of a box after shutting its lid) goes beyond representing what is available in the optic array. Unfortunately, it was not possible in the time available to meet these architectural requirements, except in restricted ad hoc fashion, though the dialogue system (Chapter 8) presupposed some of them, and the Explorer system (Chapter 10) used SLAM techniques to represent far more information about currently unperceived entities than PlayMate.

12.7 De-fusing Diversity: Towers and Layers

Designs for whole systems can vary in uncountable ways, if all possible alternatives for every design decision are considered. This raises the urgent question: is there any way a science of intelligent systems can impose some intelligible and useful structure on the space of possibilities, or is it just a morass composed of an enormous collection of special cases?

12.7.1 Generative Frameworks

One way of trying to impose structure is to specify a generative framework by specifying a collection of basic building blocks for behaviours and ways of combining them into arbitrarily complex systems. That was the approach adopted by Turing, which produced a class of machines that was later shown could be generated in alternative ways, e.g. Turing machines, production systems, lambda calculus, Curry combinators, logical inference mechanisms, etc.

Another approach is to take some supposedly general architectural framework such as SOAR or ACT-R (both summarised in [54]) and then show how all other architectures of interest can be subsumed by the chosen one.

There are two problems with this generative approach, despite its great power and usefulness in computer science. First the set of machine-types generated is too restricted, being composed entirely of systems whose behaviour consists only of components in the selected framework, e.g. discrete serial steps in a Turing machine. However that limitation can sometimes be overcome by allowing the components to be combined in different ways, to generate types of machine with possibly asynchronous concurrently active components, e.g. as in digital circuit design formalisms, or Milner's pi calculus [55], or Hewitt's actor formalism [56]. The remaining restriction to discrete processes can be removed by allowing additional analog mechanisms that support continuous variation, e.g. conductors, oscillators, capacitors, etc, or mechanical devices, such as gears, pulleys, springs, strings, etc. and analog to digital and digital to analog converters. The use of generative representations has the problem of being "bottom up", making it hard to get a high-level, top-down view of the space of designs.

A closely related problem is that for the kinds of behaving systems we are interested in there is always an external environment, and we need to find ways of characterising systems not merely in terms of how they are built, from processing components, or what goes on inside them, but in terms of what sorts of environments they can interact with, and how they interact with them – i.e. they are characterised in terms of combinations of types of *functionality* rather than combinations of types of *mechanism*. A framework for presenting those ideas was loosely inspired by Nilsson's discussion in [57, Ch 25] of architectural "towers" and architectural "layers".

12.7.2 Subdivision into Towers of Functionality

Systems that act in an environment can be described as having three major (possibly overlapping) sub-systems with different functions:

- a *perceptual* sub-system that gains information from the environment, processing information derived from physical transducers that produce internal signals from incoming energy in various ways;
- an *action* sub-system that emits energy in various forms (especially applied forces) using transducers controlled by internal signals;
- and between those sub-systems an arbitrarily complex collection of "central" mechanisms that interact with the perceptual and action sub-systems but may also do many other things.

This gives us Nilsson's three towers, which he called the perception tower, the action tower and the model tower. For our purposes, the label "model" is too narrow, though our label "central" is also inadequate.

We can further subdivide types of tower according to which kinds of sensory information they use (visual, auditory, haptic, magnetic, etc.), which kinds of outputs their effectors produce (luminescence, acoustic, pressure-applying, squirting, throwing, blowing, changing shape, etc.), what kinds of information they can derive from the sensors, and what changes they can produce in the environment – including cases where acting and sensing are tightly linked (Gibson, [58]). e.g. altering gaze in order to obtain different visual information or squeezing something in order to gain haptic information. (See Chapter 3.)

12.7.3 Subdivision into Layers of Functionality

A different way of subdividing systems or sub-systems "top-down", i.e. in terms of their functionality, is by distinguishing different ways of mediating sensing and acting. This is orthogonal to the previous divisions, and can be thought of as providing three layers, though our partly biologically-inspired division into layers is not exactly the same as Nilsson's (or most others, e.g. [59], Gat [60], and Minsky [61]).

Reactive layer: The evolutionarily oldest, and easiest to implement mechanisms are systems that respond to sensory input by immediately generating a short term response (externally or internally) without representing any consequences of that response, and without reasoning about multiple possibilities before selecting that response. We can call those "reactive" mechanisms, while acknowledging that they can vary enormously in complexity, including the number of intermediate processing stages between sensing and acting, and also whether some of the things sensed or altered are internal, e.g. sensing internal energy levels and damping down internal levels of activity, or switching sub-systems on or off. Such reactive systems may be either discrete or continuous, and may or may not include feedback control, adaptive learning, or other kinds of sophistication. In more complex cases (often implemented in neural net mechanisms) the inputs and corresponding outputs are *intput and output patterns*, where a collection of sensors acting concurrently trigger a collection of coordinated outputs. There are also intermediate cases. Other possibilities include temporally extended triggering inputs (e.g. using thresholds) and temporally extended outputs (e.g. running away from something). Moreover, reactive systems can include "proto-deliberative" mechanisms, described in [10] where competing output actions are stimulated at the same time, but only one wins on the basis of some mechanism for evaluating alternatives. (Unfortunately, some researchers confusingly label this "deliberative".)

Deliberative layer: A second type of layer can provide various kinds of deliberative sub-system, which vary in the sophistication of the predictive, or control functions they support, as discussed in [10]. The common feature involves the ability to respond to input, or the formation of a goal, by considering not only alternative responses but also the consequences of those responses before selecting an action. More sophisticated versions can consider alternative action sequences before deciding – as many of the earliest AI systems did. The ability to explore branching futures depends crucially on sensory information being "chunked" into discrete categories, so that associations can be learnt between discrete cases, avoiding the need to handle infinitely branching futures (see [10]). (Nilsson's top two layers are included in our second layer.)

Meta-management layer (with meta-semantic competences): The third, biologically most recent, type of layer, not discussed in Nilsson's chapter, is able to do two important kinds of things, namely monitor some of its own (semantically rich) internal states and processes, including characterising them in some explicit, structured formalism (unlike hierarchical analog control systems), and representing other individuals as also having such internal states, describable using meta-semantic competences (explained above in Section 12.6.1).

Although three layers have been distinguished, there are intermediate cases not yet mentioned, as well as other useful ways of dividing up functionality (e.g. Minsky [61] has six layers). A possible source of confusion is that all mechanisms must ultimately be implemented in reactive mechanisms of some kind.

12.8 The CogAff Architecture Schema – One Small Step

By the time the CoSy project started, we had attempted in previous work (mentioned in Chapter 1) to make the task of exploring design- and niche-space more tractable by using a generic schema, in which the tower and layer classifications presented above were superimposed, thus forming the nine cell CogAff architecture schema, depicted in Figure 12.4 and mentioned in Chapter 1 of this book. This uses a modified version of Nilsson's distinction between layers and towers, where differences between towers arise from different relationships to the environment, and differences between layers are defined in terms of differences in evolutionary age, kind of computational and representational sophistication, and types of mechanism used – though I now prefer to emphasise types of functionality rather than types of mechanism. For instance, the bottom left box could include several types of low level sensory processing, and the top left perception of communicative actions, intentions, moods, etc. Likewise the bottom right box could contain low level motor outputs of various kinds and the top right box gestures, linguistic utterances, and expressive behaviours. The schema is clearly an oversimplification, but provides a crude initial framework for comparing a wide range of architectures, according to which boxes are used, what they contain, which forms of representation are used in the boxes, and how the components are connected. Some of the problems of combining different sorts of functionality, using different forms of representation are discussed in Chapter 2. In contrast with the "generative" approach to diversity, this framework is neutral as to what the smallest processing components are. Examples of systems that do not fit this schema are systems composed of large numbers of autonomous individuals, so-called "multi-agent systems".

It is not claimed that this is the only way of dividing architectural components or that the divisions are sharp. Different AI researchers who use layered architectures describe their layers differently (e.g. Brooks [59], Gat [60], and Minsky [61]) though some of those differences can be subsumed within the CogAff Schema. Sun [63], describes an architecture (CLARION) made of several components each with two layers corresponding roughly to the two bottom layers of CogAff. This can be re-described as a two-layer architecture, where the two layers are subdivided into components, each closely linked to a component in the other layer. Exploring all these ways of characterising architectures, to see whether there is some useful over-arching form of representation is a topic for further research.

To a first approximation the CogAff schema provides a sort of "grammar" for architectures. There are many special cases of that schema, including special cases that use only the reactive layer e.g. purely reactive subsumptive

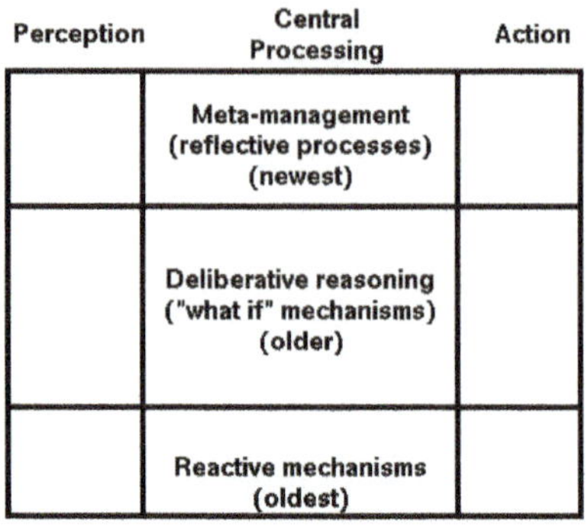

Fig. 12.4. The CogAff schema [62] loosely based on a combination Nilsson's ideas of towers and layers, provides a framework for comparing a wide range of architectures. However, as explained in the text, it is only crude beginning.

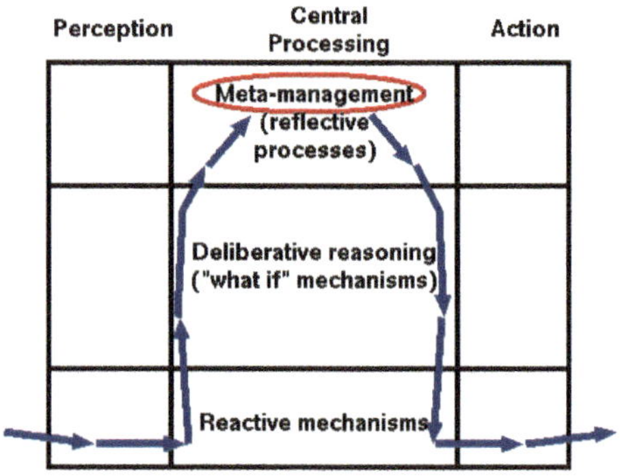

Fig. 12.5. A popular type of architecture, which could be called the "Omega" architecture, depicted here, is a special case of the CogAff schema

architectures [59], and reactive insect-like architectures that include a reactive "global alarm" mechanism receiving inputs from all parts of the system and capable of modulating or redirecting all parts very quickly.

12.8.1 Omega Architectures

A popular type of architecture incorporates a sequential multi-layer pipeline: sensory information comes in via low level sensors ('bottom left'), gets abstracted as it goes up through higher central layers, until action options are proposed near the top, where decisions are taken, after which control information flows down through the layers and out to the motors ('bottom right'). This can be called an 'Omega' architecture because the pattern of information flow is shaped like a the Greek letter Ω, as shown in Figure 12.5. Many models in AI and psychology have this style e.g. Albus in [64]. The influential 'contention scheduling' model of Shallice and collaborators [65] is a variant in which the upward information flow activates a collection of competing units where winners are selected by a high level mechanism, on the basis of possibly learnt preferences. Their later work added a further layer the "Supervisory Attentional System" (SAS), for dealing with novel situations. Such systems are examples of a general three step cyclic pipeline model for processing information:

REPEAT: (1) Sense. (2) Think and decide. (3) Act.

The CogAff schema accommodates such sequential pipeline architectures, as a special case, but also permits alternatives where the mechanisms in different layers are *concurrently* active, and various kinds of information constantly flow within and between the sub-systems in both directions. An important feature of embodied animals and robots with multiple physical sub-systems concerned with locomotion, perception, manipulation, communication, and internal functions, is that they illustrate the need for such concurrency. For example, the CoSy Explorer needs to control its movements at the same time as it uses its visual and other perceptual systems to check whether it has reached its target and simultaneously processes some speech input. In general the sorts of concurrency required in a robot with multiple interacting sub-systems refute philosophical functionalist theories that use a finite state machine model of mind, e.g. Block in [66]. This functionality is also inconsistent with sequential "sense think act" models.

Many researchers do not understand the need for anything but Omega architectures: they think of perception and action as essentially low level processes of transduction, so they use small boxes for them in architecture diagrams, possibly connected to low level internal processes. That is roughly how perception and action are currently implemented in the CoSy robots, (except perhaps for the linguistic components). Section 12.12 below explains why towers are needed for vision and action in more advanced robots, though only primitive forms have so far been implemented in CoSy, described in previous chapters.

12.9 Beyond the CogAff Schema

The subdivisions in the CogAff schema are not offered as the only subdivisions to be used in describing designs. A study of the products of evolution will reveal many intermediate cases, requiring a finer-grained subdivision of types of component. The task of designing working systems, as in CoSy, can also drive further development of such a conceptual framework, identifying discontinuities in requirements and designs as illustrated in [12] and below, challenging philosophical and other theories that propose a small number of major steps in evolution (e.g. expanding the useful overview in [67]). Some finer-grained distinctions between layers are described in [10].

One of high level features of a design not represented in the CogAff schema is the extent to which it learns or develops. It is clear that the human architecture is not fixed at birth but develops over time, including acquiring new layers of competence, new forms of representation (including use of new languages, mathematical and scientific notations, musical notations, technical diagrams, maps, circuit diagrams, etc.), new ontologies and new reactive skills related to the other developments. There may be still unknown *internal* forms of representation that humans and other animals develop after birth.

In a more detailed survey we would need to divide up systems in terms of different patterns of change and development, with at one extreme a completely fixed system, followed by a system that is fixed except for parameters that can be adjusted, followed by more and more complex forms of learning and development, as illustrated in Figure 12.3. It is not clear that researchers in AI and Cognitive Science have so far produced a comprehensive taxonomy of the sort that would be required. The CogAff schema may turn out to provide a useful way of indicating that different sub-systems develop in different ways – for instance many reactive systems learning by parameter adaptation, the high level systems growing by developing new languages, building explanatory theories, extending their ontologies, and absorbing new values from the surrounding culture, and all of them developing by acquiring new links between subsystems (e.g. compiling new reactive versions of pre-existing deliberative capabilities). Fido, the domestic robot in Chapter 1 may need many of these forms of development.

Another set of divisions between boxes is concerned with forms of representation. AI researchers have frequently noted the importance of choosing forms of representation that are suited to particular problems, e.g. Minsky in 1961 [68] and McCarthy and Hayes in 1969 [69]. The latter extended Chomsky's ideas about adequacy of grammatical formalisms, by distinguishing metaphysical adequacy, epistemological adequacy and heuristic adequacy. Following ideas of Marr, it is now commonplace to distinguish viewer-centred, object-centred and room-centred forms of representation, among others. The functional differences between the layers and the towers in the CogAff schema will be only loosely connected with differences in form of representation. Further work is needed to work out which sorts of representation fit where. A special case is the use of linguistic forms of representation.

12.9.1 Where Are the Linguistic Mechanisms?

Analysis of requirements for our robots raised the question whether linguistic competences fit naturally into a small subset of the boxes in the CogAff schema, or whether they need additional boxes, or whether they are to be distributed in all the main portions of the architecture.

The answer seems to be that linguistic competence is distributed through many parts of the architecture. It is obvious that linguistic perception of speech requires multi-level, processing dealing concurrently with acoustic, phonetic, morphological, syntactic, semantic and pragmatic information. So for speech the requirement for a *tower* of perception is clear. Similar comments apply to reading printed or hand-written text. Likewise speech production requires multiple levels of processing as ideas, sentences, phrases, word-selection, and production details such as tempo, intensity and intonation are determined, along with self-monitoring that can lead to self-correction. Similar remarks can be made about production of written communications. So producing linguistic utterances requires an action tower, rather than just a simple system for feeding signals to a transducer. These two points rule out an Omega architecture for a system with human language capabilities.

Use of language internally or externally can also function as an enhancer for other sub-systems, e.g. concerned with planning, reasoning, hypothesis formation, prediction, motivation, and conflict resolution. Moreover, while explicitly learning a new foreign language seems at first to exercise mostly the two upper levels of the schema, as expertise develops that seems to make increasing use of automatic reactive sub-systems, in the lowest layer.

Further development of these ideas and their implications for imposing structure on the space of possible designs, remains a topic for further research, especially if a future project can develop an intelligent pre-verbal toddler-robot as a basis for acquiring linguistic competence, so that we can more clearly understand what difference language makes, an issue that was not in CoSy because linguistic competence was integrated from the start. Although that appeared a reasonable strategy at the time, and may be an appropriate strategy for particular engineering applications of AI, it is arguable that the failure to produce a robot that behaved intelligently before adding linguistic abilities has seriously distorted our research because proper human-like linguistic competence should be a later addition to more basic general animal competences, extending sophisticated functionality, including communicative capabilities, that existed without language.

CoSy does have a number of such capabilities, including perceptual and planning capabilities, so in principle they could have been used to generate a kind of animal intelligence (e.g. producing play, exploration and learning). However that would have required a different way of putting things together so that motives need not be derived from a human linguistic communication. We did discuss mechanisms for "architecture-based" motivation, as opposed to the kinds of "reward-based" motivation often assumed to be necessary. Architecture-based motivation, which, it is arguable, is characteristic of most biological organisms, involves having one or more portions of the architecture where possible future states or processes, or constraints on states and processes can be described, and which, if present will tend to generate and modulate planning, decision-making, and the prioritising of actions. In that case other modules in the architecture can be triggered by various occurrences to create these motivational representations and insert them where they can become effective: for example an auditory mechanism detecting a strange noise, and automatically generating a motive to investigate the source of the noise and adding it to the store of current motives (along with other relevant information [70]). If several mechanisms cause conflicting motives to be generated, that could be detected and might trigger a conflict resolution mechanism to deal with the conflict. All such mechanisms for architecture-based motivation are probably innate in simplest animals (and machines), while products of learning and development play a role in more sophisticated types.

Pressure of time, and our commitment to integrate the subsystems developed by all the partners, prevented us investigating this kind of possibility further. It might in future demonstrate how a robot might learn by play and exploration without being reward-driven.

12.9.2 Varieties of Compositional Semantics

One of the major differences between the reactive layer and other layers in the CogAff schema is that the upper two layers can make use of formalisms supporting structural variation (e.g. predictions, plans and percepts of varying complexity) whereas the reactive level mainly uses atomic symbols (e.g. measures) or fixed dimensional vectors of symbols.

Where there is structural variation, and old structures can be combined to form new more complex ones (like sentences with sub-clauses), the semantic content of complex structures is usually assumed to be based on compositional semantics: the meaning of any complex expression is a function of the meanings of the parts and the structure of the expression. A problem addressed in CoSy led to an extension of this idea, as follows.

A question arose as to how certain words and phrases such as 'to the left of' used in an instruction to place something should be interpreted when there is a large area to the left. The solution at first proposed was to use a probabilistic semantics to select the "best" target location in the region under consideration, on the assumption that the probabilities would be derived from previous experiences. A probabilistic mechanism was therefore implemented.

Despite the empirical evidence, I found the arguments for the probabilistic interpretation unconvincing because they did not take account of the importance of context. I thought the empirical, probabilistic, data were likely to be a side-effect of deeper, more powerful and general mechanisms.

So an alternative theory was proposed, according to which Gricean principles of communication can be combined with a form of compositional semantics that allows context to play a role anywhere in a semantic structure. So on this view, the proper way to obey the request "Put the pen at the left of the book" is to use an understanding of what the pen is needed for, and what it can interact with, along with the perceived set of spatial relationships, to select a target location, rather than using a probability calculation based on previously observed placings. Where the context does not determine a selection, any location within reach of the person who asked for the pen will do.

This turned out to be a special case of an important general idea, that transforms many supposedly vague expressions (e.g. "heap", "big", "long", "efficient"), into expressions with a gap to be filled by the context of use, based on general knowledge and intelligence. For example, the number of stones required for a heap, depends on why a heap is required: to hold down a tarpaulin in a strong wind, to provide a base for one end of a bridge, to provide a platform on which to stand to see over a wall, and so on. These ideas led to a long and complex discussion paper [8] still under development.

An implication of this, is that the process of sentence comprehension can intrinsically include deep integration with reasoning, memory and visual perception. For this to work, not only the language sub-systems, but also the reasoning, memory and perceptual sub-systems must be designed so as to support the integration. That is a powerful challenge to system designers working on those sub-systems in isolation.

It is a powerful challenge anyway, and I expect it will be many years before machines can handle such context sensitivity in a useful way: it will first require development of a great deal of knowledge about the world and what people can want or intend or fear or prefer to happen in it.

12.10 The H-CogAff Special Case

One special case of the CogAff schema, labelled H-CogAff (Human-Cogaff), developed at Birmingham in the decade before CoSy [62], and also mentioned in Chapter 1, assumes all the boxes are occupied and that there are many connections, including connections to a reactive alarm system, and connections linking various layers in multi-level perception and action sub-systems to different parts of the central "tower", as explained in [62].

This special case and the diversity allowed by the CogAff schema were mentioned in our proposal. It was hoped that some of the ideas could be tested and the specifications refined and extended as a result of work on the robots. We did develop some new ideas about requirements for such an architecture as a result of the requirements analysis done during the project and work on some of the sub-systems. In particular [10] showed the need for several important subdivisions in the deliberative layer, and there were changes to our ideas about vision, language and the types of compositional semantics required in an intelligent system, as mentioned above.

Implications for meta-management were related to requirements for motive generation and selection in the robot, but so far only very simplified versions of those mechanisms have been implemented, as they sufficed for the scenarios chosen. Another important function of meta-management is mentioned below in connection with mathematical learning. These ideas are still under development, and have not yet fed into working systems (so they are not mentioned in

Chapter 2). In some cases, this will require major advances. In general, insofar as meta-management involves the ability to monitor, represent, modulate or extend internal processes, it will require meta-semantic competence: namely the ability to represent things that have semantic content.

This will require dealing with referential opacity, and many other things that have so far not arisen in the CoSy project because the scenarios addressed have been relatively simple. However, as some of the discussions among the team working on integration revealed, the problem is already just below the surface in the current system because we are close to problems where the robot needs to think about and reason about what a human knows, wants, intends, can see, etc. Future work will need to address these problems.

Some researchers, e.g. John McCarthy, favour dealing with such cases by introducing new modal logics with special modal operators for Knows, Wants, etc. I think a better solution uses architectural mechanisms providing a kind of 'encapsulation' e.g. where supposed/possible beliefs (one's own or someone else's) are treated using the usual forms of representation but not allowed to have the normal causal powers of beliefs. For instance, imagining a situation where you believe a hungry lion is running towards you should not make you decide to run away. On the other hand if you know that John thinks the bush in the shadows is a hungry lion you should be able to work out that he may wish to run away, and therefore decide to tell him that what he thinks is a lion is a bush. It is well known that young children take some time to develop the ability to handle referential opacity. It is not clear whether a future robot will also have to grow the ability, as opposed to having it pre-programmed. This may be related to whether pre-verbal competences need to be well developed before language learning starts.

12.11 Study Trade-Offs Not Special Cases

The diversity of designs and sets of requirements mentioned above is very daunting. One way of trying to make sense of this diversity is to consider divisions at a fairly high level of abstraction to start with, as proposed in previous sections.

Another strategy for imposing structure on the problem is to think in terms of *trade-offs*. Instead of arguing about how things must be, or trying find ways of categorising whole systems using evaluation functions, we can look at costs and benefits of different options within a single design problem. These costs and benefits can be thought of as niche-relative disadvantages and advantages, without assuming that these must all have numerical values. Often the trade-offs need to be expressed descriptively rather than numerically: e.g. design X makes certain kinds of learning impossible, but enables predators of type Y to be avoided – not-unlike the style of consumer reports on multi-functional objects such as cars, cookers, computers, etc. (These issues are addressed in this old paper on "better" [71], which suggests a semantic structure with a component for context, not unlike the analysis of vague words, above.)

12.11.1 Nature-Nurture Trade-Offs

A particularly interesting case is the trade-off between pre-programming all the behaviour required in a system (as evolution in effect does for the majority of species, a feature for which the label "precocial" is often used) and instead producing an "altricial" organism or machine that starts off superficially highly incompetent but has a sophisticated meta-competence for developing itself in a certain class of environments, including acquiring many new competences through interactions with the environment [41, 72, 6, 1, 27] This seems to be the option evolution has developed for many mammals, especially hunting mammals and primates, as well as many birds).

In between the two extremes are different combinations of "precocial" (or preconfigured) competences and "altricial" (or meta-configured in the terminology of [1]) competences. Jackie Chappell and I argued at IJCAI'05 [6] that both extremes and intermediate cases will be required in robots.

The CoSy robots have a particular combination of pre-programmed and learnt competences dictated largely by limitations of our time and available techniques: future work should explore more possibilities in a more principled way, after a deep analysis of trade-offs. There is work in progress in building a version of the PlayMate that discovers things about the world and effects of its actions in Birmingham[8] and Freiburg.[9] We are still a long way from the kind of bootstrapping mentioned in Section 12.5.4.

12.11.2 Image-Scene Tradeoffs in Visual Processing

One trade-off that provoked much discussion, especially in the context of the PlayMate scenario was the trade-off between trying to perform visual tasks by using and learning about 2-D image features, relationships, models and processes, as opposed to using and learning about 3-D structures and processes, which could be either inferred from image information or used to project to image information, in a mixture of bottom up and top-down processing.

[8] Work by Marek Kopicki, partially reported in `http://www.cs.bham.ac.uk/~msk/report8/report8.final.pdf`, `http://www.cs.bham.ac.uk/~msk/report9/report9.pdf`

[9] See Jürgen Sturm, Christian Plagemann, Wolfram Burgard. Adaptive Body Scheme Models for Robust Robotic Manipulation. In *Proceedings of Robotics: Science and Systems (RSS)*, Zürich, Switzerland. 2008.
Jürgen Sturm, Christian Plagemann, Wolfram Burgard. Unsupervised Body Scheme Learning through Self-Perception. In *IEEE International Conference on Robotics and Automation (ICRA)*, Pasadena, CA, USA, 2008.
`http://www.informatik.uni-freiburg.de/~sturm/media/resources/public/zora-7dof-demo.avi`

An illustration of the need for a 3-D ontology is shown by the ambiguity of the Necker cube, in which the experience of a pattern of straight lines flips between seeing a cube with one of two spatial orientations, and where the relative distances change. Since *nothing* changes in the image when the cube flips it is clear that something not in the image must be represented in each of the two states.

It is a small step from there to argue that even in unambiguous images where a cube is seen the percept must represent 3-D structure, not just identify a portion of the image and its features and relationships. This kind of representation using an exosomatic ontology (i.e. referring to something outside the perceiver, that can exist independently of whether it is sensed or not) will be crucial in a robot with the competences we aimed for in CoSy. However at present the vision systems deployed are very limited in their understanding of 3-D structures.

Another tradeoff whose potential importance emerged in the analysis of requirements was the tradeoff between representing only static structures and then attempting to represent processes, including actions in terms of collections of representations of static structures, *vs* constructing representations of processes that could be used during the perception of processes and also when reasoning about possible past or future or unperceived processes. This topic is discussed in [23] and some of the online presentations.

12.11.3 Trade-Offs Related to Noise and Uncertainty

Another trade-off discussed at various times and especially during the final year is the trade-off between (a) dealing with noise and uncertainty by making use of probability distributions and performing probabilistic inferences, as opposed to (b) finding a form of representation that avoids noise and uncertainty (in a particular situation) by using a high level of abstraction. For example if it is impossible to decide for certain whether a curved line has constant curvature or not (i.e. is a circular arc), then instead of working with a probability distribution over a range of possible curvatures, simply describe it as curved. In some cases where there is uncertainty the robot could avoid reasoning about probabilities by getting new information. E.g. if the size of an object is uncertain because it is partly occluded, instead of dealing with probable lengths choose a different viewpoint, or temporarily move the occluding object. Another important possibility seems to be to notice that there may be regions of definiteness where an answer is "yes" and regions of definiteness where an answer is "no", and a phase boundary where the answer is uncertain. In that case, it may be possible to move away from the phase boundary when it is encountered. E.g. if you can't tell whether your trajectory will or will not cause you to bump into the edge of a wall, it will typically be possible to ensure that you will definitely not bump into it if you aim more to one side. These points are discussed in this draft document [13] mentioned below in connection with changing affordances.

There are other trade-offs that surfaced during the project, most of which still need further work.

12.12 Requirements for Visual Systems

12.12.1 Why Do Perception and Action Need Towers?

There are many different sorts of requirements that could be considered for visual systems. For example, work on systems that can be trained to recognise objects in images of cluttered scenes is challenging and useful for many application domains, but it does not necessarily provide the kind of visual competence required for a robot with a movable hand to be able to work out a good way to pick up an object with a complex shape, subject to varying constraints – e.g. it may be fragile, or full of liquid, or very hot, or too large to be grasped by one hand, or partially obstructed by another object. In some sense it is obvious that the ability to perceive 3-D shape, including distinguishing parts with different features and relationships and seeing their relationships to one another and to other objects must, for humans and other animals, and presumably also future intelligent robots, have priority over recognition of objects, since it is possible to see and interact with (e.g. picking up, climbing over, disassembling) something you do not recognise. This is related to the theories of Gibson mentioned later.

Humans can perceive and interact with objects in poor light, can see spatial structures and potential for actions of various sorts even in low resolution noisy images, as illustrated in Figure 12.6. A more detailed analysis of some of the requirements can be found in [73].

Fig. 12.6. Despite low resolution, poor lighting, and noise in this image, people easily perceive a collection of objects with definite spatial relationships, even though what is perceived (including shapes, orientations, curvature, relative thickness, etc.) is not perceived with very great precision. A challenge is to devise forms of representation that (a) are derivable from images despite poor image quality, and (b) have sufficient definiteness to allow actions to be planned and executed reliably. (From [73]).

12.12.2 Multi-strand Process Perception

Analysis of requirements for PlayMate in year 1 [30] revealed the importance, for a robot doing 3-D manipulations, of being able to perceive and think about "multi-strand relationships" (relationships holding between different parts of two or more objects, in addition to relations between the whole objects). Some of the relationships will be metrical relationships of size, distance, angle, volume, curvature, etc. Others will be qualitative relationships, such as containment, contact, being above, overlapping, being nearer, being between, etc. The latter may change discontinuously while the former change continuously. Other relationships may be causal or functional, e.g. pushing, supporting, stretching, compressing, etc. All of these need to be perceived by a robot with the capabilities we were aiming for in the PlayMate, and to some extent also in the Explorer. However, the work on perception in CoSy was not able to address most of this.

When actions are performed the different sub-relations can change in parallel, producing "multi-strand processes". Some of the changes are continuous and others discrete (e.g. topological). This is related to the need for perception of static structures to use multiple ontologies in parallel (not just part-whole layers). Examples of the use of different ontologies in seeing a static scene include seeing lines on a page, seeing the lines forming a face with parts seen as eyes, nose, mouth, cheeks, etc., and seeing the face as happy, or sad. When the well known ambiguous duck-rabbit figure (Figure 12.7) flips the perceived 2-D image features do not change, but different ontologies are involved in the two percepts, e.g. ears vs bill. Furthermore there is a meta-semantic ontology involved insofar as the duck is seen as looking one way and the rabbit as looking the other way. Around 30 years ago, the Popeye program described in [39, Ch 9] used a mixture of top-down and bottom-up processing, and stored knowledge, to interpret messy pictures in terms of different ontology layers, with distinct part-whole relationships in each layer. Although it was noticed many years ago that seeing static *structures* could involve perception of structures at different levels of abstraction, using different ontologies, the need for perception of *processes* at different levels of abstraction went unnoticed, though something like this was noted by Grush [74] in 2004.

This generates requirements for human, animal and robot vision that appear not to have been widely appreciated. This is because when things change there can also be changes going on simultaneously at different levels of abstraction. For instance looking at a video of a rotating wire frame cube, involves seeing changing light and dark portions of the image and simultaneously seeing edges, corners and faces of the cube moving around in 3D space, changing positions, orientations and relative distances. That is a relatively simple case compared with what is required for the PlayMate robot to see what it is doing when it manipulates objects and far more complex shapes than wire frame cubes are involved, in addition to changing causal and functional relationships. At present there is nothing in CoSy that represents multi-strand processes,

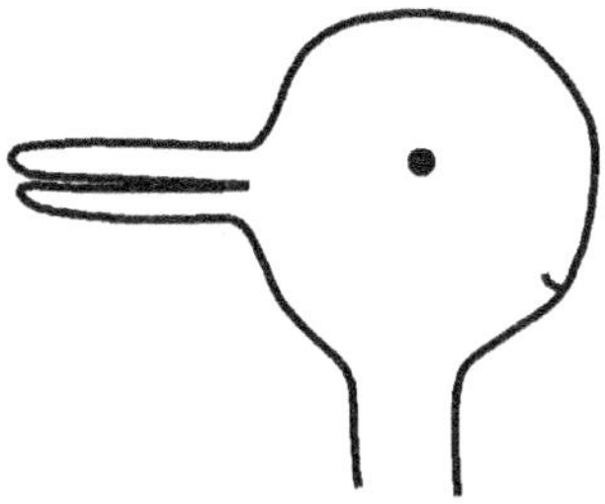

Fig. 12.7. Many ambiguous figures flip between two percepts without anything in the image changing. This often gives important clues to the multi-layered ontologies that can be involved in visual perception. What changes when this flips between the two views?

although some fragments are there. However the visual challenge of seeing our robot's arm and hand seem to be well beyond the state of the art in vision, especially seeing it moving. These ideas were presented in an invited talk at a multidisciplinary workshop in May-June 2007 and will be published in the proceedings: [23].

12.12.3 How to Acquire Useful Ontologies

If the above is correct, then that requires a visual system not merely to be able to learn to see part-whole hierarchies on different levels but also more abstract interpretation layers. This is related to the need for an intelligent system to use amodal exosomatic ontologies. How the ontologies develop and how the semantics of newly defined symbols are generated if they are not definable in terms of old symbols or sensorimotor patterns is a complex issue, discussed in more detail in [17].

12.12.4 Varieties of Complexity Reduction

Chapter 3 discusses varieties of learning/development that involve reducing complexity by reducing dimensionality of sensory motor information acquired by active exploration and experiment.

The complexity reduction involves moving to a new vector space with fewer dimensions, but preserving the "space-occupancy" form of representation. (An analogical form of representation in the sense of [51].)

We also need to understand other processes that are important in humans and some other animals and which could be essential for some future robots. A very different kind of transformation is the move from representing some portion of the world in a pixel/voxel based form (i.e. occupancy of portions of some vector space) to representing discrete, enduring, re-identifiable objects with features and relationships – possibly changing features and relationships, where the same object can occupy different spatial regions/volumes at different times.

Consider this example:
A 1000x1000 video with 1000 frames, containing changing blue red and white pixels might be described either in terms of a huge three-valued array, or like this:

- A big blue roughly circular pulsating blob moves right with increasing speed against a white background.
- A smaller red triangular blob rotates clockwise about its centroid while moving left with decreasing speed.
- The second blob starts to the right of the first blob and is temporarily obscured by the first blob as it moves.

These descriptions summarise the contents of a 1000x1000x1000 array. The example illustrates a trade-off between simplicity of conceptual apparatus and simplicity of descriptions:

The conceptual apparatus required to describe all the array cells is very simple: just a uniform formalism based on three coordinates and a colour label, whereas the object-based summary description above achieves massive complexity reduction by using a much more sophisticated ontology, including a distinction between space and time dimensions, the notion of a spatially extended, temporally enduring, but changing object, (with complex criteria for identity of the object), etc.

Further, the three sentence, object-based, summary above describes not only the one scenario with 10^9 cells, but a huge number of different scenarios with slightly different contents (different blob sizes, slightly different shapes, different speeds and accelerations, etc.) though the boundaries between what is and what is not included are somewhat vague because of the vagueness of 'big', 'roughly circular', 'pulsating', etc. and the fact that speeds and accelerations are not specified. This kind of vagueness is related to the point about context-sensitive compositional semantics in Section 12.9.2.

I think biological evolution somehow discovered the need for something like the object-based form of representation in a subset of organisms. This seems to be the basis of human abilities to use logic and to use verbal descriptions. I don't know how many organisms can use the object-based type of representation. It seems very likely that cats, monkeys, nest-building birds, primates, need it. I don't know about frogs, flies, paramecia, etc. These questions could simulate new kinds of research in animal cognition, and new questions to be asked about cognitive development in humans. Issues like this are discussed in Brian Cantwell Smith's book [75], though without specific implementation recommendations.

Much work in robotics (including CoSy) assumes the need to use an ontology of enduring but changeable objects – though often the ontology is assembled in a piecemeal unprincipled way. There is much work still to be done defining the long term ontological requirements first of all for a pre-verbal child-like robot and then for a robot learning to use a human language for communication. It is sometimes assumed that there must be some innate ontology on the basis of which everything else is constructed. However, that assumption ignores the possibility of an ongoing process of testing and debugging of the current ontology which could lead both to rejection of some of the innate components and to to construction of extensions that are not definable in terms of the starting ontology. Something like that happens in the history of science, so it is possible in principle. How to get a robot to do that is a topic for future research – one of many long term requirements for human-like robots.

12.12.5 Beyond J.J. Gibson's Affordances

J.J. Gibson [49] drew attention to the requirement for a perceiver's ontology to include positive and negative affordances: namely features of the environment that are relevant to enabling or obstructing actions that the perceiver can perform that might be relevant to achieving goals. Analysis of requirements for a robot has shown that Gibson's ideas need to be extended:

- An agent can perceive the possibility of processes that are not produced by the agent, and also perceive things that will enable or prevent such processes. (For more details see [76].) I call this perception of *proto-affordances*. Affordances will then map onto a small subset of proto-affordances.
- There is a kind of affordance that Gibson did not explicitly distinguish from an action affordance, namely an *epistemic affordance*: concerned with aspects of the environment that support or obstruct the perceiver's acquisition of new information. E.g. turning a face of a cube towards you creates the epistemic affordance that is the possibility of getting information about features of tha face. Likewise the rotation will have moved at least one other face out of sight, creating a negative epistemic affordance for that face. One of the important things a child has to learn is what sorts of epistemic affordances the environment provides, and also what actions it can perform to increase the epistemic affordances (gain access to more information).

The ability to perceive and reason about both proto-affordances, including affordances for other individuals, and epistemic affordances is relevant to many of things a future domestic robot may be required to do. In particular the epistemic affordances are very relevant to visual servoing. A partial analysis is in [13], which was written in response to difficulties in getting the PlayMate to grasp things reliably at the CoSy review in 2007.

The ability to detect that one lacks some information requires self-monitoring and a meta-semantic competence and therefore belongs in the meta-management architectural layer. The ability to perceive epistemic affordances and action affordances, and to work out which action affordances will change which epistemic affordances (e.g. moving nearer the door to a room will enable you to see more of the contents of the room), seems to be a feature of development in very young children and some animals that has not been studied except for very restricted contexts using verbal interactions, e.g. asking a child whether looking at or feeling a hidden object will provide information about its shape or its colour. In some animals the actions required to alter epistemic affordances may be genetically compiled into reactive subsystems, but animals that have to learn to manipulate epistemic affordances in environments that evolution could not anticipate will need mechanisms for acquiring such competences. For example, as you walk through a car park there are vast amounts of visual information available in the changing optic array, about changing colours, angles subtended, altering occlusions, various kinds of optical flow, and moving highlights and reflections on curved and planar surfaces.

Using that information to compute, in parallel, shapes, orientations, curvatures, distances, spatial relationships, surface properties, etc. is unlikely to be based entirely on genetically determined competences since nothing like car-parks existed to influence the selection of our distant ancestors. So there must be processes of learning, still to be studied. I suspect it will require massively parallel constraint propagation mechanisms operating at different levels of abstraction, on different time scales. It will also involve perception operating in parallel at different levels of abstraction, with different levels in the perceptual systems connecting to different levels in central systems, as indicated in the CogAff perception tower.

12.12.6 Implications of Speed of Human and Animal Visual Perception

Some experiments help to demonstrate familiar but not often noticed features of human visual competence: the speed at which very high level percepts seem to be constructed even when there are no expectations about what the next scene will be. For example, see the demonstration in [18].

Considerations of the sort presented there led to the conjecture that the architecture of a human-like visual system includes a multilayer collection of dynamical systems linked by a constraint propagation network, with different layers operating in parallel at different speeds, performing different sorts of tasks, some of them representing only sensorimotor ontologies whereas others, more remote from the sensorimotor interface can also represent ontologies referring to unperceivable, relatively inaccessible parts of the universe, including future actions and events. These ideas are discussed further in [23] See Figure 12.8.

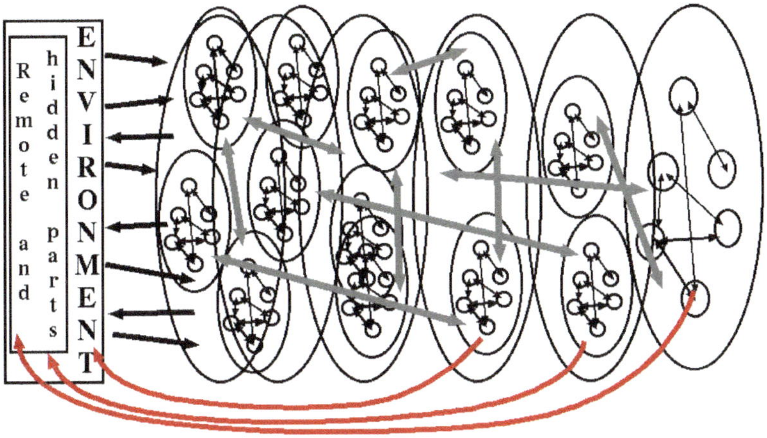

Fig. 12.8. A perceptual system composed of multiple dynamical systems linked in a constraint network, some of them dormant while others operate concurrently more and less remote from the sensorimotor interface, with more remote sub-systems able to refer to un-sensed aspects of reality. Some involve continuous dynamics, others discrete changes. The various perceptual (e.g. visual) sub-systems would also be connected with more central processing sub-systems and in some cases also with action sub-systems, e.g. for reflexes to work. Dynamical systems further from the sensorimotor interface can use semantic contents referring to things in the environment that are more remote from the sensorimotor interface, as indicated by the long arrows.

12.13 Learning to Be a Mathematician

The ability to see proto-affordances and also epistemic affordances has very deep implications. By studying requirements for a robot to be able to cope with novel configurations, we see that some things that at first are learnt as empirical generalisations, can later be regarded as mathematical (i.e. necessary, not empirical) truths, as appears to happen in young children, though this has not been noticed by developmental psychologists as far as I know. An example was mentioned in Section 12.6, namely a child coming to realise that counting a fixed set of objects in different orders must necessarily give the same result, even though initially this was learnt as an empirical generalisation.

There are many unsolved problems about how this transition happens, but it seems to be closely connected with learning about affordances and how they are related to structures of objects in a principled way that is not just empirical. When these matters are fully understood by the child, animal or robot they enable novel problems to be solved by creative reasoning about action and epistemic affordances. This depends on coming to realise that some true generalisations are not just empirical generalisations that might one day be tested. When fully understood they can be seen to be, in effect, mathematical theorems, even if the learner does not explicitly notice this fact. This topic was discussed, with more examples, in a recent paper [21] and further expanded in this slide presentation: `http://www.cs.bham.ac.uk/research/projects/cogaff/talks/#mkm08`

If we can make progress with modelling this kind of learning, not only will it be a contribution to robotics, and to developmental psychology, it will also favour Kant's philosophy of mathematics [77] over Hume's (and Russell's [78]).

12.13.1 Two Kinds of Causation

The kind of learning that uses a transition from empirical to non-empirical understanding also supports a notion of causation that is more like Kant's than Hume's conception of causation (which is essentially statistical and is the precursor of modern Bayesian notions of causation). In work done with Chappell a start has been made in using these ideas to analyse kinds of causal competence in other animals as well as in humans competences.[10] It is very likely that future robots of many kinds will need that kind of causal understanding. However, our robots were not confronted with problems requiring the ability to think about causation.

This discussion points to a need for a form of learning that is very different from the heavily Bayesian (probabilistic/statistics-based) forms of learning that currently attract the most attention. A possible initial mechanism for this would be to allow some features of what has been learnt empirically to trigger a change in the way structures or processes in the environment are represented – e.g. a change from lots of sensorimotor conditional probabilities to representing 3-D objects moving around in a locally euclidean space. That form of representation of processes will have strong implications for what is and is not possible. If the distinctions between kinds of material are included in the representations (e.g. some things are impenetrable others not, some are rigid, others not) then properties of matter can play a role in some of the reasoning. For example if one end of a rigid rod is rotated in a plane then the far end *must* move in a circular arc. If one of two meshed gear wheels made of rigid impenetrable material is rotated, the other *must* rotate in the opposite direction. It is often thought that there are only two ways a young child or animal can discover useful affordances, namely either by empirical trial and error, or by learning from what someone else does (through imitation or instruction). However, our discussion shows that there is a third way, namely by *working out* the consequences of combining spatial processes in advance of their occurrence. This point seems to be missed by many developmental psychologists, e.g. [41].

[10] See the slide presentations
`http://www.cs.bham.ac.uk/research/projects/cogaff/talks/#wonac`

Some of these implicit theories with strong implications may have been pre-preprogrammed genetically in some animals, as a precocial or pre-configured competence. In others, a genetically pre-programmed tendency to perform experiments, discover generalisations, and then build a new layer of theory to make sense of the results may reach similar competences more indirectly and more slowly, but with more scope for subsequent modification. (An example of a nature-nurture trade-off.) It seems that a new born human is already predisposed to try to interpret perceived structures and processes as inhabiting a 3-D space, as might an "altricial" robot be. However, various observations could trigger a process of abduction leading to an enriched explanatory theory about the nature of the environment, e.g. allowing that not only are there movable objects, but some are rigid and some are impenetrable. These notions would be understood in such a way as to disallow representations of certain things bending, being dented, breaking, etc. and also disallowing the representation of part of one object passing through another. (Compare McCarthy on "The well designed child" [50].)

More subtle theories would need to be developed by a child to allow it to learn facts about the nature of mappings between two structures or two processes that allow the discovery that it is not just an *empirical* fact that counting a row of objects left to right produces the same result as counting the same row right to left. Contrast the theories of Rips et al. [79].

There is still much work to be done on the architectural and representational underpinnings for these layered processes of learning and theory construction. Originally it was hoped that some such forms of discovery (e.g. in relation to counting) could be made by a version of PlayMate that was able to point at a set of objects in sequence and ask questions about what it had done.[11] But so far we do not have vision systems, episodic memory or action sub-systems capable of being used for this sort of task.

12.14 Confusions about the Role of Embodiment

What's important about embodiment (e.g. what drove the most significant evolutionary developments in primate and bird cognition) seems to have been the need to be able to perceive and interact with 3-D structures and processes (including manipulating, assembling and disassembling 3-D structures) and the need to be able to think about spatially located events, processes and entities in the past, remote spatial regions, and the future.

In contrast, much of the work on embodied cognition in robots, and much of the philosophical concern with the importance of embodiment has focused on the terribly narrow problem of learning about sensorimotor relationships. (There are some exceptions.) For a detailed critique of these ideas see [25].

[11] As proposed in
`http://www.cs.bham.ac.uk/research/projects/cosy/PlayMate-start.html`

The single most important reason why embodiment influences cognition in humans is that we are part of a very complex 4-D world that extends way beyond what we can experience or interact with at any time, but which we can think about, plan about, learn about, ask questions about, find or construct routes to, use to explain what we perceive, build and test theories about, etc. People born blind, or without limbs (like Alison Lapper, the artist), or with four legs instead of two, or born as conjoined twins (two heads sharing a torso and legs) can develop those human cognitive competences.

That contrasts with disembodied AI systems that interact only with and think only about, some abstract information structure, such as a financial database, the internet, mathematical proofs, or a board game whose physical implementation is irrelevant. They don't need to be embodied or even to know anything about the 4-D environment and where they are in it or which of its occupants could affect them.

12.15 Developing the Revolution in Philosophy

Work in progress, partly inspired by the above work related to CoSy shows that old ideas about conceptual analysis in philosophy (the study of "logical geography" in Ryle's terminology [80]) needs to be reconsidered as a special subset of a larger task: investigation of complex aspects of reality that generate a rich "logical topography" that can be divided up in different ways, supporting different "logical geographies". A long paper explaining these ideas is under development [81].

It is not common for an AI project to be promoted as a contribution to philosophy, although there have been philosophers involved in AI projects (e.g. Daniel Dennett in COG, Bruce Buchanan in Dendral, Selmer Bringsjord, John Pollock, etc.) and at least one philosopher, Margaret Boden, has made major contributions to the history and philosophy of AI.

My concern was not merely to contribute to philosophy, but also to clarify problems in biology, psychology and eventually brain science.

There have been many AI projects involving philosophers, though usually as collaborators helping with engineering goals. Some philosophers interested in technical philosophical issues, for example, issues concerned with how scientific theories relate to evidence, or issues concerned with how a modal logic could be used in reasoning about permissions and obligations have tried using AI languages, tools and techniques to develop and test their philosophical theories (Herbert Simon, Paul Thagard).

A recent Edinburgh PhD thesis by Alison Pease took the philosophical analysis of the history of Euler's theorem by Lakatos as the basis for a model of mathematical exploration.[12]

In some cases a particular philosophical viewpoint, for example a philosophical theory about meaning (e.g. symbol-grounding theory), or a theory

[12] Available at `http://homepages.inf.ed.ac.uk/apease/research/phd.html`

concerning the nature of emotions, or of consciousness, or of the importance of embodiment, has influenced the design of a working AI system, especially, in recent years a flood of work influenced by the bad philosophy of symbol-grounding, (criticised in Section 12.5).

If the claims in Section 12.13 about the processes of transforming empirical discoveries to something like mathematical theorems can be substantiated and modelled, this will have deep significance for several aspects of philosophy, including philosophy of mind, philosophy of mathematics, philosophy of causation, and philosophical questions about evolution.

12.16 Further Documentation on These Ideas

For anyone interested in finding out the extent of the impact of CoSy on thinking about the themes presented here there are three sources of further information (still growing):

- The online repository of papers, discussion notes and presentations at the Birmingham CoSy site:
  `http://www.cs.bham.ac.uk/research/projects/cosy/papers/`
- The collection of online presentations at seminars, workshops and conferences, available here:
  `http://www.cs.bham.ac.uk/research/projects/cogaff/talks/`
- The disorganised collection of html discussion notes in this web site
  `http://www.cs.bham.ac.uk/research/projects/cogaff/misc/`

12.17 Why Other Disciplines Need AI

Very often philosophers and psychologists who attempt to think about possible cases in order to express their theories lack the experience of designing, implementing, testing and debugging working systems, so that they use levels of description that no engineer could take as a specification for a working system: the verbal descriptions used (sometimes with accompanying diagrams) are so non-specific as either to determine no possible working implementation, or as to determine very many different implementations with very different properties that the original proposers would never have considered.

One of the ways of changing this is for AI researchers to involve more people from other disciplines both in the detailed analysis of sets of requirements, and also in the processes of designing, implementing and testing, so that the depth and precision of future theories in the other disciplines can be improved over time.

12.18 Conclusion: The Future

Work on the follow on EU-funded CogX project (2008-2012)[13] will provide opportunities to develop a subset of the ideas presented here, though limitations of AI technology will remain a constraining factor for some time to come.

In parallel with that a proposal is under development for a collaborative project, with two biologists, Jackie Chappell and Susannah Thorpe, to investigate some of the cognitive competences displayed by orangutans moving through trees. Unlike most other animals, including other apes, Sumatran orangutans are able to use the compliance of branches intelligently to overcome difficulties caused by large gaps and the inability of some of the branches to support their weight. These problems and the achievements of the animals will be analysed from the standpoint of a robot designer and related to controversies about the evolutionary origins of ape intelligence, in which too much weight is sometimes given to social requirements.

Attempts will also be made to use some of the lessons learnt during the CoSy project, mentioned in Section 12.13, in collaborative research with developmental psychologists (yet to be identified) investigating the role in young children of the ability to transform empirical generalisations into proto-mathematical knowledge that can be used with confidence in solving novel problems. A start is being made by developing a collection of "toddler theorems".[14]

References

1. Chappell, J., Sloman, A.: Natural and artificial meta-configured altricial information-processing systems. International Journal of Unconventional Computing 3(3), 211–239 (2007), `http://www.cs.bham.ac.uk/research/projects/cosy/papers/#tr0609`, `http://cognitivesystems.org/cosybook/chap12.asp#Chappell/etal:2007`
2. Sloman, A.: How to Put the Pieces of AI Together Again, Tech. Rep. COSY-TR-0608, University of Birmingham, School of Computer Science. Poster summary for AAAI 2006 Members Poster Session, Boston (July 2006); 2-Page abstract at, `http://www.cs.bham.ac.uk/research/projects/cosy/papers/#tr0608`, Poster at `http://www.cs.bham.ac.uk/research/projects/cosy/papers/#pr0603` (2006), `http://www.cognitivesystems.org/cosybook/chap12.asp#Sloman:2006a`

---

[13] Described in `http://www.cs.bham.ac.uk/research/projects/cogx/`

[14] See this recent workshop paper [21] and two related online slide presentations: `http://www.cs.bham.ac.uk/research/projects/cogaff/talks/#math-robot` `http://www.cs.bham.ac.uk/research/projects/cogaff/talks/#toddlers`

3. Sloman, A.: Introduction to Symposium GC5: Architecture of Brain and Mind Integrating high level cognitive processes with brain mechanisms and functions in a working robot. In: Proceedings of the AISB 2006 Adaptation in Artificial and Biological Systems, Bristol (2006), http://www.cs.bham.ac.uk/research/projects/cosy/papers/#tr0602, http://cognitivesystems.org/cosybook/chap12.asp#Sloman:2006
4. Sloman, A.: Putting the Pieces Together Again. In: Sun, R. (ed.) Cambridge Handbook on Computational Psychology, ch. 26, pp. 684–709. Cambridge University Press, New York (2008), http://www.cs.bham.ac.uk/research/projects/cogaff/07.html#710, http://cognitivesystems.org/cosybook/chap12.asp#Sloman:2008a
5. Sloman, A.: The Well-Designed Young Mathematician. Artificial Intelligence 172(18), 2015–2034 (2008), http://www.cs.bham.ac.uk/research/projects/cosy/papers/#tr0807, http://cognitivesystems.org/cosybook/chap12.asp#Sloman:2008
6. Sloman, A., Chappell, J.: The Altricial-Precocial Spectrum for Robots. In: Proceedings IJCAI 2005, IJCAI, Edinburgh, pp. 1187–1192 (2005), http://www.cs.bham.ac.uk/research/cogaff/05.html#200502, http://cognitivesystems.org/cosybook/chap12.asp#Sloman/etal:2005
7. Sloman, A., Schiele, B. (eds.) Tutorial on Learning and Representation in Animals and Robots. In: IJCAI 2005, Edinburgh (2005), http://www.cs.bham.ac.uk/research/projects/cosy/conferences, http://cognitivesystems.org/cosybook/chap12.asp#Sloman/etal:2005c
8. Sloman, A.: Spatial prepositions as higher order functions: And implications of Grice's theory for evolution of language. Research Note COSY-DP-0605, School of Computer Science, University of Birmingham, Birmingham, UK (2005), http://cognitivesystems.org/cosybook/chap12.asp#Sloman:2005
9. Sloman, A.: Polyflaps as a domain for perceiving, acting and learning in a 3-D world. In: Position Papers for 2006 AAAI Fellows Symposium. AAAI, Menlo Park (2006), http://www.aaai.org/Fellows/fellows.php, http://www.aaai.org/Fellows/Papers/Fellows16.pdf, http://cognitivesystems.org/cosybook/chap12.asp#Sloman:2006b
10. Sloman, A.: Requirements for a Fully Deliberative Architecture (Or component of an architecture). Research Note COSY-DP-0604, School of Computer Science, University of Birmingham, Birmingham, UK, http://www.cs.bham.ac.uk/research/projects/cosy/papers/#dp0604 (May 2006), http://cognitivesystems.org/cosybook/chap12.asp#Sloman:2006c
11. Sloman, A.: Sensorimotor vs objective contingencies. Research Note COSY-DP-0603, School of Computer Science, University of Birmingham, Birmingham, UK, http://www.cs.bham.ac.uk/research/projects/cosy/papers/#dp0603 (May 2006), http://cognitivesystems.org/cosybook/chap12.asp#Sloman:2006d
12. Sloman, A.: Diversity of Developmental Trajectories in Natural and Artificial Intelligence. In: Morrison, C.T., Oates, T.T. (eds.) Computational Approaches to Representation Change during Learning and Development. AAAI Fall Symposium 2007, Technical Report FS-07-03, pp. 70–79. AAAI Press, Menlo Park (2007), http://www.cs.bham.ac.uk/research/projects/cosy/papers/#tr0704, http://cognitivesystems.org/cosybook/chap12.asp#Sloman:2007a

13. Sloman, A.: Predicting Affordance Changes (Alternatives ways to deal with uncertainty). Tech. Rep. COSY-DP-0702, School of Computer Science, University of Birmingham, unpublished discussion paper, http://www.cs.bham.ac.uk/research/projects/cosy/papers/#dp0702HTML (November 2007), http://cognitivesystems.org/cosybook/chap12.asp#Sloman:2007e
14. Sloman, A.: Requirements for Digital Companions: It's harder than you think. Position Paper for Workshop on Artificial Companions in Society: Perspectives on the Present and Future Organised by the Companions project, October 25-26. Oxford Internet Institute (2007), http://www.cs.bham.ac.uk/research/projects/cogaff/07.html#711 (October 2007), http://cognitivesystems.org/cosybook/chap12.asp#Sloman:2007d
15. Sloman, A.: What evolved first and develops first in children: Languages for communicating? or Languages for thinking (Generalised Languages: GLs). Presentation given to Birmingham Psychology department, http://www.cs.bham.ac.uk/research/projects/cosy/papers/#pr0702 (2007), http://cognitivesystems.org/cosybook/chap12.asp#Sloman:2007g
16. Sloman, A.: Why Some Machines May Need Qualia and How They Can Have Them: Including a Demanding New Turing Test for Robot Philosophers. In: Chella, A., Manzotti, R. (eds.) AI and Consciousness: Theoretical Foundations and Current Approaches AAAI Fall Symposium 2007, Technical Report FS-07-01, pp. 9–16. AAAI Press, Menlo Park (2007), http://www.cs.bham.ac.uk/research/projects/cosy/papers/#tr0705, http://cognitivesystems.org/cosybook/chap12.asp#Sloman:2007b
17. Sloman, A.: Why symbol-grounding is both impossible and unnecessary, and why theory-tethering is more powerful anyway, http://www.cs.bham.ac.uk/research/projects/cogaff/talks/#models (2007), http://cognitivesystems.org/cosybook/chap12.asp#Sloman:2007f
18. Sloman, A.: A Multi-picture Challenge for Theories of Vision (2008), http://cognitivesystems.org/cosybook/chap12.asp#Sloman:2008e
19. Sloman, A.: A New Approach to Philosophy of Mathematics: Design a young explorer, able to discover "toddler theorems". Online presentation, http://www.cs.bham.ac.uk/research/projects/cogaff/talks/#toddler (2008), http://www.cs.bham.ac.uk/research/projects/cogaff/talks/sloman-toddler1.pdf
20. Sloman, A.: Architectural and representational requirements for seeing processes, proto-affordances and affordances. In: Cohn, A.G., Hogg, D.C., Möller, R., Neumann, B. (eds.) Logic and Probability for Scene Interpretation, no. 08091 in Dagstuhl Seminar Proceedings, Schloss Dagstuhl - Leibniz-Zentrum fuer Informatik, Germany, Dagstuhl, Germany (2008), http://cognitivesystems.org/cosybook/chap12.asp#Sloman:2008b
21. Sloman, A.: Kantian Philosophy of Mathematics and Young Robots. In: Autexier, S., Campbell, J., Rubio, J., Sorge, V., Suzuki, M., Wiedijk, F. (eds.) Intelligent Computer Mathematics. LNCS, vol. 5144, pp. 558–573. Springer, Heidelberg (2008), http://www.cs.bham.ac.uk/research/projects/cosy/papers#tr0802, http://cognitivesystems.org/cosybook/chap12.asp#Sloman:2008c

22. Sloman, A.: Varieties of Meta-cognition in Natural and Artificial Systems. In: Cox, M.T., Raja, A. (eds.) Workshop on Metareasoning, AAAI 2008 Conference, pp. 12–20. AAAI Press, Menlo Park (2008), http://www.cs.bham.ac.uk/research/projects/cosy/papers/#tr0803, http://cognitivesystems.org/cosybook/chap12.asp#Sloman:2008d
23. Sloman, A.: Architectural and representational requirements for seeing processes and affordances. In: Computational Modelling in Behavioural Neuroscience: Closing the gap between neurophysiology and behaviour. Psychology Press, London (2009), http://cognitivesystems.org/cosybook/chap12.asp#Sloman:2009
24. Sloman, A.: Machines in the Ghost. In: Dietrich, D., Fodor, G., Zucker, G., Bruckner, D. (eds.) Simulating the Mind: A Technical Neuropsychoanalytical Approach, pp. 124–148. Springer, Vienna (2009), http://www.cs.bham.ac.uk/research/projects/cosy/papers/#tr0702, http://cognitivesystems.org/cosybook/chap12.asp#Sloman:2009a
25. Sloman, A.: Some Requirements for Human-like Robots: Why the recent over-emphasis on embodiment has held up progress. In: Sendhoff, B., Koerner, E., Sporns, O., Ritter, H., Doya, K. (eds.) Creating Brain-like Intelligence, pp. 248–277. Springer, Berlin (2009), http://www.cs.bham.ac.uk/research/projects/cosy/papers/#tr0804, http://cognitivesystems.org/cosybook/chap12.asp#Sloman:2009b
26. Sloman, A., Chappell, J.: Altricial self-organising information-processing systems. AISB Quarterly (121), 5–7 (2005), http://www.cs.bham.ac.uk/research/cogaff/05.html#200503, http://cognitivesystems.org/cosybook/chap12.asp#Sloman/etal:2005a
27. Sloman, A., Chappell, J.: Computational Cognitive Epigenetics (Commentary on Jablonka and Lamb: Evolution in Four Dimensions (2005)). Behavioral and Brain Sciences 30(4), 375–376 (2007), http://www.cs.bham.ac.uk/research/projects/cosy/papers/#tr0703, http://cognitivesystems.org/cosybook/chap12.asp#Sloman/etal:2007
28. Sloman, A., Chappell, J., Team, T.C.: How an animal or robot with 3-D manipulation skills experiences the world. In: The tenth annual meeting of the Association for the Scientific Study of Consciousness, Oxford, ASSC, Internet (2006), http://www.cs.bham.ac.uk/research/projects/cosy/papers/#pr0602, Poster for ASSC10, Oxford (June 2006), Also at ASSC10 Eprints Archive, http://eprints.assc.caltech.edu/112/, http://cognitivesystems.org/cosybook/chap12.asp#Sloman/etal:2006
29. Sloman, A., Chappell, J., The CoSy PlayMate team.: Orthogonal Recombinable Competences Acquired by Altricial Species (Blankets, string, and plywood). Research Note COSY-DP-0601, School of Computer Science, University of Birmingham, Birmingham, UK, http://www.cs.bham.ac.uk/research/projects/cosy/papers#dp0601 (January 2006), http://cognitivesystems.org/cosybook/chap12.asp#Sloman/etal:2006c

30. Sloman, A.: Cosy-partners. CoSy deliverable DR.2.1 Requirements study for representations, Tech. Rep. COSY-TR-0507, The University of Birmingham, UK, http://www.cs.bham.ac.uk/research/projects/cosy/papers/#tr0507 (2005), http://cognitivesystems.org/cosybook/chap12.asp#Sloman/etal:2005b
31. Sloman, A., Birmingham CoSy Project Team, Chappell, J.: Poster: Acquiring Orthogonal Recombinable Competences. In: Bekkering, H. (ed.) Proceedings CogSys-II. Radboud University Nijmegen, NL (2006), http://www.cs.bham.ac.uk/research/projects/cosy/papers/#pr0601, Conference, http://www.socsci.ru.nl/CogSys2, http://cognitivesystems.org/cosybook/chap12.asp#Sloman/etal:2006b
32. Sloman, A., Wyatt, J., Hawes, N., Chappell, J., Kruijff, G.-J.M.: Long Term Requirements for Cognitive Robotics. In: Cognitive Robotics: Papers from the 2006 AAAI Workshop: Technical Report WS-06-03, pp. 143–150. AAAI Press, Menlo Park (2006), http://www.aaai.org/Library/Workshops/ws06-03.php, http://www.cs.bham.ac.uk/research/projects/cosy/papers/#tr0604, http://cognitivesystems.org/cosybook/chap12.asp#Sloman/etal:2006a
33. Sloman, A.: The SimAgent TOOLKIT – for Philosophers and Engineers (And Some Biologists, Psychologists and Social Scientists) (1996), http://www.cs.bham.ac.uk/research/projects/poplog/packages/simagent.html
34. Sloman, A., Logan, B.: Building cognitively rich agents using the Sim_agent toolkit. Communications of the Association for Computing Machinery 42(3), 71–77 (1999), http://www.cs.bham.ac.uk/research/projects/cogaff/96-9%9.html#49
35. Fodor, J.: The Language of Thought. Harvard University Press, Cambridge (1975)
36. Sloman, A.: The primacy of non-communicative language. In: MacCafferty, M., Gray, K. (eds.) The analysis of Meaning: Informatics 5 Proceedings ASLIB/BCS Conference, Oxford, Aslib, London, March 1979, pp. 1–15 (1979), http://www.cs.bham.ac.uk/research/projects/cogaff/81-95.html#43
37. Jablonka, E., Lamb, M.J.: Evolution in Four Dimensions: Genetic, Epigenetic, Behavioral, and Symbolic Variation in the History of Life. MIT Press, Cambridge (2005)
38. Warneken, F., Tomasello, M.: Altruistic helping in human infants and young chimpanzees. Science, 1301–1303 (2006), doi:10.1126/science.1121448
39. Sloman, A.: The Computer Revolution in Philosophy. Harvester Press/Humanities Press, Hassocks (1978), http://www.cs.bham.ac.uk/research/cogaff/crp
40. Ambler, A.P., Barrow, H.G., Brown, C.M., Burstall, R.M., Popplestone, R.J.: A Versatile Computer-Controlled Assembly System. In: Proc. Third Int. Joint Conf. on AI, Stanford, California, pp. 298–307 (1973)
41. Gibson, E.J., Pick, A.D.: An Ecological Approach to Perceptual Learning and Development. Oxford University Press, New York (2000)
42. Sloman, A.: A First Draft Analysis of some Meta-Requirements for Cognitive Systems in Robots, contribution to euCognition wiki, also available as, http://www.cs.bham.ac.uk/research/projects/cosy/papers/#dp0701 (2007), http://cognitivesystems.org/cosybook/chap12.asp#slomanDp0701
43. Carnap, R.: Meaning and necessity: a study in semantics and modal logic. Chicago University Press, Chicago (1947)

44. Sloman, A.: The structure of the space of possible minds. In: Torrance, S. (ed.) The Mind and the Machine: philosophical aspects of Artificial Intelligence, Ellis Horwood, Chichester (1984), http://www.cs.bham.ac.uk/research/projects/cogaff/07.html#704
45. Sloman, A.: Explorations in design space. In: Cohn, A. (ed.) Proceedings 11th European Conference on AI, Amsterdam, August 1994, pp. 578–582. John Wiley, Chichester (1994)
46. Sloman, A.: Interacting trajectories in design space and niche space: A philosopher speculates about evolution. In: Deb, K., Rudolph, G., Lutton, E., Merelo, J.J., Schoenauer, M., Schwefel, H.-P., Yao, X., et al. (eds.) PPSN 2000. LNCS, vol. 1917, pp. 3–16. Springer, Heidelberg (2000), http://www.cs.bham.ac.uk/research/projects/cogaff/00-02.html#62
47. Scheutz, M.: The evolution of simple affective states in multi-agent environments. In: Cañamero, D. (ed.) Proceedings AAAI Fall Symposium 2001, pp. 123–128. AAAI Press, Falmouth (2001)
48. Neisser, U.: Cognitive Psychology. Appleton-Century-Crofts, New York (1967)
49. Gibson, J.J.: The Ecological Approach to Visual Perception. Houghton Mifflin, Boston (1979)
50. McCarthy, J.: The Well-Designed Child. Artificial Intelligence 172(18), 2003–2014 (2008)
51. Sloman, A.: Interactions between philosophy and AI: The role of intuition and non-logical reasoning in intelligence. In: Proc. 2nd IJCAI, pp. 209–226. William Kaufmann, London (1971), http://www.cs.bham.ac.uk/research/cogaff/04.html#200407
52. Pylyshyn, Z.: Is vision continuous with cognition? The case for Cognitive impenetrability of visual perception. Behavioral and Brain Sciences 22(3), 341–423 (1999)
53. Trehub, A.: The Cognitive Brain. MIT Press, Cambridge (1991)
54. Langley, P., Laird, J.: Cognitive architectures: Research issues and challenges, Tech. rep., Institute for the Study of Learning and Expertise, Palo Alto, CA (2006), http://cll.stanford.edu/~langley/papers/final.arch.pdf
55. Milner, R.: The Polyadic pi-Calculus: A Tutorial, Tech. Rep. LFCS report ECS-LFCS-91-180, The University of Edinburgh
56. Hewitt, C., Bishop, P., Steiger, R.: A Universal Modular ACTOR Formalism for Artificial Intelligence. In: Proc. Third Int. Joint Conf. on AI (1973)
57. Nilsson, N.: Artificial Intelligence: A New Synthesis. Morgan Kaufmann, San Francisco (1998)
58. Gibson, J.: The Senses Considered as Perceptual Systems. Houghton Mifflin, Boston (1966)
59. Brooks, R.: A robust layered control system for a mobile robot. IEEE Journal of Robotics and Automation RA-2, 14–23, 1 (1986)
60. Gat, E.: On three layer architectures. In: Kortenkamp, D., Bonnasso, R.P., Murphey, R. (eds.) Artificial Intelligence and Mobile Robots. AAAI Press, Menlo Park (1998)
61. Minsky, M.L.: The Emotion Machine. Pantheon, New York (2006)
62. Sloman, A.: The Cognition and Affect Project: Architectures, Architecture-Schemas, And The New Science of Mind. Tech. rep., School of Computer Science, University of Birmingham (2003), http://www.cs.bham.ac.uk/research/projects/cogaff/03.html#200307

63. Sun, R.: The CLARION cognitive architecture: Extending cognitive modeling to social simulation. In: Sun, R. (ed.) Cognition and Multi-Agent Interaction, pp. 79–99. Cambridge University Press, New York (2006), http://www.cogsci.rpi.edu/~rsun/sun.clarion2005.pdf
64. Albus, J.: Brains, Behaviour and Robotics, Byte Books. McGraw Hill, Peterborough (1981)
65. Cooper, R., Shallice, T.: Contention scheduling and the control of routine activities. Cognitive Neuropsychology 17(4), 297–338 (2000)
66. Block, N.: On a confusion about the function of consciousness. Behavioral and Brain Sciences 18, 227–247 (1995)
67. Dennett, D.: Kinds of minds: towards an understanding of consciousness. Weidenfeld and Nicholson, London (1996)
68. Minsky, M.L.: Steps towards artificial intelligence. In: Feigenbaum, E., Feldman, J. (eds.) Computers and Thought, pp. 406–450. McGraw-Hill, New York (1963)
69. McCarthy, J., Hayes, P.: Some philosophical problems from the standpoint of AI. In: Meltzer, B., Michie, D. (eds.) Machine Intelligence 4, pp. 463–502. Edinburgh University Press, Edinburgh (1969), http://www-formal.stanford.edu/jmc/mcchay69/mcchay69.html
70. Beaudoin, L., Sloman, A.: A study of motive processing and attention. In: Sloman, A., Hogg, D., Humphreys, G., Partridge, D., Ramsay, A. (eds.) Prospects for Artificial Intelligence, pp. 229–238. IOS Press, Amsterdam (1993), http://www.cs.bham.ac.uk/research/projects/cogaff/81-95.html#16
71. Sloman, A.: How to derive "better" from "is". American Phil. Quarterly 6, 43–52 (1969), http://www.cs.bham.ac.uk/research/cogaff/sloman.better.html
72. Rochat, P.: The Infant's World. Harvard University Press, Cambridge (2001)
73. Sloman, A.: Perception of structure: Anyone Interested?, http://www.cs.bham.ac.uk/research/projects/cosy/papers/#pr0507, http://cognitivesystems.org/cosybook/chap12.asp#sloman-cosypr0507 (2005)
74. Grush, R.: The emulation theory of representation: Motor control, imagery, and perception. Behavioral and Brain Sciences 27, 377–442 (2004)
75. Smith, B.: On the Origin of Objects. MIT Press, Cambridge (1996)
76. Sloman, A.: Actual possibilities. In: Aiello, L., Shapiro, S. (eds.) Principles of Knowledge Representation and Reasoning: Proceedings of the Fifth International Conference (KR 1996), pp. 627–638. Morgan Kaufmann Publishers, Boston (1996), http://www.cs.bham.ac.uk/research/cogaff/96-99.html#15
77. Kant, I.: Critique of Pure Reason. Macmillan, London (1929); 1781, translated by Norman Kemp Smith
78. Russell, B.: Mysticism and Logic and Other Essays. Allen & Unwin, London (1917)
79. Rips, L.J., Bloomfield, A., Asmuth, J.: From Numerical Concepts to Concepts of Number. The Behavioral and Brain Sciences
80. Ryle, G.: The Concept of Mind. Hutchinson, London (1949)
81. Sloman, A.: Two Notions Contrasted: 'Logical Geography' and 'Logical Topography' (Variations on a theme by Gilbert Ryle: The logical topography of 'Logical Geography'.). Tech. Rep. COSY-DP-0703, School of Computer Science, University of Birmingham, Birmingham, UK (2007), http://cognitivesystems.org/cosybook/chap12.asp#cosy-dp-0703

13

Lessons and Outlook

Henrik I. Christensen

Robotics and Intelligent Machines, Georgia Institute of Technology,
Atlanta, Ga. USA
hic@cc.gatech.edu

13.1 Introduction

The CoSy project had a very ambitious set of goals from the outset. The effort was driven forward by scientific and technical goals as outlined in Chapter 1. Early in the project integration efforts were undertaken to ensure that both component and systems issues could be addressed. Already after 12 months early demonstrators were available for empirical studies of cognitive systems. Obviously, at that stage, the systems were brittle and of limited functionality but they framed the problems in a nice way.

In some respects CoSy is about the marriage of traditional artificial intelligence, computational perception, human computer interaction and robotics. In each of the participating disciplines good progress has been reported over the last few decades, but the expected break through has not been achieved. A major emphasis of CoSy has been a change in emphasis. By framing the problem in the context of an embodied or situated agent many of the challenges change. First of all the context in terms of task and spatial location in many respects simplifies the problem and makes it much more tractable. The embodiment also allow for direct interaction with the world. The system is no longer a passive *observer* that is trying to understand what is going on. The system has the ability to participate in the world and change it to make intractable problem more tractable. The embodiment is thus a key part both in terms of mobility and manipulation. It must, however, be recognized that the situated embodied context also poses a number of additional challenges. As the agent moves through the world the view point changes, the illumination varies, the human-robot relation changes over time. The situated context comes as a cost. Nonetheless as essential part of CoSy has been the study of embodied systems. Another aspect that has played an important part in CoSy is the assistive role. The systems studied are not expected to be fully autonomous but to operate in cooperation with humans. This implies that humans can be queried when ambiguities arise, which again changes the problem.

H.I. Christensen et al. (Eds.): Cognitive Systems, COSMOS 8, pp. 485–489.
springerlink.com

Over the duration of the project significant progress was made across all areas and many of the problems have been well framed with in the individual chapters and some of the more philosophical issues have already been discussed as part of Chapter 12. It is clear that there are a number of important lessons coming from an effort such as CoSy and obviously the number of problems discovered by the project are likely to be a numerous as the lessons. In Section 13.2 a number of the major lessons will be discussed and correspondingly some of the new challenges will be outlined in Section 13.3.

13.2 Lessons

Architectures: An important part of CoSy has been the study of systems. A key aspect is here the consideration of architectures. CoSy has performed extensive analysis of a rich variety of architectures in the literature but as important is the design of an architectural toolkit CAST, that enable flexible implementation of a variety of different applications. The architectural approach could be considered a federation of blackboard systems that are loosely synchronized. Some of the considerations related to the design are discussed in Chapters 2 and 11. Early design of systems have indicated that this might be a highly effective framework for construction of relatively complex systems.

Representations: Tied to the problem of architectures and systems design is the problem of representations, which permeates the entire systems discussion. For many of the sub areas of cognitive systems there has been progress, but a frequently encountered problem is scalability and generalization. Across the different areas of planning, geometry and perception hybrid representations have been adopted as a strategy to address locality and scalability. In visual perception a hierarchical method is used where signal level representations are trained in an unsupervised manner, and as abstractions are introduced so is the degree of supervision leading to effective models for recognition. In mapping local structures are modelling using standard Euclidean geometry where as topological models are adopted for description of the layout of large scale structures. The change of metric to topological models enables a much higher degree of scalability and learning of geometric structures can be accommodated at the local level, which implicitly enables generalization.

Interpretation for Interaction: Through adoption of hybrid models and tying them to probabilistic reasoning it is also possible to perform efficient interpretation of scenes. As part of the PlayMate scenario it has been demonstrated how the system can be taught to play a number of simple games and it is also here demonstrated how the system can generalize from one game to another. The interpretation is here closely tied to knowledge acquisition, dialog generation and dynamic re-planning. The integration of these components into a common framework has allowed for a number of compelling demonstrations. Initially it was expected that a significant portion of the PlayMate demonstrator would emphasize basic manipulative skills. In reality the set of manipulative skills needed for studies of basic interaction has been limited. The set of tasks a system can perform is directly related to the dexterity of the interaction with the world, but even with a limited functionality the richness of the world is tremendous.

Robust operation: Operating in a real-world setting requires robustness to have any degree of extended interaction or autonomy. The issue of robustness can be approached from several different directions. In CoSy at least three different approaches have been adopted. First of all processing of sensory information has largely been modelling within a Bayesian framework combined with learning for acquisition of data models, such that sources of variation and noise are explicitly considered. Another approach has been through integration of multi sensory modalities. Through redundancy and complementarity in information it is possible generate an added degree of robustness. An example is in the mapping of environments where visual and laser based information is integrated to provide more robust recognition of structures. Visual information typically has a high degree of ambiguity, where as the laser information is more specific, but it is entirely 2D so verification with visual data enables added discrimination. However both of these approach are entirely data driven. Another approach pursued in CoSy has been through performance monitoring and planning.

Monitoring and Re-planning: Within CoSy a particle based method was developed for monitoring of state. The particle approach was adopted to enable description of the high degree of variation that may happen in processes and to cope with scalability. The performance monitoring was integrated into the planning framework to allow generation of predictions about process evolution, i.e. what are the expected and unexpected evolutions in a process. The prediction of process evolution within a formal framework also allow for meta-level reasoning to understand when re-planning can be used for recovery from unexpected situations. The dynamic re-planning was used in the PlayMate scenario to recover for poor manipulative capabilities. However, it was also integrated into the dialog system for communication with humans to handle the variability dialog patterns. The added degree of robustness demonstrated was essential in the design of demonstrator systems that exhibited some degree of autonomy.

Situated dialog generation for HRI: A very important part of the CoSy project was the design of situated dialog systems. Few systems within Human-Robot Interaction have actually demonstrated dialog behaviour where is there is any level of flexibility. As such most of these systems exhibit a high degree of brittleness. Open-ended dialog systems are obviously a major challenge, but by casting the dialog problem as situated it is possible to limit the complexity of the overall dialog and consequently to make it robust and meaningful to users with a limited degree of training. It has further been demonstrated that it is possible to generate dialog structures that generalize across mapping and manipulation. The availability of a common framework allow for easy adaptation to a variety to applications. At the same time it must be recognized that the dialog today is almost entirely based on spoken utterances and auditory feedback. The integration of a riches set of spatial gestures and body postures would enrich the interaction significantly.

Overall CoSy has generated a substantial body of knowledge during its lifetime as witnessed by the results reporting in this volume and by the significantly body of publications in the archival literature. Progress has been achieved across all of the areas.

13.3 Outlook

Integration: Cosy was from the outset an "integrated" project with a healthy balance between topical research and integration of methods into a systems context. In addition the deliverables where both in terms of theory and methods. It must, however, be recognized that the project in many respects merely has formulated the problems and there is tremendous room for new research to understand the problem of cognitive systems. The ambitious goals put forward by the European Commission "to construct physically instantiated or embodied systems that can perceive, understand, ... and interact with their environments and evolve in order to achieve human-like performance" are far from solved. It is, however, important to recognize that the situated context paves a way to study many of the essential problems. The embodiment allows for re-formulation of the problem(s) and to attack them from a new perspective and CoSy has clearly demonstrated an initial strategy for such studies.

Categorization: It is important to recognize that CoSy by no measure has solved the categorization problem in perception. CoSy has generated new methods for multi-categorical recognition but the more general problem of categorization is still a major challenge. It is clear that through adoption of a categorical approach it would be possible to provide an added degree of scalability. Traditionally categories have been defined by linguistic concepts, which is a straight forward approach, but it is not immediately clear for example visual categories have a one to one mapping to linguistic concepts. As an example the "chair" maps to a number of different visual categories. A stable and sitable surface can be provided in a number of different ways. Also categories may be defined in several different ways.

Affordances: During the setup of CoSy it was envisaged that object affordances would play a crucial role in the design of a system and in organization of representation. In many respects affordances were considered a fundamental building block in the study of cognitive systems. One could consider affordances perception-action invariants. However, there are several other possible interpretations of the affordance concept as already outlined in Chapter 12. The exact role of affordances is still largely a open problem.

Physical Interaction: In relation to the affordance discussion it is important to recognize that CoSy only deployed relatively modest methods for physical interaction with the environment. The mobile robots had no way to push or move objects around and the manipulators have simple / limited functionality grippers. In addition the physical systems were not designed to allow for learning by demonstration. These limitations allow for studies of a particular set of problems. However problems such as skill and task learning for interaction with objects were largely ignored. This limitation was necessary due to lack of hardware but also due to lack of resources to address this much more complex problem.

CogX: Some of the open problems will be addressed in a continuation of the CoSy project. A consortium that is composed of a sub-set of the CoSy partners and complemented with a few new partners have embarked on new research effort that is building on the results obtained in CoSy. Further information about the new project CogX is available from the project web facility `http://cogx.eu`. The emphasis of CogX is in particular on introspection/self-understanding and acquisition of new skills and tasks.

Author Index

Cognitive Systems Monographs

Edited by R. Dillmann, Y. Nakamura, S. Schaal and D. Vernon

Vol. 1: Arena, P.; Patanè, L. (Eds.):
Spatial Temporal Patterns for
Action-Oriented Perception
in Roving Robots
425 p. 2009 [978-3-540-88463-7]

Vol. 2: Ivancevic, T.T.; Jovanovic, B.;
Djukic, S.; Djukic, M.; Markovic, S.:
Complex Sports Biodynamics
326 p. 2009 [978-3-540-89970-9]

Vol. 3: Magnani, L.:
Abductive Cognition
534 p. 2009 [978-3-642-03630-9]

Vol. 4: Azad, P.:
Visual Perception for Manipulation
and Imitation in Humanoid Robots
270 p. 2009 [978-3-642-04228-7]

Vol. 5: de de Aguiar, E.:
Animation and Performance Capture
Using Digitized Models
168 p. 2009 [978-3-642-10315-5]

Vol. 6: Ritter, H.; Sagerer, G.;
Dillmann, R.; Buss, M.:
Human Centered Robot Systems
216 p. 2009 [978-3-642-10402-2]

Vol. 7: Levi, P.; Kernbach, S. (Eds.):
Symbiotic Multi-Robot Organisms
approx. 480 p. 2010 [978-3-642-11691-9]

Vol. 8: Christensen, H.I.;
Kruijff, G.-J.M.; Wyatt, J.L. (Eds.):
Cognitive Systems
491 p. 2010 [978-3-642-11693-3]